Lecture Notes in Networks and Systems

Volume 75

The series "Lecture Notes in Networks and Systems" publishes the latest developments in Networks and Systems—quickly, informally and with high quality. Original research reported in proceedings and post-proceedings represents the core of LNNS.

Volumes published in LNNS embrace all aspects and subfields of, as well as new challenges in, Networks and Systems.

The series contains proceedings and edited volumes in systems and networks, spanning the areas of Cyber-Physical Systems, Autonomous Systems, Sensor Networks, Control Systems, Energy Systems, Automotive Systems, Biological Systems, Vehicular Networking and Connected Vehicles, Aerospace Systems, Automation, Manufacturing, Smart Grids, Nonlinear Systems, Power Systems, Robotics, Social Systems, Economic Systems and other. Of particular value to both the contributors and the readership are the short publication timeframe and the world-wide distribution and exposure which enable both a wide and rapid dissemination of research output.

The series covers the theory, applications, and perspectives on the state of the art and future developments relevant to systems and networks, decision making, control, complex processes and related areas, as embedded in the fields of interdisciplinary and applied sciences, engineering, computer science, physics, economics, social, and life sciences, as well as the paradigms and methodologies behind them.

**** Indexing: The books of this series are submitted to ISI Proceedings, SCOPUS, Google Scholar and Springerlink ****

More information about this series at http://www.springer.com/series/15179

Sheng-Lung Peng · Nilanjan Dey ·
Mahesh Bundele
Editors

Computing and Network Sustainability

Proceedings of IRSCNS 2018

Springer

Editors
Sheng-Lung Peng
Department of Computer Science
National Dong Hwa University
Hualien, Taiwan

Nilanjan Dey
Techno India College of Technology
Kolkata, West Bengal, India

Mahesh Bundele
Poornima University
Jaipur, Rajasthan, India

ISSN 2367-3370 ISSN 2367-3389 (electronic)
Lecture Notes in Networks and Systems
ISBN 978-981-13-7149-3 ISBN 978-981-13-7150-9 (eBook)
https://doi.org/10.1007/978-981-13-7150-9

Library of Congress Control Number: 2019934365

This Springer imprint is published by the registered company Springer Nature Singapore Pte Ltd.
The registered company address is: 152 Beach Road, #21-01/04 Gateway East, Singapore 189721, Singapore

Organization

Organizing Chairs

Mr. Bharat Patel, Vice Chairman, ASSOCHAM Western Region
Mr. Peter Kent, Chief Executive Officer (CEO), UKEI, UK

Organizing Co-Chair

Mr. Nakul Sharedalal, Advisor, G R Foundation, India

Organizing Secretary

Amit Joshi, SITG, Ahmedabad; Global Knowledge Research Foundation

Members

Dr. Vijay Singh Rathore, Department of CSE, JECRC, Jaipur, India
Mr. Aman Barot, G R Foundation, India
Dr. Mahipal Singh Deora, BN University, Udaipur, India
Dr. Nisheeth Joshi, Banasthali University, Rajasthan, India
Mr. Nilesh Vaghela, Electromech, Ahmedabad, India
Mr. Vinod Thummar, SITG, Ahmedabad, Gujarat, India
Dr. Chirag Thaker, L. D. College of Engineering, Gujarat, India
Prof. S Rama Rao, Goa University, Goa, India
Dr. Nitika Vats Doohan, Indore, India

Dr. Parikshit Mahalle, Sinhgad Group, Pune, India
Dr. Priyanks Sharma, RSU, Ahmedabad
Dr. Nitika Vats Doohan, Indore
Dr. Mukesh Sharma, SFSU, Jaipur
Dr. Manuj Joshi, SGI, Udaipur, India
Dr. Bharat Singh Deora, JRNRV University, Udaipur
Mr. Ricky, P Tunes, Goa University, India
Prof. L. C. Bishnoi, GPC, Kota, India
Dr. Vikarant Bhateja, Lucknow, India
Dr. Satyen Parikh, Dean, Ganpat University, Ahmedabad, India
Dr. Puspendra Singh, JKLU, Jaipur, India
Dr. Aditya Patel, Ahmedabad University, Gujarat, India
Mr. Ajay Choudhary, IIT Roorkee, India

Technical Program Committee

Prof. Milan Tuba, Singidunum University, Serbia
Dr. Durgesh Kumar Mishra, Chairman, Div IV, CSI

Technical Program Committee Co-Chairs

Dr. Mahesh Bundele, Dean Research, Poornima University, Jaipur
Dr. Nilanjay Dey, Techno India College of Engineering, Kolkata, India

Advisory Committee

Shri Nitin Kunkolienker, President, MAIT
Prof. S Rama Rao, Goa University, Goa
Mr. P N Jain, Add. Sec., R&D, Government of Gujarat, India
Prof. J. Andrew Clark, Department of Computer Science, University of York, UK
Mr. Vivek Ogra, President, GESIA
Prof. Mustafizur Rahman, Endeavor Research Fellow, Australia
Dr. Kalpdrum Passi, Laurentian University, ON, Canada
Mr. Chandrashekhar Sahasrabudhe, ACM India
Dr. Pawan Lingras, Saint Mary's University, Canada
Dr. S C Satapathy, Professor, CSE, AP, India
Mr. Niko Phillips, Active Learning, UK, India
Dr. Bhagyesh Soneji, Chairperson, ASSOCHAM Western Region
Dr. Dharm Singh, NUST, Windhoek, Namibia

Members

Preface

Second International Research Symposium on Computing and Network Sustainability (IRSCNS 2018) targeted theory, development, applications, experiences, and evaluation of interaction sciences with fellow students, researchers, and practitioners.

Conference may concern any topic within the conference scope. Workshops may be related to any topics within the conference scope. The conference is devoted to increase the understanding role of technology issues, how engineering has day by day evolved to prepare human-friendly technology. The conference provided a platform for bringing forth significant research and literature across the field of ICT for sustainable development and an overview of the technologies awaiting unveiling. This interaction was the focal point for leading experts to share their insights, provide guidance, and address participant's questions and concerns.

The conference was held during August 30–31, 2018, at Hotel Vivanta by Taj, Panaji, Goa, India, organized by Global Knowledge Research Foundation, state chamber partner—Goa Chamber of Commerce & Industry, and incubation partner—Centre for Incubation and Business Acceleration, and supported by The Institution of Engineers (India)—CEDB, IE(I); Department of Electronics and IT, Ministry of Communications & Information Technology, Government of India, Asian-African Chamber of Commerce & Industry; Unified Brainz; and World Peace & Diplomacy.

The research submissions in various advanced technology areas were received, and after a rigorous peer review process, with the help of program committee members and 56 external reviewers for 400 papers from 8 different countries including Algeria, USA, United Arab Emirates, Serbia, Qatar, Mauritius, Egypt, Saudi Arabia, Ethiopia, and Oman, 55 were accepted with an acceptance ratio of 0.13.

Technology is the driving force of progress in this era of globalization. Information and communication technology (ICT) has become a functional requirement for the socioeconomic growth and sustainable development of any country. The influence of information and communication technology (ICT) in shaping the process of globalization, particularly in productivity, commercial, and financial spheres, is widely recognized. The ICT sector is undergoing a revolution that has momentous implications for the current and future social and economic situation of all the countries in the world. ICT plays a pivotal role in empowering people for self-efficacy and in understanding how it can facilitate this mission to reach out to the grassroots level. Finally, it is concluded that ICT is a significant contributor to the success of the ongoing initiative of Startup India.

In order to recognize and reward the extraordinary performance and achievements by ICT and allied sectors and promote universities, researchers, and students through their research work adapting new scientific technologies and innovations, the two-day conference had presentations from the researchers, scientists, academia, and students on the research work carried out by them in different sectors.

IRSCNS & Startup Summit was a flagship event of G R Foundation. This was the third edition. The earlier two series were a grand success with participation from various universities, academia, scientists, scholars, researchers, students, industries, stakeholders, and R&D institutions from within and outside the country. The researchers presented their research papers through presentations at the conference. All the earlier conferences witnessed the presence of more than 250 delegates and eye-catching presence from government institutions and ministries. The second edition of IRSCNS & Startup Summit along with Startup & ICT Awards focused on the new innovations using scientific technologies in the ICT sector. The conference-cum-awards marked the presence of national and international stakeholders, universities, R&D institutions, academics, students from Goa University and BITS Pilani—Goa Chapter, representatives from union and state governments, investors, policymakers, industry leaders, trade bodies and scientific consultants, and ICT experts to share their knowledge in this area.

Burdwan, India Nilanjan Dey

Contents

Editors and Contributors

About the Editors

Dr. Sheng-Lung Peng obtained his M.Sc. (Computer Science) from National Chung Cheng University, Taiwan, and Ph.D. (Computer Science) from National Tsing Hua University, Taiwan. He is currently a Professor in the Department of Computer Science and Information Engineering, National Dong Hwa University, Taiwan. He is also Director of Library and Information Center in the same University. His research interests are in Designing and analysing algorithms for combinatorics, bioinformatics, data mining and networks. He has about one hundred papers in Journal and Conference proceedings.

Dr. Nilanjan Dey was born in Kolkata, India, in 1984. He received his B.Tech. degree in Information Technology from West Bengal University of Technology in 2005, M.Tech. in Information Technology in 2011 from the same University and Ph.D. in digital image processing in 2015 from Jadavpur University, India. In 2011, he was appointed as an Asst. Professor in the Department of Information Technology at JIS College of Engineering, Kalyani, India followed by Bengal College of Engineering College, Durgapur, India in 2014. He is a visiting fellow of University of Reading, UK. He is now employed as an Associate Professor in Department of Information Technology, Techno India College of Technology, India. His research topic is signal processing, machine learning, and information security. Dr. Dey is an Associate Editor of IEEE Access and is currently the Editor-in-Chief of the *International Journal of Ambient Computing and Intelligence*, and Series co-editor of Springer Tracts of Nature-Inspired Computing (STNIC).

Dr. Mahesh Bundele is currently working as Professor in Computer Engineering and Dean (Research & Development) at Poornima University, Jaipur, India. He is heading Advanced Studies Research Center and responsible for design, development and execution of research curriculum and processes for Masters in Engineering and Doctoral Program in School of Engineering, Sciences and

Management. He has total 31 years experience in teaching and research. He has developed many unique research methodology concepts and implemented. He is the mentor and controller of quality research and publications at the University. He is also responsible for inculcation of innovative and critical analysis concepts across the University and across the Poornima Foundation involving three other campuses. He is working on various research projects sponsored by various agencies. He did his doctoral degree in Wearable Computing and guiding research in Pervasive & Ubiquitous Computing, Computer Networks, Software Defined Networking. His areas of interests are also Wireless Sensor Networks, Algorithmic research, Mathematical Modeling, and Smart Grids. He has worked on many government and non- government research projects and guided bachelors and master degree students for their research. He has more than 30 publications in reputed journals and conferences. He has been reviewer of few IEEE Transactions. He is actively involved in IEEE activities in Rajasthan Subsection and Delhi Section and holding the responsibility on Standing Committee of IEEE Delhi Section for Technical & Professional activities for controlling quality of conferences and publications in IEEE. He is also holding the position of membership development chair at IEEE Rajasthan Subsection.

Contributors

Naveen Aggarwal University Institute of Engineering and Technology, Panjab University, Chandigarh, India

Akashdeep University Institute of Engineering and Technology, Panjab University, Chandigarh, India

Shabir Ali Computer Science and Engineering Department, Motilal Nehru National Institute of Technology Allahabad, Allahabad, India

André Alves Departamento de Informática, Escola de Engenharia, Universidade do Minho, Braga, Portugal

Victor Alves Centro Algoritmi, Universidade do Minho, Braga, Portugal

R. Anand Department of Electronics and Communication Engineering, Sona SIPRO, Salem, India;
Sona Signal and Image PROcessing Research Center, Advanced Research Centre Block, Sona College of Technology, Salem, Tamil Nadu, India

M. Anitha Dayananda Sagar College of Engineering, Kumaraswamy Layout, Bangalore, India

J. E. Anusha Linda Kostka Noorul Islam Centre for Higher Education, Kumaracoil, India

S. Aparna Lakshmanan Department of Mathematics, St. Xavier's College for Women, Aluva, Kerala, India

Parveen Arora NEC Technologies India Pvt. Ltd., Chennai, India

A. Arunvel CDAC, Pune, India

Pallavi Asthana Amity School of Engineering and Technology, Amity University Lucknow Campus, Lucknow, India

T. S. Aswin Data Science Group, Bengaluru, India

Anurag Babel Dayananda Sagar College of Engineering, Kumaraswamy Layout, Bangalore, India

Shashwati Banerjea Computer Science and Engineering Department, Motilal Nehru National Institute of Technology Allahabad, Allahabad, India

Sivasankari Bhagavathiperumal Faculty of Engineering and Information Technology, Centre of Artificial Intelligence, University of Technology Sydney, Broadway, NSW, Australia

Parag Bhalchandra S.C.S, Swami Ramanand Teerth Marathwada University, Nanded, Maharashtra, India

Neha Bharti Department of CSE, Jaipur Engineering College and Research Center, Jaipur, India

Nityesh Bhatt Institute of Management, Nirma University, Ahmedabad, India

Ruby Bhatt Department of CSE, Rabindranath Tagore University, Bhopal, India

Gananatha Bhuyan School of Computer Engineering, Kalinga Institute of Industrial Technology University, Bhubaneswar, Odisha, India

António Capita Instituto Superior Técnico Militar, Luanda, Angola

Priyadarshini Chatterjee Department of Computer Science and Engineering, Vardhaman College of Engineering, Hyderabad, India

Naresh Kumar Chaudhary Dr. Ram Manohar Lohia Awadh University, Faizabad, UP, India

Chandresh Kumar Chhatlaniand JRN Rajasthan Vidyapeeth, University, Udaipur, Rajasthan, India

Kirti Choudhary Department of CSE, Jaipur Engineering College and Research Center, Jaipur, India

Nilam Choudhary Jaipur Engineering College and Research Centre, Jaipur, India

Keren Lois Daniel St. Xavier's College, Jaipur, India

Deepak Dashora JRN Rajasthan Vidyapeeth University, Udaipur, India

V. G. Deepa Department of Mathematics, Sree Krishna College, Guruvayur, Kerala, India

Nilesh Deshmukh S.C.S, Swami Ramanand Teerth Marathwada University, Nanded, Maharashtra, India

Sarita Deshpande Department of Information Technology, PES's Modern College of Engineering, Pune, India

B. K. Dhanalakshmi SJB Institute of Technology, Bangalore, Karnataka, India

P. M. Dinesh Department of Electronics and Communication Engineering, Sona SIPRO, Salem, India;
Sona Signal and Image PROcessing Research Center, Advanced Research Centre Block, Sona College of Technology, Salem, Tamil Nadu, India

Sanjay K. Dwivedi Babasaheb Bhimrao Ambedkar University, Lucknow, India

Filipa Ferraz Centro Algoritmi, Universidade do Minho, Braga, Portugal

Lavisha Garg University Institute of Engineering and Technology, Panjab University, Chandigarh, India

Sanjay Gaur Jaipur Engineering College and Research Centre, Jaipur, India

Geetika Gautam Department of CSE, Jaipur Engineering College and Research Center, Jaipur, India

N. Geetha Department of CSE, SOET, CMR University, Bangalore, India

Ravindra Ghritlahare National Institute of Technology, Raipur, Chhattisgarh, India

Madhu Goyal Faculty of Engineering and Information Technology, Centre of Artificial Intelligence, University of Technology Sydney, Broadway, NSW, Australia

Archana Gulati Computer Engineering, Datta Meghe College of Engineering, Airoli, India

Deepti Gupta Banasthai Vidyapeeth University, Vanasthali, Rajasthan, India

Varun Gupta Department of Electronics and Communication, Amity University, Uttar Pradesh, India

Kailas Hambarde S.C.S, Swami Ramanand Teerth Marathwada University, Nanded, Maharashtra, India

Fouziah Hamza Noorul Islam Centre for Higher Education, Kumaracoil, Kanyakumari, India

Shaikh Husen S.C.S, Swami Ramanand Teerth Marathwada University, Nanded, Maharashtra, India

Lavika Jain Aishwarya Institute of Management and Information Technology, Udaipur, India

Anal K. Jha Aryabhatta Knowledge University, Patna, India

V. Jyothsna Sree Vidyanikethan Engineering College, Tirupati, India

Moin Khorajiya CDAC ACTS, Pune, India

Nishita Kindo School of Computer Engineering, Kalinga Institute of Industrial Technology University, Bhubaneswar, Odisha, India

P. Krishna Kishore Chadalawada Ramanamma Engineering College, Tirupati, India

V. Rama Krishna Sathyabama Institute of Science and Technology, Chennai, India

Govind Kulkarni S.C.S, Swami Ramanand Teerth Marathwada University, Nanded, Maharashtra, India

Anil Kumar Amity School of Engineering and Technology, Amity University Lucknow Campus, Lucknow, India

Anushil Kumar Dayananda Sagar College of Engineering, Kumaraswamy Layout, Bangalore, India

Gardas Naresh Kumar CDAC ACTS, Pune, India

T. N. R. Kumar Ramaiah Institute of Technology, Bengaluru, India

Vijay Kumar Aryabhatta Knowledge University, Patna, India

Santosh S. Lomte College of Engineering Matoshri Pratishthan, Nanded, India

Priti Maheshwary Department of CSE, Rabindranath Tagore University, Bhopal, India

Chetan Mali Department of CS & IT, JRN Rajasthan Vidyapeeth, University, Udaipur, Rajasthan, India

Ch Mamatha Department of Information Technology, Vardhaman College of Engineering, Hyderabad, India

Pratima Manhas ECE Department, FET, Manav Rachna International Institute of Research & Studies, Faridabad, India

S. Maria Celestin Vigila Noorul Islam Centre for Higher Education, Kumaracoil, Kanyakumari, India

Pratistha Mathur Manipal University, Jaipur, Rajasthan, India

Ankur Meena B N University, Udaipur, India

Gaurav Mishra Development Management Institute, Patna, India

Sumita Mishra Amity School of Engineering and Technology, Amity University Lucknow Campus, Lucknow, India

Puneet Misra Department of Computer Science, University of Lucknow, Lucknow, UP, India

Girish Mohan Department of Information and Communication Technology, Manipal Institute of Technology, Manipal Academy of Higher Education, Manipal, Karnataka, India

D. Mukesh Chadalawada Ramanamma Engineering College, Tirupati, India

K. Munivara Prasad Chadalawada Ramanamma Engineering College, Tirupati, India

Darshana A. Naik Department of CSE, MSRIT, Banglaore, India

João Neves Mediclinic Arabian Ranches, Dubai, United Arab Emirates

José Neves Centro Algoritmi, Universidade do Minho, Braga, Portugal

G. Nischal B V Bhoomaraddi College of Engineering and Technology, Hubli, Karnataka, India

Garima Ojha Jaipur Engineering College and Research Centre, Jaipur, India

Antara Durgesh Oza Gujarat Technological University, Gujarat, India

Ronali Padhy School of Computer Engineering, Kalinga Institute of Industrial Technology University, Bhubaneswar, Odisha, India

Naraharisetty Padmaja National Institute for Smart Government (NISG), Hyderabad, India

Smitha N. Pai Department of Information and Communication Technology, Manipal Institute of Technology, Manipal Academy of Higher Education, Manipal, Karnataka, India

M. Pallavi Department of CSE, SOET, CMR University, Bangalore, India

Mudra Panchal Department of Information Technology, PES's Modern College of Engineering, Pune, India

Mayank Pandey Computer Science and Engineering Department, Motilal Nehru National Institute of Technology Allahabad, Allahabad, India

Shraddha Pandit University Institute of Technology, RGPV, Bhopal, Madhya Pradesh, India

Darshanaben D. Pandya Madhav University Abu Road, Sirohi, India

Namita Pandya Department of Human Resource, Geetanjali University, Udaipur, India

Rajeev Pandya B N University, Udaipur, India

M. E. Paramasivam Department of Electronics and Communication Engineering, Sona SIPRO, Salem, India

Ajeet Ram Pathak Department of Information Technology, PES's Modern College of Engineering, Pune, India

Savitri Patil Ramaiah Institute of Technology, Bengaluru, India

Nishant S. Pawar Gujarat Technological University, Ahmadabad, India

Sofia K. Pillai G.H. Raisoni College of Engineering, Nagpur, Maharashtra, India

Marco Prata Departamento de Informática, Escola de Engenharia, Universidade do Minho, Braga, Portugal

Shalini Puri Birla Institute of Technology, Mesra, Ranchi, Jharkhand, India

M. M. Raghuwanshi G.H. Raisoni College of Engineering, Nagpur, Maharashtra, India;
YCCE, Nagpur, Maharashtra, India

M. Rajesh Babu Sri Ramakrishna Engineering College, Coimbatore, Tamil Nadu, India

S. Ramakrishna Department of Electronics and Communication Engineering, B V Bhoomaraddi College of Engineering and Technology, Hubballi, Karnataka, India

Jayesh Ranjan Information Technology Department, Government of Telangana, Hyderabad, India

GT Venkateswara Rao Information Technology Department, Government of Telangana, Hyderabad, India

Vagvala Satyajit Rao National Institute for Smart Government (NISG), Hyderabad, India

Vijay Singh Rathore Department of CSE, Jaipur Engineering College and Research Center, Jaipur, India

Aman Rauniyar Dayananda Sagar College of Engineering, Kumaraswamy Layout, Bangalore, India

Jorge Ribeiro Escola Superior de Tecnologia e Gestão, ARC4DigiT—Applied Research Center for Digital Transformation, Instituto Politécnico de Viana do Castelo, Viana do Castelo, Portugal

Mário Ribeiro Departamento de Informática, Escola de Engenharia, Universidade do Minho, Braga, Portugal

S. H. Rohini B V Bhoomaraddi College of Engineering and Technology, Hubli, Karnataka, India

R. S. Sabeenian Department of Electronics and Communication Engineering, Sona SIPRO, Salem, India

A. N. Sahil Sri Ramakrishna Engineering College, Coimbatore, Tamil Nadu, India

Hemant Sahu Geetanjali Institute of Technical Studies, Udaipur, Rajasthan, India

S. Sai Satyanarayana Reddy Department of Computer Science and Engineering, Vardhaman College of Engineering, Hyderabad, India

V. Samba Siva Chadalawada Ramanamma Engineering College, Tirupati, India

S. D. Sawarkar Computer Engineering, Datta Meghe College of Engineering, Airoli, India

Ashutosh Saxena CR Rao Advanced Institute of Mathematics Statistics and Computer Science, University of Hyderabad Campus, Hyderabad, India

S. Seema Department of CSE, MSRIT, Banglaore, India

R. S. Shaktawat B N University, Udaipur, India

S. Sharanraj Sri Ramakrishna Engineering College, Coimbatore, Tamil Nadu, India

Lokesh Sharma JRN Rajasthan Vidyapeeth University, Udaipur, India

Manish Sharma Department of Electronics and Communication, Amity University, Uttar Pradesh, India

Nonita Sharma Dr. B. R. Ambedkar National Institute of Technology, Jalandhar, Punjab, India

Praveen Sharma Mewar University, Chittorgarh, Rajasthan, India

Rajashekar B. Shettar Department of Electronics and Communication Engineering, B V Bhoomaraddi College of Engineering and Technology, Hubballi, Karnataka, India

Suhas B. Shirol Department of Electronics and Communication Engineering, B V Bhoomaraddi College of Engineering and Technology, Hubballi, Karnataka, India

Urmila Shrawankar G.H. Raisoni College of Engineering, Nagpur, Maharashtra, India

Pooja Shrimali Informatics & Computational Sciences, Mohanlal Sukhadia University, Udaipur, Rajasthan, India

Piyush Shukla Department of CSE, UIT, Bhopal, India

Piyush Kumar Shukla University Institute of Technology, RGPV, Bhopal, Madhya Pradesh, India

Shobhit Shukla Department of Computer Science, University of Lucknow, Lucknow, UP, India

Abhinav Singh CloudxLab Inc., Bangalore, India

Geetika Singh Department of CSE, MSRIT, Banglaore, India

Krishna Vijay Singh Sikkim Manipal University, Tadong, Gangtok, India

Satya Prakash Singh Birla Institute of Technology, Mesra, Ranchi, Jharkhand, India

Aditya Kumar Sinha DAC, Pune, India

M. S. Smitha Rao Department of CSE, SOET, CMR University, Bangalore, India

A. N. Sreedhar Chadalawada Ramanamma Engineering College, Tirupati, India

V. N. Sreeja Department of Mathematics, Sree Krishna College, Guruvayur, Kerala, India

M. Sreenivasulu Chadalawada Ramanamma Engineering College, Tirupati, India

K. C. Srikantaiah SJB Institute of Technology, Bangalore, Karnataka, India

Josyula Srinivas National Institute for Smart Government (NISG), Hyderabad, India

R. Subhasini Sathyabama Institute of Science and Technology, Chennai, India

G. A. Sumalatha SJB Institute of Technology, Bangalore, Karnataka, India

S. Suresh Kumar Sri Ramakrishna Engineering College, Coimbatore, Tamil Nadu, India

Preetam Tamsekar S.C.S, Swami Ramanand Teerth Marathwada University, Nanded, Maharashtra, India

Peeyush Tewari Birla Institute of Technology, Noida, India

B. R. Thakar Raksha Shakti University, Ahmedabad, India

Shaveta Thakral ECE Department, FET, Manav Rachna International Institute of Research & Studies, Faridabad, India

Akhilesh Tiwari Madahv Institute of Technology & Science, Gwalior, Madhya Pradesh, India

Vineeta Tiwari CDAC ACTS, Pune, India

Sanket G. Torambekar Yogeshwari Polytechnic, Ambajogai, India

Neeraj Tyagi Computer Science and Engineering Department, Motilal Nehru National Institute of Technology Allahabad, Allahabad, India

Sourabh Tyagi Department of Information and Communication Technology, Manipal Institute of Technology, Manipal Academy of Higher Education, Manipal, Karnataka, India

Priyanka Anurag Urla Department of Information and Communication Technology, Manipal Institute of Technology, Manipal Academy of Higher Education, Manipal, Karnataka, India

S. Varun Sri Ramakrishna Engineering College, Coimbatore, Tamil Nadu, India

M. D. Vasan Raksha Shakti University, Ahmedabad, India

G. S. Veena Ramaiah Institute of Technology, Bengaluru, India

Sirisha Velampalli CR Rao Advanced Institute of Mathematics Statistics and Computer Science, University of Hyderabad Campus, Hyderabad, India

K. R. Venugopal Bangalore University, Bangalore, Karnataka, India

Henrique Vicente Centro Algoritmi, Universidade do Minho, Braga, Portugal
Departamento de Química, Escola de Ciências e Tecnologia, Centro de Química de Évora, Universidade de Évora, Évora, Portugal

Shweta Vikram Babasaheb Bhimrao Ambedkar University, Lucknow, India

R. Vineeth Sri Ramakrishna Engineering College, Coimbatore, Tamil Nadu, India

Pratiksha D. Warule Computer Engineering, Datta Meghe College of Engineering, Airoli, India

Sourabh Yadav Gautam Buddha University, Uttar Pradesh, India

Kashif Zahid Dayananda Sagar College of Engineering, Kumaraswamy Layout, Bangalore, India

Snaring Cyber Attacks on IoT Devices with Honeynet

Antara Durgesh Oza, Gardas Naresh Kumar, Moin Khorajiya and Vineeta Tiwari

Abstract Internet of things is a buzzword of this era, which tries to correlate digital gadgets that we have currently. Internet became the backbone of numerous things which interconnect uniquely in already prevailing Internet infrastructure called as the Internet of things. Due to wider implementation of IoT, devices are prone to cyber attacks. To beat those attacks, concept of a Honeypot is implemented in IoT. The Honeypot is the security resource frame up to act as decoy to allure attackers. But this mechanism focuses only on a single system, whereas Honeynet is the network of high-interaction Honeypots. Therefore, we will implement concept of Honeynet in field of IoT. This research concentrates on how we can prevent the man-in-the-middle attack in IoT devices and still be capable of analyzing it using deception. In this research, we have implemented an authorization mechanism, OAuth, in Honeynet using which we will be capable to solve the issue of MITM attacks in IoT.

Keywords Internet of things · Honeynet · OAuth · Man-in-the-middle attack

1 Introduction

Internet of things (IoT) is just what it sounds like, things that have Internet. It is all about how one device is communicating to another. Things can talk to each other, and it is not about just computers, each and every IP-enabled device such as AC, cars, and city trashcans can interconnect with one another. IoT is simply "a network

A. D. Oza (✉)
Gujarat Technological University, Gujarat, India
e-mail: antaraoza@gmail.com

G. N. Kumar · M. Khorajiya · V. Tiwari
CDAC ACTS, Pune, India
e-mail: guru.jasper@gmail.com

M. Khorajiya
e-mail: khorajiyamoin@gmail.com

V. Tiwari
e-mail: vineetat@cdac.in

S.-L. Peng et al. (eds.), *Computing and Network Sustainability*, Lecture Notes in Networks and Systems 75, https://doi.org/10.1007/978-981-13-7150-9_1

of Internet-connected things that can gather and exchange data." The catchphrase "Internet of things" contains mainly two core parts, Internet and things. The Internet is a backbone of connectivity, and things mean objects or devices.

In a couple of decades, you are going to wake up in an absolutely different world where almost each and every device people interact with is having some sort of "sense." However, there is a price for providing devices the ability to talk with each other, to be precisely named as critical security vulnerabilities. Even if a device is separately secure, when it gets connected to other devices forming a network, it becomes vulnerable to numerous attacks [1].

1.1 Architecture of IoT

IoT can be seen as a huge network consisting of not only computers but also different devices connected via series of miscellaneous technologies, e.g., barcode, RFID, Bluetooth, NFC, etc. Wired and wireless mediums are the enablers of its connectivity. Basic architecture of IoT is presented in Fig. 1.

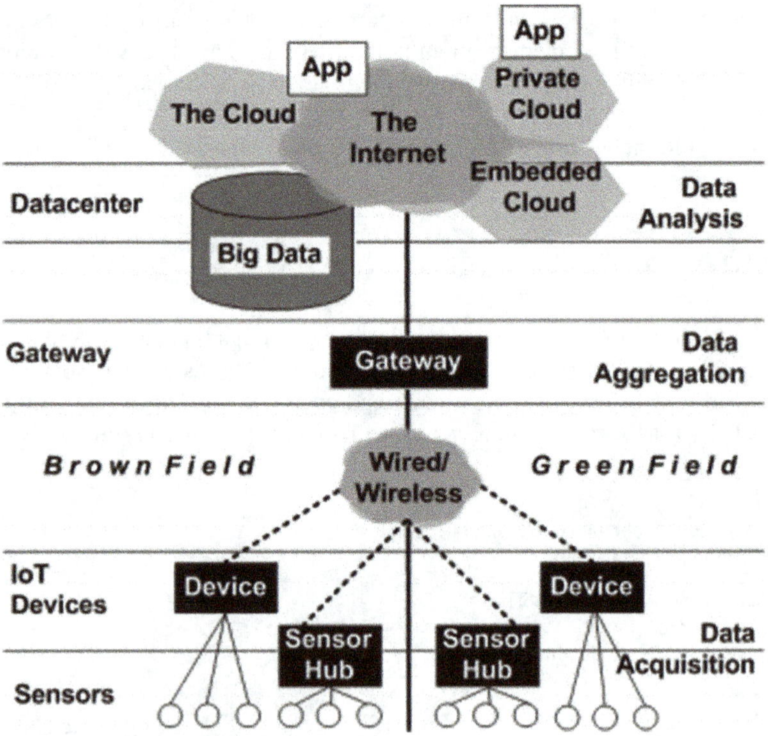

Fig. 1 Architecture of Internet of things

IoT devices mainly involve sensors and actuators. Sensors are responsible for sensing data. On the basis of sensing data, it has to make some action which can be performed by actuators [2]. Its control point is known as a gateway. Data is collected and forwarded to the gateway. The gateway connects to the cloud. Before sending collected data to the cloud, edge computing is performed. Preprocessing of aggregated data before entering into the cloud is known as edge computing. To control and monitor data remotely, applications can be used which will be handled by a user. The term "Greenfield" refers to creating new products from the scratch rather than the term "Brownfield" refers to linking the already existing devices, systems, and infrastructures [1].

The recent developments to improve security in IoT include ways to provide confidentiality, integrity and access control in the IoT, trust and privacy among users and smart things, and enforcement of security and privacy policies. Nevertheless, even after implementing all these methods, field of IoT is vulnerable to numerous attacks intended to disrupt network [3]. Due to this reason, designing another line of defense aimed to detect and defend attacks is necessary. Even though there are many ways to evade those attacks, there is a crucial need for one which can turn table on attacker by using active approach. The best solution is to use Honeypot and Honeynet [1].

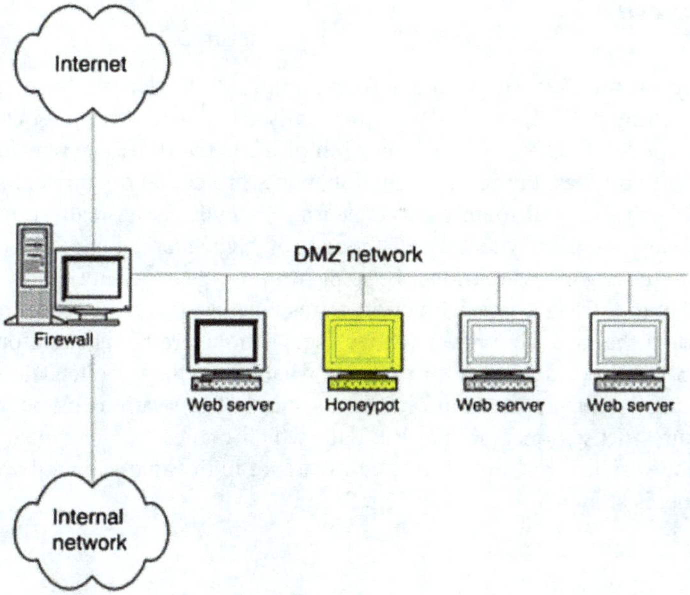

Fig. 2 Simple Honeypot

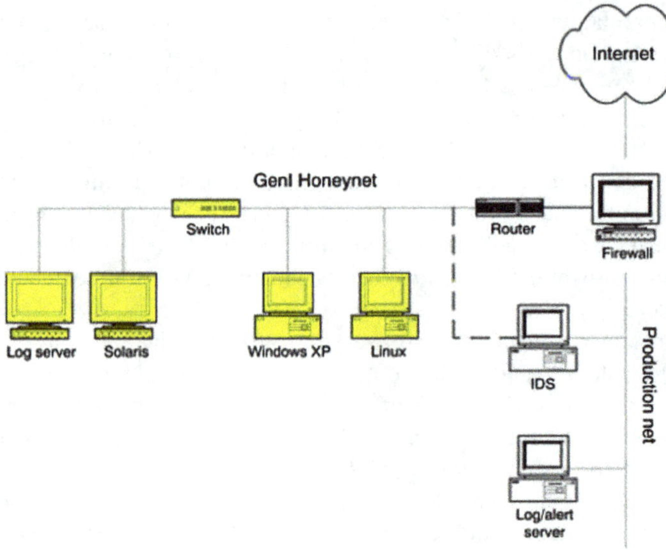

Fig. 3 Architecture of Honeynet

1.2 Honeynet

The Honeypot is the security resource frame up to act like decoy whose importance resides in getting probed, attacked, or compromised [4]. It contains no sensitive data; however, it pretends to be a valuable portion of the network. Honeypots are divided into mainly two types, the production Honeypots protect an organization, whereas research Honeypots will mainly use to learn. Honeynet is basically a network of several Honeypots. Honeynet is well-thought-of high-interaction Honeypot. There are three requirements which must be kept in mind: data capture, data collection, and data control. Data control function is responsible for controlling the flow of data ensuring that attackers never realize their actions are being monitored and to offer them sufficient flexibility to perform whatsoever they want. Data collection focuses on collecting captured information securely from various individual nodes. Data capture focuses on grabbing the data which comes in Honeynet. Honeynet captures the very less amount of data but that is of high value. General architecture of Honeynet is shown in Fig. 3 [1] (Fig. 2).

2 Problem Identification

As development of IoT is catching high peak, MITM attacks in IoT are going to turn into a much bigger concern. MITM attacks often remain undiscovered and unrecorded in the statistics. Traditional security measures like blocking traffic, stop attackers from achieving access to network, but it also blocks the analysis of attack. So we need a strong mechanism where we can prevent the MITM attacks in IoT network and still be capable enough to analyze it.

If we focus on authentication part, then typical techniques in Honeypot and Honeynet authentication involve the comparison of incoming requests with patterns of existing viruses or worms. Traditional authentication methods are not suitable for use where attackers can monitor network traffic and intercept passwords. The use of strong authentication methods that do not disclose passwords is imperative.

Traditional Honeynet mechanisms use IDS to identify whether the traffic is malicious or not. But IDS has one disadvantage; if traffic load increases, it starts generating false positives. Due to that, extra processing time and power of Honeynet will get wasted to analyze traffic which was never really malicious and it will lead to inaccurate results.

3 Implemented Framework

Main idea of this framework is that we have to prevent the MITM attacks in IoT environment using Honeynet, but prevention does not mean that we have to block the traffic. As we are going to design Honeynet, we should be able to capture the attacks and analyze it for further use. So this framework provides deception as well as authentication mechanism to evade the MITM attack. Framework of IoT Honeynet pillared with authentication mechanism is portrayed in Fig. 4.

User sends message request to the IoT device to execute some events remotely by logging in. Request will be checked into the system. If request is new, then its details are not stored in signature database, so it will directly get forward to authenticator. Authenticator provides the interface where it sends message to the valid user (database contains valid user's mail ID and MAC address) along with sender's message request. Here we are using OAuth2 for authentication and authorization. Here Python-OAuth2 library is used for implementing OAuth. If user approves the request, then and only then it will pass to the actual IoT devices and request will get executed. If request message was altered by attacker, then user will get to know it using this authentication mechanism and user will discard the request. If user discards the request, then traffic will be redirected to Honeynet to verify that request discarded by user is actually malicious or not. Sometimes user can discard the request by mistake so it should not be considered as malicious. Then request will be redirected to IDS to check port and protocol. If it is malicious, then details of malicious traffic and signature will get saved in Signature Verification Database. So that next time if same attack happens, then it will able to redirect request to Honeynet directly.

Main working of Signature Verification Database is that if any request is getting matched with the database, then it will not pass that request to authenticator but directly redirect it to the Honeynet for further analysis and sends notification to user that "your original request is altered." As a result, it will reduce the processing time. Using this framework, we will able to prevent the MITM attacks in IoT environment and still be able to analyze it by providing deception. Implemented Framework reduces the problem of false-positive generation and stops Honeynet to lead to inaccurate results. It is explained in pictorial diagram given in Fig. 5.

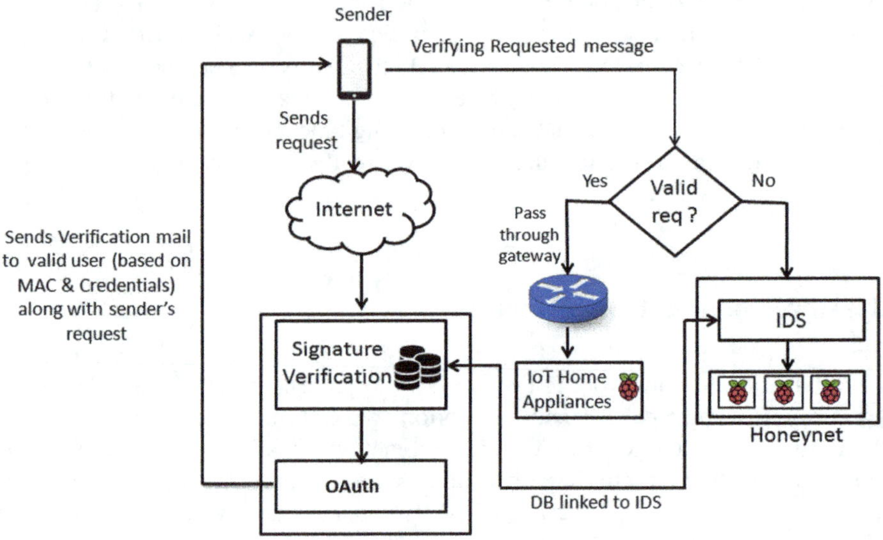

Fig. 4 Architecture of Implemented Framework

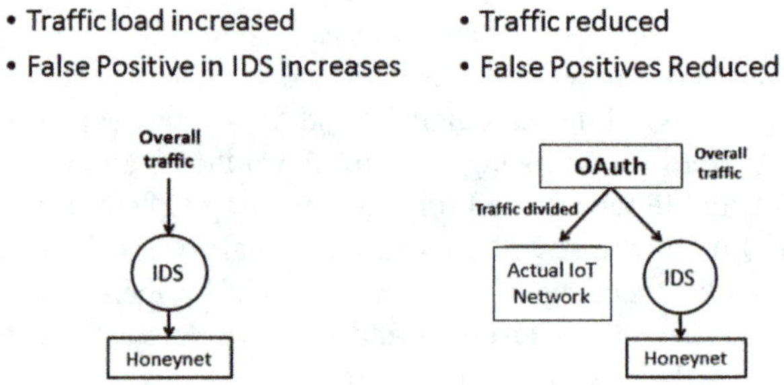

Fig. 5 Reduced network traffic due to OAuth

3.1 Process Flow of Implemented System

See Fig. 6.

3.2 Virtual Honeynet Architecture

See Fig. 7.

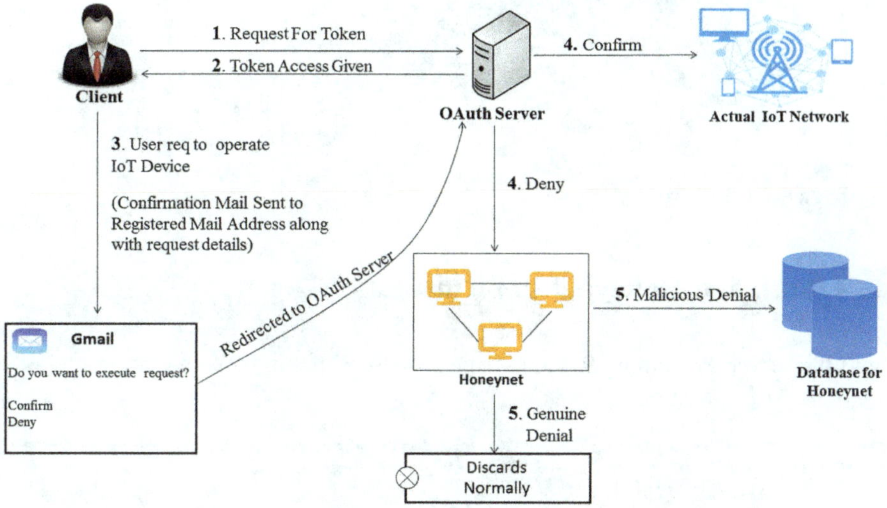

Fig. 6 Process flow of Implemented System

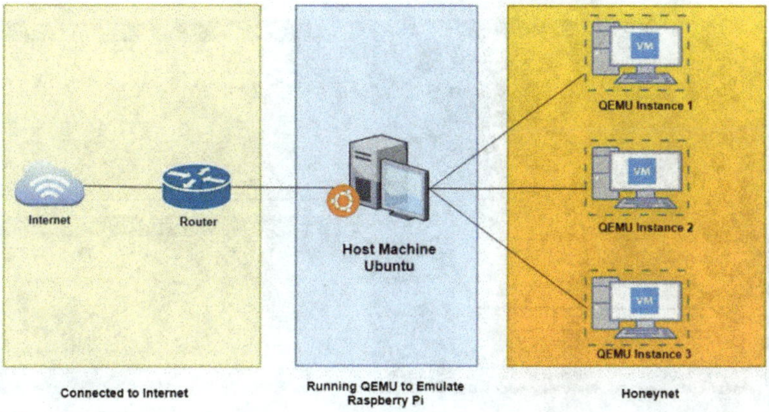

Fig. 7 Architecture of virtual Honeynet

Fig. 8 Actual IoT network

4 OAuth and Honeynet Implementation

Start the server by running Python script. Now server is in listening mode.

```
>>>
============== RESTART: C:\Users\Antara\Desktop\server_oauth.py ==============
Starting OAuth2 server on http://192.168.1.53:8081/...
Visit http://192.168.1.53:8081/ in your browser
```

Run client script to operate IoT device from other machine.

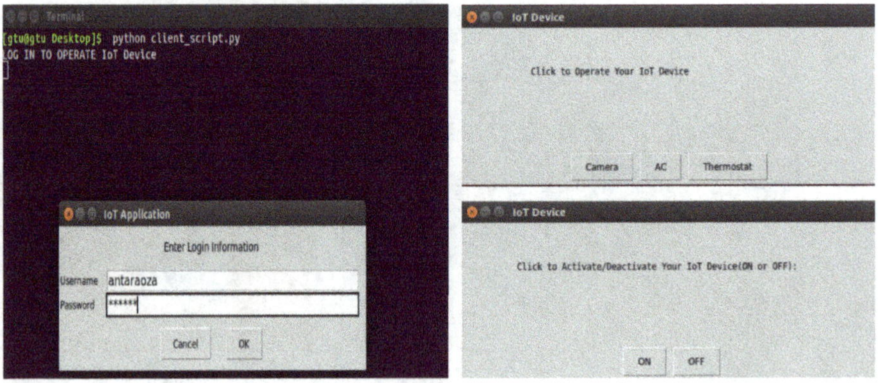

Verification mail will be sent to valid user.

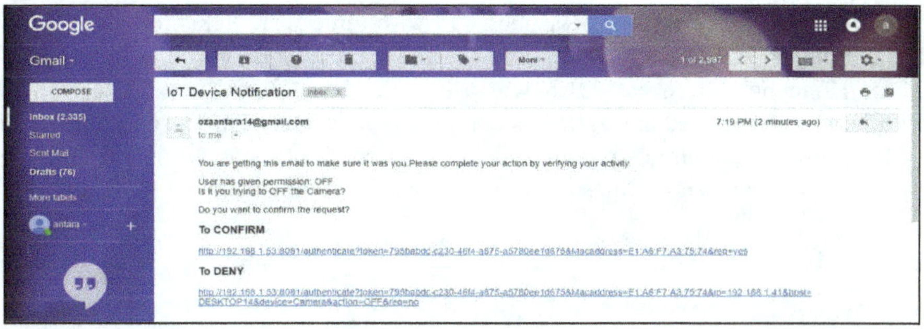

Details captured by Honeynet is getting saved in database for analysis.

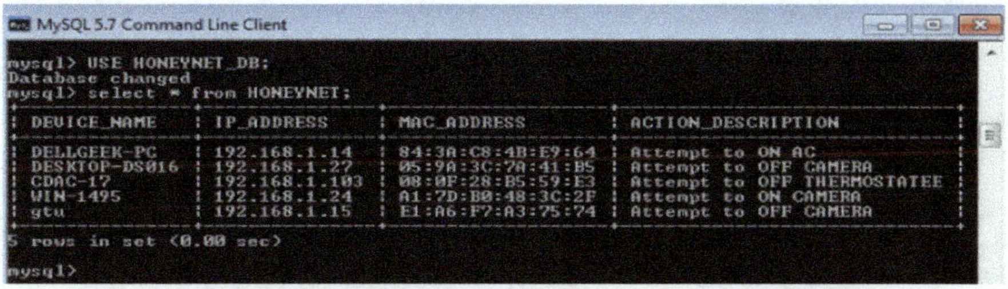

MySQL 5.7 Command Line Client

```
mysql> USE HONEYNET_DB;
Database changed
mysql> select * from HONEYNET;
```

DEVICE_NAME	IP_ADDRESS	MAC_ADDRESS	ACTION_DESCRIPTION
DELLGEEK-PC	192.168.1.14	84:3A:C8:4B:E9:64	Attempt to ON AC
DESKTOP-DS016	192.168.1.27	05:9A:3C:7A:41:B5	Attempt to OFF CAMERA
CDAC-17	192.168.1.103	08:0F:28:B5:59:E3	Attempt to OFF THERMOSTATEE
WIN-1495	192.168.1.24	A1:7D:B0:48:3C:2F	Attempt to ON CAMERA
gtu	192.168.1.15	E1:A6:F7:A3:75:74	Attempt to OFF CAMERA

```
5 rows in set (0.00 sec)

mysql>
```

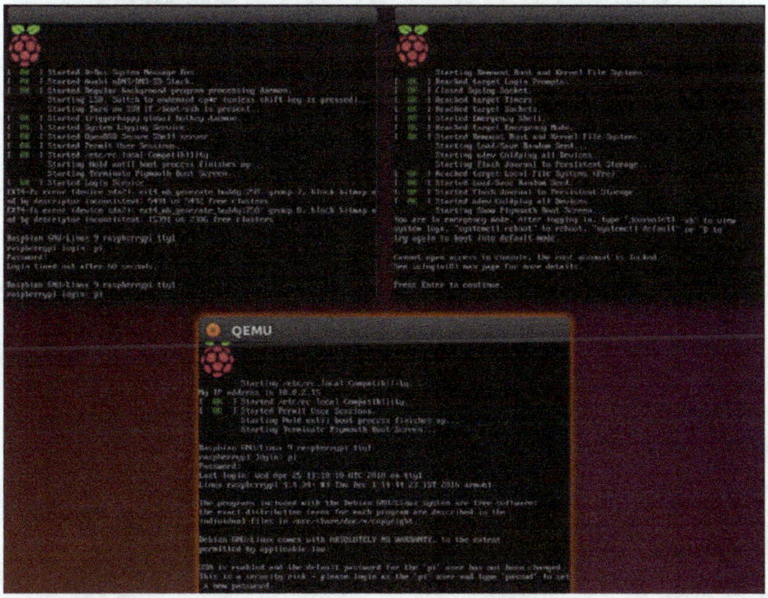

Fig. 9 Honeynet instances running on QEMU

Here actual IoT network is Raspberry Pi with three LED lights for camera, AC, and thermostat, respectively, and QEMU is used to create virtual Honeynet instances. Three Honeynet instances are shown in Fig. 9. Here QEMU is used to emulate Raspberry Pis. We need multiple Raspberry Pis to make the network so virtualization is the best and economic solution in this case. Benefit of using QEMU is that, it is lightweight because it nullifies the weight of any extra management tools (Fig. 8).

5 Result Parameter Analysis

See Figs. 10 and 11.

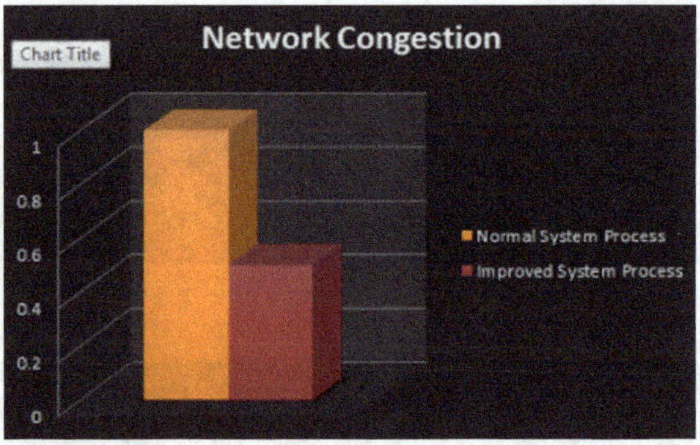

Fig. 10 Network congestion comparison

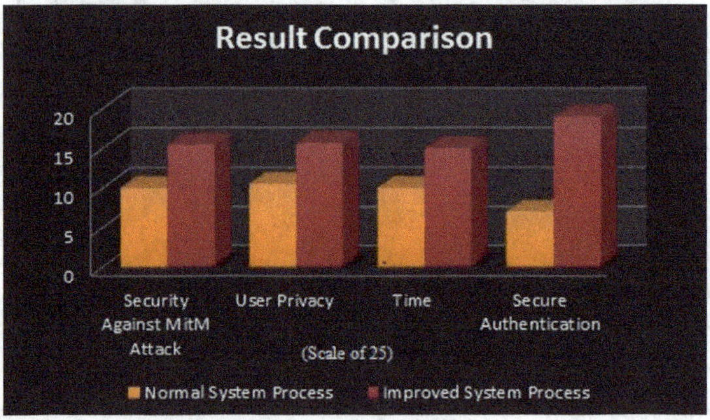

Fig. 11 Parameter comparison

6 Conclusion and Future Work

You have to initiate your footsteps with knowledge that nothing in this world is perfect. Our purpose is not in direction of hunting down the perfection and to live within it. It is to take whatsoever broken pieces we found and sew them together in best possible way we can. This exact rule is applicable to the defense strategy in cyberspace. Security in IoT is like swimming in a sea along with sharks. It is a never-ending war. As new defenses begin to prevent older threats, attackers adapt by way of developing new tricks and tactics to break the security. But then it does not mean that we stop making efforts. This research addressed the issue of how can we prevent the man-in-the-middle attack in IoT devices and still be able to analyze it. Common security measures, e.g., blocking, prevent the attacker from accessing the network but also prevent an analysis of the attack. The redirection of the malicious traffic to the Honeynet for further analysis is a suitable solution. This framework provides the OAuth mechanism that is open authorization standard. OAuth is implemented in Python language. Reinforcing authenticity of user by implementing OAuth into IoT makes it truly sound secure. Implementation of OAuth also reduces network congestion that in turn reduces the false positives. As a result, it increases accuracy in Honeynet analysis. Future directions of our research include focusing on other surface attack areas of IoT.

References

1. Oza AD, M.E. Student, Gujarat Technological University, Kumar GN, Project Engineer, CDAC-ACTS INDIA, Khorajiya M, Project Engineer, CDAC-ACTS INDIA (2018) Survey of Snaring Cyber Attacks on IoT devices with Honeypots and Honeynet. In: 3rd International conference for convergence in technology (I2CT), IEEE
2. Drrajivdesaimd.com (2017) INTERNET OF THINGS (IoT)—Dr Rajiv Desai. [online] Available at: http://drrajivdesaimd.com/2016/07/19/internet-of-things-iot
3. Sherasiya T, Upadhyay H (2016) Intrusion detection system for internet of things. IJARIIE-ISSN(O)-2395-4396
4. Dial.uclouvain.be (2017) Measurements of compromised IoT devices from blackhole and honeypot|Mémoire UCL. [online] Available at: https://dial.uclouvain.be/memoire/ucl/en/object/thesis%3A10671
5. Honeynet.org (2017) Know your enemy: Honeynets|The Honeynet project. [online] Available at: https://www.honeynet.org/papers/honeynet
6. Anas Abd Almonim Nour Albashir (2015) Detecting unknown vulnerabilities using Honeynet, IEEE
7. La QD, Member, IEEE, Quek TQS, Senior Member, IEEE, Lee J, Member, IEEE, Jin S, Member, IEEE, Zhu H (2015) Deceptive attack and defense game in Honeypot-enabled networks for the internet of things, IEEE
8. Zhang W, He H, Kim T (2013) Xen-based virtual honeypot system for smart device. Springer Science+Business Media, New York
9. IEEE Spectrum: Technology, Engineering, and Science News (2017) Internet-exposed energy control systems abound. [online] Available at: https://spectrum.ieee.org/energywise/energy/thesmartergrid/thousands-of-control-systems-connected-to-the-internet

10. Anirudh M, Thileeban SA, Nallathambi DJ (2017) Use of Honeypots for mitigating DoS attacks targeted on IoT networks. In: IEEE international conference on computer, communication, and signal processing, IEEE
11. Sans.org (2017) [online] Available at: https://www.sans.org/readingroom/whitepapers/detection/designing-implementing-honeypotscada-network-35252
12. Miloslavskaya N, Tolstoy A (2017) Ensuring information security for internet of things. In: 2017 IEEE 5th international conference on future internet of things and cloud, IEEE

Image Enhancement Using Morphological Operations: A Case Study

Pratima Manhas, Parveen Arora and Shaveta Thakral

Abstract Morphology consists of two words: morph+ology. Morph means transformation, and ology means a branch of study. Image transformation may take place due to various factors. One of such factors is noise. Hence, morphology is considered to be an image transformation technique which is used to remove imperfections in an image caused by noise. Morphology finds its application in the representation of shape of a region, describing skeletons, surface deviation analysis, and other surface metrological applications. Morphological operations are performed using erosion and dilation algorithms, and the results are verified using MATLAB R2013a.

Keywords Pixel · Morphology · Erosion · Dilation · Transformation

1 Introduction

Morphology is basically an image enhancement technique based on mathematical set theory which with the help of structuring element removes noise while retaining useful objects under interest. Basic morphological operations are erosion and dilation. Certain compound operations can also be implemented under morphology. Opening and closing are such operations. Erosion followed by dilation leads to opening, yet reverse operation, i.e., dilation followed by erosion, leads to closing. These composite operations act as filter and probe an image by smoothing of corners from inside or outside.

P. Manhas · S. Thakral
ECE Department, FET, Manav Rachna International Institute of Research & Studies, Faridabad, India
e-mail: pratima.fet@mriu.edu.in

S. Thakral
e-mail: shaveta.fet@mriu.edu.in

P. Arora (✉)
NEC Technologies India Pvt. Ltd., Chennai, India
e-mail: parveen.arora@india.nec.com

S.-L. Peng et al. (eds.), *Computing and Network Sustainability*, Lecture Notes in Networks and Systems 75, https://doi.org/10.1007/978-981-13-7150-9_2

Erosion operation is based on shrink principle where areas of foreground pixels get shrink and background holes within them expand, while dilation operation is based on expanding principle where areas of foreground pixels grow and background holes within them shrink [1].

2 Erosion Algorithm

Morphology is a collection of nonlinear operations related to shape, and it does not change the pixel numerical value. The erosion and dilation are two types of operations performed on image. The algorithm of erosion has the following steps:

(1) A set of '9' pixels is chosen to perform this operation where one pixel is called point pixel under operation, and rest '8' pixels are called its neighboring pixels. As erosion is based on mathematical operation, AND means output is '1' only if all the inputs are one, indicating if point pixel belongs to foreground and all its neighboring pixels are also belonging to foreground and are set, then point pixel remains set and maintains its position of foreground.
(2) As erosion is based on mathematical operation, AND means output is '1' only if all the inputs are one, indicating if point pixel belongs to foreground and any one of its neighboring pixels is not belonging to foreground and is reset, then point that needs to reset its position is shuffled to background [2].
(3) If point pixel belongs to background, then it maintains its position to background.

In terms of grayscale images, the above algorithm can be summarized as:

(a) Compute the numerical values of the point pixel under operation and its eight neighboring pixels.
(b) Then, find out minimum value among '9' pixels.
(c) Then, the numerical value of the point pixel will be replaced by minimum value found out among 9 pixels.

2.1 *MATLAB Coding: Erosion*

```
clc
clear all;
close all;
I1 = imread('img.jpg');
SE = strel('arbitrary', eye(5));
I2 = imerode(I1,SE);
imshow(I1); title('original');
figure,imshow(I2); title('Eroded')
```

$$\begin{bmatrix} 1 & 0 & 0 & 0 & 0 \\ 0 & 1 & 0 & 0 & 0 \\ 0 & 0 & 1 & 0 & 0 \\ 0 & 0 & 0 & 1 & 0 \\ 0 & 0 & 0 & 0 & 1 \end{bmatrix}$$

Fig. 1 Structuring element

The structuring element is taken as 5 × 5 square matrix where all diagonal elements are one, as shown in Fig. 1.

3 Dilation Algorithm

The other operation is known as dilation. The algorithm of dilation has the following steps:

(1) A set of '9' pixels is chosen to perform this operation where one pixel is called point pixel under operation and rest '8' pixels are called its neighboring pixels. As dilation is based on the mathematical operation, OR means output is '1' only if any (one or more) of the inputs is one, indicating if point pixel belongs to background and at least one of its neighboring pixels is belonging to foreground and is set, then point pixel is shuffled to foreground [3].
(2) As dilation is based on mathematical operation, OR means output is '1' only if any of the inputs are one, indicating if point pixel belongs to background and none of its neighboring pixels is belonging to foreground and is reset, then point needs to maintain its position to background.
(3) If point pixel belongs to foreground, then it maintains its position to foreground.

In terms of grayscale images, the above algorithm can be summarized as:

(a) Compute the numerical values of the point pixel under operation and its eight neighboring pixels.
(b) Then, find out maximum value among '9' pixels.
(c) Then, the numerical value of the point pixel will be replaced by maximum value found out among '9' pixels.

Fig. 2 Structuring element for dilation

$$\begin{bmatrix} 1 & 1 & 1 \\ 1 & 1 & 1 \\ 1 & 1 & 1 \end{bmatrix}$$

3.1 MATLAB Coding: Dilation

```
clc
clear all;
close all;
I1 = imread('img.jpg');
SE= strel('square',3)
I2 = imdilate(I1,SE);
imshow(I1); title('original');
figure,imshow(I2); title('Dilated')
```

The structuring element is taken as 3×3 square matrix where all elements are one, as shown in Fig. 2.

4 Opening Operation

It is one of the morphological operations in which erosion is followed by dilation. In this operation, two images are used—one is the actual image and next is structuring image. It can be useful for separating out of particular objects from the background. For example, suppose an image consists of lines and dots in the same time, and if someone wants to count the dots, then choosing the structuring element as disk-shaped can be used to remove the lines.

4.1 MATLAB Coding: Opening

```
clc
clear all;
close all;
i=imread('leaf.jpg');
subplot(2,2,1),imshow(i);xlabel('original image');
se=strel('disk',1);
y=imopen(s,se);
subplot(2,2,2),imshow(y);xlabel('open image');
```

5 Closing Operation

It is one of the morphological operations in which dilation is followed by erosion. It can be used to remove the noise. MATLAB software tool is used to apply closing operation by using different structuring elements.

5.1 MATLAB Coding: Closing

```
clc
clear all;
close all;
i=imread('leaf.jpg');
subplot(2,2,1),imshow(i);xlabel('original image');
se=strel('disk',1);
y=imclose(s,se);
subplot(2,2,2),imshow(y);xlabel('closed image');
```

6 Results

The results of the following operations are shown in Figs. 3, 4, 5, 6, and 7. The original image and eroded image are shown in Fig. 4. It shows the shrinkage. The white noise is reduced in erosion operation. By performing the operation of dilation, the image

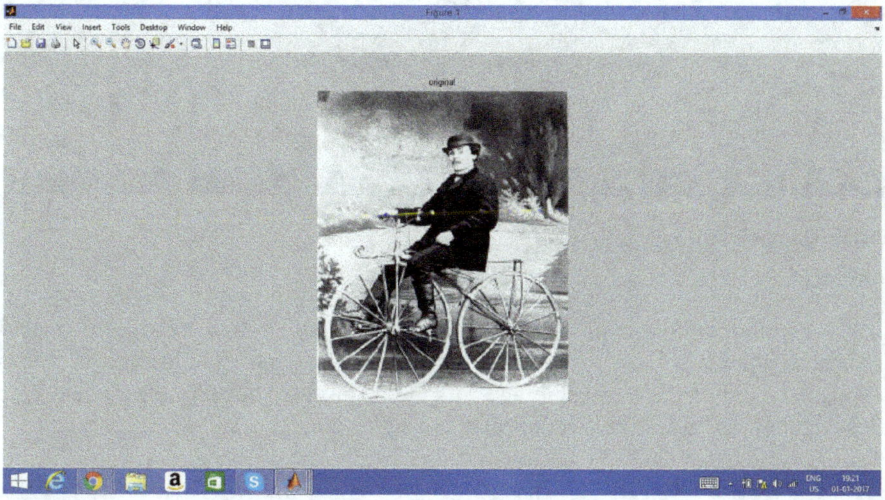

Fig. 3 Original image

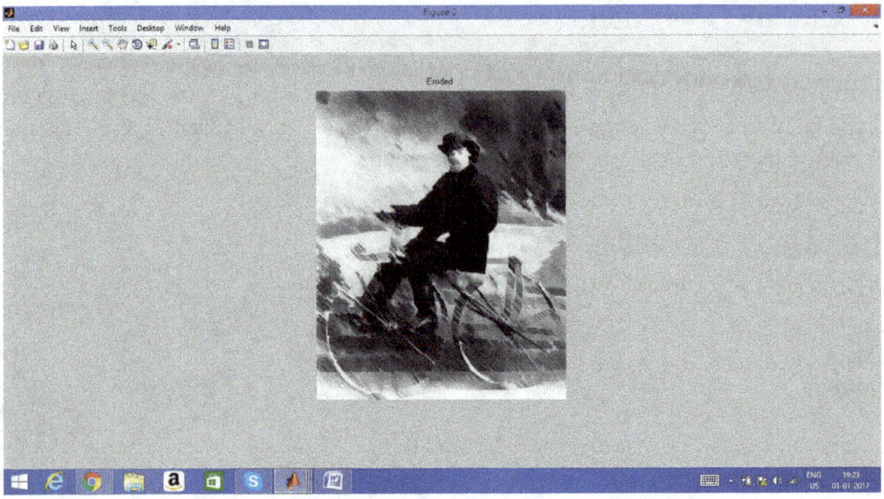

Fig. 4 Eroded image

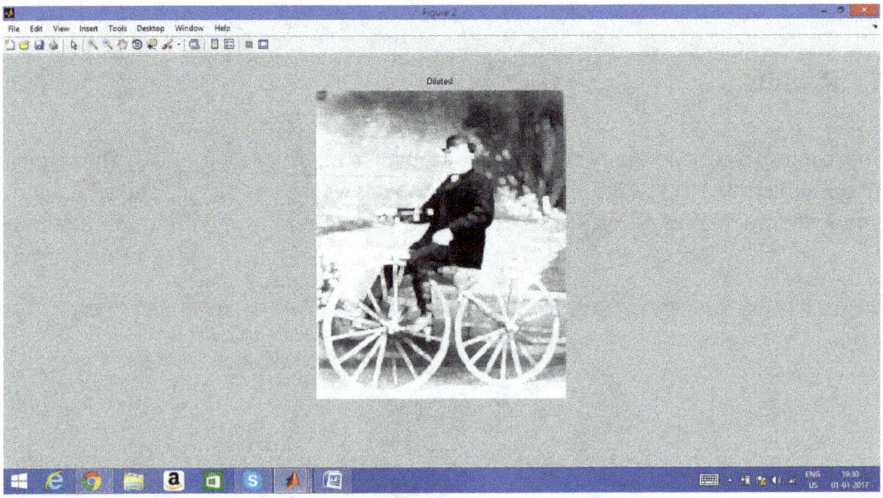

Fig. 5 Dilated image

gets wider and white region is enhanced. The original, eroded, and dilated images have been shown in Figs. 3, 4, and 5. Further, the opening and closing operations are also applied on the original image, and the results are shown in Figs. 6 and 7. In these operations, the noise is reduced.

Fig. 6 Opening operation with the original image

Fig. 7 Closing operation with the original image

7 Applications of Morphology

Morphological operations can be used in image restoration, edge detection, shape recognition, and feature extraction. Image restoration is used to discover the damaged regions of an image. It can be done by extracting the skeletons and boundaries using various morphological operations. Morphological operations can be used to obtain the attributes of the image, and it is a very simple process. This process can be further used for image segmentation and noise filtering. By applying dilation, erosion, opening, and closing to the image, the quality of the image can be further enhanced.

8 Conclusion

Morphology is a technique which is used on binary images, and it requires two inputs such as the original image and the structuring element which is also known as kernel. The kernel decides the mode of operation. In this paper, two modes are shown such as erosion and dilation which provide the shrinkage and enlargement of image. They help in the reduction of white noise, and they were used in extracting the image components (such as boundaries). Further, the work can be modified by using the operation of both erosion and dilation (opening) to encrypt the image, and further decryption can be done by using dilation and erosion (closing).

References

1. Droske M, Rumpf M (2007) Multiscale joint segmentation and registration of image morphology. IEEE Trans Pattern Anal Mach Intell 29(12):2181–2194
2. Gu Y, Liu T, Jia X, Benediktsson JA, Chanussot J (2016) Nonlinear multiple kernel learning with multiple-structure-element extended morphological profiles for hyperspectral image classification. IEEE Trans Geosci Remote Sens 54(6):3235–3247
3. Pratima M, Soni MK (2015) Performance analysis of DWT-OFDM and FFT-OFDM using various digital modulation techniques and channel coding. Int J Comput Appl (0975 –8887) 128(11): 34–39
4. Sternberg SR (1986) Grayscale morphology. Comput Vis Graphics Image Process 35(3):333–355
5. Serra J, Soille P (eds) (2012) Mathematical morphology and its applications to image processing, vol 2. Springer Science & Business Media
6. Batool N, Chellappa R (2015) Fast detection of facial wrinkles based on Gabor features using image morphology and geometric constraints. Pattern Recogn 48(3):642–658
7. Thakral S, Manhas P, Kumar C (2010) Virtual reality and m-learning. Int J Electron Eng
8. Li H, Wang R, Cao S, Chen Y, Tian N, Chen X (2016) Weak signal detection using multiscale morphology in microseismic monitoring. J Appl Geophys 133:39–49
9. Manhas P, Thakral S, Kumar C (2010) Performance analysis of GPRS/EDGE. Int J Electron Eng Res 655–658
10. Hu Z, Wang C, Zhu J, Liu X, Kong F (2016) Bearing fault diagnosis based on an improved morphological filter. Measurement 80:163–178
11. Thornton H (2016) Interaction of morphology and syntax in American Sign Language. Routledge

Evaluation of e-NAM Adoption: A Case of *Jetalpur Mandi*, Gujarat

Gaurav Mishra and Nityesh Bhatt

Abstract Electronic-National Agriculture Market (e-NAM) is being implemented in India to promote uniformity, streamlining of procedures across the integrated agricultural markets. e-NAM aims to reduce information asymmetry between buyers and sellers and facilitate the provision of real-time price discovery. Through a pilot study of the project in *Jetalpur mandi*, Gujarat, the authors intend to bring out the issues and challenges in adoption and implementation by various stakeholders of e-NAM. The study concludes that for e-NAM's successful adoption, aspects related to observability, relative advantage, compatibility, trialability and complexity are important and should be considered in the design of existing and future e-NAM centres.

Keywords IT in agriculture · ICT for development · e-NAM

1 Introduction

Sustained food and nutritional security is a big challenge for the growing population in today's world. According to Wani et al. [17], in India, around 1.4 billion people have to be fed by 2025. Hence, the ever-increasing demand for food puts immense pressure on the agriculture production. The land availability for agriculture is limited, and therefore, better crop production strategies have to be in place to match the ever-increasing food demand. There are many challenges being faced by farmers in the Indian context, for example, crop failures, animal illness lack of credit, lack of irrigation facilities and reliable agriculture-related information. In the literature (e.g. [4]), lack of infrastructure and connectivity technologies has also been cited as the factors which prevent the reach of the right information at the right time to farmers.

G. Mishra
Development Management Institute, Patna, India
e-mail: gmishra@dmi.ac.in

N. Bhatt (✉)
Institute of Management, Nirma University, Ahmedabad, India
e-mail: nityesh@nirmauni.ac.in

S.-L. Peng et al. (eds.), *Computing and Network Sustainability*, Lecture Notes in Networks and Systems 75, https://doi.org/10.1007/978-981-13-7150-9_3

According to Sachan et al. [13], marginal reductions in the post-harvest losses may bring great relief on the food security front as well as improve the income level of the farmers. Marketing of most farm products generally involves many operations and processes through which the food and raw materials move from the cultivated farm to the final consumers. Agricultural markets have grown in size and complexity. Information provision to farmers can play a significant part in improving the livelihoods of farmers. Agriculture could benefit tremendously with the applications of information and communication technologies (ICTs), especially in enhancing socio-economic conditions of farmers. New marketing channels are coming up by both public and private stakeholders and these stakeholders are playing new and evolving roles [16].

In India, Agriculture Produce Marketing Committee (APMC) Act was enacted to enhance market efficiency, increase transparency and protect rights of farmers [7]. On the whole, the APMC Act has helped in getting rid of several malpractices and imperfections in agricultural markets. In addition, it has aided in the creation of orderly and transparent marketing conditions and ensured a fairer deal to farmers selling their produce [1].

However, in the literature, there are indications that the APMC Act has not been able to overcome some of the challenges faced by farmers. Few researchers believe that selling at a designated place, i.e. yard of a *mandi,* has its own disadvantage [2]. Once agricultural produce has been brought to it, it is seldom taken back in the event of an unfair deal. Hence, loss to farmers is difficult to avoid. According to Kumar et al. [5], few 'licensed' or 'registered' participants/agents/traders have a control on the markets and the Act became a deterrent to the development of competitive markets.

2 Electronic-National Agriculture Market (e-NAM)

e-NAM is an electronic trading portal which links the existing APMC *mandis* to create a unified national market for agricultural commodities. All the APMC-related information and services can be gathered from the e-NAM portal. Types of information which can be accessed from portal are: commodity arrivals and prices, buy and sell trade offers, provision to respond to trade offers, among other services. In addition, the portal helps to reduce transaction costs and information asymmetry. According to the information available on the e-NAM website, e-NAM's objective is to provide an integrated market through an electronic platform for efficient flow of agricultural commodities, both at state and at national level. The platform also aims to improve the agricultural supply chain by reducing information distortion between farmers and traders, providing real-time commodity price to farmers and uniformity of procedures across the integrated markets. Through e-NAM, the auction process will be made transparent with the provision of online payments. In addition, it will help the consumers to avail better quality commodity at a reasonable price. e-NAM software is developed by National Informatics Centre (NIC), Hyderabad.

Small Farmers' Agribusiness Consortium (SFAC) at the national level provides guidelines for setting e-NAM. At the state level, State Agriculture Marketing Board implements e-NAM. The monitoring of e-NAM implementation is done by the Board of Governors and Director of APMC. According to the project officials, Ministry of Agriculture pays for the integration costs, customization of software, training, etc., for local *mandis*. Rs. 30 lakhs is provided by the central government and provision of Rs. 30 lakhs is made by the respective state government for setting each e-NAM centre. e-NAM software is installed at each centre by technical support of Nagarjuna Fertilizers and Chemicals Limited (NFCL). The government has entitled SFAC as the lead promoter of e-NAM. At every *mandi*, a laboratory will be established by the government for quality assessment of the commodity to be traded. The laboratory will provide a certificate and potential price of the commodity based on the quality of the commodity.

Through an open tender, NFCL has been selected as a strategic partner (SP) for development, maintenance and support of the e-NAM platform. SP's role is to write software, customize it based on the requirements of *mandis* in states and run the platform. Parameters for quality testing of 69 commodities have been set by SFAC. Initially, the laboratories will do testing of a single commodity, and in future, testing for more than one commodity will be done. Presently, 455 *mandis* across 13 states are live on e-NAM platform.[1] To facilitate commodities trading on e-NAM, common tradable parameters have been developed for 25 commodities. Total 585 e-NAM *mandis* have to be set by the year 2018. Presently, 250 APMC of 10 states have been integrated into NAM platform. After registration in e-NAM, traders are provided username and password. Using these credentials, traders are able to do buying and selling of commodities. According to the e-NAM website, it is necessary for the states to have a single unified licence which is valid across the state, electronic price discovery mechanism and a single point charge of market fee. The states/UTs which follow the above criterion may benefit under the scheme. In addition, the states have to carry marketing reforms as per the provisions laid in the APMC Act and rules and develop requisite ICT infrastructure with respect to e-NAM integration. Also, the State Marketing Boards/APMCs must create awareness for the e-auction platform.

3 Research Methodology

As there is not much information on e-NAM available in the literature, this study is exploratory in nature. Through a pilot study of the project in *Jetalpur mandi*, Gujarat, the authors intend to bring out the issues and challenges in adoption and implementation of the electronic platform by farmers and APMCs. *Jetalpur mandi* was selected for the study as it is the only *mandi* which has adopted e-NAM in Ahmedabad district. Ahmedabad district was chosen because it is the biggest city in

[1] Source: http://www.enam.gov.in/NAM/home/implemented_progress.html, accessed on 03/02/2017.

Gujarat and has convenience of access to the project site. For the study, the authors conducted semi-structured interviews with various project stakeholders, i.e. project officials based at *Jetalpur* as well as from state headquarters based at Gandhinagar, farmers, *mandi* agents and traders. Authors also interacted with officials (local and state levels) from the implementation agency, i.e. NFCL. In addition, other infrastructure support providers (ex. Internet bandwidth) were also contacted for technical inputs.

4 e-NAM in Gujarat

The Gujarat Agricultural Produce Markets Act, introduced since 1963, is implemented in Gujarat for the regulation of agricultural marketing for the development of existing markets and establishment of new market yards. In Gujarat, there are 225 *talukas* (administrative division of India within a district) having *mandis* where agricultural produce is brought for sale. Under the provision of the Act, 207 market committees have been established in Gujarat. There are 190 principal market yards and 222 sub-market yards under the market committees. Presently in Gujarat, 40 *mandis* have enrolled under e-NAM, and according to the project officials, 25 new e-NAMs are to be set in Gujarat. The selection of e-NAMs is based on the volume of transaction done in various APMCs and also on the type of commodity traded. APMCs where fruits and vegetables are traded are not integrated with e-NAM. In Gujarat, major e-NAM centres are located in *Jetalpur, Jasdan*, *Botad*, *Himmatnagar*, *Jamnagar*, *Vijapur*, *Deesa*, *Gondal*, *Tharad*, *Junagadh*, *Mauva* and *Visnagar*. The implementation of e-NAM will be completed in three phases. In the first phase, local farmers will have the provision to trade locally. In the second stage, they will be in a position to trade at the state level. In the third phase, farmers will be able to trade at the national level.

4.1 About Jetalpur Mandi

Jetalpur mandi was started in the year 2007 and is semi-government in nature. e-NAM project in *Jetalpur mandi* was adopted in September 2016. Rice and wheat are the major commodities traded at *Jetalpur* market yard. During the season, 600–700 tractor loads of rice are brought to market yard. There are three seasons of rice, and hence, *mandi* trading for rice is more compared to wheat. On an average, 100–150 tractor loads of commodities come to the market yard daily. The *mandi* works 24 h, i.e. commodity may come and go out the *mandi* at any time. However, an auction takes place from 10.00 a.m. from Monday to Saturday. All the payments to the APMC are done during working hours. Presently, there are 12 employees at the *mandi*.

There are 72 registered traders for e-NAM out of 117 traders (registered by APMC office). Farmers from other districts also come to *Jetalpur* for trading because of two reasons. First, they get a better price due to the presence of a large number of traders in this *mandi*. And second, traders are prompt in giving the agreed amount to farmers, usually in the form of a cheque for a larger amount and cash for a smaller amount. Traders pay a charge of Rs. 0.80 for every Rs. 100.00 of the transaction to the market yard. For gate pass, traders pay Rs. 50 for small vehicles and Rs. 100.00 for larger vehicles. Farmers need to pay Rs. 20 for the gate pass. With respect to the IT infrastructure, e-NAM in *Jetalpur* has: 10 personal computers (8 working) with configuration—Operating System: Windows 10, RAM: 10 GB, Processor: AMD A8 PRO 7600B, 10 cores, 3.10 GHz, 64 bit and X64-based processor; two laser printers; five tablets; one laptop; battery backup for the systems of 8 h; and Internet connection with a bandwidth of 6 Mbps. e-NAM operation at *Jetalpur* is manned by three employees, of which two (on the payroll of NFCL) are working as an analyst. They manage e-NAM transactions, handle challenges with software and send periodic reports to NFCL. The third employee looks after the overall administration of e-NAM. Ishaan Group, a private IT company, provides technical support to e-NAM in Gujarat.

4.2 Process Flow for the Commodity at e-NAMs

According to the e-NAM project officials, e-NAM follows a definite process for trading. The step-wise process flow in e-NAM is:

1. **Gate Entry and lot management**: When a farmer comes to a *mandi*, he/she is given a gate pass. The gate pass has details such as time and date of arrival, the name of the farmer, type of vehicle, weight of the vehicle type of produce and gate pass number. When the vehicle passes through the gate, the commodity brought by farmers has to placed/stored correctly. This is important as there should be no loss of commodity to farmers due to mishandling.
2. **Sampling and Assaying**: During sampling and assaying, a sample of the commodity is taken and is tested for quality. The price of the commodity is set based on the quality parameters. It is time-consuming process and is implemented only in few states like Uttar Pradesh, Rajasthan and Chhattisgarh.
3. **Approval for trade**: After sampling and assaying, based on assessed quality approval is given for trade. However, in Gujarat, the sampling and assaying do not happen. Hence, trade is approved for all the commodities which have gate entry pass.
4. **Bidding and weighment**: There are three parts to bid declaration, i.e. bid creation, actual bidding and bid declaration. Bid creation is of two types—open bid and closed bid. In an open bid, both farmers and traders know the final price. In a closed bid, there is a minimum bid set by APMC and farmers know the price. Last bid cannot be seen by the traders. Multiple bids can be placed by traders, and through auto assign functionality of the e-NAM software, the winner is declared by the APMC staff. Bidding is done on the trader's portal. Bid declaration is of

three types: publish the result, bid extension and exit bid. In publish the result, the final result of a bid is displayed. In bid extension, farmers or traders may extend bid due to several reasons like a technical failure in bidding platform or farmer is not happy with the price. In exit bid, the farmer may exit the bidding process. Once a trader wins the bid, he/she can deny the final bid price to the farmer. The objective of weighment is to know the net movement of commodity entered and exited through the gate of *mandi*. APMC has to keep track and see that there is no imbalance in net inflow and outflow quantities of any commodity.

5. **Sale agreement and settlement**: Once the final bid is made and weighment is done, a sale agreement is made. The sale agreement usually consists of the commission charges, market cess and *hamali* (labour) charges. For settlement, all the stakeholders need to have a bank account which has to be registered with e-NAM.

5 Findings

The *Jetalpur mandi* provided a glimpse of the implementation of e-NAM in Gujarat. However, some important observations were made during the study. It is evident in the literature that most of the ICT initiatives with respect to the empowerment of rural citizens have a low success rate [6]. Innovation is an idea, behaviour or object that is perceived to be new by its audience [10]. e-NAM is a new IT-based initiative which aims to empower farmers by providing a platform where they could sell their produce at a competent price. The authors view e-NAM as an IT artefact. The decision to adopt or reject any new technology is affected by five attributes, namely observability, relative advantage, compatibility, trialability and complexity [11]. Observability, relative advantage, compatibility and trialability show a positive correlation with adoption, while complexity can be found to show an inverse relationship.

5.1 Observability

Observability is the way people see the benefits or results of an innovation. In theory, e-NAM proposes various benefits to its stakeholders. Few farmers and traders opine that bidding on e-NAM takes more time than the traditional bidding process. The time taken for each e-auction is 20–25 min which is more than normal auctioning (which takes maximum of 5 min). Traders believe that current e-NAM platform will not be suitable and useful for trading during peak time when 600–700 paddy tractors come every day for trading. As the project is in its initial phase of life cycle, the stakeholders, i.e. traders, farmers and APMC officials, are not able to verify the benefits of e-NAM. Its only functioning aspect at the *Jetalpur mandi,* i.e. computerized gate pass, will lead to huge queue of vehicles during peak time. Traders do not trust the tradable quality parameters as specified by e-NAM. They have their own quality checking techniques which are less time-consuming. Further, traders perceive that quality checks by e-NAM facility may not even reflect the true quality. In normal auction, they can see the commodity as a whole. But, in sampling, they will only get to know about a small sample.

5.2 *Relative Advantage*

Relative advantage is the added benefit, which an adopter perceives over the old practice. Theoretically, the e-NAM project has many advantages over the traditional way. However, most stakeholders find the added advantage of online transaction painful. For example, in auctioning, it is necessary for all the stakeholders to have bank accounts. However, most farmers do not have bank accounts. Most farmers are small farmers with less produce for trading. They prefer cash transactions over cheques. During our research, one farmer shared his concern, '*I want cash when I sell my produce. How will I verify if trader has put the required amount in my bank account*'? Interestingly, another farmer said '*If money is transferred to bank then I have to withdraw from the bank. For small transactions cash is better. I don't want to run to banks for such sum of money. It costs us a day and money to visit bank also*'. Many farmers do not have bank accounts and they do not want to have them. Some have bank account but they use it rarely. During discussion, traders were found reluctant to share their bank details.

5.3 *Compatibility*

Compatibility is the extent to which any new technology complies with the present social norms and beliefs, past experiences, existing ideas and needs of adopters. Compatibility is one of the key facilitating conditions for the adoption of technology artefacts [3]. e-NAM seems to be incompatible within the social context. First, it entails the use of mobile devices, tablets, computer systems for trading. Almost, all farmers and traders in the study were not savvy in using such electronic gadgets. In addition, use of English as a language in the application adds to the woes. It is observed in the literature (e.g. [15]) that lack of digital skills, access to the Internet, lack of digital experience, language, etc., may lead to 'digital divide'. Those people who are proficient in using digital technology will benefit the most from e-NAM. e-NAM promotes cashless transaction, and therefore, it will be a formidable challenge to have bank accounts opened for all the farmers interested in trading. A farmer said '*I don't have necessary papers to open a bank account*'. Farmers and traders were also found to be apprehensive of sharing their personal details such as bank account, Permanent Account Number (PAN) and Aadhaar (it is a twelve-digit unique identity number based on biometric and demographic data) (in case they have) for trading purpose.

5.4 *Trialability*

Trialability is the extent to which a technology may be tested or experienced with on a limited basis [12]. The trialability of e-NAM is not evident in the data. More or less, it is compulsory for the farmers to use computerized gate pass. However, the farmers were found to be sceptical of the project until they find the real difference between the traditional trading and electronic trading. Digital skills related to finding and assessing information are crucial for using such electronic platform [14]. Trialability will provide the stakeholders to learn more about the services. In addition, they would be in position to know about the digital skills from their peers who have digital skills. Farmers can tap into and benefit from the experiential knowledge of fellow farmers and experts in the farming community [8].

5.5 *Complexity*

Complexity is about ease in using an innovation. Given the social context of farmers, the complexity of the process flow at e-NAM and digital skills of users, the stakeholders will find it difficult to use e-NAM services. The use also becomes difficult due to the lack of standardization of measure in the electronic platform. Farmers are not used to the metric system of measurement. They are more comfortable with *Adha* (half kg), *Sava* (1.250 kg), *Paseri* (5 kg), *Daseri* (10 kg), etc., as units of measurement. However, in the e-NAM system, the price of a commodity is given for 20 kg. Such differences bring in not only complexity but also incompatibility with the existing social belief and norms. The complexity of the e-NAM system increases with IT infrastructure-related challenges like lack of the Internet/server connectivity, electricity, lack of computers, human resource during peak trading seasons.

6 Discussion and Conclusion

e-NAM is a unique project for farmers, and it has the prospect of providing economic benefits to farmers by connecting them to *mandis* across the country in future. However, as discussed in the earlier sections, a lot of work is to be done at the behavioural, strategic and implementation levels. Training programmes and workshops should be conducted for farmers and traders to build awareness and trust towards the services. Also, bottlenecks with respect to the IT infrastructure, digital illiteracy, standardization of measurement, quality checks, etc., have to be taken into consideration for successful implementation. Project officials should continuously drive efforts towards creating an environment where all stakeholders trust each other and have positive attitude towards the project. Authors concur with Potnis [9] that mobile phones can be efficiently used for trading as they have relatively low learning curve, affordability and ease of availability. Employees of *mandi* also need appropriate training as well as positive attitude for digital drive. In addition, use of different models to reduce time in processing of tasks could also be considered. For example, project officials may explore the possibility of linking e-NAM portal with Aadhaar system to reduce time taken for making gate pass at *mandis*.

References

1. Acharya SS (2004) Agricultural marketing in India, vol 17. Millennium Study of Indian Farmers, Government of India, Academic Foundation, New Delhi
2. Chand R (2012) Development policies and agricultural markets. Econ Polit Wkly 47(52):53
3. Chengalur-Smith I, Potnis DD, Mishra G (2016, January) The adoption of IBM's spoken web in information poor communities: a pilot study with farmers in Gujarat, India. In: 2016 49th Hawaii international conference on system sciences (HICSS), IEEE, pp 3878–3887
4. Dwivedy N (2011) Challenges faced by the agriculture sector in developing countries with special reference to India. Int J Rural Stud 18(2)
5. Kumar G, Gulati A, Cummings Jr A (2007) Foodgrains policy and management in India, responding to today's challenges and opportunities. International Food Policy Research Institute, New Delhi and Indira Gandhi Institute of Development Research, Mumbai
6. Madon S (2009) e-Governance for development. In: e-Governance for development. Palgrave Macmillan, UK, pp 53–70
7. Parikh KS (2012) Development policies and agricultural markets. Econ Polit Wkly 47(52):53
8. Patel N, Shah K, Savani K, Klemmer SR, Dave P, Parikh TS (2012, March) Power to the peers: authority of source effects for a voice-based agricultural information service in rural India. In: Proceedings of the fifth international conference on information and communication technologies and development, ACM, pp 169–178
9. Potnis DD (2011) Cell-phone-enabled empowerment of woman earning less than $1/day. IEEE Technol Soc 30(2):39–45
10. Robinson L (2009) A summary of diffusion of innovations. In: Enabling change
11. Rogers EM (1995) Lessons for guidelines from the diffusion of innovations. Jt Comm J Qual Improv 21(7):324–328
12. Rogers EM (2003) Elements of diffusion. In: Diffusion of innovations, 5th edn, pp 1–38
13. Sachan A, Sahay BS, Sharma D (2005) Developing Indian grain supply chain cost model: a system dynamics approach. Int J Prod Perform Manag 54(3):187–205
14. Van Dijk J (2005) The deepening divide: inequality in the information society. Sage, Thousand Oaks
15. Van Dijk J, Hacker K (2003) The digital divide as a complex and dynamic phenomenon. Inf Soc 19(4):315–326
16. Vijayshankar PS, Krishnamurthy M (2012) Understanding agricultural commodity markets. Econ Polit Wkly 47(52):35
17. Wani SP, Bergvinson D, Raju KV, Gaur PM, Varshney RK (2016) Mission India for transforming agriculture (MITrA). Research report IDC-4

A New Technique for Extrinsic Text Summarization

Nishita Kindo, Gananatha Bhuyan and Ronali Padhy

Abstract Text summarization is the process where a text document is downsized, with a computer program, so as to create a summarized text and possesses the most essential points of the primary text. The present work is one new type of extrinsic technique of summarization, which finds an exemplary subset of the original sentence set, which possesses the precise erudition of the primary text.

Keywords Text summarization · NLP · Intrinsic summarization · Extrinsic summarization · Text mining · Term frequency · Inter document frequency · Sentence weight · Redundant sentence · Clustering · OCA

1 Introduction

Before knowing what text summarization is, first we need to understand what is a summary? A summary [1] is nothing but the important information extracted from the primary text and is shorter than that of the primary ones. The aim of automatic or machine-generated text summarization is constricting the primary text into a stipulated version without losing the overall meaning or gist of the content. Some portions of the primary text can be treated as the summary in an descriptive way that covers all congruent information of the text. The summarized text has the advantage of reduced reading time which is quite efficient for readability.

We can classify the overall text summarization [2, 3] methods into extractive-based summarization and learning-based summarization. In extractive summarization method, important sentences, paragraphs, etc., are being selected from the pri-

N. Kindo (✉) · G. Bhuyan · R. Padhy
School of Computer Engineering, Kalinga Institute of Industrial Technology University, Bhubaneswar, Odisha, India
e-mail: nishita.kindofcs@kiit.ac.in

G. Bhuyan
e-mail: gananatha.bhuyanfcs@kiit.ac.in

R. Padhy
e-mail: ronali.padhyfcs@kiit.ac.in

S.-L. Peng et al. (eds.), *Computing and Network Sustainability*, Lecture Notes in Networks and Systems 75, https://doi.org/10.1007/978-981-13-7150-9_4

mary text and integrating them into briefer form by eliminating irrelevant words. The importance of sentences is decided based on the weight of sentences. This paper focuses on neoteric techniques based on extractive text summarization methods.

Text document summarization is the technique of condensing a text. The primary text is given as input to the system, and a condensed or briefed text is resulted, which is an extract from the primary text with no redundancy [4, 5].

This technique started in the 1960s and has been advancing in the last 40 years and has become more important with the extensive usage of the Internet. Since 1997, Microsoft Word has been using text summarizer. Text document summarization is based on the weight of the sentences where the system calculates how frequently a particular keyword appears in the text file. The keywords belong to the soi-disant open class words. The frequency of the keywords is calculated by the summarization system, i.e., which keyword present in which sentence and which sentence present where in the text. It checks whether the tagged text has bold letter, numerical values, etc. Finally, these data are collected and processed to summarize the primary text.

Text document summarization can be used:

- To reduce the size of news to a small-sized SMS or WEB format for mobile phones.
- To allow the system synthetically read the compressed text as the whole text can be lengthy and boring.
- Easy to represent the compact illustration in various search engines like Google, Yahoo, Bing, etc.
- While searching for the meaning of foreign languages, the summarized translated text helps rather than the lengthy description.
- It reduces time while reading multiple texts in a compressed way.

2 Basic Ideas and Theory of Operation

2.1 Text Mining

Text mining is also known as text data mining, a part of data mining where it digs out meaningful words as information from the text document. Meaningful information is primarily obtained through casting of patterns and trends by virtue of statistical pattern learning. Customarily text mining is nothing but the process of configuring the given text (disintegrating the text by adding some obtained linguistic features and removing some other features and then inserting into the database subsequently), obtained patterns from the processed data, and finally evaluating and interpreting the result. Meaningfulness in text mining refers to combinations of pertinence, uniqueness, and nosiness.

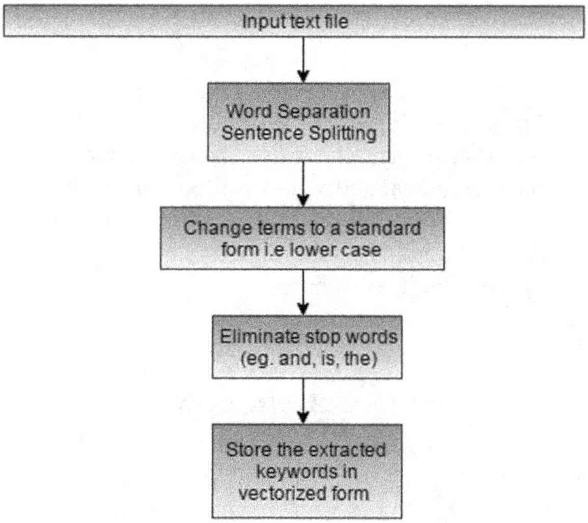

Fig. 1 Image of text preprocessing

2.2 *Text Preprocessing*

The first thing to do in text preprocessing is to alter documents, which basically is a character string, into something that is suitable for the learning algorithm and the classification task. The text alteration (text transformation) basically is of the following types:

- Word separation, sentence splitting;
- Remove HTML or other tags;
- Remove stop words, punctuations.

The stop words are those repetitive words that don't hold any useful information, i.e., pronouns, conjunctions, prepositions, etc. (Fig. 1).

2.3 *Sentence Extraction*

Sentence extraction is a technique used for summarizing the text automatically. In this superficial approach, for identifying the most intrusive sentences of a text statistical heuristics are used. There are many approaches that are more knowledge-concentrated which mainly requires added knowledge such as ontologism or linguistic knowledge or weight of sentence, whereas sentence extraction is quite cost-effective. In brief what this current technique does is filters the important sentences to the end users making it more user-friendly.

2.4 *Multi-text Summarization*

Multi-text summarization is a programmed method designed for extracting information from multiple text files drafted about the same topic. The summary resulted from this technique allows users, who use the same for professional work, to make themselves acquainted with information comprised in a large collection of texts. This is how a multi-text summarization system consummates the news combiners to transform the lengthy news into tiny SMS without reducing the information and is easily readable in any phone.

3 Overview of Current Approach

In this work, to get the summary of the input texts, the extractive method has been employed. Below-mentioned features are used to get the summary extracted:

1. *Content words or keywords* (w_k): The keywords are the leftover words that have been obtained after eradicating all the stop words such as pronouns and prepositions. The total number of keywords has been taken during the assignment of weight to each term.
2. *Term frequency* (t_f): The count of a particular keyword or term in the text.
3. *Sentence feature*: In a text document, the first sentence of the first paragraph has more importance and hence more likely to have been included in the summary. So here the starting sentence has been included in the summary and counting the number of sentences in the text. And assigning weights to the keywords to find the weight of the sentence. Sentence weight is calculated according to original text and then sorted in descending order of sentence weight.
4. *Normalized term frequent* (t_f): Normalized term frequency of the ith keyword k_i,

$$t_{f_i} = \left(\frac{f_i}{f_1 + f_2 + \cdots + f_n}\right) \times w_k$$

   where f_i = frequency of the keyword k_i for $1 \le i \le n$
   w_k = total number of keywords.
5. *Inter document frequency* (*IDF*): Inter document frequency IDF_i of the keyword k_i is the ratio of the number of text files in which that keyword k_i is occurring (n_i) and the total number of text files (n).

$$\text{Therefore, } \mathrm{IDF}_i = \frac{n_i}{n}$$

6. Weight of the keywords k_i in a sentence $s_i = \left(t_{f_i} + \mathrm{IDF}_i\right)$ weight of a sentence $= \sum_{i=1}^{j} t_{f_i} + \mathrm{IDF}_i$ where there are j different keywords in sentence s_i. Now, we arrange the sentences in descending order of their weights.
7. *Sentence grouping*: Here, each sentence from the texts is compared with every other sentence in that text file to form a group or cluster [6–8] of similar sentences having same meaning. If a cluster is having a cardinality greater than one, then the normalized weight of all sentences of the clusters is computed by the formula:

$$w_{s_i} = \frac{\text{weight of the sentence}}{\text{lenght of the sentence}}$$

   where w_{s_i} is the weight of the sentence s_i in the cluster or group.
   Then, the sentence s_i in the cluster for which w_{s_i} is maximum is selected and all the others are deselected and thus are regarded as redundant sentences.
8. *Evaluation*: Compare the system-generated summary with the same benchmark or standard summary usually human-generated summary.

4 Algorithm for Present Technique

a. Algorithm for single text file

Input: Text file for summarization, stop words file for extracting keywords and threshold.

Output: Sorted sentences in descending order of the sentence weight (only keywords).

Steps:

i. Extract keywords k_i
ii. Find occurrence of keywords k_i in the text
iii. Determine the keywords (most frequent words) in the text greater than threshold (th), i.e., k_i>th
iv. Determine weight of keyword k_i,

$$w_i = \frac{t_f}{n}$$

v. Determine the weighted sum of all the sentences

$$w_{t_s} = \frac{w_{t_i}}{n} \quad i = 1$$

where w_{t_s} = effective weight of the sentence.
$w_{t_1}, w_{t_2}, w_{t_3}, \ldots, w_{t_n}$ are the weights of individual terms in that sentence.
n = total number of terms in that sentence.

vi. Finally, arrange the sentences in descending order of weighted sum.

b. Algorithm for multiple text file

Input: Text file for summarization, stop words file for extracting keywords and threshold.

Output: Sorted sentences in descending order of the sentence weight (only keywords).

Steps:

i. Extract keywords k_i
ii. Find occurrence of keywords k_i in the text
iii. Determine the keywords (most frequent words) in the text greater than threshold (th), i.e., k_i>th
iv. Determine weight of keyword k_i,

$$w_i = \frac{t_f}{n}$$

v. Determine the weighted sum of all the sentences

$$w_{t_s} = \frac{w_{t_i}}{n} \quad i = 1$$

where w_{t_s} = effective weight of the sentence.
$w_{t_1}, w_{t_2}, w_{t_3}, \ldots, w_{t_n}$ are the weights of individual terms in that sentence.
n = total number of terms in that sentence.

vi. Cluster all the redundant sentences by comparing each sentence from the text with every other sentence in that text file to represent the sentence having the maximum ratio of the cluster, i.e.,

$$\max\left(\frac{w_i}{l_i}\right)$$

where w_i = weight of the keywords
l_i = length of the sentences.

vii. *Inter document frequency* (*IDF*): Inter document frequency of word w_i is the ratio of the number of text files in which that word is occurring (n_i) and total number of text files (n).

$$\text{IDF} = \frac{n_i}{n}$$

viii. Then, the weight of the sentence in that text file is $[(tf + idf)]$. This is performed in all the text files in the folder.
ix. Step 1 to step 8 is repeated for all the text files in the folder.
x. Finally, sequentially arrange the sentences in descending order of weighted sum in all the text files of the folder to get the summarized text file for each file of the folder (displays only 25% of the summarized text).

c. Algorithm for clustering meaningful sentences
Input: Text files for clustering, pre-defined threshold
Output: A set of sentences with maximum ratio
Steps:

i. Extract weighted sentence w_{ts} in a text file
ii. Grouping sentences based on commonality of keywords between the sentences, given a pre-defined threshold 'th'
iii. For $1 \leq i \leq n \,\forall\, s_i \in G_i$
iv. Then, redundancy is removed by keeping the sentence which has maximum ratio, i.e.,

$$\max\left(\frac{w_i}{l_i}\right)$$

v. Finally, meaningful sentences are obtained from the text file.

5 Experimental Results

The platform is evaluated using an Intel(R) Xeon(R) CPU E5645 @2.40 GHz having 8 GB RAM and Windows 7 64-bit OS.

The English text file data of different subjects like accounting, geography, biology, economics, AI, etc., are collected from the Internet (www.worldscientific.com, en.wikipedia.org, scienceworld.scholastic.com www.ncert.nic.in/ncerts/textbook/textbook www.textbooksonline.tn.nic.in/books, and many more) (Table 1).

Table 1 Performance evaluation

S. No.	Input folder	Input text file	No. of input texts files	Time for summarization (s)	Accuracy of summarized text	Average accuracy
1.	ACC TXT	ACC1.TXT	7	348.36	54.44	38.6
		ACC2.TXT		171.66	64.55	
		ACC3.TXT		1135.5	36.67	
		ACC4.TXT		1054.27	37.78	
		ACC5.TXT		1371.41	28.18	
		ACC6.TXT		184.38	31.42	
		ACC7.TXT		816.52	17.14	
2.	AI TXT	AI1.TXT	6	628.88	54.44	40.96
		AI2.TXT		1167.64	45.71	
		AI3.TXT		1287.47	29.05	
		AI4.TXT		1763.79	41.57	
		AI5.TXT		1358.62	37.27	
		AI6.TXT		1443.6	37.77	
3.	BIO TXT	BIOLOGY1.TXT	5	501.36	75	48.9
		BIOLOGY2.TXT		293.76	75	
		BIOLOGY4.TXT		100.84	32.22	
		BIOLOGY5.TXT		386.33	35	
		BIOLOGY6.TXT		1160.42	27.65	
4.	ECO TXT	ECO1.TXT	8	527.31	60	46.34
		ECO2.TXT		931.48	55.45	
		ECO3.TXT		585.78	68.6	
		ECO4.TXT		299.69	26.67	
		ECO5.TXT		618.78	50	
		ECO6.TXT		507.99	35	
		ECO7.TXT		526.06	35	
		ECO8.TXT		575.26	40	
5.	GEO TXT	GEO1.TXT	6	374.44	55.45	53.68
		GEO2.TXT		1791.71	50	
		GEO3.TXT		491.83	60	
		GEO4.TXT		365.29	55.45	
		GEO5.TXT		1445.95	38.57	
		GEO10.TXT		935.77	62.63	

6 Conclusion

Since this summarization technique follows the extraction method, it may happen in some cases that one sentence contains the noun and the next sentence contains the pronoun referring to the noun in the previous sentence. While summarizing, if the system considers the second sentence only, then it may look inappropriate and incomprehensible. The above issues are unavoidable and at the same time a major drawback in text summarization. So to give a meaning to such type of summarized text, we are in process to resolve this type of anaphoric problems occurring in text summarization.

Here, we have illustrated a generic analysis of text summarization executing the extractive method. The status and state of summarizing a text have entirely changed through the years. It can be more beneficial from the work of other tasks, e.g., information extraction, information retrieval, or text categorization. As text summarization task is not finished yet and there is still a lot of effort to put, much to investigate and to improve, therefore, research on this field will still abide. Different approaches and evaluation techniques along with feature, types, and definition of text summarization have already been exposed and developed. In the future, we are planning to contribute toward the improvement of this domain by ameliorating the standards of summaries and analyzing the effect of other related tasks on summarization.

References

1. Nalini Dr K, Jaba Sheela L (2014) Survey on text classification. Int J Innovative Res Adv Eng (IJIRAE) 1(6). ISSN: 2349-2163
2. Salton G, Singhal A, Mitra M, Buckley C (1997) Automatic text structuring and summarization. Inf Process Manage 33(2):193–207
3. Jones KS (2007) Automatic summarizing: the state of the art. Inf Process Manage 43(6):1449–1481
4. Sarker G, Roy K (2013) An RBF network with optimal clustering for face identification. In: International conference on information & engineering science-2013 (ICIES-2013)
5. Sarker G (2010) An unsupervised natural clustering with optimal conceptual affinity. J Intell Syst 19:289
6. Shah K, Chaterjee N, Saha B (2004) A new approach to a single pass optimal clustering algorithm in data mining
7. Sarker G, Besra M, Dhua S, An optimal clustering for fuzzy categorization of cursive handwritten text with weight learning in textual attributes. In: Accepted for publication in International conference on advances in computer engineering and applications
8. Deng J, Hu J, Chi H, Wu J (2010) An improved fuzzy clustering method for text mining. In: 2010 Second international conference on networks security wireless communications and trusted computing (NSWCTC), vol 1, pp 65–69
9. Sarker G (2007) A heuristic based hybrid clustering for natural classification. Int J Comput Inf Technol Eng (IJCITAE) 1(2):79–86
10. Alguliev RM, Aliguliyev RM (2005) Effective summarization method of text documents. In: Proceedings of the 2005 IEEE/WIC/ACM international conference on web intelligence (WI'05), France, pp 264–271, 19–22 September 2005

11. Sarker G, Besra M, Dhua S, A Malsburg learning Bp network combination for handwritten alpha numeral recognition. In: Accepted for publication in International conference on advances in computer engineering and applications
12. Yan Y, Chen L, Tjhi W-C (2013) Semi-supervised fuzzy co-clustering algorithm for document categorization. Knowl Inf Syst 34(1):55–74

Dynamic Provisioning of Cloud Resources Based on Workload Prediction

Sivasankari Bhagavathiperumal and Madhu Goyal

Abstract Most of the businesses nowadays have started using cloud platforms to host their software applications. A cloud platform is a shared resource that provides various services like software as a service (SAAS), infrastructure as a service (IAAS) or anything as a service (XAAS) that is required to develop and deploy any business application. These cloud services are provided as virtual machines (VM) that can handle the end-user's requirements. The cloud providers have to ensure efficient resource handling mechanisms for different time intervals to avoid wastage of resources. Auto-scaling mechanisms would take care of using these resources appropriately along with providing an excellent quality of service. The researchers have used various approaches to perform auto-scaling. In this paper, a framework based on dynamic provisioning of cloud resources using workload prediction is discussed.

Keywords Auto-scaling · Horizontal scaling · Vertical scaling · Virtual machine · Cloud server · Load balancer · Load measurer · Load predictor · Load detector

1 Introduction

Cloud computing is the popular technology that provides the distributed infrastructure or services that include environments to host vendor applications, network resources, build and deploy applications. The users like online stores, webmasters tend to prefer the cloud services like Amazon Elastic Compute Cloud (Amazon EC2), Microsoft Azure that offers resources in terms of virtual machines (VM) instead of setting up their infrastructure [1]. Scaling is one of the most prominent features which allocate resources dynamically based on the volume of the request to the server. The cloud servers offer two types of scaling namely horizontal scaling and vertical

S. Bhagavathiperumal (✉) · M. Goyal
Faculty of Engineering and Information Technology, Centre of Artificial Intelligence, University of Technology Sydney, PO Box 123, Broadway, NSW 2007, Australia
e-mail: Sivasankari.bhagavathiperumal@student.uts.edu.au

M. Goyal
e-mail: madhu.goyal-2@uts.edu.au

S.-L. Peng et al. (eds.), *Computing and Network Sustainability*, Lecture Notes in Networks and Systems 75, https://doi.org/10.1007/978-981-13-7150-9_5

scaling. Horizontal scaling is connecting more servers at the same time to do the same work increasing the speed or availability of the logical units. Vertical scaling is the ability to raise the power of the server like increasing the RAM of the computer to improve the speed and performance of the computer.

Auto-scaling is provisioning of resources through virtual machines ensuring the quality of service in accordance with the service level agreements (SLA). For example, consider a situation where X number of machines are serving N number of customers at the peak time of stock exchange. Suddenly, the unanticipated number of customers uses the same application increasing the number of nodes. This is when the performance of the application slows down violating the SLA. At this time, an efficient auto-scaling system would be required to manage the load and balance the service [2].

Moreover, auto-scaling ensures that the required sources of services are supplied seamlessly when the demand is high and reduce the supply when the demand decreases. The automated solution to this horizontal and vertical scaling would benefit both the cloud providers and the users of this cloud services concerning with utilization of resource wisely and cost-effectively along with increasing the performance. However, identifying the required resources would be challenging as the demand fluctuates from time to time [2]. Therefore, it is important to build the frequently occurring events in the system so that the cloud systems predict the requirements of the users [3].

The cost of the service is reduced if less resources are leased resulting in the performance getting affected at the peak hours [4]. As shown in Fig. 1 [1], the auto-scaling mechanism starts with the end-user sending a request to the application through a device connected with the Internet. The application forwards this request to the virtual machines which is connected to the load balancer. The applications has the auto-scaling mechanism when the decision of the virtual machines required are decided and request from the user is served [1].

2 Related Works

The cloud community has proposed various researches to perform the auto-scaling. An efficient process of auto-scaling is proposed based by IBM called as MAPE loop [1]. The authors suggested a cost-efficient scale down commands which is more focused towards selecting redundant virtual machines. The authors presented an efficient auto-scaling mechanism to adapt to adequate services with lesser operational expenses. The performance of this model was based on four phases of monitoring, analysing, planning and execution. According to the authors, the implementation of the virtual machines will take place after careful monitoring and analysing only. This model suggested an executor named as Suprex that executed the plan of auto-scaling that can handle the resources based on the requests from the end-users. Another theory proposes the implementation of autonomic computing by the model suggested by

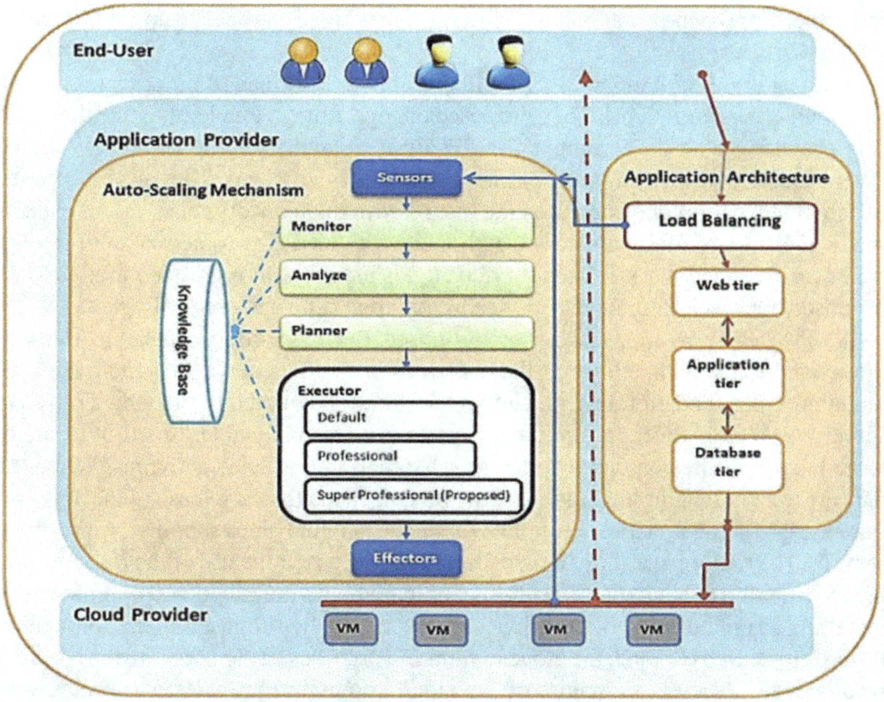

Fig. 1 Architecture of cloud-based Web applications [1]

IBM called as MAPE-K loop to manage the elements of any software and hardware resources [3].

An approach implemented by Amazon that aimed at selling their unused capacities based on auction like mechanism which were called as spot instances was suggested [5]. These spot instances were suitable mainly for fault-tolerance flexible Web applications. The cost of this instance was low compared with on-demand price. Moreover, this approach can be used for applications that can be interrupted. Applications like background processing, batch jobs can utilize this approach compared to the critical applications. On the other hand, the spot instances also take more time to boot when compared to the on-demand instances. So, the authors suggested a heterogeneous approach where a mix of both spot and on-demand instances can be used to meet the end-users' demand.

Another model suggested a repacking approach where they investigated how to repack the virtual machines to provide the required services most efficiently [6]. In addition, another theory proposed a concept of the efficient auto-scaling scheme (EAS) which minimizes the processing time by implementing a huge number of cores in the Internet of Things [7]. Other studies researched on the workload prediction approach to enhance the service to the users reducing the cost by forecasting the expected load [2]. While the workload prediction model was successful, there were few other approaches that focused more on workload and came up with a linear model [4].

A linear regression model which will predict the workload of the services used in the cloud was proposed [4]. They proposed an algorithm to efficiently scale the cloud services along with developing a cloud scaling architecture. The authors discussed the various cloud service providers and cost of service cloud. They have proposed different algorithms and compared the results with related works and various aspects like REMICS and FP7 project. The authors investigated the problem of auto-scaling based on predicted workloads in service clouds using linear regression model. They proposed approach to scale the service in both real time and pre-scaling.

Although various approaches were proposed, the main idea of all the theories was reducing the cost with delivering the resources in an efficient way to benefit both the cloud providers and cloud users. Unlike the traditional method of having a physical server, the cloud servers are reliable, secured and maintainable. Small business to large business relies on cloud computing for their data-related activities. When the data are transmitted through the cloud servers, the need for the release of the data also increases. Moreover, data are supplied on the virtual machines supplied by the cloud servers. The major benefit of auto-scaling is paying as per the use of the resource and using as per the demand of the request. Compared with the traditional approaches of installing an application on their physical servers and then connecting the application through a secured channel, the cloud computing provides easier access to the required resources [8]. The elastic nature of the cloud computing provides the resources in proportion to the volume request [9].

3 Proposed Framework for Auto-scaling Based on Workload Prediction

In this paper, an n-tier architecture is proposed for the auto-scaling framework based on the workload prediction as shown in Fig. 2.

The framework consists of four layers, i.e. end-users' layer, application layer, VM layer and predictor layer. The predictor layer consists of four different process namely load detector, load measurer, load analyser and load predictor.

It takes the responsibility of delivering the required VMs making the application run based on the user demands. All these processes contain a storage base that will store the data of load. The processes of predictor layer are described in Fig. 3.

3.1 End-User Layer

The end-user layer is where the business customers use Internet to access the business application to perform their required task. The request for accessing the applications

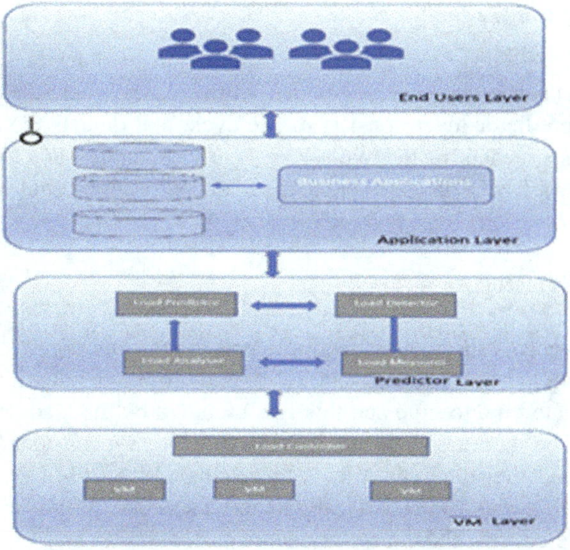

Fig. 2 Framework for auto-scaling based on workload prediction

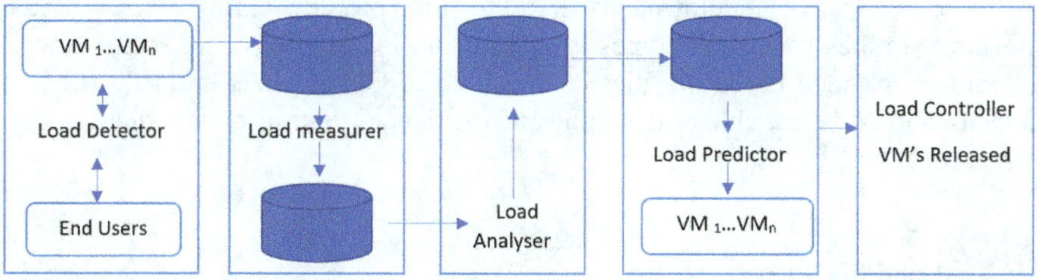

Fig. 3 Flow chart for the proposed framework

varies at different time intervals and thus based on the number of end-users' request, the VMs should be released. At certain times, there might be enormous number of users trying to reach the application layer and there might be less or no requests also.

3.2 *Application Layer*

This is the layer where the business applications are hosted. This layer may consist of numerous services, applications and databases. The end-users would send the request to this layer. To maintain a quality of service, this layer should be active always irrespective of the demands.

3.3 *Predictor Layer*

The prediction layer is the main component of this paper where the scaling process happens. The predictor layer consists of life cycle that detects the signal received from the end-user measuring the request, analysing the request and finally predicting the required VMs. These data are sent to the load controller present in the VM layer.

3.3.1 Load Detector

This process receives the request from the end-user to the application along with detecting the number of virtual machines released to the handle the demands. The load will fluctuate time to time and this gets recorded by the load detector actively. The results will be shared with the load measurer. When we consider S_i as the service provided by cloud, then the VMs utilized are calculated for the t-th interval along with calculating the requests [3]. Auto-scaling can be successful only when the metrics of the load is detected with suitable granularity.

In general, there are two possible approaches which are homogeneous approach where the resource pool is of same size and heterogeneous approach in which distinct size of resources is allowed. Most of the cloud providers offer different virtual machine families for diverse types of applications. Currently, there scaling mechanism are provided based on rules or threshold of CPU utilization [10]. The load detector should be capable of detecting such variances to ensure the request always is recorded.

3.3.2 Load Measurer

A linear regression model is used to measure the load. Linear regression model calculates the current values of series against the prior values in the series. The general form of linear regression is given as follows [3]:

$$Y_{t+1} = \beta_1 + \beta_2 * X_1$$

where t is indexes, Y_1 is the incoming workload and X is the actual value of the instance at that moment. The load can also be measured using the moving average method or exponential smoothing methods. In moving average, the mean of the n last values is calculated; on the other hand, the exponential smoothing method decrease values in each value of time series [11].

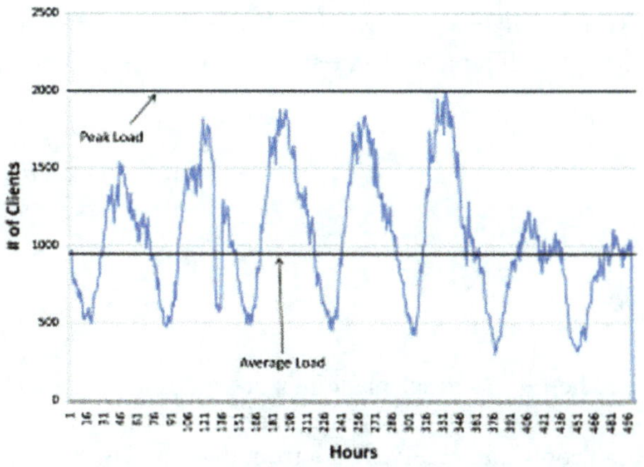

Fig. 4 World Cup Soccer 1998 workload [2]

3.3.3 Load Analyser

The load analyser task is to analyse the load that is measured. This will help the controller control the release or ceasing of the VM. The analyser will take the responsibility of analysing both resource release and time. The end-user's request is analysed and validated by the load analyser. After care analysis, the data are supplied to the load controller. The load analyser uses the auto-regressive or linear regression method to analyse the load so that the data of analysed load help the load controller predict the future loads.

Consider the example of World Cup Soccer 1998 workload [2] shown in Fig. 4.

In the above figure, the workload is calculated to granularity of hours versus number of clients. The graph shows the fluctuation of the resources; while at peak time, the load is huge and at some normal hours, the load is low, the load analyser should impose these variations and save the data to be provided for the load controller.

3.3.4 Load Predictor

The load predictor uses the Auto-regressive integrated moving average (ARIMA) to release the VMs based on the prediction calculated by ARIMA method. ARIMA pattern follows a very popular time series model which is a sequence of measurements over time, usually obtained at equally spaced intervals which can be—daily–monthly–quarterly–yearly. This can also be calculated based on seasonal trends. Seasonal trends are the patterns in time series algorithm where the sequence of measurement is done seasonally. For example, a year could have a seasonal trend of four seasons which typically repeats for every season.

Given a time series of data X_t, where t is an integer index of the time interval and X_t are real numbers which represents the number of end-users an ARMA (p', q) is given by

$$X_t - \alpha_1 X_{t-1} - \cdots - \alpha_{p'} X_{t-p'} = \varepsilon_t + \theta_1 \varepsilon_{t-1} + \cdots + \theta_q \varepsilon_{t-q}$$

3.4 VM Layer

The VM layer is where the virtual machines are ready to auto-scale based on the requirement. This layer consists of the n number of virtual machines and the load controller. The load controller receives data from the predictor layer and ensures that the required VMs are released to the meet the end-users' demand.

4 Conclusion and Future Works

In this paper, a framework for dynamically provisioning of cloud resources is proposed. This would help in handling the increasing volume of data throughput of enterprises and businesses which are moving towards the cloud computing services. It will also be helpful with analysis of configuration issues on provisioning these resources dynamically especially when the applications are running during the peak hours. The future studies will specifically focus on prediction of provisioning vertical or parallel processing of resources using data mining techniques specially to meet the requirements of big data.

References

1. Aslanpour MS, Ghobaei-Arani M, Nadjaran Toosi A (2017) Auto-scaling web applications in clouds: a cost-aware approach. J Netw Comput Appl 95:26–41
2. Roy N, Dubey A, Gokhale A (2011) Efficient autoscaling in the cloud using predictive models for workload forecasting. In: 2011 IEEE 4th international conference on cloud computing
3. Ghobaei-Arani M, Jabbehdari S, Pourmina MA (2016) An autonomic resource provisioning approach for service-based cloud applications: a hybrid approach. Future Generation Comput Syst
4. Yang J et al (2014) A cost-aware auto-scaling approach using the workload prediction in service clouds. Inf Syst Front 16(1):7–18
5. Qu C, Calheiros RN, Buyya R (2016) A reliable and cost-efficient auto-scaling system for web applications using heterogeneous spot instances. J Netw Comput Appl 65:167–180
6. Sedaghat M, Hernandez-Rodriguez F, Elmroth E (2013) A virtual machine re-packing approach to the horizontal vs. vertical elasticity trade-off for cloud autoscaling. In ACM international conference proceeding series
7. Kim HW, Jeong YS (2016) Efficient auto-scaling scheme for rapid storage service using many-core of desktop storage virtualization based on IoT. Neurocomputing 209:67–74

8. Dutreilh X et al (2010) From data center resource allocation to control theory and back. In: Proceedings—2010 IEEE 3rd international conference on cloud computing, CLOUD 2010
9. Mogouie K, Mostafa Ghobaei A, Shamsi M (2015) A novel approach for optimization auto-scaling in cloud computing environment. Int J Mod Educ Comput Sci 7(8):9–16
10. Sahni J, Vidyarthi DP (2017) Heterogeneity-aware adaptive auto-scaling heuristic for improved QoS and resource usage in cloud environments. Computing. Archives for Informatics and Numerical Computation, 2017 99(4):351–381
11. Lorido-Botrán T, Miguel-Alonso J, Lozano JA (2013) Comparison of auto-scaling techniques for cloud environments

Conceptual Machine Learning Framework for Initial Data Analysis

M. S. Smitha Rao, M. Pallavi and N. Geetha

Abstract This century has witnessed the emergence of new branch of science—data science that facilitates the analysis of large amounts of data which in turn helps in taking model-based data-driven decisions. The prelude to any successful analytical model building and implementation phase is a properly conducted initial data analysis stage. IDA encompasses laborious tasks of data cleansing: missing value treatment, outlier detection, checking the veracity of data, data transformation, and thus preparing data for model building. A systematic, disciplined, and non-personalized approach to IDA reduces the probability of incorrect and inaccurate results from the model. The amount of data presented for model building today makes the IDA stage a very crucial task which cannot be manually conducted. Machine learning can be applied to analyze complex and bigger data, find patterns accurately, etc. Hence, it could also be used for data preparation prior to model building. This paper tries to reduce the ad hoc nature of IDA by providing a conceptual framework using machine learning.

Keywords IDD · EDA · Machine learning · Feature engineering · Support vector machines · Regression · Principal component analysis

1 Introduction

Today humongous amounts of data are being collected from innumerable sources in numerous formats. This data when analyzed could provide: excessive insights to comprehend the past behavior, understand the current performance, and predict the future trends. This century has witnessed the emergence of new branch of

M. S. Smitha Rao (✉) · M. Pallavi · N. Geetha
Department of CSE, SOET, CMR University, Bangalore 562149, India
e-mail: smitha_rao_ms@yahoo.com

M. Pallavi
e-mail: pallupavana@gmail.com

N. Geetha
e-mail: geetha.vtu@gmail.com

S.-L. Peng et al. (eds.), *Computing and Network Sustainability*, Lecture Notes in Networks and Systems 75, https://doi.org/10.1007/978-981-13-7150-9_6

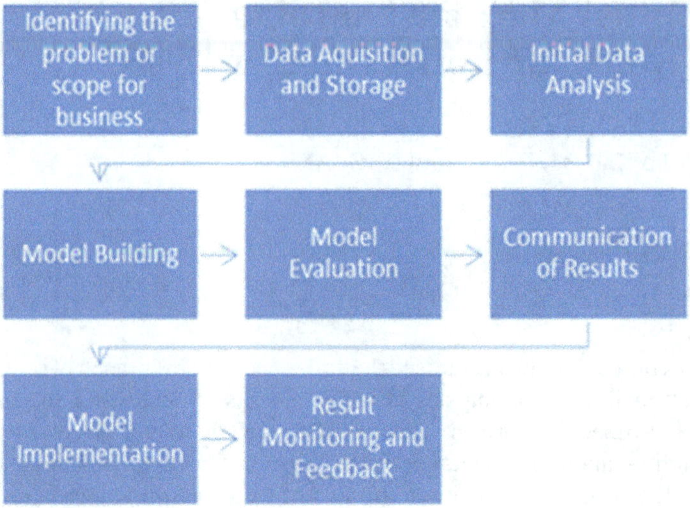

Fig. 1 Data science process

science—data science that facilitates the analysis of large amounts of data and helps in taking data-driven effective decisions. Data science is a large umbrella encompassing data cleansing, preparation, data analytics, machine learning, and everything related to gaining insights from data. With a massive increase in the volume of data captured and stored today, corporates are investing in data science to optimize business processes, to find solutions to business problems, and also to find new avenues for business. Business intelligence (BI) uses data analytics to provide actionable information to various business stakeholders that in turn help them to take effective strategic and tactical decisions. Data analytics is a multidisciplinary approach that has mathematical foundation, uses statistical tools and machine learning algorithms to logically analyze data in order to find patterns in them. The data science process through an elaborate and exhaustive process is the fundamental need for the corporate today. Major stages in the data science process (in Fig. 1) are identifying the problem or scope for business, data acquisition and storage, initial data analysis (IDA), exploratory data analysis (EDA), model building, model evaluation, visualization and communication of results, model implementation, and finally result monitoring and feedback. To make this data suitable for pattern mining and pattern analysis, it has to be passed through data preprocessing phase [1]. Initial data analysis is the prelude to successful model building and implementation. This phase is conducted independently of the fundamental analysis phase post-phrasing the research statement identified by the problem faced.

A systematic, disciplined, and non-personalized approach to IDA reduces the probability of incorrect and inaccurate results from the model. In [2], data envelopment analysis (DEA) model might be useful to screen training data so a subset of examples that satisfy monotonicity property can be identified. In [3], the authors have created database of cleaned data post-data cleaning, user identification, and session identification. In [4], authors present several data preparation techniques of access stream before the mining process can be started and these are used to improve the performance of the data preprocessing to identify the unique sessions and unique users.

The central idea in this paper is to bring out an approach to IDA that uses machine learning to perform various tasks involved at this stage.

Organization of the paper is as follows. Concepts and significance of machine learning and machine learning for initial data analysis are presented in Sects. 2 and 3, respectively. A conceptual machine learning framework for initial data analysis is presented in Sect. 4. Finally, conclusions are drawn in Sect. 5.

2 Machine Learning

Machine learning is a strategy of data analysis that automates analytical model building. It comes under artificial intelligence based on the thought that systems can learn from information and recognize designs and patterns followed by decision making with less human intervention. Models in machine learning are exposed to new data, able to adapt independently which leads to have high priority to the iterative aspect which learns from previous computations to produce repeatable reliable decisions and results. The iterative viewpoint of machine learning is vital as models are uncovered to modern information; they are able to adjust. Widely publicized illustrations of machine learning applications are self-driving Google car, online recommendation, sentiment analysis, detecting fraudulent transactions, etc. Today machine learning has been enhanced with statistical techniques of Bayesian analysis and data mining. Machine learning can be applied to analyze complex and bigger data, find patterns accurately, etc. Hence, it could also be used for data preparation prior to model building. Algorithms in machine learning are categorized as supervised, unsupervised, semi-supervised, and reinforcement learning [5].

- **Supervised Machine Learning**: These algorithms use labeled examples to predict future new data events from data which has been learned in the past. These algorithms can also be used to compare generated outputs with expected proper and intended output and hence to find errors and abnormalities which needs to be modified accordingly.
- **Unsupervised Machine Learning**: These algorithms are used when the data is not labeled, and we need to find patterns and classify them accordingly. Though they are not used for predictive analytics, these algorithms like *K*-means and hierarchical clustering use the inherent nature of the data to draw inferences and prepare the data for other algorithm implementations.
- **Semi-supervised Machine Learning**: These algorithms use both labeled (small amount) and unlabeled (large amount) data for training.
- **Reinforcement Machine Learning**: These algorithms interact with its environment by producing actions and discover errors or rewards. It is a type of learning

which makes decisions based on which actions to take such that the outcome is more positive. The learner has no knowledge which actions to take until it has been given a situation [6].

3 Machine Learning for Initial Data Analysis (IDA)

Data analytics, characterized as the automated part of data science, involves the process of obtaining raw data, preparing it for modeling, and finally applying various modeling algorithms to this data that results in information which helps in effective decision making. García et al. [7] explore the various data preprocessing in data mining. Initial data analysis (IDA) is one of the very crucial pre-modeling stages in data analytics. It ensures that the data presented to the modeling stage is accurate, complete, concise, and unambiguous. The impact of different preprocessing steps on the accuracy of machine learning algorithms for various aspects like sentiment analysis [8] has also been extensively researched. There are no formal set of activities and procedures defined for IDA due to which the demarcation between IDA, exploratory data analysis (EDA), and modeling has become blurred. Today due to the increase in the number and type of data sources, varied data formats, and complexity of data sets, standardized practices for pre-modeling stage are essential to ensure stable and robust models. In addition to the various pre-modeling activities involved, these activities are susceptible to individual subjectivity and hence a disciplined approach to each of these activities is essential. EDA aims at understanding the data by trying to establish patterns, simplifying its complexity, framing initial hypothesis for modeling, etc. While EDA phase involves usage of visual and statistical tools and techniques to explore and summarize data, IDA focuses on preparing data for EDA and subsequently for modeling. Post-data acquisition and data integration, IDA phase involves data preparation activities like data cleansing, data standardization, missing value treatment, outlier treatment, data transformation, etc. In many analytics processes, EDA and IDA are iterative processes. IDA activities are guided by the extent of quality of data, measurements and transformations of data. Various activities involved in IDA can be primarily classified into two groups: data cleansing and data transformation. Data cleansing can be further classified as the process of treatment of the following deficiencies in data: inaccuracy, missing values, outliers, etc. Data transformation would involve rescaling of data and feature synthesis.

3.1 Data Cleansing

Data veracity is one of the primary requirements for any analytics process; hence, one of the important IDA activities is data cleaning. Efficient anomaly detection mechanisms are becoming an urgent and critical topic in the presence of big data applications [9]. Data cleaning involves sub-activities like identifying corrupt, incomplete, incorrect, invalid, inaccurate, irrelevant parts of the data and taking corrective actions like rejecting, validating, correcting, modifying, filtering, removing, replacing, and verifying the errors and ambiguities in the data. Data cleaning process ensures reliability, relevance, and correctness of data by accessing the quality of data at the initial stage itself. The various machine learning algorithms for the different phases of IDA are:

(1) **Inaccurate Data**: Data that has not been maintained or entered correctly would be one of the causes for bad data. Inaccurate data may generate invalid reports which in turn lower productivity and affect the revenue of a business. One of the algorithms that help to improve accuracy by reducing inaccurate data is support vector machine.
(2) **Missing Value Treatment**: Missing values occur when no data value is stored for the variable in an observation. This can have a significant effect on the conclusions that can be drawn from the data. Data could be of qualitative/quantitative type. Missing data would be treated by listwise or case deletion, pairwise deletion, mean/median/mode substitution, regression imputation, last observation carried forward, maximum likelihood, expectation-maximization, multiple imputations, sensitive analysis, etc. Missing values in the data set can also be detected by decision tree approach.
(3) **Outlier Treatment**: An outlier is an observation point that is distant from other observations. An outlier may be due to variability in the measurement or it may indicate experimental error. This may lead to poorer results. Outliers can be treated by Naïve Bayesian, Apriori techniques, capping, imputation prediction, etc.

3.2 Data Transformation

Data transformation is the process of converting data from one format or structure into another format or structure. Data transformation can be simple or complex based on the required changes to the data between the source (initial) data and the target (final) data. Tools and technologies used for data transformation can vary widely based on the format, structure, complexity, and volume of the data being transformed. The various categories of data transformations include data rescaling and feature synthesis.

(1) **Data Rescaling**

Rescaling can be used if there is a need to change the type of data measurements. It is also often used to compare two data sets that are otherwise incomparable because they use a different scale. Data rescaling in some cases is mandatory prior to modeling as it increases accuracy of the model when the data has consistent scale and distribution. The major techniques for rescaling are: normalization and standardization

- *Normalization*: It is rescaling of the data from the original range so that all values are within the required range usually 0 and 1. Various techniques used for this are min-max scaling or *Z*-score scaling. Normalization works provided the minimum and the maximum can be accurately estimated.
- *Standardization*: Standardization is centering data by rescaling the values so that the mean of the values is 0 and the standard deviation is 1. Unlike normalization which can be applied to any data set, standardization assumes that the data is normally distributed.

(2) **Feature Synthesis**

With massive amounts of data and abundant features, data modeling becomes a subjective component. Feature synthesis involves arriving at relevant feature set which improves the precision of the model. The various techniques in feature synthesis are feature engineering, feature selection, etc. Feature engineering is performed initially to create new set of features from the existing attributes, followed by feature selection to arrive at a final set of features which would be fed to the modeling algorithm.

- *Feature Engineering*: Feature engineering involves creating new features for the data that has substantial influence on the final model built. Considering the correlation and other dependencies between the attributes of the data, one can arrive at a new feature set that can be used to build a model. It involves using domain knowledge intelligently to combine or split existing raw attributes into new features which have a higher predictive power. The main techniques in feature engineering are: linear regression, various types of binning, statistical, or log transformations, decision tree, genetic programming, inductive logic programming, annotation-based approaches, generic feature construction methods like *K*-means, SVD, deep feature synthesis (DFS), etc.
- *Feature Selection*: It is the process of ranking the attributes by their value to predictive ability of a model. Algorithms such as decision trees automatically rank the attributes in the data set. Feature selection methods can be classified as filters, wrappers, embedding methods, etc. Filter methods are based on statistical tests measuring intrinsic properties of the data set like information gain, chi-square test, correlation coefficient, variance threshold, etc. Wrapper method includes recursive feature elimination, sequential feature selection algorithms, and genetic algorithms. Embedded method includes L1 (LASSO) regularization, etc. Dimension reduction techniques like principal component analysis (PCA), non-negative matrix factorization, and discriminant analysis are a few of the effective feature selection and extraction methodologies implemented.

4 Machine Learning Framework for IDA

Figure 2 provides the framework for a disciplined approach to IDA. The basic IDA framework involves three basic tasks: initial data diagnostics (IDD), data cleansing,

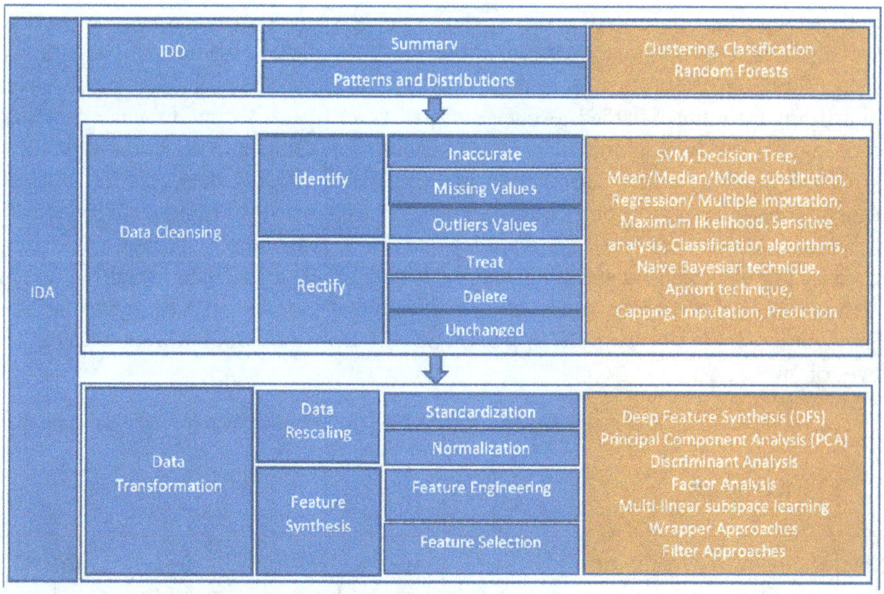

Fig. 2 IDA process

and data transformation. Data is provided to IDD post-data acquisition which itself comprises of tasks such as identifying the source, source integration, data extraction, and data storage. IDD phase identifies patterns in the data by using summary statistics or in some cases exploratory data analysis (EDA). IDD phase helps in finding abnormalities in data which is later used to resolve errors, inaccuracies, outliers, missing values, etc. IDD produces a report identifying abnormalities in the data which is used in the data cleansing task. The next phase in IDA is the cleansing task which involves identifying errors, missing values, outliers in the data and taking corrective measures for each of them. The rectification process could be treatment, deletion, or leaving the values unchanged. Machine learning algorithms like SVM, classification, Naïve Bayesian approach, Apriori technique, regression imputation, last observation carried forward, maximum likelihood, expectation-maximization, multiple Imputation, sensitive analysis, decision-tree approach, capping, imputation, prediction, etc., can be used to treat any abnormality identified. The next task in IDA is data transformation which involves data rescaling and feature synthesis. Data rescaling could be done either through standardization or normalization. Feature synthesis further involves subtasks like feature engineering and feature selection. Machine learning algorithms like principal component analysis, discriminant analysis, factor analysis, multi-linear subspace learning, wrapper approaches, filter approaches, regression, regularization, etc., could be used for feature synthesis. After each sub-phase of IDA could result in report being generated summarizing the findings, benchmarks, and corrective actions for each of the abnormal behavior identified.

Every data set presented to the model building algorithm inherently consists of two significant components: pattern and noise. If the pattern is identified accurately, the variation in the data is due to the stochastic noise. Model building involves identifying the pattern in the presence of noise. Over fitting occurs if the noise is tried to fit into the model which leads to good training set results and poor test set results. Noise component occurs due to abnormalities in data during data capture, storage, transmission, conversion, etc. These abnormalities have to be eliminated or reduced to reasonable limits to arrive at good models for prediction and classification. Machine learning algorithms used at each stage of IDA ensures that the data presented finally to the model building algorithm is a set of relevant features free of abnormalities. The data set so obtained post-IDA phase can be used for model building with higher level of confidence.

5 Conclusion

Machine learning techniques provide accurate models for data analytics, provided the data is cleansed and is presented to the model building algorithm with relevant features. This paper explains the significance of initial data analysis and presents a systematic and disciplined framework for IDA highlighting the need for various machine learning algorithms in the various tasks of this preprocessing phase of analytics.

References

1. Dwivedi SK, Rawat B (2015) A review paper on data preprocessing: a critical phase in web usage mining process. In: 2015 International conference on green computing and internet of things (ICGCIoT), Noida, 2015, pp 506–510
2. Pendharkar PC (2005) A data envelopment analysis-based approach for data preprocessing. IEEE Trans Knowl Data Eng 17(10):1379–1388. https://doi.org/10.1109/tkde.2005.155
3. Suneetha KR, Krishnamoorthi Dr R (2009) Data preprocessing and easy access retrieval of data through data ware house. In: Proceedings of the World congress on engineering and computer science 2009, vol I WCECS 2009, October 20–22, 2009, San Francisco, USA, 978-988-17012-6-8
4. Sudheer Reddy K, Kantha Reddy M, Sitaramulu V (2013) An effective data preprocessing method for Web Usage Mining. In: 2013 International conference on information communication and embedded systems (ICICES), Chennai, 2013, pp 7–10
5. Das K, Behera RN (2017) A survey on machine learning: concept, algorithms and applications. Int J Innovative Res Comput Commun Eng 5(2):2320–9801
6. Dey A (2016) Machine learning algorithms: a review. (IJCSIT) Int J Comput Sci Inf Technol 7(3):1174–1179
7. García S, Luengo J, Herrera F (2015) Data preprocessing. In: Data mining. Springer International Publishing Switzerland
8. Alam S, Yao N (2018) The impact of preprocessing steps on the accuracy of machine learning algorithms in sentiment analysis. In: Computational and mathematical organization theory. Springer, Berlin

9. Xu S, Qian Y, Hu RQ (2017) A data-driven preprocessing scheme on anomaly detection in big data applications. In: 2017 IEEE conference on computer communications workshops (INFOCOM WKSHPS), Atlanta, GA, 2017, pp 814–819. https://doi.org/10.1109/infcomw.2017.8116481
10. Kaur S, Jindal S (2016) A survey on machine learning algorithms. Int J Innovative Res Adv Eng (IJIRAE) 3(11):2349–2763
11. Khanum M, Mahboob T (2015) A survey on unsupervised machine learning algorithms for automation, classification and maintenance. Int J Comput Appl 119(13):0975–8887

The Role of Social Factors in Education: A Case Study in Social Network Perspective

V. G. Deepa, S. Aparna Lakshmanan and V. N. Sreeja

> *Literacy in itself is no education. Literacy is not the end of education or even the beginning. By education, I mean an all-round drawing out of the best in the child and man-body, mind and spirit.*
>
> —Mahatma Gandhi

Abstract Education is the process of facilitating learning or the acquisition of knowledge, skills, values, beliefs and habits. This process of acquiring, adapting and transmitting knowledge depends on so many social factors. Social network analysis is a growing body of research with a great deal of opportunities. The study of social network is a multidisciplinary area involving social, mathematical, statistical and computer sciences. The idea is to map social relations and to look for parameters which will eventually lead to a better understanding of how relations work, information flows and organizations collaborate. This paper is an attempt to understand the impact of social factors in education by the help of a weighted graph and to find how the factors, viz. parents, relatives, teachers, friends, reading materials, visual medias, social medias, homely atmosphere, institutional atmosphere, health conditions, grasping capacity and memory power, that directly or indirectly influence in their education.

Keywords Social network · SNA · Education · Correlation coefficient · Weighted graph

AMS subject classification 91D30 · 91D10 · 05C82

V. G. Deepa (✉) · V. N. Sreeja
Department of Mathematics, Sree Krishna College, Guruvayur, Kerala, India
e-mail: skcdeepa@gmail.com

V. N. Sreeja
e-mail: drsreejavn@gmail.com

S. Aparna Lakshmanan
Department of Mathematics, St. Xavier's College for Women, Aluva, Kerala, India
e-mail: aparnaren@gmail.com

S.-L. Peng et al. (eds.), *Computing and Network Sustainability*, Lecture Notes in Networks and Systems 75, https://doi.org/10.1007/978-981-13-7150-9_7

1 Introduction

The study of social relationships among a set of individuals is a quickly widening multidisciplinary area in the field of research. Social network analysis (SNA) has its own existence in almost all fields of sciences like mathematics, statistics, computer science, biology, physics, chemistry and social sciences. Among them, researchers have focused on the graphical characterization of network structure for analyzing the descriptive measures of simple networks. The work on networks in the social and biological sciences mainly focuses on many graphical parameters such as centrality, reciprocity, path lengths, diameter, cliques and statistical parameters like correlation coefficients and degree distribution. More recently, researchers have become interested in the global organization of social networks, which provides a new medium for social interaction. But the main challenge in this type of network analysis is the inconsistency of the dynamic networks. To a great extent, this can overcome by considering time interval graphs of such type networks. For more details about classical network theory, one can refer Wasserman and Faust [1], Brandes and Erlebach [2], Anheier, Gerhards and Romo [3], Wouter De Nooy [4], and Scott and Wasserman [5].

Adolescence education in India aims to provide young people with appropriate age and cultural information, promote healthy attitudes and develop skills to enable to face real-life situations meritoriously. Education should provide integrity among individuals to enable them to respond in practical situations effectively. Even though the adolescence gets almost an equal opportunity for being educated, the output shows great variability among individuals. This provides a scope for understanding the relevance and contributions of social factors in an education system. For all notations and terminology note mentioned here please refer [6–8].

1.1 Objectives and Main Contributions

This paper invites our understanding to the impact of social factors in education with the help of a weighted graph. After a pilot study, we choose 10 suitable adolescence and find how the factors, viz. parents, relatives, teachers, friends, reading materials, visual medias, social medias, homely atmosphere, institutional atmosphere, health conditions, grasping capacity and memory power, that directly or indirectly influence in their education.

2 Methodology

Education is not only literate, but also powerful for the sake of living, that's why social factors play this much important role in the field of education. When we go through the educational output; for example, CGPA—Course Grade Point Average for graduate students, of a group of adolescence, we were keen to know, how the

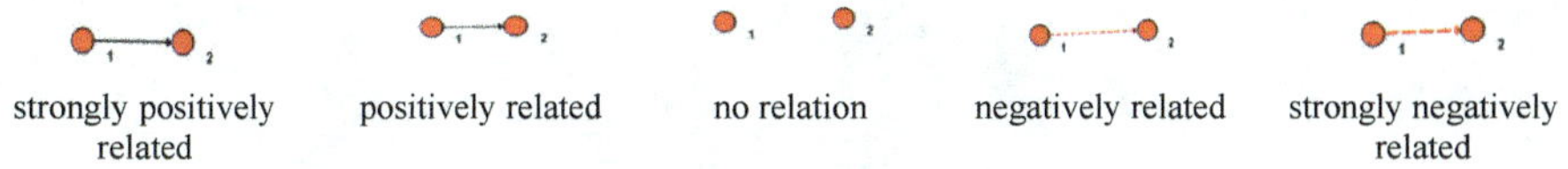

Fig. 1 Different types of relation among two vertices

social factors directly or relationship to these factors affect their educational output. So we choose 10 students from all categories having different CGPA and conducted our study.

To find the impact on social factors like parents and teachers on education of adolescence, we formulate this situation as a social network with vertex set V. Let $V = \{v_1, v_2, \ldots, v_{22}\}$, where vertices v_1 to v_{12} denote the factor parents, relatives, teachers, friends, reading Materials, visual medias, social medias, homely atmosphere, institutional atmosphere, health conditions, grasping capacity and memory power, respectively. In our study, the vertices v_{13} to v_{22} indicate tenadolescence. We have collected the data from each $v_i; i = 13, 14, \ldots, 22$, that is the factors directly or relationship to the factors $v_j, j = 1, 2, \ldots, 12$ affecting their education through questionnaire. We take weight to each of these relationships as -2, -1, 0, 1, 2 for strongly negatively related, negatively related, no relation, positively related and strongly positively, related respectively to the factors $v_j, j = 1, 2, \ldots, 12$. This forms a graph $G = (V; E)$ which consists of a set of objects $V = \{v_1, v_2, \ldots, v_{22}\}$ called vertices, and another set $E = \{e_1, e_2, \ldots\}$ whose elements are called the edges, such that each edge e_k is identified with an unordered pair (v_i, v_j) of vertices. A graph G is said to be weighted if there exists a function $w : E(G) \to Z$ where Z is the set of all integers. So we can consider the situation of adolescence education affected by various social factors as a weighted graph. A strongly positively related vertices, edge with weight 2 are connected by a thick black line, positively related vertices, edge having weight 1 are connected by a thin black line, no relation, weight 0 has no connecting edge, negatively related vertices (weight—1) are joined by thin red dotted line and finally strongly negatively related vertices (weight—2) are connected by thick red dotted lines (Fig. 1). Then the situation of factors affecting education is formulated as a weighted graph and the resulting figure is 2.

Let $G = (V, E)$ be any graph, and let $S \leq V$ be any subset of vertices of G. Then the induced subgraph $G[S]$ is the graph whose vertex set is S and whose edge set consists of all of the edges in E that have both endpoints in S [7]. For understanding the factors affected by education for each individual $v_i; i = 13, 14, \ldots, 22$, we have to consider the induced subgraph of $\{v_1, v_2, \ldots, v_{12}, v_i\}$. Note that, the induced subgraphs for v_{13} are the subgraph with vertex set $\{v_1, v_2, \ldots, v_{13}\}$ and that for v_{14} are the subgraph with vertex set $\{v_1, v_2, \ldots, v_{12}, v_{14}\}$. Similarly, we have considered the induced subgraphs of $v_{15}, v_{16}, v_{17}, v_{18}, v_{19}, v_{20}, v_{21}andv_{22}$ for analyzing the final output for each individual. The resulting induced subgraphs are in Figs. 3, 4, 5, 6, 7, 8, 9, 10, 11, 12, 13 and 14.

Analysis of the induced subgraphs of the ten individuals gives rise to the requirement of a single value for each actors which is capable of expressing the totality of

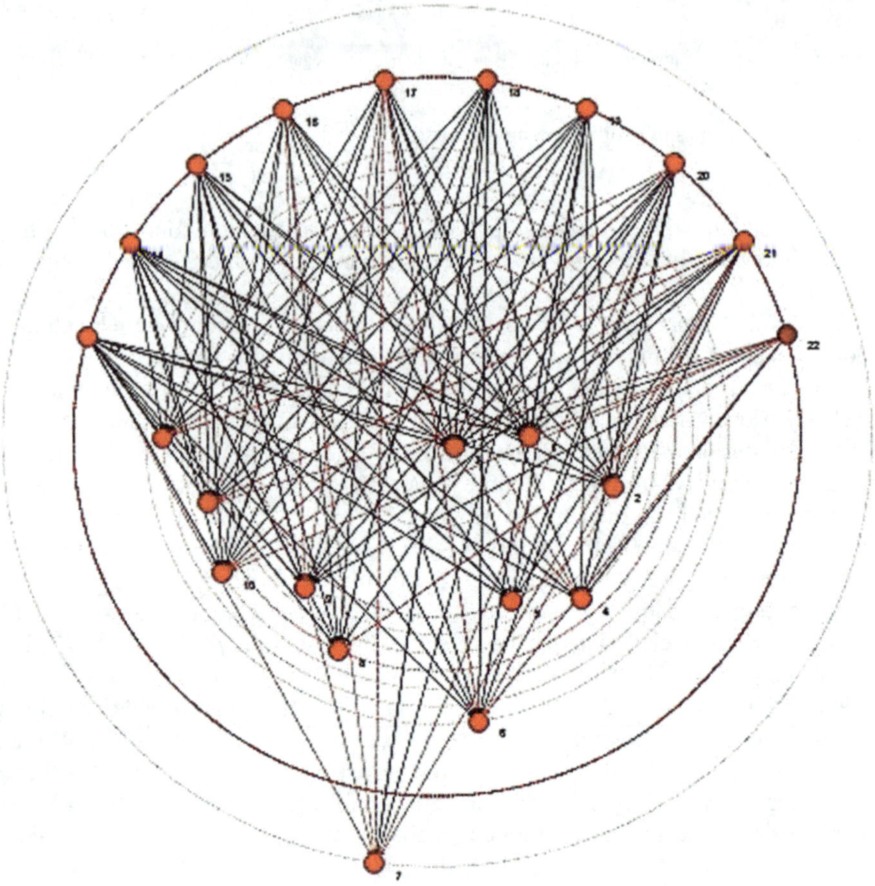

Fig. 2 Factors affected in education

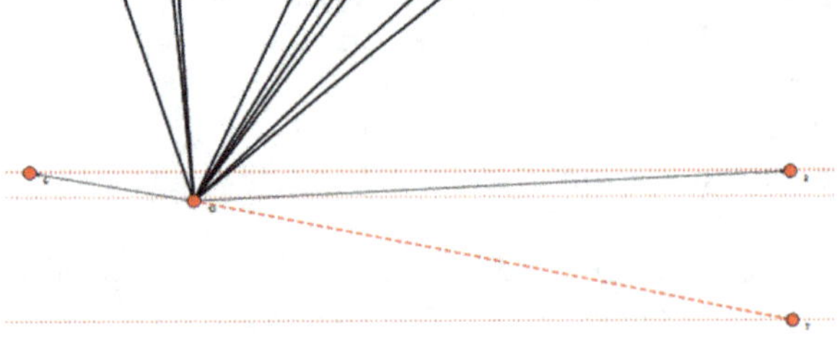

Fig. 3 Induced subgraph of v_{13}

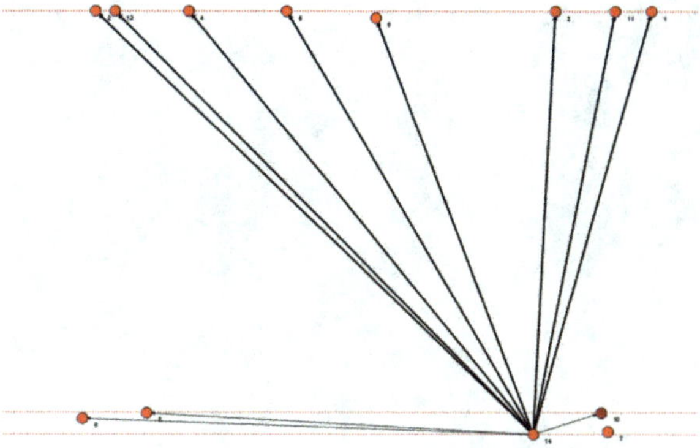

Fig. 4 Induced subgraph of v_{14}

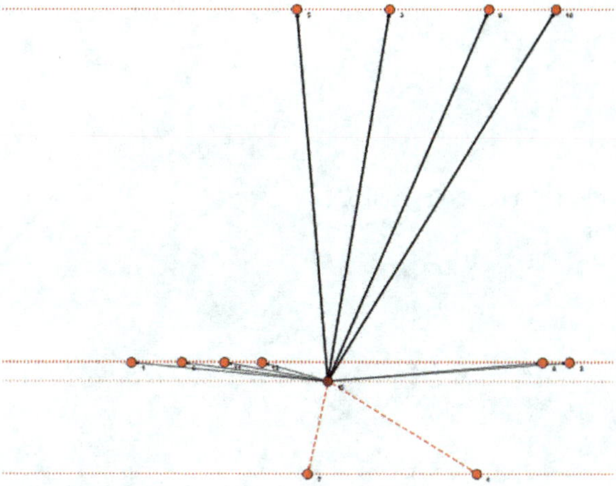

Fig. 5 Induced subgraph of v_{15}

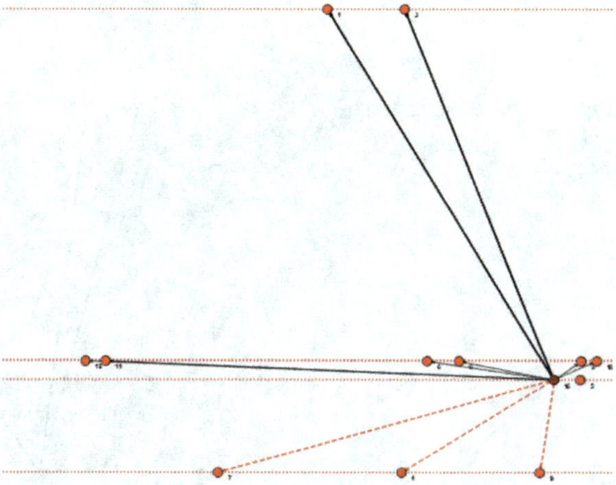

Fig. 6 Induced subgraph of v_{16}

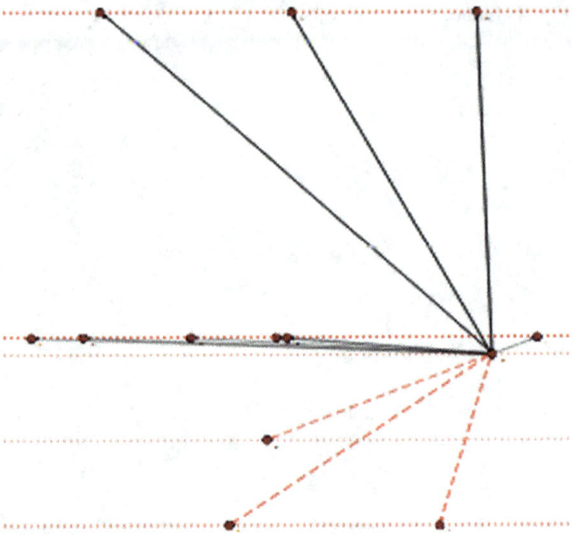

Fig. 7 Induced subgraph of v_{17}

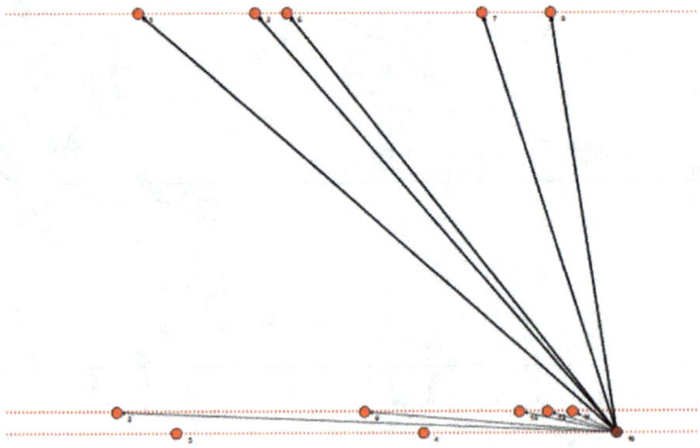

Fig. 8 Induced subgraph of v_{18}

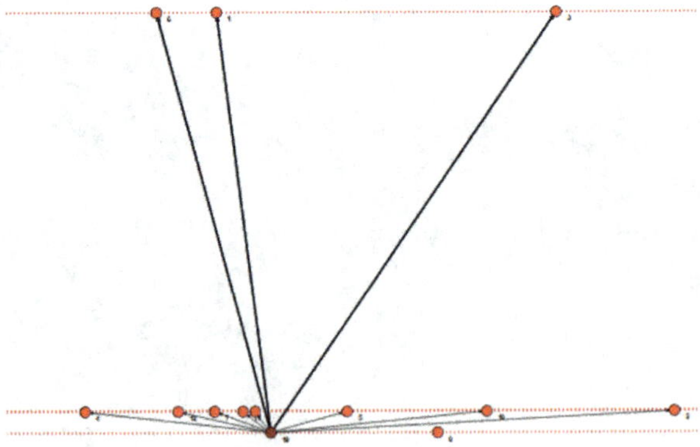

Fig. 9 Induced subgraph of v_{19}

Fig. 10 Induced subgraph of v_{20}

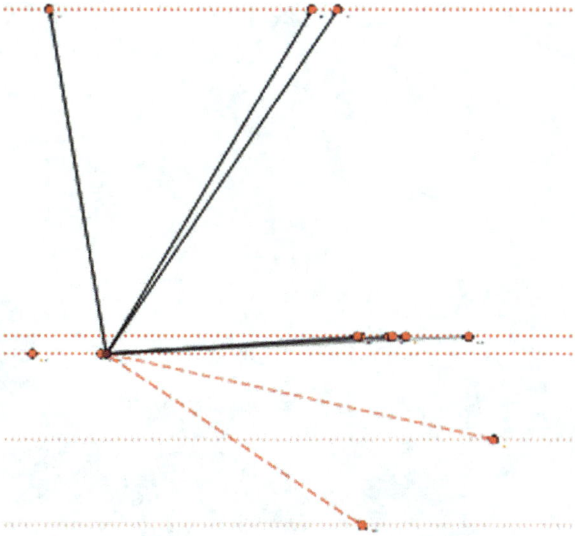

Fig. 11 Induced subgraph of v_{21}

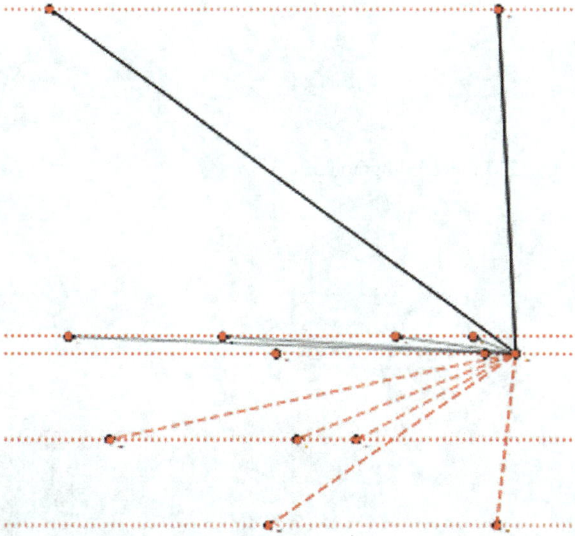

the abovementioned subgraphs. This constant for each v_i; $i = 13, 14, \ldots, 22$ can express their influence on factors v_1 to v_{12}. So we termed this single value as weight quotient (WQ) and defined as follows;

For each $i = 13, 14, \ldots, 22$; weight quotient (WQ) of v_i is defined as

$$\mathrm{WQ}(v_i) = \frac{\sum_{j=1}^{12} w(e_{ij})}{\sum_{j=1}^{12} \text{maximum } w(e_{ij})}$$

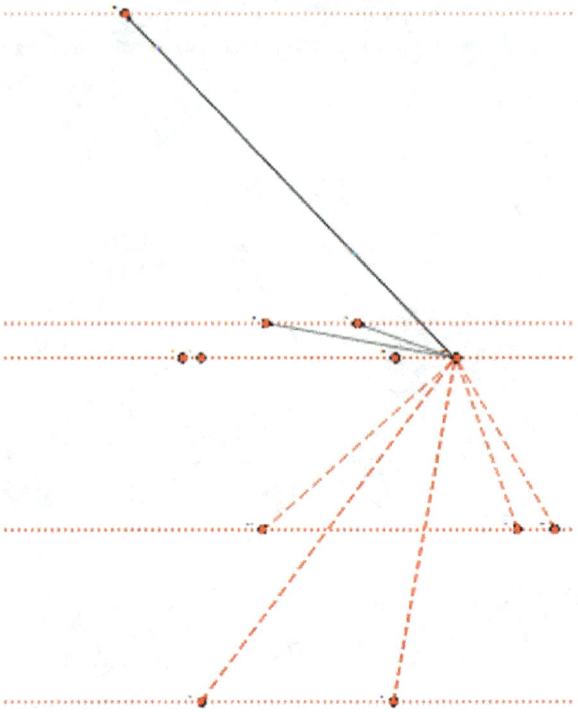

Fig. 12 Induced subgraph of v_{22}

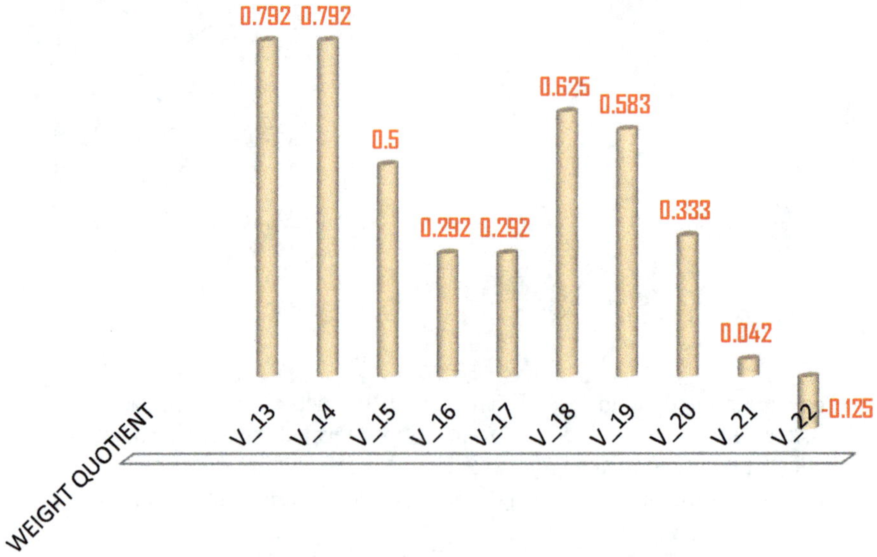

Fig. 13 Weight quotient (WQ)

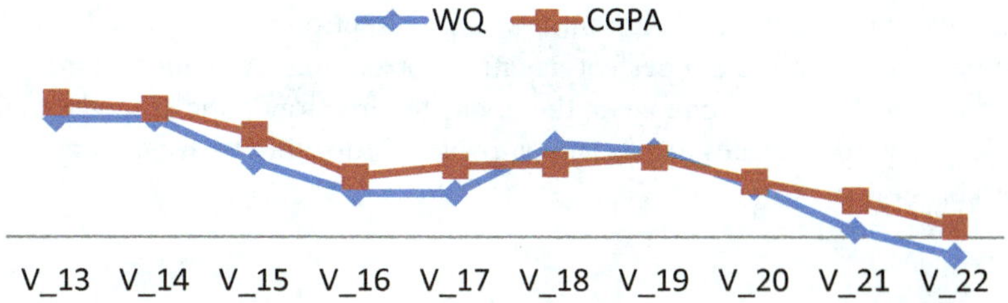

Fig. 14 Comparison between WQ and CGPA

Then clearly, $-1 \leq \text{WQ} \leq 1$.

Apart from the main theme, it is quite interesting to find which factors mostly influenced the students either negatively or positively or by what extend, each of these social factors chosen for this study was influenced. Even though the factors may change according to the social status, through this paper, we are just indented to know how much these factors influence adolescence. For this purpose, we introduced a factor quotient (FQ) for each factors under consideration. FQ is defined as follows;

For each $i = 1, 2, \ldots, 12$; factor quotient (FQ) of v_i is defined as

$$\text{FQ}(v_i) = \frac{\sum_{j=13}^{22} w(e_{ij})}{\sum_{j=13}^{22} \text{maximum } w(e_{ij})}$$

Then clearly, $-1 \leq \text{FQ} \leq 1$.

The weight quotient gives an idea about the totality of factors influenced by each actor v_i; $i = 13, 14, \ldots, 22$ and the educational output, we have mentioned earlier is also considered as the totality of influence of social factors at the time of education. So for a finer comparison, for each v_i, $i = 13, 14, \ldots, 22$; we collected the CGPA, converted as $0 \leq \text{CGPA} \leq 1$. Then we got the data as follows.

Vertex	v_{13}	v_{14}	v_{15}	v_{16}	v_{17}	v_{18}	v_{19}	v_{20}	v_{21}	v_{22}
CGPA	0.902	0.861	0.691	0.403	0.474	0.488	0.535	0.372	0.245	0.071

The correlation coefficient is a statistical measure to calculate changes of the value of one variable to the value of another. Actually, correlation is a mutual relationship between two or more variables. The correlation coefficient between the variables x and y is denoted by γ_{xy} and is defined as:

$$\gamma_{xy} = \frac{\text{cov}(x, y)}{\text{SD}(x)\text{SD}(y)}$$

where $\text{cov}(x, y)$ is the covariance between variables and SD denotes the standard deviation. Correlation coefficients are expressed as values between -1 and $+1$. A coefficient of $+1$ indicates a perfect positive correlation: A change in the value of one variable will predict a change in the same direction in the second variable. A coefficient of −1 indicates a perfect negative correlation: A change in the value of one variable will predict a change in the opposite direction in the second variable. A coefficient of zero indicates there is no palpable relationship between fluctuations of the variables.

3 Result and Discussion

The relationship between the twelve social factors and tenadolescence has been found and expressed as a weighted graph in Fig. 2. The weights of each factor are given in columns against the raw for each actor

	v_1	v_2	v_3	v_4	v_5	v_6	v_7	v_8	v_9	v_{10}	v_{11}	v_{12}
v_{13}	2	1	2	2	2	1	−1	2	2	2	2	2
v_{14}	2	2	2	2	2	1	0	1	2	1	2	2
v_{15}	1	1	2	−1	2	1	−1	1	2	2	1	1
v_{16}	2	1	2	−1	0	1	−1	1	−1	1	1	1
v_{17}	2	1	2	1	1	−2	−2	1	−1	2	1	1
v_{18}	2	1	2	0	0	2	2	2	1	1	1	1
v_{19}	2	1	2	1	1	2	1	0	1	1	1	1
v_{20}	2	1	2	2	1	1	−1	1	1	−2	1	−1
v_{21}	1	1	2	1	1	−1	−1	0	2	−1	−2	−2
v_{22}	−1	0	1	2	0	−2	0	−2	1	−1	−1	0

Using this output, we found the weight quotient (WQ) for each actors v_i where $i = 13, 14, \ldots, 22$ and the result is given in Fig. 15

By the help of the statistical tool correlation coefficient, we found the relation between the weight quotient (WQ) as X and the CGPA as Y.

Vertex	v_{13}	v_{14}	v_{15}	v_{16}	v_{17}	v_{18}	v_{19}	v_{20}	v_{21}	v_{22}
WQ (X)	0.792	0.792	0.500	0.292	0.292	0.625	0.583	0.333	0.042	−0.125
CGPA (Y)	0.902	0.861	0.691	0.403	0.474	0.488	0.535	0.372	0.245	0.071

$$\text{correlation coefficient,} = 0.930103$$

The value of γ_{xy} is nearly equal to one implies that there exists a perfect positive correlation between x and y.

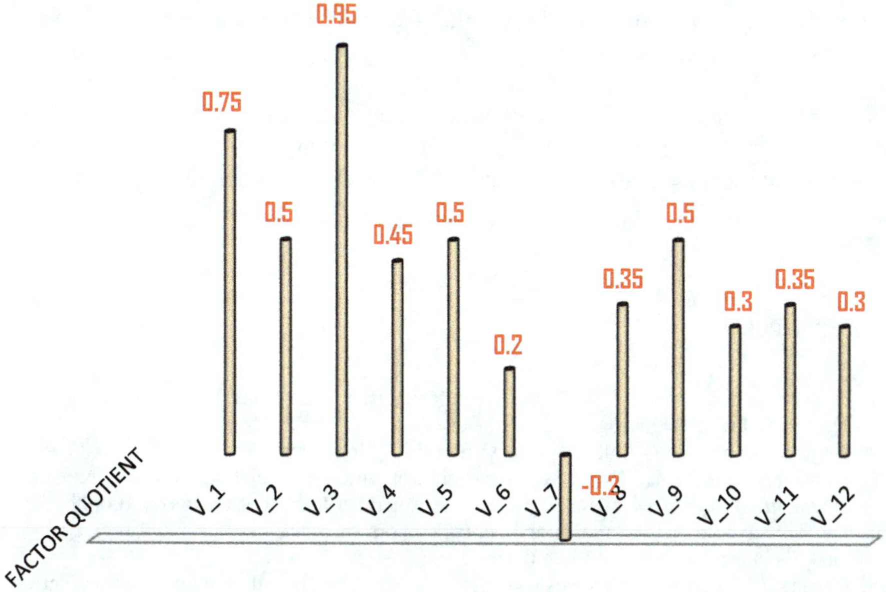

Fig. 15 Factor quotient

Even though the social factors under consideration is purely depend on the social atmosphere, geographical areas, financial stability, etc., we are keen to know the rate of dependence of factors included in this study. The factor quotient for the vertices v_i where $i = 1, 2, 3, \ldots, 12$ is given in Fig. 15.

4 Conclusion

The field of SNA is broad and growing, and a new methods and approaches are constantly in development. Since SNA mainly focuses on the relation among entities, graph plays a major role for the analysis of SNA. Through this paper, we invited our understanding to the impact of social factors in education with the help of a weighted graph and found how the factors, viz. parents, relatives, teachers, friends, reading materials, visual medias, social medias, homely atmosphere, institutional atmosphere, health conditions, grasping capacity and memory power, that directly or indirectly influenced in their education. After this study, we found that along with the parents, teachers also act as the most influential factor in education. Also found that social media possesses a negative impact. Here we conducted only a small sample study and open a scope for large sample study before arriving a general conclusion. The social factors under consideration purely depend on the social atmosphere, geographical areas, etc., so it can change according to the social status. It is very interesting to observe that the weight quotient (WQ) of each individual and the educational output (CGPA) are perfect positively correlated. So there arises an opportunity for improving the final output by improving WQ. Since every social network is dynamic, changes may happen within the period of study. Here we treated as a static situation and conducted our study. In order to overcome the dynamic nature of the network, we can take the time interval graphs of such networks for the analysis.

References

1. Wasserman S, Faust K (1994) Social network analysis: methods and applications. Cambridge University Press, Cambridge, ENG and New York
2. Brandes U, Erlebach T (2005) Network analysis. Springer, New York
3. Anheier HK, Gerhards J, Romo FP (1995) Forms of capital and social structure in cultural fields: examining Bourdieu's social topography. The American Journal of Sociology 100:859–903
4. De Nooy Wouter (2003) Fields and networks: correspondence analysis and social network analysis in the framework of field theory. Poetics 31(56):305–327
5. Carringtom PJ, Scott J, Staneywasserman S (2005) Models and methods in social network analysis. Cambridge University Press, Cambridge
6. Knoke David, Yang Song (2007) Social network analysis (II). SAGE publications Ltd., Thousand Oaks
7. Douglas B (2000) Introduction to graph theory (II). Pearson, London
8. Scott J (2017) Social network analysis (IV). SAGE, London
9. Wang C, Lizardo O, Hachenetal D (2012) A dyadic reciprocity index for repeated interaction networks. ZU064-05-FPR
10. Durkheim Emile (1895) The rules of sociological method. The first and most fundamental rule is: consider social facts as things, p. 14
11. Barcelo H, Laubenbacher R (2006) Graph theoretic tools for dynamic network analysis
12. Arumugam S, Haughton B, Vasanthi B, Zhang C (2014) Reciprocity in social networks—a case study in Tamil Nadu India. CS-BIGS 5(2):126–131

Investigating the Effect of Compression and Decompression in Video Using Fractal Technique

Shraddha Pandit, Piyush Kumar Shukla and Akhilesh Tiwari

Abstract With the advent of multimedia technology, video compression has become imperative. The high definition of video required huge amount of storage space and large amount of bandwidth for the transmission of video. The largest part of multimedia is video. There is upsurge in demand of compressed data due to excessive usage of multimedia applications on Internet. Hence in success of multimedia data, video compression and decompression are majorly used. There are already various transform functions such as wavelet transform, DCT transform and fractal transform functions which are used for compression and decompression of video. In all transform function, the fractal transforms function adhere to the rule of block symmetry. It is very proficient process, but the rate of compression is very time-consuming, however, the decompression is very fast. In this paper, we adopted fractal triangular partitioning scheme to compress and decompress the videos. Here in this work for our analysis, we used very short length videos which are of different size. The primary objective of this work is to minimize the encoding time of video and achieve better compression ratio. The process of video compression and decompression methods is simulated in MATLAB software and used some standard parameters for the evaluation of compression and decompression results.

Keywords Fractal · Video compression · Video decompression · ITP · MSE · PSNR · CR · ET · rms

---

S. Pandit (✉) · P. K. Shukla
University Institute of Technology, RGPV, Bhopal, Madhya Pradesh, India
e-mail: svpandit_pict@yahoo.co.in

P. K. Shukla
e-mail: pphdwss@gmail.com

A. Tiwari
Madahv Institute of Technology & Science, Gwalior, Madhya Pradesh, India
e-mail: atiwari.mits@gmail.com

S.-L. Peng et al. (eds.), *Computing and Network Sustainability*, Lecture Notes in Networks and Systems 75, https://doi.org/10.1007/978-981-13-7150-9_8

1 Introduction

Due to high resolution, the storage and transmission of multimedia is a big challenge. In order to improve storage and transmission of multimedia, data compression techniques are recommended. The most of the part of multimedia is video [1, 2]. The part of video data increases day to day. Thus, the utilization of computerized images is expanding quickly. Despite the fact that there is a fast advance in mass stockpiling thickness, speed of the processor and the execution of the computerized correspondence. Frameworks, the interest for information stockpiling limit and information transmission data transfer capacity remain on surpassing the abilities of close by advances.

Due to these developments, the importance of efficient compression methods of audio/video/image data has been increasing tremendously. A compression strategy comprises of meanings of two complex procedures: compression and decompression [3, 4]. Compression is a change of unique information portrayal into various portrayal described by more modest number of bits [5, 6]. Inverse process remaking of the first informational collection is termed as decompression. There can be recognized two kinds of compression: lossless and lossy [4, 7].

In lossless compression strategies, the informational collection reproduced amid decompression is indistinguishable as the first informational collection. In lossy strategies, the compression is irreversible the remade informational collection is just an estimation of the first image [8]. Video compression and decompression play a vital role in success of multimedia data [9, 10]. In a search space, finding of coefficient and blocks in video compression is important phase. For the searching of blocks and coefficient, many researchers used zig-zag and some other random searching techniques as well as various transform functions. Fractal transform is one of them which is a lossy compression technique.

Here in this paper, we adopted fractal triangular partitioning scheme (ITP) to compress and decompress the videos. Section 2 corresponds to the fundamental of fractal video compression. Section 3 speaks about the fractal triangular partitioning scheme (ITP). Results are mentioned in Sect. 4. Conclusion is being discussed in Sect. 5.

2 Fractal Video Compression

A video is a collection of frames or snapshots and has to be compressed to low bit rate without damaging the content and meaning of the frame [11–13]. For example, for a single snapshot it needs 1024 bytes to store and transfer which has to be minimized. The frames of a video segment are more related and have little amount of changes in the pixels of the frames. For a single video segment, there will be number of frames according to type of video. There are high redundancies in a similar casing, the spatial redundancies, and between progressive edges, the transient redundancies.

The greater part of video compression methods plans to misuse these connections and high redundancies keeping in mind the end goal is to accomplish better compression.

Fractal image compression has become a conventional method for compressing still images [14, 26]. It depends on iterated function system [15], in which an image is portrayed by sets of conditions that give contractive mappings. The main successful fractal coder was presented by Jacquin, who saw that a block of an image is like another block of an image [16]. The main thought behind fractal compression is that it works on the self-similarity which is present in images [17]. It suited well for videos with firm picture in its place of incessantly changing pictures. For compression of natural images, it gives best results, but also gives inferior results with videos holding textual information. But the fractal compression can attain a very high compression ratio [4, 12], which attracts researchers to work in this field.

Encoding time [18] is the most demanding issue in fractal-based video compression. Fractal compression confers high CR. But encoding time is very high due to finding best matched range and domain block and number of search points. Dividing the whole image into domain blocks and stored all domain blocks in a domain pool becomes a problem. Because domain pool blocks are overlapping blocks. If there are a large number of domain blocks, then it needed huge amount of space. Also implementing an algorithm in fractal compression becomes more intricate as it requires more encoding time (ET) to compress an image and video. There are two ways which we can apply for fractal video compression. First is cube based and second is frame based [4, 15, 19].

3 Fractal Triangular Partitioning Scheme (ITP)

Transform-based function is critical in video compression [20, 21]. Most of the multimedia comprises of the visual clips and image. The varied nature of visual clips demands more memory and bandwidth for storage and transmission. Therefore compression is 'must be' for efficient processing of video. An essential issue associated with fractal-based image compression is to split an entire image into sub-blocks for encoding. For this, in fractal-based compression there are fix point partitioning scheme, quad tree partitioning scheme, H-V partitioning scheme and triangular partitioning scheme [17]. In [18], frame group-based fractal video compression based on Hadoop computing environment is discussed. For experimentation, author implemented parallel algorithm based on DCOM,.NET remoting and Hadoop cloud computing platform. In [1] used fast fractal image compression technique to compress AVI files and performance is measured using standard parameters. This section describes the process of video encoding and decoding using fractal triangular partitioning scheme.

To reduce the compression time [22] and better searching of block symmetry here, we used frame-based approach to compress and decompress video using triangular partitioning scheme. Whenever frame is to be encoded, it is triangularly partitioned into right-angled isosceles triangles. Right-angled isosceles triangles are chosen in

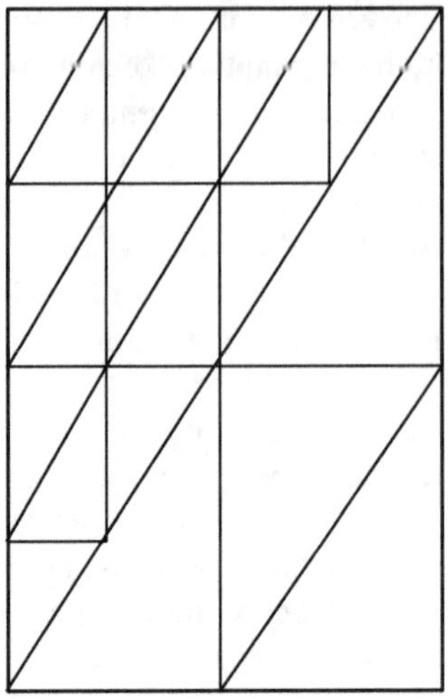

Fig. 1 Triangular partitioning scheme

order to cover the whole image in partitioning without leaving any gray area in between.

Figure 1 represents the triangular partitioning scheme.

Algorithm 1 and algorithm 2 describe the video compression and decompression process using fractal triangular partitioning scheme (ITP).

3.1 Video Encoding Process Using Triangular Partitioning Scheme

Algorithm 1: Video Decompression Input: AVI video
Output: Compressed File

1. Extract frames from the video
2. Divide each frame diagonally into two isosceles triangles by considering range and domain block
3. Set the minimum threshold value (rms)
4. Create range pool and domain pool where all range and domain blocks are stored
5. Compare first range block(R block) with the corresponding domain block (D block) and find out the best matching block with rms.
6. If match is found, store range size, domain size and rms value
7. If match is not found, apply geometric transformation and calculate the difference between the range and domain blocks through rms method and store range size, domain size, rms value and transformation angle.

8. If geometric transformation is discarded, subdivide the range block recursively into sub-isosceles triangles and repeat step 3
9. Repeat step 8
10. End.

3.2 *Video Decoding Process Using Triangular Partitioning Scheme*

The decompression of video is inverse process of compression [4, 15, 23]. The decompression of video is very complex process in terms of recovery of image and their coefficient for the restoration process. The decompression of video measures the process in terms of number of iterations and the value of PSNR of decoded image for the grouping of frames. The following steps describe the process of video decompression.

Algorithm 2: Video Decompression **Input**: Load encoded video
Output: Decompressed Frames

1. Differentiate symmetry blocks
2. Estimate the triangular blocks in symmetry
3. Generate the set of images/frames and coefficients
4. Generate frames
5. Generate original video
6. End

4 Results

The algorithm of compression and decompression is simulated in MATLAB 8.3.0.532(R2014a). All tested videos are AVI videos and obtained from computer vision video dataset. The computer vision dataset of video is freely available for research. The dataset consists of verity of videos in the form of RGB mode in different frames and cycles. For the simulation creates GUI interface and write encoding and decoding process in terms of function and script. Performance is measured using standard parameters like peak signal-to-noise ratio (PSNR), compression ratio (CR), mean square error (MSE) and encoding time(ET) (Fig. 2).

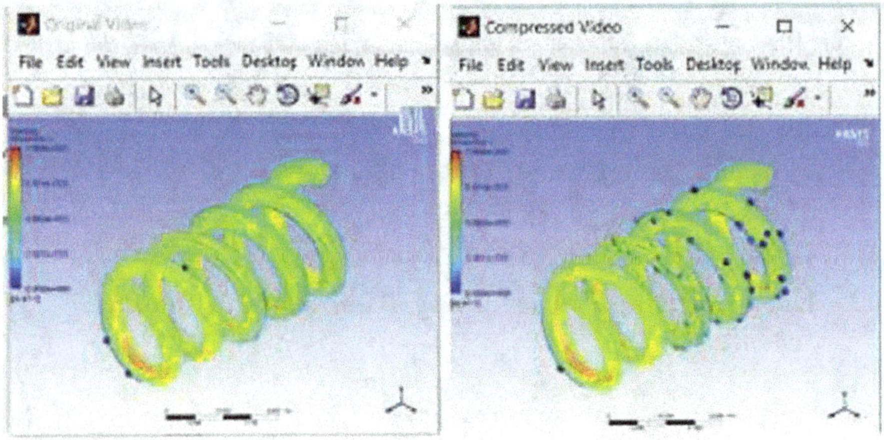

Fig. 2 Original and compressed view of coil.avi video using ITP

Table 1 Performance of ITP using coil and xylophone.avi video

	AVI video	
	Coil	Xylophone
Compression ratio	0.45	0.45
MSE	14.97	14.47
PSNR (dB)	39.99	28.89
Encoding time (ms)	1.07	0.40

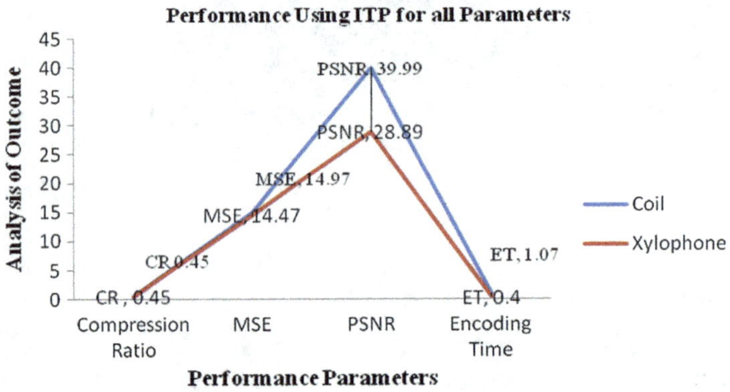

Fig. 3 Performance analysis of ITP using all parameters

Table 1 represents the performance of isosceles triangle partition (ITP) for coil and xylophone videos. Here, we observed that as the value of MSE increases, the value of PSNR also increases. ITP take less time to compress xylophone.avi video (Figs 3, 4, 5 and Table 2).

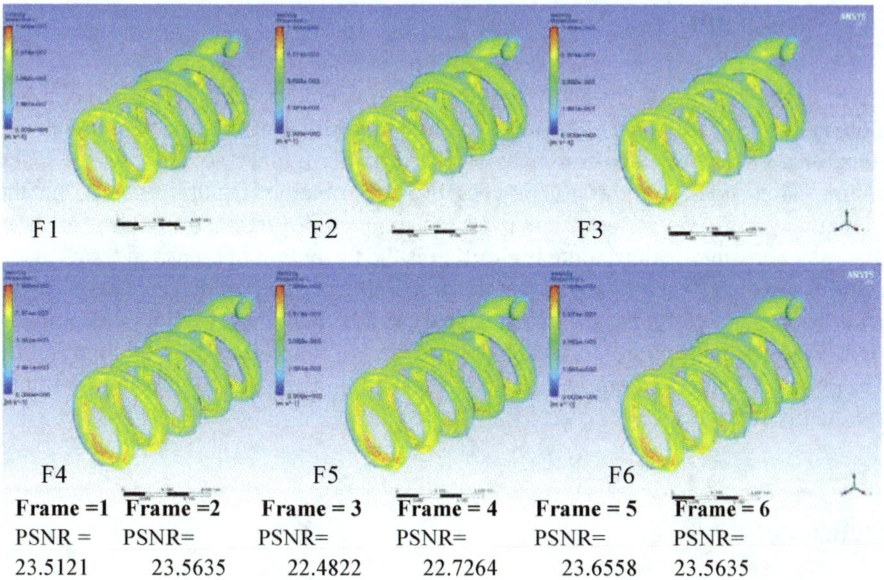

Fig. 4 Decompressed images in form of frames using triangular partitioning scheme for coil.avi video

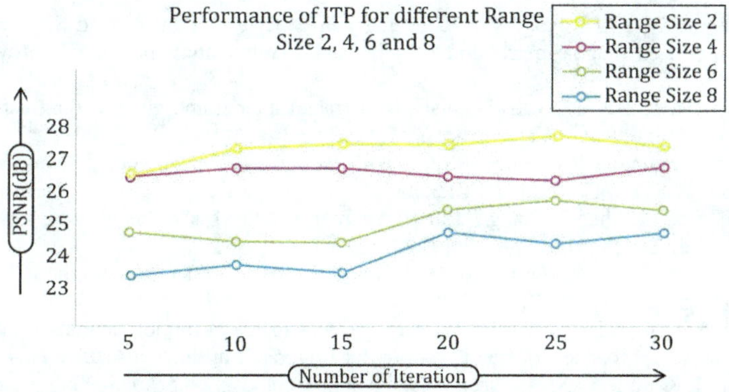

Fig. 5 Performance of ITP method with given Iteration (5, 10, 15, 20, 25, 30), range size (2, 4, 6, 8) and PSNR in decompression for coil.avi video

Table 2 Iteration and PSNR according to the range size (2, 4, 6, 8) using triangular partition scheme for coil.avi video

Iteration		5	10	15	20	25	30
PSNR	Range size 8	23.7898	23.5635	24.7898	24.4875	24.7898	27.3238
	Range size 6	24.4875	24.4831	25.4875	25.7877	25.4875	28.4458
	Range size 4	26.7877	26.7898	26.5554	26.4215	26.7898	29.4823
	Range size 2	27.4215	27.5554	27.5212	27.7898	27.4875	30.7845

5 Conclusion

Here, in this paper, we have successfully implemented process of video compression and decompression using fractal triangular partitioning scheme. The transform function is critical in video compression. The transform function improves the video compression and decompression process. From the obtained results, it is seen that the triangular partitioning scheme enhances the symmetry of fractal transform function. The major advantage of triangular partition is the formation of blocks in terms of triangle not as rectangular block so that the processing of video frames increases. In the process of video compression, we also analyzed the execution of video decompression. Video decompression is the reverse process of video encoding (compression). The process of video decompression evaluates the numbers of iteration and maintains the quality of video.

References

1. Minas A, Sediq H (2013) Compression of an AVI video file using fractal system. Int J Comput Sci Issues 10(2):182–189
2. Yuanfang X, Xia S (2014) System design for real-time image processing based on multicore DSP. J Netw 9(11):3143–3150
3. Alarabeyyat A, Al-Hashemi S, Khdour T, Hjouj Btoush M, Bani-Ahmad S, Al-Hashemi R (2012) Lossless image compression technique using combination methods. J Softw Eng Appl 5(12):752–763
4. Kulkarni M, Kulkarni D (2017) Analysis of fractal inter frame video coding using parallel approach. J Sign, Image Video Process 11(4):629–634
5. Dhok S, Deshmukh R, Keskar A (2012) Efficient fractal image coding using fast Fourier transform. GSTF Int J Comput 1(2):35–40
6. Gao H, Zeng W, Chen J, Zhang C (2016) An improved fast fractal image compression coding method. Rev Téc Ing Univ Zulia 39(9):241–247
7. Lima V, Robson W, Pedrini H (2013) 3D searchless fractal video encoding at low bit rates. J Math Imaging Vision 45(3):239–250
8. Kamble S, Thakur N, Bajaj P (2016) Review on block matching motion estimation and automata theory based approaches for fractal coding. Int J Interact Multimedia Artif Intell 4(2):91–104
9. Aswani R, Kamble S (2014) Fractal video compression using block matching motion estimation—a study. IOSR J VLSI Sign Process 4(2):82–90
10. Sudhkar R, Letitia S (2015) Motion estimation scheme for video coding using hybrid discrete cosine transform and modified unsymmetrical-cross multi hexagon grid search algorithm. Middle-East J Sci Res 23:848–855
11. Pandit S, Shukla P, Tiwari A (2018) Fractal compression of an AVI video file using DWT and particle swarm optimization. Int J Comput Sci Inf Secur 16(1):128–131
12. Pandit S, Shukla P, Tiwari A (2018) A proficient video compression method based on DWT and HV partition fractal transform function. Int J Sci Eng Technol 7(2):20–24
13. Pandit S, Shukla P, Tiwari A (2018) Enhance the performance of video compression based on fractal H-V partition technique with particle swarm optimization. Int J Comput Sci Eng 6(1):31–35
14. Andonova S, Popovic D (1994) Video coding using fractal based image compression. In: IEEE international conference on systems, man and cybernetics, pp 343–348

15. Zhu S, Li L, Chen J, Belloulata K (2014) An automatic region-based video sequence codec based on fractal compression. AEU-Int J Electron Commun 68(8):795–805
16. Jacquin A (1992) Image coding based on a fractal theory of iterated contractive image transformations. IEEE Trans Image Process 1(1):18–30
17. Fisher Y (1995) Fractal image compression: theory and application. Springer, New York
18. Huang Z (2017) Frame-groups based fractal video compression and its parallel implementation in Hadoop cloud computing environment. Int J Multi-dimension Syst Sign Process 29(100):1–18
19. Ponlatha S, Sabeenian R (2014) An artificial neural network based lossless video compression using multilevel snapshots and wavelet transform using intensity measures. Int J Eng Technol 6(4):1900–1908
20. Wu Z, Yan B (2010) An effective fractal image compression algorithm. In: Proceeding of IEEE international conference on computer application and system modeling, pp 139–143
21. Peng H, Wang M, Lai C (2007) Design of parallel algorithms for fractal video compression. Int J Compute Math 84(2):193–202
22. Distasi R, Nappi M, Riccio D (2006) A range/domain approximation error based approach for fractal image compression. IEEE Trans Image Process 15(1):89–97
23. Wang M, Lai C (2007) Grey video compression methods using fractals. Int J Comput Math 84(11):1555–1566
24. Zhao E, Dan L (2005) Fractal image compression methodology: a review. In: International conference on information, technology and applications, pp 756–759

Classifier Ensembling: Dataset Learning Using Bagging and Boosting

Santosh S. Lomte and Sanket G. Torambekar

Abstract Data mining is used to tackle the huge amount data which are preserved in the data warehouse and database, to chase desired information. Many data mining have been projected such as association rules, decision trees, neural networks, etc. from many years it takes its position in center. The efficiency of the real time or artificial is being enhanced using certain data mining techniques like bagging and boosting and AdaBoost by which the accuracy of the different classifiers is also efficient. Boosting and bagging are two widely used ensemble methods for classification. As the task is to increase the efficiency of the classifier, as per the past works it is always better to have combination of classifiers than to go for random guessing. From the available boosting techniques, the AdaBoost algorithm is one of the best techniques especially in the case when carrying out the branching type of tasks. In the following work, a classification technique is explained which considers the ensemble of classifiers for bagging and for boosting separately, where decision tree is used as the classifier for bagging and for boosting artificial neural network (ANN) is being used as the classifier. For the ensembling, the classifier MDT is being used. Sections are divided as Introduction, Literature Review, Problem Statement, Issues and Challenges, Research Methodology, and Analysis of the Work.

Keywords Classifiers · Bagging · Boosting · Meta-decision tree · Data mining · Ensemble · Errors · Learning · Training

1 Introduction

Data mining is an ongoing process which consists of various steps in it to unravel the pattern. Most important factor in data mining is the huge amount of data. Problem that grows in today's data mining is streams of data which include records of the

S. S. Lomte (✉)
College of Engineering Matoshri Pratishthan, Nanded, India
e-mail: drsantoshlomte@gmail.com

S. G. Torambekar (✉)
Yogeshwari Polytechnic, Ambajogai, India
e-mail: sankettorambekar@gmail.com

S.-L. Peng et al. (eds.), *Computing and Network Sustainability*, Lecture Notes in Networks and Systems 75, https://doi.org/10.1007/978-981-13-7150-9_9

telephonic calls, datasets of the Web pages, data in the multimedia format, transactions of the various retail chains, mediclinical records, data about analysis of the different scientific researches, and various research records of the market.

Classification technique is being defined as the tool for data mining which is quite robust in which the open-ended methodologies are being used for rising the techniques used for the variation of data from simple to the complex format of data. The unseen instances are classified by using the induced methodology.

As in the case of the classification algorithm, different levels are being assigned to the data points; hence, it can be stated as the supervised leading technique. The induced methodologies can be generated in many ways like using the concept of tree meta-algorithm, ANN, nearest neighbor technique, and rough set-based technique. The meta-classifiers are generally being utilized as classification algorithms. In the case when more than one classifier is being used in the technique, then the meta-classifier is being used for combining the results of the predictions from the each of the classifier. The meta-classifier is efficient in the case when the model types are different in their nature.

On the basis of the band-based performance, the bagging and boosting techniques are considered as the most generally used techniques for classification and also due to the fact that the theoretical performance of the techniques provides strong evaluation results. Bagging and boosting are generally used in the batch format means the training set to be used for the learning process should be made available at once and also random access of the data is being asked in many of the cases. The bagging is found useful for the weak and unstable form of data where the learning curves are non-regarding and the training data available as sample is being considered as critical part. The boosting is useful for the cases of the weak and simple classifiers where the non-regarding learning curves are considered along with the availability of large sample data for training [1].

Ensemble learning is defined as the two-stage process of learning—initial stage is all about the prediction of the every single classifier and the second stage for the overall prediction of all the classifiers used as one. In current years, different methods for making of an ensemble have been projected, and it is proved that these methods are very useful for machine learning and pattern recognition.

An ensemble method forms base for classifiers and the output to unite them, generally by voting, to get improved classification accuracy. Better classification can be accomplished by using diverse classifiers. Three most frequently used and famous ensemble creation methods are bagging, boosting, and rotation forest.

2 Related Work

In this section, brief review of different bagging and boosting techniques for the classification of data are being provided which are based on different data mining techniques; the methodologies are considered such that they provide the enhancement one over another.

Freund and Schapire, separately working for the boosting, proposed an adaptive boosting algorithm also termed as the AdaBoost algorithm [2]. In the methodology, the author has used the weighted version of the training dataset instead of using the random samples. For the reason discussed the similar training set is being used again, the training dataset used is need not to be large which was quite used in the earlier cases. Nowadays, the AdaBoost algorithm is mostly used technique for the ensemble of the classifiers where the performance of the methodology is quite comparable.

In AdaBoost, final classifiers are being obtained using the weak classifiers and weak classifiers are first trained before ensemble. The generation of the weak classifier is being done on the basis of the re-weighted version of the training dataset and also depends on the accuracy level of the previous classifier. As the process goes for several iteration and for every iteration the training dataset remains the same and instances of the dataset are weighted on the basis of the classification of the past classifier.

The patterns which are not well classified by the previous classifier are the point of focus for the weak learners at every iteration. For the generation of the base classifiers, it is quite important to make a selection of weak classifiers, and the selected weak classifiers are trained without compromising the weights of the classifiers which are trained previously. In the case when the base learner is strong enough, then it can attain high accuracy level and only leaves the instance in the dataset which are outliers or the noisy points with some specific weights for the upcoming iterations.

The concept of real AdaBoost was enhanced in gentle AdaBoost [3] which involves the steps by Newton and provides reliable and efficient ensembling, but it focuses less on the outliers in the dataset. In the case of the gentle AdaBoost, the weighted least square is being considered over the estimation of the class probability.

The gentle AdaBoost [3] is the enhancement over the real AdaBoost; the gentle AdaBoost considers the usage of the different steps by Newton, which actually helps in providing better and reliable ensembling as in the process the outliers are considers at very less extent. As the process is completely an iterative part, the gentle AdaBoost uses the weighted least squares regression instead of using the class probability. The technique is termed as gentle because it is quite stable and also efficient and also is conservative as compared to the real AdaBoost.

Log ratios are not considered in the computation of the gentle AdaBoost just due to the fact that it might be unstable as it involves the quotients or it is possible that the denominator might reach zero. As per the analysis, the performance of the gentle AdaBoost is quite similar to the real AdaBoost and logit Boost and sometime outperforms both techniques; in some cases, the analysis is being considered on benchmarked data.

The modest AdaBoost algorithm [4] is another variant of the AdaBoost, and the generalization error is quite less and the training error is quite high as compared to the other available variants of the AdaBoost. For the case of depiction of the classifiers as correct and incorrect patterns for which it uses the weighted scheme an inverted distribution. Higher weights wi are being assigned to training sample which are misclassified in the previous steps.

And similarly higher weights *wi* are assigned to those training patterns which are properly classified.

The Float Boost [5] is another variant of the AdaBoost; following are the processing steps of the technique:

1. initialization;
2. forward inclusion;
3. conditional exclusion;
4. output.

The above-mentioned stages are just similar to the processing stages of the AdaBoost; the only difference in the AdaBoost and the Float Boost techniques is in stage 3, conditional exclusion, which actually works for finding and at the same time removing the classifier which is weak among the combination of classifier. The removal of the weak classifier sometime increases the error rate for some specific threshold.

Underbagging is the special case of the bagging technique in which the examples of the majority class from every bootstrap are reduced randomly till the Nmin which actually defines the cardinality of the minority class. In the case of the Exactly Balanced Bagging (EBB), the complete minority class is being considered at once and in same the minority class is merged with the subsets of the minority class which are randomly chosen, the same is done to balance the cardinality between the classes.

Roughly balanced bagging (RBB) is the another variant of the bagging technique which actually is the enhancement over the technique EBB, where the number of majority and minority samples is just same as every bootstrap [6]. The probability of the each class is being equalized before fixing the constant sample size. The negative binominal distribution is being utilized for deciding the size of the majority class with T iterations and with S bootstrap samples.

In the case of the overbagging, which is considered as the simplest variant of the bagging techniques, the random oversampling is being used for transforming the bootstrap samples. S m_{aj} of the majority class is being replaced with respect to the replacement for balancing the majority class of the every sample. Replacement technique is used for sampling in the case of the Bagging technique.

SMOTEBagging is another variant of the bagging technique which actually rises the diversity in the case of classifier components [7]. The SMOTE technique is used as the replacement of random oversampling in the case of minority class. The re-sampling rate of the SMOTE is updated from the lowest to highest rate every of the iteration which actually varies from 10 to 100%, where the ratio actually is the representation of the number of re-samples Nmin that are to be done in the upcoming stage or iteration. The bootstrap samples are generated in the same way what is done for varying the ratio from underbagging to overbagging ensemble. On the basis of the results and analysis presented by the author in his work, it is quite clear that the SMOTEBagging provides better results as compared to the ensembles where the re-sampling is being done.

The preliminary analysis in the work presented in [8] describes that it is not quite efficient and also the processing of the technique is very much similar to the overbagging. The below section provides the list of other two different variants of the underbagging technique. First one is proposed by Chan and Stolfo where the majority class is being partitioned into subsets as they are not overlapping one another and every of the subset must have the Nmin approximately.

The component classifiers are then generated by utilizing the subsets of the majority class and the minority class as well from the bag. Stacking technique is used for combining the predictions of all the classifiers available. Classical random forest can be adopted as the replacement over balanced random forest.

The technique first considers the bootstrap samples generated from the minority class with Nmin and considering exactly similar number of majority class samples. The random subset selection technique which is being obtained from the CART is being used at the split of each and every tree which is quite similar to that of original random forest. Liu et al. analyzed that its performance is not better than that of Chan and Stolfo's method or balance cascade [9].

3 Issues and Challenges

- Boosting seems to be robustness to noise.
- Voting technique is being used in the case of the bagging algorithm where the heterogeneity of the instances space is not considered.
- Increasing the efficiency of the bagging and boosting techniques.
- Errors sometimes are just unrealistic in the case of the un-correlation of the individual models.
- Problem of overfitting.
- Problem of imbalance in dataset.

4 Problem Statement

Learning algorithms provide the prediction for the training data for any instance and also can find the optimal solution in terms of prediction if we have sufficient data. As the bagging and the boosting algorithms generate different predictions among them, many looks very much similar and accurate to one another when considering the training dataset. From all available predictions, one is selected as the final for that instance, but the problem can be more likely may be solved by combining the available classifiers for any instance in the training data. It is being realized that the group of classifiers can work more efficiently than the single classifier. So, it is quite clear that why to use a single classifier when the combination of the classifiers is providing the better outcomes, which actually is the basic reason for the replacement of the single classifier with the combination.

The boosting and the bagging are being applied on the same training data with a defined base learner as we will be using the Classification and Regression Tree

(CART) because of its simplicity in classification and also the decision tree is having the problem of overfitting (a small alteration in the training pattern results in huge change in the developed model). Different classifiers obtained from the different base learners for any instance are then ensembled using the stacking approach.

5 Proposed Algorithm

In this work of **Freund and Schapire**, the adaptive classification technique (AdaBoost) algorithm is being extended. The key term of the AdaBoost algorithm is the usage of the weights for the same training data repeatedly instead of selecting new randomly, because of which the AdaBoost is not a complex and large methodology as compared to other classification algorithms.

The efficiency of the classifiers can be increased by just averaging the decisions of all of the classifiers available in the process or in the ensemble of the classifier (Perrone et al. 1993). In the case when the individual classifiers are found inaccurate and diverse, then some more improvement can be suggested, and the same can be attained by considering the base learning methodology for several training datasets. The two different techniques incorporated can be differentiated by the way they are carrying out the construction of the training dataset, bagging (Breiman 1994) and boosting (Freund 1995).

In the case of the bagging technique, every classifier carries out the learning process by considering the duplicate of the bootstrap training set which is actually original. Considering training set S having N different samples, a replacement is being done in the training sets for generating the new training sets. Just because of the replacement, it is possible that some of the examples might occur multiple times in the single set and some might absent from the same. As per the analysis, it is seen that around two-thirds of the examples are seen as replicates in the training set, and the training sets are independent for all of their operations but the multiple classifiers can be trained simultaneously.

In the case when the classifiers are quite unstable, the bagging is quite effective, means the updation in the training set can make allowable changes in the behavior of the classifier which are being formulated with reference to the decomposition of the variance (Geman et al. 1992) [10]. As the variance is reduced in the case of the bagging which results in the better generalization performance, at the same time the bias is maintained. Note, however, that there is no unique bias/variance decomposition for classification tasks [10, 11] (Kong and Dietterich 1995; Kohavi and Wolpert 1996; Tibshirani 1996).

The learning process is all about the generation of the prediction from the different classifiers used in the overall process. The proposed methodology uses two different classifiers: artificial neural network (ANN) and decision tree. ANN is the artificial implementation of the human brain which actually considers the calculation, estimation, or prediction on the basis of the past happenings. The decision tree is the

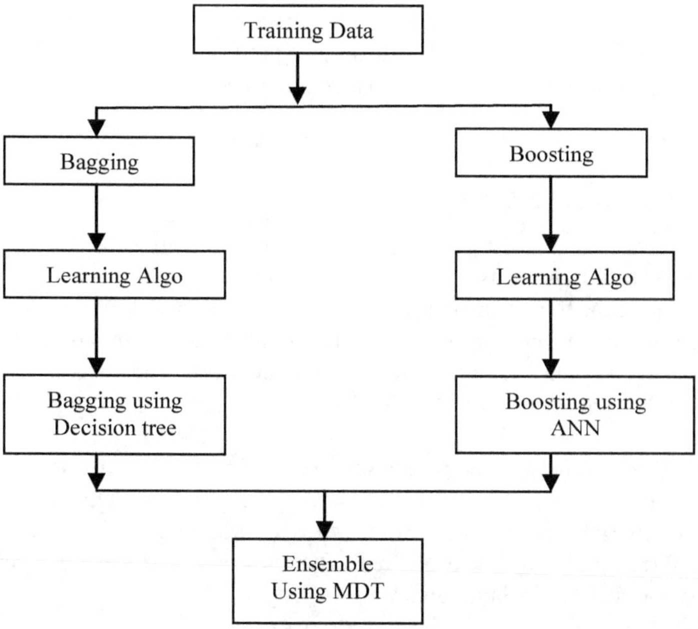

Fig. 1 Proposed system architecture

binary tree which actually works on the concept of the feature test; the splitting of the branches is done for the status of the feature test processing.

Some specific patterns are considered in the case of the AdaBoost for generating the composite classifiers, and the training of the classifiers is done sequentially. The probability distribution $D_t(i)$ is being maintained in the case of the AdaBoost for the training dataset. The same distribution is being used for training the classifier. The weighted cost function is not considered for some of the techniques, for which the concept of sampling and replacement is being used for weighted cost function approximation. The examples whose probability is higher will be seen very frequently, and similarly those having low probability will not be seen frequently; some of the examples might be completely absent even though the probability is not zero (Fig. 1).

5.1 Using Decision Tree as Base Classifier

The decision tree used as base classifier is structured in the format of a tree and works in manner of divide-and-conquer way. Feature tests are represented by the non-leaf nodes and are also termed as the split; data over the splitting node will be further divided into subnodes on the basis of the values of the feature sets. Every of the leaf node is assigned a label, which represents the instances that are at leaf node. Feature tests are carried out in series for the purpose of prediction carried out from the root node, and at the reach of leaf node output is generated.

5.2 *ANN as Base Classifier*

Neural network which is also termed as artificial neural network is the artificial representation of the biological neural system. Different models of neurons are being used to determine the function of the neural system. Neurons are also defined as the unit which is basic building blocks of the neural network. The neurons are linked to one another that the information of the data of one neuron is available at each and every neuron in the network.

Forma network is being used for the representation of the connection between the neurons which actually are weighted. In the case of the structure of the neural network, many representations are possible where the multi-layer feed-forward network is mostly preferred and used. Neurons are interconnected as layer by layer, and the connections are neither in-layer and nor cross-layer.

6 Result and Discussion

For the final evaluation of the work, certain parameters are considered which will define the classification of the dataset and the efficiency of the technique discussed. In the evaluation part, the instance is considered as the single data point and parameters are being evaluated for every single instance in the dataset. Mean absolute error, root mean squared error, relative absolute error, root relative squared error, etc., are some of the considered evaluation parameters.

```
** Decision Tress Evaluation with Datasets **

Correctly Classified Instances          2               66.6667 %
Incorrectly Classified Instances        1               33.3333 %
Kappa statistic                         0
Mean absolute error                     0.4333
Root mean squared error                 0.4726
Relative absolute error                97.5     %
Root relative squared error           100.2497 %
Total Number of Instances               3
```

The methodology outperforms the existing methodology as the mean absolute error and also root mean squared error are at minimum level in the case of the proposed methodology.

```
=== Confusion Matrix ===

 a b   <-- classified as
 2 0 | a = yes
 1 0 | b = no

=== Detailed Accuracy By Class ===

                 TP Rate   FP Rate   Precision   Recall   F-Measure   ROC Area   Class
                   1         1         0.667       1         0.8        0.5       yes
                   0         0         0           0         0          0.5       no
Weighted Avg.    0.667     0.667     0.444       0.667     0.533      0.5
```

For the better evaluation of the work, confusion matrix is been used or generated, and the confusion matrix actually provides the classification representation of the dataset. A confusion matrix is depiction of the dataset in tabular form which is used for **describing the performance of a classification model** (or "classifier") when applied over the specific set of data and in the condition when the true values for the data are known. The relative terminology of the confusion matrix is quite confusing while it is quite in depiction.

```
** Decision Tress Evaluation with Datasets **

Correctly Classified Instances           3              100      %
Incorrectly Classified Instances         0                0      %
Kappa statistic                          1
Mean absolute error                      0
Root mean squared error                  0
Relative absolute error                  0      %
Root relative squared error              0      %
Total Number of Instances                3
```

In the case when the decision tree is used as classifier, the classification of the dataset has reached to the level of 100% and other errors have been degraded to minimum like zero.

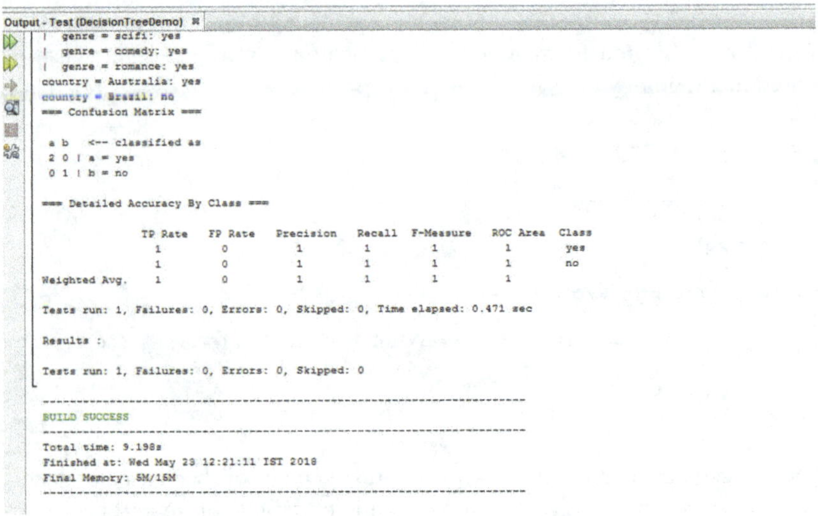

In the case when the decision tree is considered as the base classifier, the results are better than the previous methodology in the terms of the recall and precision as shown in the snapshot, and also the number of instances classified and missed is considered for computing the efficiency of the work.

7 Conclusion

Ensemble learning combines the power of multiple models into a single decision. Benefits from ensemble learning can include reduced overfitting and increased classification performance which makes ensemble learning a potential useful tool for large and dynamic datasets. However, many ensemble approaches, such as bagging and boosting, do not take into account the inherent high dimensionality found in these datasets. Thus, we developed new ensemble approaches, bagging using the decision tree as base classifier and boosting using the ANN as base classifier, and the above two are combined using the meta-decision tree approach.

In this work, we seek to determine whether proposed methodology is best suited for large and dynamic datasets. The methodology is tested and analyzed for different parameters like mean absolute error, root mean squared error, relative absolute error, etc., and results have shown that the methodology outperforms the existing work with respect to the number of instance classified and also considering the above-mentioned parameters taken as efficiency factors for the classification of the dataset.

References

1. Dietterich TG (2001) Ensemble methods in machine learning. In: Kittler J, Roli F (eds) Multiple classifier systems. Lecture Notes Computer Sciences, vol 1857, pp 1–15
2. Freund Y, Schapire R (1994) A decision-theoretic generalization of on-line learning and application to boosting. In European conference on computational learning theory—Euro-COLT. Springer, New York
3. Friedman J, Hastie T, Tibshirani R (2000) Additive logistic regression: a statistical view of boosting. Ann Stat 28(2):337–374
4. Freund Y, Schapire R (1996) Experiments with a new boosting algorithm. In Thirteenth international conference on machine learning, pp 148–156, Bari, Italy
5. Li S, Zhang Z (2004) Floatboost learning and statistical face detection. Trans Pattern Anal Mach Intell 26(9):23–38
6. Hido S, Kashima H (2009) Roughly balanced bagging for imbalance data. Stat Anal Data Min 2(5–6):412–426
7. Wang S, Yao T (2009) Diversity analysis on imbalanced data sets by using ensemble models. In IEEE symposium on computational intelligence and data mining, pp 324–331
8. Błaszczyński J, Stefanowski J, Idkowiak L (2013) Extending bagging for imbalanced data. In: Proceedings of the 8th CORES 2013, Springer series on advances in intelligent systems and computing, vol 226, pp 269–278
9. Liu A, Zhu Zh (2013) Ensemble methods for class imbalance learning. In: He H, Ma Y (eds) Imbalanced learning: foundations, algorithms and applications. Wiley, Hoboken, p 6182
10. Breiman L (1996) Bagging predictors. Mach Learn 24:123–140
11. Ho TK, Hull JJ, Srihari SN (1994) Decision combination in multiple classifier systems. IEEE Trans Pattern Anal Mach Intell 16(1):66–75

Toward m-Governance: T-Wallet Application for Financial Inclusion

Jayesh Ranjan, GT Venkateswara Rao, Naraharisetty Padmaja, Vagvala Satyajit Rao and Josyula Srinivas

Abstract New digital platforms offer convenient and less expensive ways to make payments and transfer funds. Owing to the requirement of the citizens during demonetization period, Telangana State Government came out with a unique initiative called the T-Wallet, enabling its citizens in making cashless electronic payments to both government and private services. T-Wallet is the official digital wallet of Telangana State, available as an Any Time Any Where digital payment option. T-Wallet is distinctive as a digital payment system for the banked and unbanked citizens, customized for citizens with a smartphone, a feature phone or no phone. The paper describes T-Wallet case study, an initiative taken by the government to encourage digital financial inclusion.

Keywords m-Governance · Financial inclusion · Mobile wallet · Digital payments · Security standards

1 Introduction

The adoption of mobile technologies to support and enhance government performance and foster a more connected society—can help improve government per-

J. Ranjan · G. V. Rao
Information Technology Department, Government of Telangana, Hyderabad, India
e-mail: dir_eseva@telangana.gov.in

G. V. Rao
e-mail: secy_itc@telangana.gov.in

N. Padmaja (✉) · V. S. Rao · J. Srinivas
National Institute for Smart Government (NISG), Hyderabad, India
e-mail: padmaja@nisg.org

V. S. Rao
e-mail: satyajit@nisg.org

J. Srinivas
e-mail: srinivas.j@nisg.org

S.-L. Peng et al. (eds.), *Computing and Network Sustainability*, Lecture Notes in Networks and Systems 75, https://doi.org/10.1007/978-981-13-7150-9_10

formance and strengthen public good governance is termed as m-Governance. The Sustainable Development Goals, otherwise known as the Global Goals, built on the Millennium Development Goals (MDGs), eight anti-poverty targets that the world committed to achieve by 2015, address the root causes of poverty and the universal need for development that works for all people [1].

The 10th 5-year Plan of India acknowledged that good governance is one of the most crucial factors required if the targets of the plan were to be achieved. e-Governance and m-Governance distinctly promote good governance by amplifying the levels of transparency, accountability and accessibility.

1.1 Evolution of Mobile Technology

Globally, mobile technology has emerged as a primary engine of communications. Mobile has become the fastest adopted technology of all time [2]. Nearly every fundamental human pursuit has been touched by mobile. In less than 15 years, 3G and 4G technologies have reached 3 billion subscribers as per Ericsson, making mobile the most rapidly adopted consumer technology in history. The key drivers for mobile governance are:

- Government mandate for inclusive growth;
- Increasing mobile penetration;
- Increasing availability of smartphones at low costs;
- Increasing consumer demand or awareness;
- Automation due to Information Communication and Technology;

1.1.1 Mobile Standards and Specifications

Main standards bodies involved in today's mobile device and network regulations are International Telecom Union, 3GPP, IEEE and IETF [3]. These bodies define the voice and video standards, standards for radio communication, Wi-fi standards and Internet protocols.

Department of Telecommunications is the holding organization of policies, regulatory bodies (like TRAI) and many Public Sector and Corporates are included under its umbrella. The mobile ecosystem and the standard bodies provide the possibilities and limitations of technology when designing mobile applications and deployment in the field of use. Security and privacy aspects of e-Governance and m-Governance applications have been defined under IT Act 2000, IT (Amendment) Act 2008, National Cyber Security Policy (2013) and privacy law enumerating nine privacy principles [4–7].

1.2 Important Factors for m-Governance Applications

There are fundamentally four mobile service delivery channels namely, voice (native voice, IVR), messaging (SMS, USSD, MMS), Web browsing and apps. The type of usage should be mapped to needed service and to needed type of device while designing mobile applications for m-Governance. The mobile application may need to operate on a multitude of platforms, supporting different operating systems of smartphones. Data Security and identity security are important as sensitive information about the end user can be an issue of concern.

1.3 Mobile Service Delivery Initiatives

An integrated service delivery platform can provide seamless service across the manual, Web and mobile channels. There have been several initiatives across the globe to build and deploy mobile applications to enable good governance. Table 1 provides a snapshot of some mobile applications included under UNDP Good Governance Parameter [8] with corresponding governance functions and sectors [9].

2 Research Methodology

The paper is presented as a case study to depict the main themes identified. Merriam [10] talking of case study research, defines a study case as a holistic and intensive description of a delimited program, institution, a process, a social unit [10]. This case study attempts to explain m-Governance initiative of Telangana State Government. The data used in the case study comprises of primary and secondary data. The primary sources comprise meetings with policy level government officials involved in leading the project, meetings with the department head who supported project implementation, technical staff who were part of the project deployment. The secondary data has been sourced from the articles on the project and managerial records

3 T-Wallet

Demonetization had given the country the impetus required to go cashless, supported by Prime Minister's call for cashless transactions and reciprocated by State Governments. It was during this cash crunch that the Telangana State Government announced its support for cashless transactions in the state. It also wanted its citizens to effortlessly make payments to both government and private services. As a result, Telangana State Government came out with a new initiative called the T-Wallet,

Table 1 m-Gov projects and UNDP good governance parameters

UNDP good governance parameter	m-Governance initiative	Domain/sector	Region
Inclusion	Bangladesh Health Service through Mobile Phone	Health	South Asia
	Kerala Kissan Project	Agriculture	India
	Aakash Project	Education	India
Consensus oriented	National Farmers Information Service (NAFIS)	Agriculture	Africa
Transparency	Karnataka MGNREGA m-platform	Employment	India
	SAKALA Project	Government	India
	Supreme Court of India Smartphone App	Courts	India
Responsiveness	Himmat App	Safety	India
	CELEPAR	Employment	South America
	Disaster Alert	Disaster Management	Global
Efficiency and effectiveness	Movilidad 2.0	Transport	Europe
	Tiruvallur m-Government Project	Governance	India
	Western Railway Ticketing System	Transport	India
Accountability	Tax challan status enquiry	Taxes	India
Participation	GRAAM Arogyashreni	Governance	India

which was launched on 1st June, 2017. Given that all the mobile payment systems were dependent on smartphones, T-Wallet was required to be accessible to citizens who owned just a feature phone, or had no phone at all (targeting the rural population) and all this without any financial burden on the citizen users.

T-Wallet is the official digital wallet of Telangana State, where citizens can use T-Wallet to make payments for both government and private transactions to avail services. T-Wallet was conceptualized with the objectives of providing a digital payment system to the banked and unbanked citizens, irrespective of whether the citizen owned a smartphone, feature phone or didn't own a phone. It also aims at ensuring the "Direct Benefit Transfers" for state government welfare schemes reaching the beneficiaries directly and gives the citizens a government payment system which they can trust in.

3.1 Conceptualization of T-Wallet

In the cash crunch actuated by the demonetization of November 2016, the Government of Telangana had made the decision to roll out T-Wallet. However, the Reserve Bank of India, the administrator for all financial transactions in cash or in e-form, is required to give a license [11], called the Prepaid Payment Instruments license (PPI license), to the wallet providers based on whether the provider is able to meet required criteria. The RBI needs documentation on expected benefits to the financial systems, operating risks, systems in place to mitigate the risks, amount of finance required, sources of finance, amount of own capital proposed, amount of borrowings expected from banks, the rate of return on investment, etc.

Officials designated to take up this task along with NISG team evaluated the application for its advantages such as its wider reachability, mobility, and ubiquity which would enable citizens' access government services from anywhere and anytime. The officials contemplated on the lines of increased democracy, green government, faster information flow, personalization of services and cost effectiveness to citizens. Government of Telangana envisaged cost optimization, improved communications, expanded service delivery and progress toward digital equality with T-Wallet.

The government also has taken care of limitations. Rural areas still support only 2.5 G (GPRS) which can provide only low-speed Internet. So the application was designed to be low in multimedia so that downloads are possible on slow-speed networks. Limitations such as low levels of literacy, computer illiteracy and affordability of the device for the rural or BPL population have led them to design this application for citizens without phone as well, and who can transact with the help of MeeSeva centers. The team also analyzed the rural citizen-centric service delivery aspect of the application, at a granular level (reaching out to migrant labor) and conducted Aadhaar seeding drives.

3.2 Development and Deployment of T-Wallet

The Government of Telangana had envisioned a wallet where the payment instruments, i.e., the preloaded balance were redeemable at a group of clearly identified merchant locations/establishments which contract specifically to accept the payment instrument. These instruments, called Semi-Closed System Payment Instruments, do not permit cash withdrawal or redemption by the holder. Given the time and the resources which would be required in order to create a completely self-owned payment gateway, in addition to the fact that RBI has put a hold on the issue of PPI licenses at the time, the Government of Telangana made the decision to own and co-brand the wallet with the firm "Transaction Analysts (India) Pvt. Ltd.", who own the PPI license. TA's team collaborated with the Government of Telangana to create the T-Wallet app and incorporate the payment gateway into to the app as delineated by the RBI guidelines for PPI license holders.

A ground infrastructure comprising of Government citizen service centers, public distribution centers and local entrepreneurs who act as digital partners provide the backbone of this cashless service. The Government Citizen Service Centres—Mee-Seva Centers—were enabled to transact using the T-Wallet for the creation of new user IDs, loading balance and making transfers on behalf of the citizen. The T-Wallet pilot was rolled out in two districts in Telangana initially—Ibrahimpur and Siddipet—post the integration of the T-Wallet with the Civil Supplies outlets.

3.3 Technical Details of Application

The T-Wallet is available on both Android- and iOS-based smartphones as well as accessible via feature phones. Furthermore, the wallet is operable via Web browsers from the official Web site. It also supports multilingual presentation in Telugu and Urdu languages besides English.

The data storage is done on Azure Cloud storage with the servers based in Pune, Maharashtra. For further network requirements, the systems are dependent on the State Data Centres (SDCs) and the SWANs.

The payments and transactions are all routed through the third party payment gateway of PayU. The major advantage that the T-Wallet provides for is the scalability of operations, i.e., it is able to handle a massive number of transactions with relative ease. In addition, the application design and user interfaces are easy to understand and operate. While Bharat Interface for Money (BHIM) is an application to enable fast, secure, reliable payments through mobile has a national footprint, T-Wallet also comes with a loading facility and also works for a feature phone and no phone segment of population.

3.4 T-Wallet Dashboard

Citizens can use T-Wallet to make payments for both government and private transactions to avail services and is integrated with Government departments such as Mee-Seva, GHMC, HMWSSB, TSNPDCL, TSSPDCL, TSRTC, RTA, TASK, CDMA, HMDA ORR Tolls and the Hyderabad Metro Rail Limited (HMRL). The Streenidhi SHG loan repayments are being enabled through T-Wallet.

T-Wallet serves through online Web browser, smartphone, feature phone and even no phone. Citizens with feature phone or no phone can use MeeSeva centers to open T-Wallet, load money into their wallet and make payments. All of the transactions are recorded, tracked, and analyzed at the backend by the app creation partner team from Transaction Analysts. Updated information on the registrations and usage can be found on the T-Wallet official Web site (https://twallet.telangana.gov.in/Dashboard/Dashboard.aspx). Figure 3 provides a glimpse of T-Wallet dashboard [12] as on 30th July 2018.

T-Wallet Dashboard

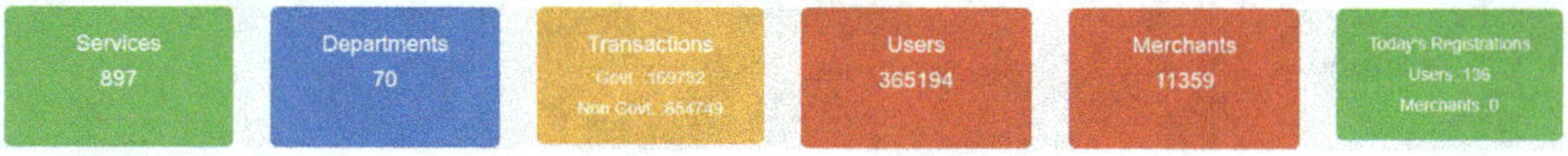

Fig. 1 T-Wallet dashboard. *Source* T-Wallet Official Web site

Table 2 Transactions of T-Wallet

Payments to departments	Total Trxns	Total Trxn amount
Traffic police	3238	900,712
HMWSSB	22,240	14,117,123
TSNPDCL/TSSPDCL	145,014	106,116,185
HMDA-ORR	2486	112,146
MeeSeva/E-Seva	79,854	61,699,916
Pvt. bill payments	291,660	35,462,469
CDMA	480	1,046,435
GHMC	1956	5,417,376
IFFCO Bazar	3	150
IGRS-TSCHITS	1	2
LABOUR	326	172,648
TSPSC	340	69,924
T-app folio	6945	1,600,566
TASK	450	444,382
TSRTC	3	1078
Total		227,161,115

A glimpse of dashboard from the official Web site of T-Wallet (Fig. 1) provides us with details of the users, number of government and non-government transactions along with departments and services which have T-Wallet included as a payment option.

Table 2 gives us an insight into the department-wise transactions done through T-Wallet. While utility bills have the highest number of transactions, transactions enabled through common service centers (MEESEVA/E-SEVA) rank second and the new initiative of government T-App folio and private bill payments are also done using T-Wallet. This indicates the reach of the wallet. Figure 2 indicates the mobile penetration, where the registrations of T-Wallet are done largely through mobile phones.

Figure 3 with transaction amount and numbers highlight the segment of population who reached out to CSCs to make payments through T-Wallet, while other transactions using debit/credit card and net banking are also made.

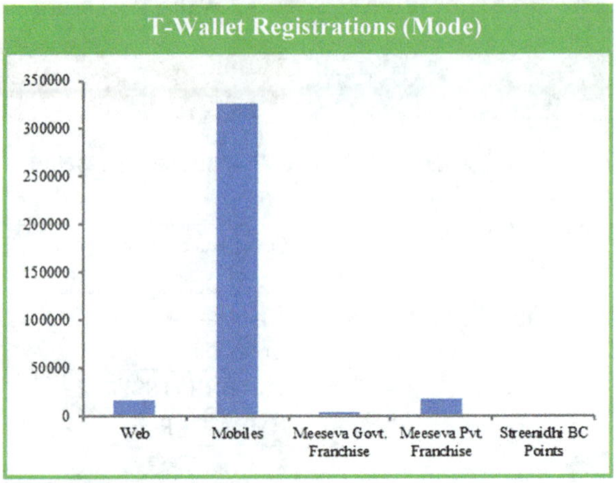

Fig. 2 Mode of T-Wallet registration. *Source* T-Wallet Official Web site

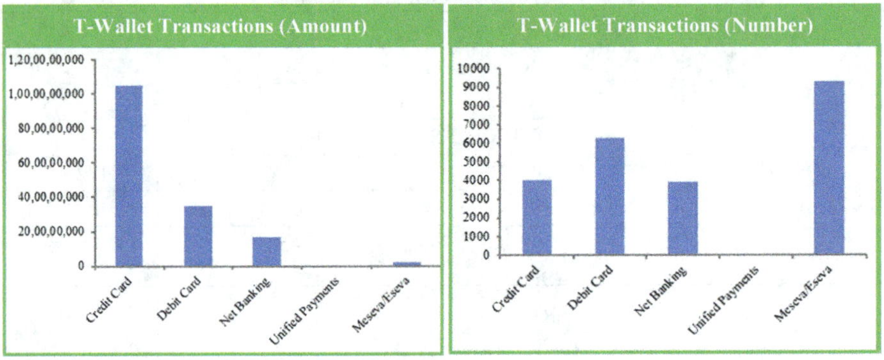

Fig. 3 T-Wallet dashboard. *Source* T-wallet PMU

3.5 Collaboration with Private Agency and Security Audits

The recording of the transaction details and the analyses required alongside the recurrent reconciliation of transaction amounts is all handled and managed by the TA Team and reported to the Government of Telangana. As per the RBI guidelines for PPI operators, the app is back-ended by Vijaya Bank for all fund transfer activities.

Security Audits are carried out by the TA Team every quarter and the reports are delivered to both the Government of Telangana as well as the RBI as part of the due diligence.

Against the norm, where the service provider charges the user a nominal transaction fee for every transaction, there is no fee levied on the citizen for the services provided by T-Wallet. All transaction costs are being absorbed by the government at present.

As part of the adherence to the RBI guidelines for PPI license holders/applicants, the banking of all the funds parked with the digital payment gateway is with the banking partner—Vijaya Bank. In compliance with the same guidelines, there is no interest accrued on the funds deposited in the PPI licensed accounts.

The Government of Telangana is preparing to apply for RBI license and develop its own application for T-Wallet within the next two years. After the Cabinet Sub Committee recommendations are accepted by the government, the implementation of the same by all the departments would be periodically reviewed by the Chief Secretary's office for the next six months to a year.

3.6 Authentication of Accounts and Storage of Data

There are three distinct cadres in the users based on the kind of authentications given for the account—low KYC, e-KYC and full KYC. As per the RBI guidelines, low KYC accounts do not need an Aadhaar number affixed to them. However, the usage too is limited to approximately rupees five thousand per month. Where the e-KYC is done, Aadhaar number is authenticated via OTP and the account is coupled to the authenticated Aadhaar number. The usage limit for such e-KYC is rupees fifteen thousand per month. Where accounts have the full KYC done, the account is created at a MeeSeva Center, uniting the biometric authentication as well as Aadhaar authentication and connecting them to the T-Wallet account. For such accounts, the usage limit is the highest at rupees one lakh per month. For each of these accounts, the maximum balance in the account at any instance is the same as the maximum amount for transactions, i.e., rupees five thousand, fifteen thousand and one lakh.

The MeeSeva Centers are authorized to act on behalf of the citizens using a two—factor authentication of Aadhaar number and biometric validation OR Aadhaar and OTP to Aadhaar linked mobile number for no phone and feature phone users, respectively.

In case of feature phones, the balance in the T-Wallet can be topped up via SMS as well. Options to do transactions via USSD and IVRS are in the development phase to be integrated into the T-Wallet.

Top-up or loading the wallet balance can be done in multiple methods—at any MeeSeva Center by paying cash at the counter, through the smartphone-based app, net banking, mobile banking or an SMS from the linked mobile number on a feature phone.

3.7 Role of a Consultant

NISG has been chosen as a consultant for this project. NISG has played a pivotal role in conceptualizing the solution end to end, and in seeking proposals from industry with a reference from government of Telangana and in finalizing the vendor for development of the mobile application. Telangana Information Technology Institute was engaged for capacity building and a PMU is set up to govern, monitor and evaluate the project.

3.8 Challenges Faced During Implementation of T-Wallet

The biggest challenge faced by the team in implementation of the T-Wallet project has been the branding—convincing the citizens that it is a trustworthy, efficacious and easy way of online money transfers even for the non-banked, no phone citizens. While the privately owned e-wallets have been doing massive promotional activities for the awareness and usage of their e-wallets, the T-Wallet team had budget constraints on advertisement and promotional activities. A collaboration with the Telangana Information Technology Institute (TITA) effectuated into a campaign for the awareness and digital literacy in a 10-day drive titled "DIGITHON" across districts in September of 2017. They achieved a massive 1.8 lakh registrations happened from rural areas. Digital literacy programs launched by state and central government, online training material provided in the official Web site of T-Wallet contributed to increasing awareness of the application.

Furthermore, the expectations from a state-owned e-wallet are much higher than those from a privately owned one. The T-Wallet's effectiveness is increased exponentially when the benefits of the various direct benefit transfers from government are linked directly to the appropriate citizens' T-Wallet account. With the constant comparison between such disparate e-wallets, it is tougher on the team to meet such expectations.

The risk factor in the operation of the T-Wallet is extremely high and rests for the major part with the technology partner, Transaction Analysts. In the face of the rising rate of cybercrime, risk mitigation warrants have been given the utmost attention.

3.9 RBI Guidelines (for Wallet)

The wallet completely follows the RBI master guidelines and the wallet application is built as per RBI guidelines. Right from RBI approval for opening of PPI license, defining information security and privacy policies, conducting risk assessment and establishing a mechanism for monitoring cyber security incidents, it is built in accordance with guidelines. Any changes proposed or done were to be in compliance with RBI guidelines.

3.10 Payment Reconciliation Mechanism Adopted

All transactions are done online and so when the wallet account is debited, the amount debited will be credited to the Merchant (Government department or private merchant) account in T-Wallet System. Settlements are done in two models. In both the cases, Escrow account is debited and merchant pre-registered account is credited. Programme Management Unit is set up by government to take care of the governance of the application.

In the first model, the government department/s provide a URL through which success transactions data is pulled automatically at configured time after the end of the day. Successful transactions are matched at both the end and reverse the failed transactions at government department end. Then net amount is settled on the next business day ($T + 1$) by way of NEFT/RTGS to the government pre-registered account. In the second model, successful transactions data is not shared by government departments. Hence on the next working day, TA team shares the successful transactions data from their end to government departments and once confirmed, settlement is done.

3.11 Privacy Policy

To ensure privacy of data, only required information is captured from T-Wallet as per RBI guidelines. Critical data is stored in encrypted format in T-Wallet database. User information is not shared to any external agencies.

For internal staff operations such as helpdesk, reconciliation and settlement teams, only the needed information about the end users is provided such as masked Aadhaar id with last 4 digits is only shown. Debit or credit card information is captured in the payment gateway Web site which is PCI DSS compliant.

3.12 Networking and Security Standards Adopted

T-Wallet is hosted in cloud environment (Microsoft Azure) and Internet is used for wallet operations. So T-Wallet supports networking standards such as IPv6, https. Data transferred from mobile or Web app to server is encrypted. HTTPS is used with encrypted data for server connectivity and firewall is used at the perimeter of T-Wallet servers.

Critical data in T-Wallet servers is encrypted and external communication with Payment Gateway service providers and other agencies is encrypted and always using https connectivity. Public and private key pair is used for securing data with external entities. Data transmission security standards are followed using an encryption mechanism with https connectivity when the data gets transferred from either the mobile or Web app to the server and similarly from server to externals servers. PCIDSS security compliance certification, ISO certification for DC and DRC is done.

3.13 Interoperability of Wallets

Transfer of balances from one T-Wallet account to another is one of the benefits. Interoperability between wallets such as T-Wallet to PayTM is not yet a reality. Such interoperability would be realized when the much anticipated RBI Master Direction for interoperability is rolled out.

3.14 Impact of the Application

In the span of a year, post the launch of the wallet, there have been 3.5 lakh registrations (i.e., new accounts) and more than 265 Crores worth of transactions. About 5.2 Lakh payment transactions are done.

3.15 Further Enhancements to the System

Cash cannot be withdrawn directly from the T-Wallet account. Once the monies have been loaded into the T-Wallet account, they may be used only to make payments, transfer to another T-Wallet account or to a bank account. A more noteworthy facet is that the RBI has recently given the approval for withdrawal of cash deposited in a T-Wallet account via MeeSeva Centers (which have a deeper reach into the villages than even Nationalized Banks) to be rolled out before the end of July 2018. The approval has been given for a pilot rollout across 5 (five) districts over a span of 3 (three) months.

3.16 Conclusion

This makes T-Wallet unique among all wallets in the country. If the cash out from T-Wallet works well, government may consider routing the entitled benefits to the citizens through T-Wallet than through a bank account. Given that there is no set precedence for a state-owned e-wallet with such unique and convenient features, the convenience of withdrawing cash directly from the T-Wallet account, only work toward enhancing the credibility of the application, increasing the confidence of the citizen in the government.

References

1. Sustainable Development Goals. Retrieved from http://www.undp.org/content/en/home/sdgoverview/post-2015-development-agenda.html
2. Boston Consulting Group (2015) The mobile revolution, how mobile technologies drive a trillion dollar impact
3. ITU presentation on Competition, Intellectual Property and Standard Setting, Slideshare http://www.slideshare.net/OECD-DAF/ip-standard-settingdoreinternationaltelecomunion17dec2014
4. Section 43 A, IT Act 200. Retrieved from http://www.itlaw.in/section-43a-compensation-for-failure-to-protect-data/
5. Section 72 A, IT Act 2000. Retrieved from http://www.itlaw.in/section-72-a-punishment-for-disclosure-of-information-in-breach-of-lawful-contract/
6. Shri Kapil Sibal releases National Cyber Security Policy. Retrieved from http://www.pib.nic.in/newsite/erelease.aspx?relid=96971
7. Report of the Group of Experts on Privacy. Retrieved from http://planningcommission.nic.in/reports/genrep/rep_privacy.pdf
8. Characteristics of Good Governance. Retrieved from https://www.gdrc.org/u-gov/g-attributes.html
9. Mobile Governance Handbook, p. 10, Table 2, 2017, m-Gov cases and UNDP Good Governance Parameters, National Institute for Smart Government training hand books
10. Merriam SB (1988) Case study research in education: a qualitative approach. Jossey Bass, San Francisco (CA)
11. Master Direction on Issuance and Operation of Prepaid Payment Instruments (Updated as on December 29, 2017). Retrieved from https://rbidocs.rbi.org.in/rdocs/notification/PDFs/58PPIS11102017A79E58CAEA28472A94596CFA79A1FA3F.PDF
12. T-Wallet Website. Retrieved from https://twallet.telangana.gov.in

Predicting Diabetic Foot Maturing Through Evolutionary Computation

José Neves, André Alves, Marco Prata, Mário Ribeiro, Victor Alves, Filipa Ferraz, João Neves, Jorge Ribeiro, António Capita and Henrique Vicente

Abstract It is a twenty-first-century disease, its numbers are still growing exponentially. This brings one to the subject of this work, the *Maturing of Diabetic Foot* which, like diabetes, rises to values never seen before. It is envisaging the development of an *ImageJ* plug-into extract relevant feature from diabetic foot images and, in conjunction with the patient's clinical and lifelong data, a computational system to

J. Neves · V. Alves · F. Ferraz · H. Vicente (✉)
Centro Algoritmi, Universidade do Minho, Braga, Portugal
e-mail: hvicente@uevora.pt

J. Neves
e-mail: jneves@di.uminho.pt

V. Alves
e-mail: valves@di.uminho.pt

F. Ferraz
e-mail: filipatferraz@gmail.com

A. Alves · M. Prata · M. Ribeiro
Departamento de Informática, Escola de Engenharia, Universidade do Minho, Braga, Portugal
e-mail: a69381@alunos.uminho.pt

M. Prata
e-mail: a68516@alunos.uminho.pt

M. Ribeiro
e-mail: a67657@alunos.uminho.pt

J. Neves
Mediclinic Arabian Ranches, 282602, Dubai, United Arab Emirates
e-mail: joaocpneves@gmail.com

J. Ribeiro
Escola Superior de Tecnologia e Gestão, ARC4DigiT—Applied Research Center for Digital Transformation, Instituto Politécnico de Viana do Castelo, Viana do Castelo, Portugal
e-mail: jribeiro@estg.ipvc.pt

A. Capita
Instituto Superior Técnico Militar, Luanda, Angola
e-mail: antoniojorgecapita@gmail.com

S.-L. Peng et al. (eds.), *Computing and Network Sustainability*, Lecture Notes in Networks and Systems 75, https://doi.org/10.1007/978-981-13-7150-9_11

predict and evaluate its severity. The applied problem-solving method is based on a symbolic/sub-symbolic line of logical formalisms that make complex systems easier to develop and analyze, where solutions to new problems are based on answers to previous ones, and itemized as a Case-Based Reasoning/Artificial Neural Network approach to computing.

Keywords Diabetic mellitus · Case-based reasoning · Artificial neural networks · ImageJ · Many-valued machines

1 Introduction

Diabetes stands for a chronic disease that spreads throughout the world. The *International Diabetes Federation* estimates that by the year 2030, approximately 438 million people will develop diabetes, an increase of around 54% to present data. On the one hand, *Diabetes Mellitus* (*DM*) occurs when the body becomes resilient to the effects of insulin or does not produce enough insulin to keep blood sugar levels within the normal range. It can develop at any age and is often preventable [1, 2]. *DM* can be silent, especially in the early stages, when the patient feels comfortable and presents almost none of the above symptoms. But just as quietly, diabetes causes chronic complications in various organs, e.g., heart, blood vessels, nerves, eyes, and kidneys. Effective control of blood sugar levels can obviate these complications. Though they develop slowly and gradually, these chronic complications of diabetes can lead to disability and even death [3–5]. On the other hand, wounds that appear on the feet due to poor blood circulation are easily infectious and difficult to treat and may result in amputation of the fingers, feet, or legs [6]. Therefore, the development of Decision Support Systems (DSS) for predicting, analyzing, and evaluating the outcome of diabetic foot ulcers can be an asset to healthcare workers, particularly to identify the best type of abnormal treatment. The DSS proposed in this paper is based on features obtained from diabetic foot images according to a series of historical data that focus on a *Proof Theoretical* approach to problem-solving [7] and implemented through a symbiosis between *Artificial Neural Networks* (*ANNs*)/*Case-Based* approaches to computing [8–10]. Medical images are analyzed by a developed *ImageJ* plug-in [11]. It is divided into five sections. The introduction makes the first and defines the goals of the work and its context. Following is a section that explores the theoretical foundations and describes the basic concepts used throughout the article. The third and fourth sections define the case study. It ends in a partition that takes into account the results and outlines the upcoming work.

H. Vicente
Departamento de Química, Escola de Ciências e Tecnologia, Centro de Química de Évora, Universidade de Évora, Évora, Portugal

2 Theoretic Fundamentals

2.1 *Knowledge and Argumentation*

Knowledge and Argumentation (*KA*) aims to understand the complexity of information and its implications. In fact, automated closing allows the system to fill in the gaps, enabling it to deal with defective information. In this study, the data items must be understood as if one finds something less inside when one occupies something separately, i.e., it is essentially made up of different elements, namely the *Interval* in which is supposed to be, its *Quality of Information* (*QoI*), the *Degree of Confidence* (*DoC*), and its *Entropy State* (*ES*) [9]. These are just four of the number of infinite items. No doubt you can do almost anything you think by combining different elements. Therefore, the universe of discourse may be eliciting in the form, viz.

$$\text{predicate}_{1 \le i \le n} - \bigcap_{1 \le j \le m} \text{clause}_j\big(\big(\big[A_{x_1}, B_{x_1}\big]\big(QoI_{x_1}, DoC_{x_1}, ES_{x_1}\big)\big), \ldots, \big(\big[A_{x_m}, B_{x_m}\big]\big(QoI_{x_m}, DoC_{x_m}, ES_{x_m}\big)\big)\big)::QoI_j::DoC_j::ES_j$$

where n, $\cap$, and m refer to the predicates' set cardinality, conjunction and extension of the predicate, respectively. On the other hand, with the aim of quantifying qualitative data and facilitating the process understanding, it is introduced graphically. Thus, considering a series of N issues regarding a certain topic, with K options per issue (e.g., *low* · *medium* · *high* and *unknown*), itemized as a circle divided into three (three) slices, with the markers in the axis corresponding to each of the possible choices per issue, one may have (Fig. 1) [12], viz.

2.2 *Case-Based Reasoning*

An approach based on *CBR* to problem-solving is aimed at solving new problems by using knowledge from unraveling similar ones, where modularity on top of the desirable as it supports reuse of parts of the application or data logics and also facilitates maintenance by allowing repair or replacement of segments of it without requiring extensive stand-ins. So, in order to allow *CBR* systems to work under these settings where default data or information is presented, an extended *CBR* cycle that considers the cases' *QoIs*, *DoCs,* and *ESs* metrics are presented [10, 13] (Fig. 2).

Issues' Data		
Issue 1	Issue 2	Issue 3
Low/Medium	High	Unknown

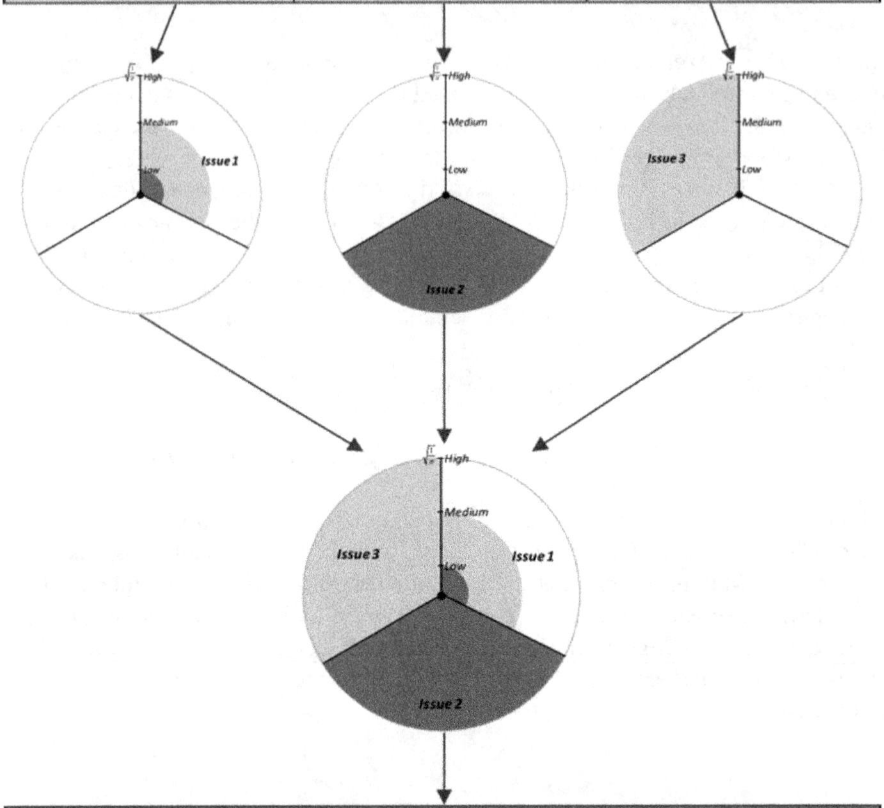

Issues' Entropic Ranges			
Issue 1	Issue 2	Issue 3	Global View
[0.11, 0.22]	[0.33, 0.33]	[0, 0.33]	[0.44, 0.88]

Fig. 1 Going from *Qualitative* to *Quantitative* data

2.3 Artificial Neural Networks

ANNs are powerful problem-solving techniques emanating from the artificial intelligence arena whose capacity and applicability in modeling of a widely types of problems and phenomena is clearly demonstrated. *ANNs* are able to capture the specificity of the problems under observation, using its architecture and nonlinearity nature, presenting better performances when compared with the classical methodologies for problem-solving [9, 12].

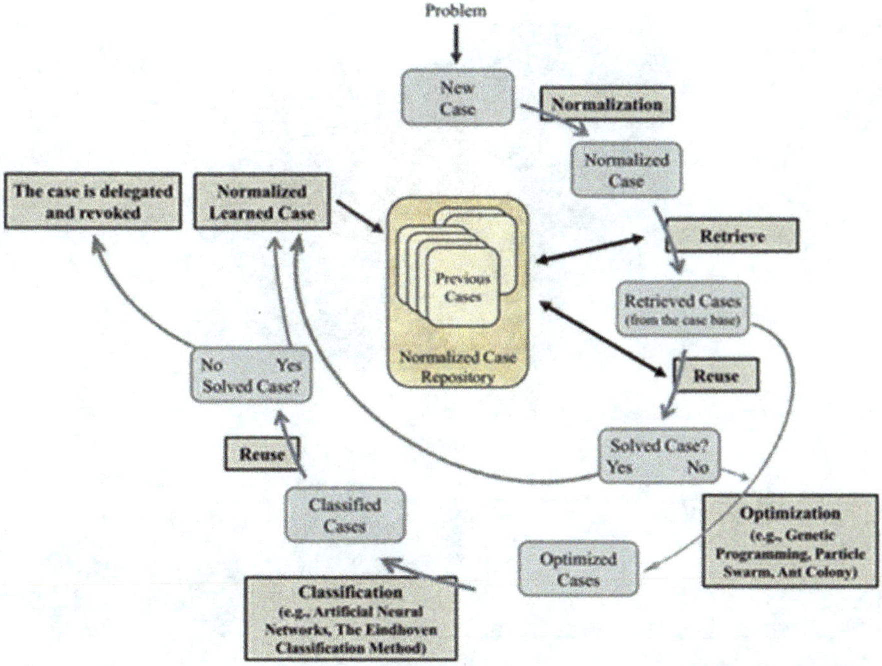

Fig. 2 A comprehensive vision of the embraced *CBR* cycle [10, 13]

3 Methods

3.1 Plug-in

Using the *ImageJ's* framework, a plug-in was developed to extract features from images of diabetic foot wounds. The area under scrutiny and the type of tissue to be extracted are handset in order to assess its severity [11, 14] (Fig. 3).

3.2 Dataset and Feature Extraction

The original dataset contained diabetic foot images, characterized by a wide variety of attributes, such as image location and the existence of bleeding, bubbles, ulcers, and other abnormalities, making the way to the respective knowledge database. Table or relation arguments follow the degrees *low* · *moderate* · *high* · *very high* (Fig. 4). The extensions of these relations make the *Case Base* for the problem in study, leading to the definition of predicate *dfs*, whose extension is given in table *Diabetic Foot State*,

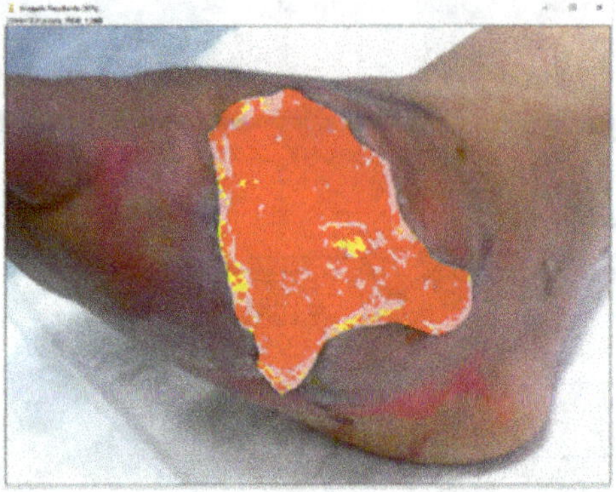

Fig. 3 Area from where the tissue features are extracted

Health Habits			
#	Smoking Status	Drinking Habits	Physical Exercise
1	None	High	High/Very High
2	High	None	Unknown
...	...	...	...
100	Unknown	Very High	Very High

Case 1

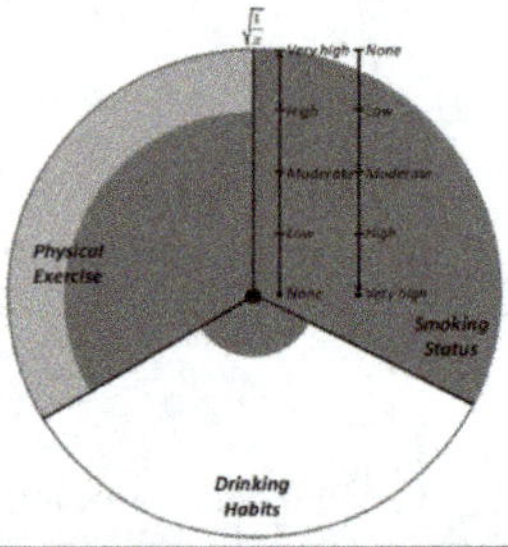

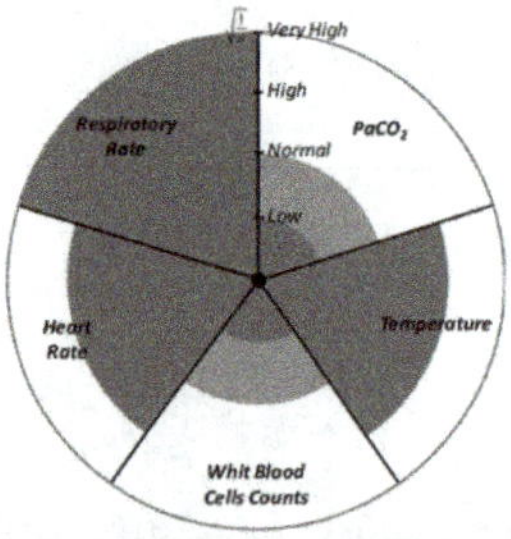

Diabetic Foot State			
#	Health Habits	Patients' Data	Entropic State
1	[0.67, 0.75]	[0.40, 0.70]	[0.54, 0.72]
2	[0.42, 0.75]	[0.25, 0.70]	[0.33, 0.72]
...	...	...	...
100	[0.33, 0.67]	[0.35, 0.75]	[0.34, 0.71]

Case 1

Patients' Data					
#	$PaCO_2$	Temperature	White Blood Cells Count	Heart Rate	Respiratory Rate
1	Low/Normal	High	Low/Normal	High	Unknown
2	Unknown	Normal/High	Normal	Unknown	Low
...	...	...	...	...	...
100	High	Unknown	Low	Unknown	High

Fig. 4 An extract of the relational data base for *Diabetic Foot State* prediction

which is used to train the *ANNs* that will work out *Diabetic Foot Evolution*, viz [9]. Predicate's *dfs* is to be understood as setting problem's objective function, viz.

$$dfs : \mathrm{H_{ealth}H_{abits}}, \mathrm{P_{atients}D_{ata}} \rightarrow \{\text{true, false}\}$$

4 Case Study

Now, in a presence of a new case, for example, the one with feature vector *Health Habits (HH) = [Moderate, Low/Moderate, High], Patients Data (PD) = unknown*, one may have, viz.

$$dfs_{new\ case}\left(\frac{([0.50, 0.58](1, 0.99, [0.50, 0.58])), \ldots, ([0, 1](1, 0, [0, 1]))}{\begin{array}{c}\text{attribute's values ranges once normalized}\\ \text{and respective QoI, DoC and ES values}\end{array}}\right)$$

whose data, once delivered to the *ANN*, returns the patient diagnosis score for *Diabet Foot Severity* (*DFS*) and its *State of Degradation* (*SD*) (the truth values are in the interval [0, 1]) (Fig. 5). Last but not least, the *ANN* may also be understood as a *Many-Valued Empirical Machine*, i.e., given the logical formula, viz.

$$dfs(([A_{x_1}, B_{x_1}], QoI_{x_1}, DoC_{x_1}, ES_{x_1}), \ldots, ([A_{x_m}, B_{x_m}], QoI_{x_m}, DoC_{x_m}, ES_{x_m}))$$

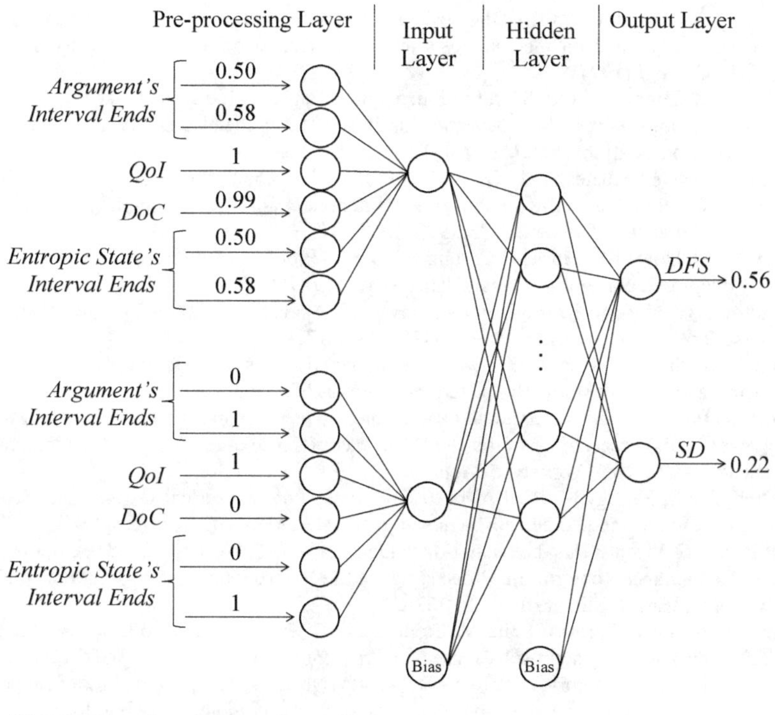

Fig. 5 Diabetic Foot Severity wound's prediction

returns an assignment of truth values *(i.e., DFS and SD)* under which the logical formula just referred to above takes the truth value *true*.

5 Conclusions

Syntheses and characterization of a workable method for problem-solving that allows for the classification of foot wound's severity were set as a *Many-Valued Empirical Machine*, i.e., it returns a patient's diagnostic to diabetic foot wound evolution in terms of a *Truth Value Valuation*. On the other hand, it must be noted that the meaning of the word *Empirical*, in this context, is *based on/concerned with/verifiable by observation or experience* rather than *theory/pure logic*. In upcoming work, we intend to move from *Empirical* to *Logic Machines*, showing the way to *Many-Valued Logic Machines* [15].

References

1. The Expert Committee on the Diagnosis and Classification of Diabetes Mellitus: Report of the expert committee on the diagnosis and classification of diabetes mellitus. Diab Care 20(7):1183–1197 (1997)
2. Alberti KG, Zimmet PZ (1998) Definition, diagnosis and classification of diabetes mellitus and its complications. Part 1: diagnosis and classification of diabetes mellitus. Provisional report of a WHO consultation. Diab Med 15(7):539–553
3. UK Prospective Diabetes Study Group (1998) Intensive blood-glucose control with sulphonylureas or insulin compared with conventional treatment and risk of complications in patients with type 2 diabetes. The Lancet 352(9131):837–853
4. New JP, McDowell D, Burns E, Young RJ (1998) Problem of amputations in patients with newly diagnosed diabetes mellitus. Diabet Med 15(9):760–764
5. Boulton AJ, Vileikyte L, Ragnarson-Tennvall G, Apelqvist J (2005) The global burden of diabetic foot disease. The Lancet 366(9498):1719–1724
6. Gale L, Vedhara K, Searle A (2008) Patients' perspectives on foot complications in type 2 diabetes: a qualitative study. Br J Gen Pract 58(553):555–563
7. Neves J (1984) A logic interpreter to handle time and negation in logic databases. In: Muller R, Pottmyer J (eds) Proceedings of the 1984 annual conference of the ACM on the 5th generation challenge. ACM, New York, pp 50–54
8. El-Sappagh S, Elmogy M, Riad AM (2015) A fuzzy-ontology-oriented case-based reasoning framework for semantic diabetes diagnosis. Artif Intell Med 65(3):179–208
9. Fernandes F, Vicente H, Abelha A, Machado J, Novais P, Neves J (2015) Artificial neural networks in diabetes control. In: Proceedings of the 2015 science and information conference (SAI 2015), IEEE Edition (2015), pp 362–370
10. Neves J, Vicente H, Ferraz F, Leite AC, Rodrigues AR, Cruz M, Machado J, Neves J, Sampaio L (2018) A deep learning approach to case based reasoning to the evaluation and diagnosis of cervical carcinoma. In: Sieminski A, Kozierkiewicz A, Nunez M, Ha QT (eds) Modern approaches to intelligent information and database systems, studies in computational intelligence, vol 769. Springer International Publishing, Cham, pp 185–197
11. ImageJ https://imagej.net/. Last accessed 2018/05/25

12. Fernandes A, Vicente H, Figueiredo M, Neves M, Neves J (2016) An adaptive and evolutionary model to assess the organizational efficiency in training corporations. In: Dang TK, Wagner R, Küng J, Thoai N, Takizawa M, Neuhold E (eds) Future data and security engineering, vol 10018. Lecture Notes on Computer Science. Springer International Publishing, Cham, pp 415–428
13. Silva A, Vicente H, Abelha A, Santos MF, Machado J, Neves J, Neves J (2016) Length of stay in intensive care units—a case base evaluation. In: Fujita H, Papadopoulos GA (eds) New trends in software methodologies, tools and techniques, frontiers in artificial intelligence and applications, vol 286. IOS Press, Amsterdam, pp 191–202
14. International best practice guidelines: wound management in diabetic foot ulcers. http://www.woundsinternational.com/media/best-practices/_/673/files/dfubestpractice forweb.pdf, last accessed 2018/06/28
15. Fernandes B, Vicente H, Ribeiro J, Analide C, Neves J (2018) Evolutionary computation on road safety. In: Cos Juez F, Villar J, de la Cal E, Herrero ÁH, Quintián H, Muñoz J, Corchado E (eds) Hybrid artificial intelligent systems, Lecture Notes in Artificial Intelligence, vol 10870. Springer International Publishing, Cham, pp 647–657

Exploration and Implementation of Classification Algorithms for Patent Classification

Darshana A. Naik, S. Seema, Geetika Singh and Abhinav Singh

Abstract Data mining techniques have seen tremendous increase in their usage in the past few years. Patent mining is one of the domains that utilize data mining techniques to a great extent. Patent mining consists of various tasks such as retrieval of patent, classification, patent valuation, patent visualization and detecting infringements. Among these, patent classification is an important task. It deals with the classification of patents into various categories. A common bottleneck in this task has been related to the automated classification of patents with better accuracy. The rapid increase in the number of patents being filed every year and the increasing complexity of the patent documents demand for advanced and revolutionized tools/machines to assist in performing patent classification in automated manner. Usually, the patents are examined thoroughly by patent analysts from various domains, who possess respective expertise and are well aware of the domain jargons. The main objective of such systems is to get rid of the time-consuming, laborious manual process and to provide patent analysts a better way for classifying patent documents. Also it helps in better management, maintenance and convenient searching of patent documents. Here, two prominent classification algorithms—Naïve Bayes and support vector machines (SVM)—are explored and implemented. Additionally, some preprocessing steps such as stop word removal, stemming, and lemmatizing are also done to obtain better accuracy. TF-IDF feature is also incorporated to obtain precise results.

Keywords Data mining · Patent mining · Patent classification · Classification algorithms · Naive Bayes · Bag of words · Support vector machine · TFIDF

D. A. Naik (✉) · S. Seema · G. Singh
Department of CSE, MSRIT, Banglaore, India
e-mail: darshananaik@msrit.edu

S. Seema
e-mail: seemas@msrit.edu

G. Singh
e-mail: geethika.singh20@gmail.com

A. Singh
CloudxLab Inc., Bangalore, India
e-mail: abhinav.foss@gmail.com

S.-L. Peng et al. (eds.), *Computing and Network Sustainability*, Lecture Notes in Networks and Systems 75, https://doi.org/10.1007/978-981-13-7150-9_12

1 Introduction

Patent documents play crucial role in the protection of invention of an individual or organization. Patent documents are hugely different from ordinary documents. Patent document has a specified format and consists of front page, description section, references and diagrams as shown in Fig. 1. However, patent documents are huge and contain myriads of technical jargons which require an extensive amount of efforts for analysis. Therefore, in order to assist patent analysts to perform certain processes such as patent data examination, processing, classifying and analyzing, a separate stream called as patent mining is created. Patent mining forms a part of data mining. Just the way data mining assists in exploring patterns in the data so does patent mining in patent content [1].

The drastic increase in the number of patents calls for the emergence of a need for designing a tool that automatically categorizes the patents. This would decrease the extensive amount of effort to be spent on performing manual patent classification. Hence a sophisticated tool developed in this direction will be of great advantage. A key bottleneck for such systems is the ability of discovering and utilizing the data stored in databases [2]. Apart from the above concern, a system must also have the ability to analyze the complexities involved in interlinking the patent information.

Patent mining comprises various tasks such as retrieval of patent, classification, patent valuation, patent visualization and detecting infringements. [1]. Patent classification is an important task. It helps in flexible management and easy maintenance of patent documents.

Usually, patent undergoes thorough examination by patent analysts from various domains, who possess respective expertise and are well aware of the domain jargons. WIPO (World Intellectual Property Organization) is one of the standard agencies that help in preserving and promoting the patents. WIPO designates patent analysts who examine the patent data and classify manually. Such manual classification is quite laborious and time-consuming.

Therefore, to classify patents efficiently, there emerges a requirement for a sophisticated system which will automatically classify the patents.

In past, many classification algorithms have been designed. However, the results are still far from obtaining the accuracy. There are many hurdles faced during the patent classification process such as (1) complex structure—patent is usually filled with over-usage of various jargons which makes classification unsuccessful because of difficulty in fetching useful features, (2) complex hierarchy of schema of patent classification and (3) increasing volume of patent documents [1]. To address these hurdles, researchers have worked on developing powerful classification algorithms. The research work was carried out with emphasis on (1) using various kinds of information provided in patent for performing classification and (2) analyzing how the different kinds patent documents behave when tested against different classification algorithms.

Fig. 1 Sample front page of patent

This paper makes an effort to explore, implement and test certain classification algorithms for patent classification. Several optimizations are also done gradually and results are recorded.

2 Bag-of-Words Model

Bag-of-words (BOW) model is used to represent textual data in machine learning domain. Similarly, BOW is used here to represent patent content. The unstructured text document is often represented in the form of BOW model. Some of the parameters in the dataset such as title, abstract and references can be represented in BOW model. Most of the classification algorithms use BOW model in their underlying implementation.

3 Dataset Generation

To perform the classification process in unsupervised manner, one requires dataset. If done manually, the dataset creation will require enormous effort. Hence, an attempt of implementing a crawler is done that will crawl the desired website or repository containing patent information.

The web crawler would crawl the pages or repository and will fetch the relevant information such as patent title, patent abstract and IPC code. One can choose to download the data in CSV format, XML or JSON, etc. Here, CSV format is chosen. The web crawler is written in python language. Table 1 shows the template of dataset collected.

4 Classification Process

The classification process generally consists of various phases such as retrieval of document, tokenization, stemming, stop word removal, vector representation and feature selection [4]. Each phase augments the process and leads to the direction of accuracy. Figure 2 illustrates all the phases involved in classification process as represented below.

Table 1 Template of dataset

Name:	PatentDataset.csv
Columns	Title—contains the title of the patent Abstract—contains the brief summary of the patent Class—represents the IPC class code of the patent Class_code—represents the equivalent of class column in numerical format

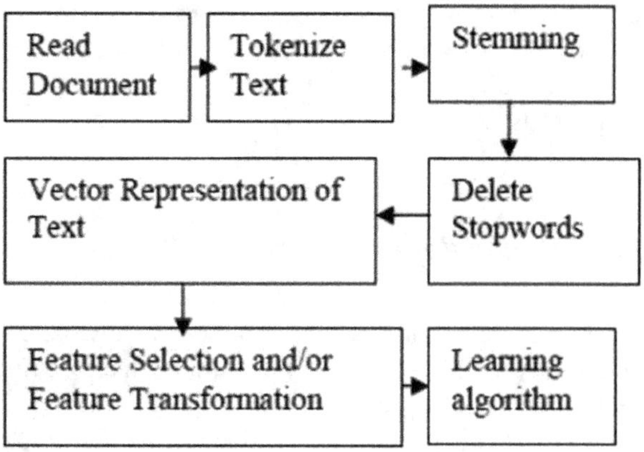

Fig. 2 Classification process

4.1 *Tokenization*

In this phase, a stream of text is divided into an indivisible element called as token. The tokens produced are fed into the subsequent phases of classification. A token can be defined as a continuous sequence of alphabetic characters or numerical characters. Tokens are delimited by whitespace characters such as single space, tab space, line break or special punctuation marks. Many algorithms contain inbuilt support to perform tokenization. Tokenization might seem an easy task in English language but is difficult for the languages that lack word boundaries like Chinese language.

4.2 *Stemming*

Stemming takes care of diminishing the deviation of inflectional forms and derivational related words to the original form. Words with similar base do not contribute toward classification process and must be disregarded.

4.3 *Stop Word Removal*

Stop words do not contribute anything toward classification process and hence must be filtered out in prior. Instead for few systems, such words cause difficulties and degrade the performance. There is no pre-defined list of stop words. A collection of words can be considered as stop words. There are certain common words that are used in some search engines.

Python provides various APIs to facilitate stop word removal. One can simply specify whether to ignore the stop words or not by mentioning the language to be considered for stop words and setting the appropriate flag. The list of stop words is pre-decided but can be customized as per one's desire. Here, the accuracy obtained was 18% before removing the stop words but was increased to 59% after removal of stop words using naive Bayes algorithm.

4.4 Vector Representation

Often the documents are represented in an algebraic model. The text documents are often represented as vectors of identifiers. Such identifiers can be index terms that can be used in various processes like retrieval, indexing, ranking on relevancy, etc. [5].

An array of words forms a document. Vocabulary or feature set is defined as group of words of a training group [6]. Python again provides support for such representation.

4.5 Feature Selection and Transformation

Feature selection is a process of diminishing the dimensionality of input dataset by removing the irrelevant data or by fetching useful information from the given data.

This helps in increasing the performance.

5 Classification Algorithms

There are several text classification algorithms available. Some of the algorithms are Naive Bayes, SVM, Winnow, TRIZ-based algorithm and KNN (K-nearest Neighbor). This paper attempts to consider Naive Bayes and SVM algorithms. It also performs the comparison of their results. An observation is also done by considering stop words, stemming, lemmatizing, tfidf, etc. The overall classification architecture can be depicted as shown in Fig. 3.

5.1 Naive Bayes Algorithm

Naïve Bayes algorithm has been a proven traditional classification algorithm and has been used since many years. Naïve Bayes classifiers are a group of classification algorithms that uses Bayes theorem. It is generally a collection of algorithms with a common principle. Naïve Bayes is consistently used in various text classification applications due to its simplicity and effectiveness [7].

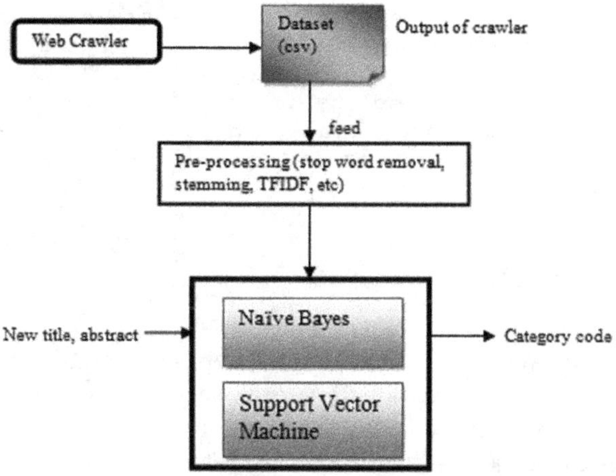

Fig. 3 Overall architecture

Here, an attempt is made to perform classification using Naïve Bayes for a dataset of 530 records in Python language. Python provides support for Naïve Bayes by offering certain packages and libraries. Following is the pseudo code:

```
Import packages
Read the dataset
Split the dataset - Testing and Training Dataset (Testing
is 20% of original Dataset)
Apply SnowBallStemmer with ignore_stopwords = true
Build the vectorizer for the training dataset
#Build the pipeline by passing vectorized dataset, Tfidf
#transformer, navie bayes classifier
Build the Pipeline (vectorizer, transformer, classfier)
Fit the input dataset and output features
Test the testing dataset against this final model
```

5.2 *SVM*

Support vector machine is categorized as supervised learning model. This algorithm uses kernel trick to transform the data and then considering these transformations as base, it distinguishes among the possible outputs. Technically, it produces a hyper plane which classifies the new dataset using the knowledge of the original dataset given [8]. Python again provides great support for implementing SVM. The pseudo code is as follows:

```
Import packages
Read the dataset
Split the dataset - Testing and Training Dataset (Testing
is 20% of original Dataset)
Apply SnowBallStemmer with ignore_stopwords = true
```

```
Build the vectorizer for the training dataset
#Build the pipeline by passing vectorized dataset, Tfidf
#transformer, clf svm classifier
Build the Pipeline (vectorizer, transformer, classfier)
Fit the input dataset and output features
Test the testing dataset against this final model
```

6 TFIDF

Term weighting is an useful concept in determining the accuracy of classification process. In a particular context, every word carries different level of importance. Such importance is denoted by term weight which is tagged with every word. This is called as term frequency (TF). There is another related phrase inverse document frequency (IDF) which is a weight that relies on the distribution of every word in the repository.

Python provides Tfidf transformer, which is responsible for calculating the tf-idf weights for our term frequency matrix.

7 Results

The classification of patents is achieved by using two algorithms—Naïve Bayes, SVM and the results of both are compared. SVM proves to be better. But the dataset upon which these algorithms are applied is obtained using the web crawler. The web crawler is written to crawl WIPO Web site and the resultant dataset is as shown in Fig. 4.

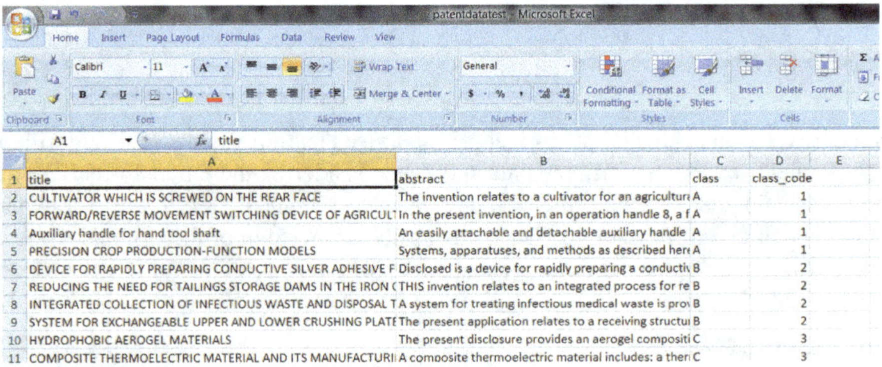

	A	B	C	D	E
1	title	abstract	class	class_code	
2	CULTIVATOR WHICH IS SCREWED ON THE REAR FACE	The invention relates to a cultivator for an agricultur	A	1	
3	FORWARD/REVERSE MOVEMENT SWITCHING DEVICE OF AGRICUL	In the present invention, in an operation handle 8, a f	A	1	
4	Auxiliary handle for hand tool shaft	An easily attachable and detachable auxiliary handle	A	1	
5	PRECISION CROP PRODUCTION-FUNCTION MODELS	Systems, apparatuses, and methods as described her	A	1	
6	DEVICE FOR RAPIDLY PREPARING CONDUCTIVE SILVER ADHESIVE F	Disclosed is a device for rapidly preparing a conducti	B	2	
7	REDUCING THE NEED FOR TAILINGS STORAGE DAMS IN THE IRON (	THIS invention relates to an integrated process for re	B	2	
8	INTEGRATED COLLECTION OF INFECTIOUS WASTE AND DISPOSAL T	A system for treating infectious medical waste is pro	B	2	
9	SYSTEM FOR EXCHANGEABLE UPPER AND LOWER CRUSHING PLATE	The present application relates to a receiving structu	B	2	
10	HYDROPHOBIC AEROGEL MATERIALS	The present disclosure provides an aerogel compositi	C	3	
11	COMPOSITE THERMOELECTRIC MATERIAL AND ITS MANUFACTURI	A composite thermoelectric material includes: a ther	C	3	

Fig. 4 Resultant dataset

Table 2 Performance of different classifiers

Classifiers	Accuracy (%)
Naive Bayes	59
SVM	74

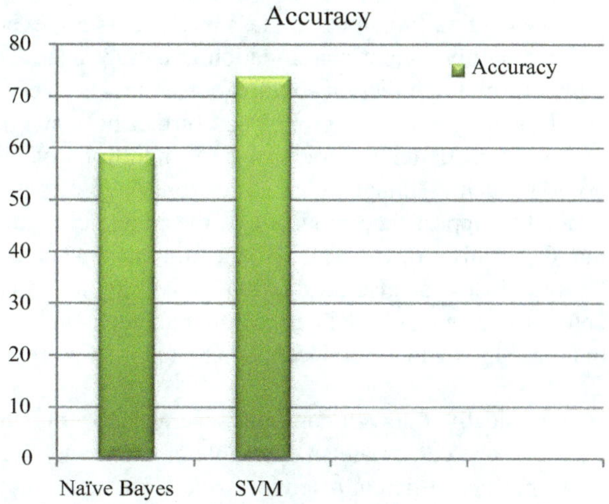

Fig. 5 Performance of dataset on different classification algorithms

Naïve Bayes algorithm gave an accuracy of 59% with the dataset having 530 records. And support vector machine presented an accuracy of 74% as shown in Table 2.

Figure 5 shows the performance of classifiers considering accuracy as parameter.

8 Conclusion

Patent classification task is a crucial task as it helps in management and further maintenance of patent documents. But because of drastic increase in number of patent documents every year, there emerges a need for a sophisticated system that automatically categorizes the patents. Some of the classification algorithms such as Naïve Bayes and SVM that are adopted in text classification and natural language processing were explored, implemented and compared for patent classification task. These classification algorithms are further tweaked or advanced by adding some pre-processing steps such as stop words removal, stemming, adding TF-IDF to provide better accuracy. Eventually, SVM proves to be better.

9 Future Enhancements

The classification of patents is done at higher category level that is section level. The hierarchy contains section at highest level followed by class level, subclass level and group level. One can perform patent classification further down the hierarchy. That is one can classify patents at class level, subclass level and group level also. As per IPC taxonomy, patent classification is hierarchical. There are eight sections at the higher level. Each section contains some classes, every class has subclasses and each subclass has some main groups that further have subgroups.

To obtain better accuracy one can additionally include complex attributes (as criteria) such as image depicted in front page of patent. The images can be transformed to appropriate models and the image classification process can be carried out. Generally, color, texture, shape, dimensions, etc., are fetched.

References can also be used in patent classification as one of the criteria to obtain better accurate results. The irrelevant character such as whitespace, tab space, punctuation marks and other special characters can be removed before performing classification.

Technically, one can combine several algorithms to obtain better accuracy. For example, one can create an ensemble of SVM and Naïve Bayes algorithms.

To perform efficient patent classification, neural networks can also be applied on larger dataset.

References

1. Zhang L, Li L, Li T (2014) Patent mining: a survey. ACM SIGKDD explorations newsletter archive, 16(2), Dec 2014
2. Abbas A (2013) A literature review on the state-of-the-art in patent analysis . Elsevier 37, Dec 2013
3. Sumathy KL, Chidambram M (2013) Text mining: concepts, applications, tools and issues—an overview. Int J Comput Appl 80(4), Oct 2013
4. Dasari DB, Rao VGK (2012) Text categorization and machine learning methods: current state of the art. Global J Comput Sci Technol Softw Data Eng 12(11) 1.0 Year 2012
5. Ikonomakis M, Kotsiantis S, Tampakas V (2005) Text classification using machine learning techniques 4(8) Aug 2005
6. https://towardsdatascience.com/document-feature-extraction-and-classification-53f0e813d2d3
7. https://en.wikipedia.org/wiki/Naive_Bayes_classifier
8. https://en.wikipedia.org/wiki/Support_vector_machine

Application of Fruit Fly Optimization Algorithm for Single-Path Routing in Wireless Sensor Network for Node Capture Attack

Ruby Bhatt, Priti Maheshwary and Piyush Shukla

Abstract Wireless sensor network (WSN) (Kaur and Kang in Int J Comput Sci Eng IJCSE 6(4):157–162, 2016, [1]) is susceptible to different types of physical attacks. It is collection of tiny-sized sensor nodes. The reason behind these attacks is its limited resource capacity. It is screened to external atmosphere for circulating data. Node capture attack is supposed to be severe attacks in WSN (Butani et al. in Int J Comput Appl IJCA 95(3):32–39, 2014, [2]). In this type, the node is substantially captured by an assailant and eradicates the secret information from the node's storage. This paper proposes a fruit fly optimization algorithm (FFOA) (Wang et al. in Mob Inf Syst 1–14, 2017, [3]). It is based on multiple objectives (Lin and Wu in J Supercomput 1–19, 2013, [4]) node capture attack algorithm. Proposed algorithm serves these objectives: maximum node contribution (Lin and Wu in J Supercomput 1–19, 2013, [4]), maximum key contribution (Lin and Wu in J Supercomput 1–19, 2013, [4]), and least resource expenses (Lin and Wu in J Supercomput 1–19, 2013, [4]). The simulation result illustrates that FFOA obtains a lower energy cost as compared with matrix algorithm (MA) (Lin et al. in J Supercomput 71:3181–3212, 2015, [5]) and other node capture attack algorithms when it is calculated for single-path routing algorithm. There are two variations in which this could have served. First is single path and the other is multi-path. But, in this paper, it is only the first case that has been discussed and calculated.

Keywords Vertex · Seizure · Fruit fly · Optimization · Capturing · Vulnerable

R. Bhatt (✉) · P. Maheshwary
Department of CSE, Rabindranath Tagore University, Bhopal, India
e-mail: ruby_15@rediffmail.com

P. Maheshwary
e-mail: pritimaheshwary@gmail.com

P. Shukla
Department of CSE, UIT, Bhopal 462023, India
e-mail: pphdwss@gmail.com

S.-L. Peng et al. (eds.), *Computing and Network Sustainability*, Lecture Notes in Networks and Systems 75, https://doi.org/10.1007/978-981-13-7150-9_13

1 Introduction

Sensor technology has evolved several applications, like catastrophic, health, and defense monitoring. Sensor networks are highly susceptible to node capture attack because of vulnerable nature. It is a practically imitative and inclusive attack in which opponent physically seizures the node by excerpting cryptographic keys [1] and confidential information. Node seizure [1] is the most incommodious problem that ventures the discretion, consistency, and protection of sensor nodes.

In these technologies, the intruders arbitrarily confine the node to cooperation in the communication of WSN. However, susceptibility assessment theory has been dignified whereby an intruder can choose a node smartly to cooperate the network using susceptibility metric [28].

To surmount the difficulties of susceptibility-based approaches, the node capture attack approach is developed by combining multiple objectives like large amount of node contribution, highest key contribution, and least resource expenses to discover an optimal node using fruit fly optimization algorithm (FFOA) [3].

2 Literature Review

Several researchers have described numerous modeling node capture techniques using vulnerability evaluation, epidemic theory, and probabilistic analysis. The intruder smartly captures sensor nodes and removes the keys from their storage to devastate the protection, consistency, and secrecy of the WSN. Matrix algorithm (MA) [4] which is matrix-based node capture attack is projected to stipulate nodes and paths correlation along with maximum destructiveness and least resource expenditure. The results represent that the MA [5] can decrease the rounds used for confronting and time required for accomplishment with the increase in the confronting competence and energy cost [4].

Greedy node captured based on route minimum key set (GNRMK) [6] was designed to find the route minimum key set [6]. It was calculated by the maximum flow of the network. The overlapping value was allocated to each node on the foundation of route minimum key set. The node with maximum overlapping value was captured in every round of attack. Results of simulation reveled that, compared with other node capture attack schemes, GNRMK [6] could conceal the network. Because of the pseudo-random key-predistribution scheme and convoluted network design, minimum resource expenditure node capture Attack (MREA) [7] is developed. It is a heuristic method. It is used to minimize energy cost along with maximum destructiveness for node capture attack.

3 Models Used

3.1 Network Model

Wireless sensor network model is represented by directed network graph $G = (N, L)$, where N is the nodes number and L is the links number.

3.2 Key-Predistribution Model

In WSN, the cryptographic keys [4] represent a key group set K and every sensor node $N_a \in N$ is randomly assigned a subset of keys $K_a \subset K$ from a key group set. Two nodes N_a and N_b, which share a set of keys $K_{a,b} = K_a \cap K_b$.

3.3 Link Model

In WSN, several links are controlled by the paths and routes. A link $L_{a,b}$ is consistent and protected if it is encoded by key.

3.4 Adversary Model

This model is described from the view of an attacker and it is supposed that the intruder has latent to spy on the information transmitting through the WSN [2].

4 FFOA [3]-Based Multi-Objective Node Capture Attack Algorithm [4]

Pacification of the whole network comes in the major task. Different routes, which contain multiple paths, are therefore confined for compromising the complete network. The following matrices are calculated:

4.1 Key-Route Matrix

$$KeR = \left[KeR_{a,b}\right]_{KeXR}, \text{ where}$$

$$KeR_{a,b} = \begin{cases} 1 \text{ if } Ke \text{ can compromise} \\ 0 \text{ Otherwise} \end{cases} \tag{1}$$

4.2 Vertex (Sensor Node)-Key Matrix

$$VeKe_{a,b} = \begin{cases} 1 \text{ if } Ka \in N_b \\ 0 \text{ Otherwise} \end{cases} \tag{2}$$

4.3 Key-Number Matrix

$$KeN_{a,b} = \begin{cases} \sum_{a=1}^{K} VeKe_{a,b} \text{ if few keys } Ka \in N_b \\ 0 \qquad\qquad\quad \text{Otherwise} \end{cases} \tag{3}$$

4.4 Vertex (Sensor Node)-Route Matrix

$$\text{Where } VeR = VeKe * KeR \tag{4}$$

We combine the values of these two matrices into a single matrix $MM = \left[MM_{b,a}\right]_{N\times R}$ as:

$$MM = \beta \times \text{VR} + (1 - \beta) \times \text{VLR} \tag{5}$$

where β is a parameter decided from (0, 1):

$$MM_{a,b} = \begin{cases} \sum_{a=1}^{K} MM_{a,b} \text{ Participation of Node } N_b \\ 0 \qquad\qquad\quad \text{Otherwise} \end{cases} \tag{6}$$

4.5 Cost Seizing Matrix

The Cost seizing matrix $CS = \left[CS_{b,a}\right]_{N\times R}$ is denoted as follows:

$$CS_{a,b} = \frac{MM_{a,b}}{W_b} \tag{7}$$

where W_b is the seizing cost of compromising N_b.

$$CS_{a,b} = \begin{cases} \sum_{a=1}^{K} CS_{a,b} & \text{Capturing Cost of Node } N_b \\ 0 & \text{Otherwise} \end{cases} \tag{8}$$

CS_b shows the resource expenses for vertex N_b.

4.6 Multi-Objective Function

The multi-objective function is denoted as follows:

$$Fn = \sum_{a=1}^{N} \sum_{b=1}^{R} \left\{ \frac{1}{MM_b} + \frac{1}{KeN_b} + CS_b \right\} \tag{9}$$

4.7 Fruit Fly Optimization Algorithm (FFOA)

After estimating multi-objective function, FFOA is instigated. It discovers optimal vertex (sensor nodes) [1] from the existing vertex, which minimizes the objective function [4] to generate the best results in order to discover the optimal nodes [4], that causes extreme ferocity in WSN [1] using FFOA [8].

Start
Step 1. Initialize population, generation, function, position, and smell of each vertex (sensor node).
Step 2. Compute the fitness of each vertex based on distance and smell and generate optimal value of individual and population.
Fitness = Function (smell)
Step 3. Update position and best index of each vertex.
Step 4. Find optimal solution, if not, repeat step 2.
If found, the optimal solution gives nodes ID's.
End

5 Results and Analysis

The performance of FFOA is analyzed on the basis of based multi-objectives node capture attack algorithm that is analyzed under following parameters listed in Table 1.

During simulation, 200 vertices (sensor nodes) [1] are distributed in the WSN [1]. 10 nodes act as source and three nodes act as destination vertices. These are

Table 1 Experimental specifications

Bounds	Standards
Vertices (sensor nodes) [1] number	200
Region-wise size	100 * 100
Source vertices	10
Sensing range	20
Destination vertices	3
Keys group	200
Allotted keys to node	20
Sender and receiver	*R*
Population size	50
Number of rounds	200

arbitrarily selected. Single-path routing protocol is used for communication between sensor vertices in the range of 20 m. The proposed algorithm is analyzed for single path and run over 200 repetitions. We evaluate the recital of the FFOA in terms of fraction of compromised traffic, energy cost, attacking rounds, and execution time and compare the results with an matrix algorithm (MA) and other node capture attack algorithms like random attack (RA) [42], maximum key attack (MKA) [8], maximum traffic attack (MTA) [8], maximum link attack (MLA) [8], and greedy node capture approximation using vulnerability evaluation (GNAVE) [8].

5.1 Energy Cost

In this simulation, the energy consumption is analyzed. The energy cost of node capturing is distributed in $U(0, 1)$. Capturing cost of network is enhanced by enhancing the number of capturing nodes. It is therefore very obvious that MA [4] and other algorithms have higher energy cost than FFOA [3]. MA and other node capture attack algorithms use greater number of nodes to compromise the whole network. It is calculated for single-path routing algorithm. These can be done for two variations in which this could have been served. First is single-path and the other is multi-path. But, here only the first case is computed (Fig. 1).

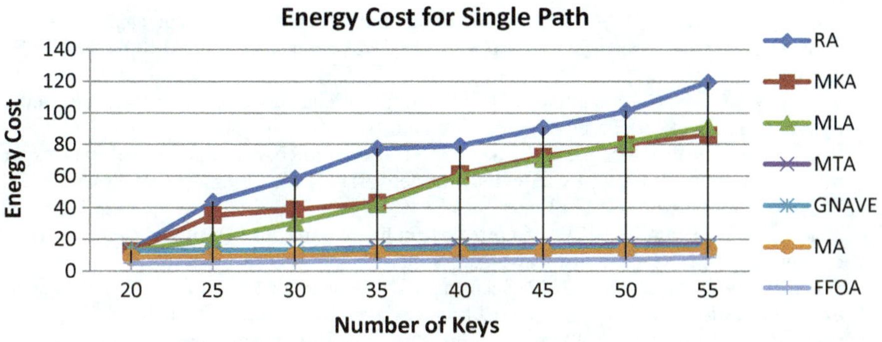

Fig. 1 Energy cost for single-path routing protocol

6 Conclusion

The proposed algorithm—fruit fly optimization algorithm (FFOA) [3] is initiated to increase the efficiency of attack using multi-objective node capture attack in WSN [2]. FFOA describes three objectives: (1) maximum key contribution (2) least resource expenses, and (3) maximum node contribution to discover optimal nodes for overall devastation of network. These nodes form the best combination of the objectives and create extreme harmfulness. The simulation result illustrates that FFOA obtains a maximum fraction of compromised traffic, lower attacking rounds, and lower energy cost compared with a matrix algorithm (MA) [4] and other node capture attack algorithms. Therefore, FFOA gives maximum attacking efficiency than matrix algorithm (MA) and other algorithms like random attack (RA) [42], maximum key attack (MKA) [8], maximum traffic attack (MTA) [8], maximum link attack (MLA) [8], and greedy node capture approximation using vulnerability evaluation (GNAVE) [8]. This is by the observation of capturing minimum nodes that compromise the whole network. Therefore, it is summed up this way that, FFOA gives maximum attacking efficiency among many node capture algorithms.

References

1. Kaur A, Kang SS (2016) Attacks in wireless sensor network—a review. Int J Comput Sci Eng IJCSE 6(4):157–162
2. Butani B, Shukla PK, Silakari S (2014) An exhaustive survey on physical node capture attack in WSN. Int J Comput Appl IJCA 95(3):32–39
3. Wang Z, Zhou C, Liu Y (2017) Efficient hybrid detection of node replication attacks in mobile sensor networks. Mob Inf Syst 1–14. https://doi.org/10.1155/2017/8636379
4. Lin C, Wu G (2013) Enhancing the attacking efficiency of the node capture attack in WSN: a matrix approach. J Supercomput 1–19. https://doi.org/10.1007/s11227-013-0965-0

5. Lin C, Wu G, Yu CW, Yao L (2015) Maximizing destructiveness of node capture attack in wireless sensor networks. J Supercomput 71:3181–3212. https://doi.org/10.1007/s11227-015-1435-7
6. Guowei W, Xiaojie C, Mohammad S, Chi L, A high efficient node capture attack algorithm in wireless sensor network based on route minimum key set
7. Lin C, Qiu T, Obaidat MS, Yu CW, Yao L, Wu G (2016) MREA: a minimum resource expenditure node capture attack in wireless sensor networks. Secur Commun Netw 9:5502–5517. https://doi.org/10.1002/sec.1713
8. Xiao C, Hao K, Ding Y (2017) An improved fruit fly optimization algorithm inspired from cell communication mechanism. Hindawi Corporation
9. Bhoopathy V, Parvathi RMS (2012) Securing node capture attacks for hierarchical data aggregation in wireless sensor networks. Int J Eng Res Appl IJERA 2(2):466–474
10. Lu C, Hu D (2016) A Fault-Tolerant routing algorithm for wireless sensor networks based on the structured directional de Bruijn Graph. Cybern Inf Technol 16(2):46–59. https://doi.org/10.1515/cait-2016-0019
11. Kim D, Kim D, An S (2016) Source authentication for code dissemination supporting dynamic packet size in wireless sensor networks. Sensors 16:1–22. https://doi.org/10.3390/s16071063
12. Kim D, Kim D, An S (2016) Communication pattern based key establishment scheme in heterogeneous wireless sensor networks. KSII Trans Internet Inf Syst 10(3):1249–1272. http://dx.doi.org/10.3837/tiis.2016.03.017
13. Kaur H (2017) Node replication attack detection using Dydog in Clustered sensor network. Computer Science and Engineering Department, Thapar University Patiala, pp 1–71
14. Qasemzadeh Kolagar I, Haj Seyyed Javadi H, Anzani M (2017) Hypercube bivariate-based key management for wireless sensor networks. J Sci Islam Repub Iran 28(3):273–285
15. Vijayarajeswari R, Rajivkannan A, Santosh J (2016) Survey of malicious node detection in wireless sensor networks. Int J Emerg Technol Innov Eng 2(6):335–338
16. Tague P, Identifying, modeling, and mitigating attacks in wireless adhoc and sensor networks

SDN-Based Fast Handover Approach to Improve the QoS of Video Streaming over Wi-Fi Networks

Krishna Vijay Singh and Mayank Pandey

Abstract In this paper, we have proposed a design and implementation of software-defined network-based fast handover approach to improve the QoS for video streaming over Wi-Fi networks. During the handover of mobile stations, two major bottleneck are experienced towards providing fast transition from one access point to another. These bottlenecks are scan latency and authentication latency. These two add up and result in a high handover latency which is undesired for video streaming applications. Using our SDN-based approach has been able to drastically reduce this latency which resulted in better QoS for video streaming application over Wi-Fi networks. We have implemented our SDN-based approach using Mininet emulator and results suggest the applicability of our approach.

Keywords Software defined networks · OpenFlow · Mobility · Handover

1 Introduction

In a recent survey [1], Cisco has estimated that IP video traffic would be 82% of all consumer internet traffic by year 2021, up from 73% in 2016. Further, a significant percentage of the end hosts receiving this IP video traffic is mobile devices such as smartphones. Although mobile radio technologies such as 3G, 4G are getting more prevalent and cheaper, Wi-Fi (based IEEE 802.11x media access control) still remains a preferable communication alternative for mobile users. However, mobile stations have to perform handovers from one access point to another during their movement. The access point from where the mobile station gets disassociated is termed as source access point (SAP) and one with which it gets associated after handover is termed as target access point (TAP). The standard handover mechanism

K. V. Singh (✉)
Sikkim Manipal University,Tadong, Gangtok, India
e-mail: krishna.si@smit.smu.edu.in

M. Pandey
Motilal Nehru National Institute of Technology, Allahabad, India
e-mail: mayankpandey@mnnit.ac.in

S.-L. Peng et al. (eds.), *Computing and Network Sustainability*, Lecture Notes in Networks and Systems 75, https://doi.org/10.1007/978-981-13-7150-9_14

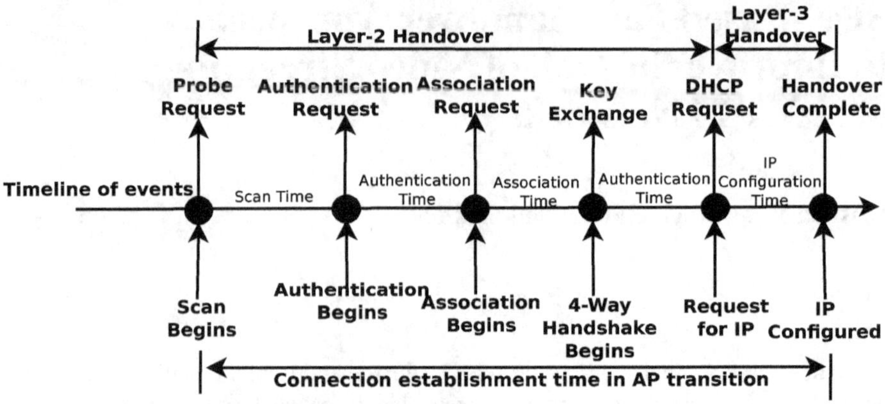

Fig. 1 Events during handover

of IEEE 802.11x introduces handover delay of 2 to 6 s which is undesirable for video streaming applications.

The handover occurs when mobile station crosses the radio range of one AP and enters in the radio range of others. In order to regain the connectivity, the stations have to undergo different phases depicted in Fig. 1. The first phase of handover is scanning in which suitable target access point (TAP) is determined. There are two techniques to perform scanning which are known as active scan and passive scan. Passive scan is not much utilized as it yields latency of 1150 ms [2, 3] for scanning all the 11 channels present in 2.4 GHz band of IEEE 802.11 standard. Active scan is preferred over passive scan as its latency is dependent on number of channels to be scanned for finding the desired TAP where the time required to scan each channel is 8 ms [2]. However, further optimizations are needed in the process of active scanning to support desired QoS for video streaming.

The next important phase is authentication to ensure that only legitimate users with verified credentials are permitted to use the wireless network. The latency is dependent upon the security standard used to protect the network. The IEEE 802.11i security standard proposed in year 2004 suffers from very high latency [4]. Later on, this security standard was optimized based on the fast transition mechanism proposed in IEEE 802.11r. Still the latency of authentication phase in IEEE 802.1x standard is 30–60 ms [4].

Generally, mobile station acquires a new IP address on association with a new access point. This is obtained through DHCP server which may be running at access points. In some cases, if both SAP and TAP are in same subnet, the mobile station can keep its IP address. However, if SAP and TAP are in different subnets, then mobile station has to acquire a new IP address as per the network prefix of the visited network. This change of IP address completely stops the streaming session running on the mobile station.

Further, when mobile station moves from one AP to another and regains connectivity from new access point, it is unable to send/receive data packets for some time.

This is because of the delayed update of MAC tables in the infrastructure switches in the wired part of the network according to the new location of mobile station. The problems of updating MAC table entries in infrastructure switches and mobility across different subnets have already been handled in our previous work [5].

The other two major bottlenecks (a) scan latency and (b) authentication latency are required to be handled properly. Some research efforts [3, 6–8] have been made to reduce the latency during handover by restricting the number of channels to be scanned. These solutions are able to reduce the scan time. However, they need additional infrastructure and modification in the access points. Also, the reduction in latency is not sufficient to meet the QoS requirement of video streaming applications. There is a need of an optimized solution which can reduce the scan latency and authentication latency without compromising the security requirements of the system.

In last few years, SDN [9] has emerged as a new networking paradigm which gives flexibility to solve several networking problems in a more innovative way [10]. This flexibility is a result of decoupling between data plane and control plane provided by SDN. In traditional networking devices (such as routers and switches), data plane and control plane are bundled together. On the other hand, in SDN, data plane devices (routers and switches) are responsible to forward the data as per the directive given from the control plane (SDN controller). The control plane is connected to all the data plane devices over the secure channel to guide them for the forwarding data. Decisions about where to forward the data is taken at the control plane and then in turn given to the data plane for implementation. Since the control plane is directly connected to the data plane devices, it has complete view of the network topology. Hence, the forwarding decision taken by the control plane is more accurate and optimized. Control plane in this case can apply forwarding rule more flexibly and dynamically based on underlying changes in the topology during mobility.

In this work, we have utilized the flexibility of SDN environment to tune the network in such a way so that data packets are delivered to mobile station at its changed location in real time. The channel to be scanned during the handover is computed by SDN controller. The bgscan technique [11] is used to find the potential available AP before mobile station associates. Moreover, the optimized 802.11 r pre-authentication technique with integration of SDN controller is used to reduce the latency of authentication on wireless access point (AP).

Our proposed methodology is capable to do this modification with very low latency. The SDN controller takes up to 2–10 ms to update the route in the data plane. The time variation from 2 to 10 ms is due to change response time of controller depending upon the load. However, the total latency required in our proposed method is presented in Table 2. The remaining paper is organized as follows. Section 2 describes the related work, Sect. 3 describes the proposed methodology and Sect. 4 describes the results along with discussion. Finally, conclusions are given in Sect. 5.

2 Related Work

Significant research is done in the field of handovers which has made phenomenal changes in QoS. However, with the increasing demand of live video streaming over Wi-Fi, latency of handover is major challenge. Majority of the existing solutions provide extensions and/or improvements only in certain phases such as scanning, authentication, IP configuration. Few proposed solutions are presented here and explained briefly. In order to reduce the probe latency, various efforts have been made in past to restrict the number of channels to be scanned at the time of handover [3, 6–8]. In [3], the authors propose an algorithm in which an AP informs the mobile station about neighbouring APs. In [6], advocates adaptive scanning based on application requirements. In [7], some masking is used based on previous probe history to optimize the scan latency. Nah et al. [8] presented a procedure for scanning the target access point (TAP), while considering the buffer level of the video player. Dely et al. [12] presented a software-defined-based scheme which talks about maintaining multiple connections with neighbouring APs. This proposal is based on out-of-band signalling where station can talk outside its coverage area. Chi et al. [4] proposed a design for fast handoff which focus is on inter-domain mobility where station can visit in a domain which is controlled by another radius server. Brik et. all states a solution [13], where multiple channels are used where one channel is dedicated administrative purpose and other is used for data transfer. The above schemes are covering some of the aspects and leaving the other. Some of them required code modification at the APs. They are also not available as open source implementations. Thus, we concluded to build a non-proprietary, vendor neutral, single radio handoff solution which improves the latency handover.

3 Proposed Solutions

3.1 Solution for Scan Latency

The scan phase is considered as a main bottleneck for fast handoff as it incurs comparatively high delay [2, 3, 12, 14]. This latency is due to scanning of all the supported channels of Wi-Fi. There are eleven to fourteen channels that are supported in the 2.4 GHz band and 40 channels supported in 5 GHz band [3]. Therefore, scanning all these channels for finding the potential APs yields significant delay. In our work, we have proposed an improved technique for scanning with the assistance of SDN controller. The number of channel to be scanned is restricted to the number of channel used by neighbouring APs only.

The SDN controller typically maintains a topology of network in its information base. We have made an additional information base at the SDN controller to maintain a proper neighbour list of APs along with its operating channel.

Table 1 Neighbours channel information base

dpid	Interface	Neighbours dpid and Channel
00:00:00:00:01	Wlan0	00:00:00:00:02, ch6
00:00:00:00:02	Wlan0	00:00:00:00:01, ch1 00:00:00:00:03, ch11
00:00:00:00:03	Wlan0	00:00:00:00:02, ch6

The main attributes maintained are depicted in Table 1. Hence, the SDN controller shares its information base with respective station for finding the potential neighbouring APs. Thereafter, the station uses bgscan technique to formulate the list of potential TAP in its vicinity. The formulated list later on is used by the station for association with TAP on handover. This technique alleviates the scanning latency completely on initiation of handover, since bgscan is applied to maintain the scan cache of mobile station. The scan cache refresh rate is dependent upon the mobility and it varies from 1 to 10 min., depending upon high to low mobility, respectively. Here, high mobility refers to the frequent handover due to mobility.

Therefore, in our proposed design the station will make use of scan cache to associate with appropriate TAP, instead of scanning all the channels and finding potential AP for association. Hence, the effective delay involve in our design in only for association as the potential list APs has been prepared in well and advance before the actual handover.

3.2 Solution for Authentication Latency

Enterprise wireless LANs are now adopting 802.1 X based EAP security solutions. IEEE 802.1 X is an IEEE standard framework for port-based access control that has been adopted by the 802.11 i security workgroup as the means of providing authenticated and secure access to wireless LAN. Our proposed design is based on 802.11 r, which is an amendment to the IEEE 802.1 X for fast handoff. Where, once the station is connected to an access point, their neighbouring access points are allowed to cache the top level key PMK. Hence, re-authentication at neighbouring access point is not necessary. But, the communication port for the new station is closed at target access point (TAP) and station is not permitted to send/receive IP datagram. The access of communication port is granted by SDN controller, upon next level of authentication known as four-way handshake. In four-way handshake, station and access point credentials are verified. The exchange of messages between TAP and SDN controller during four-way handshake is through OpenFlow protocol. SDN controller works as an authentication server, because Authentication, Authorization and Accounting (AAA)-based service module is included to serve the purpose. When a mobile station roams from one access point to another, it uses the established PMK to get the connectivity from TAP. The establishment of connectivity over PMK saves the time, and however, next level of four-way handshake prevents it from

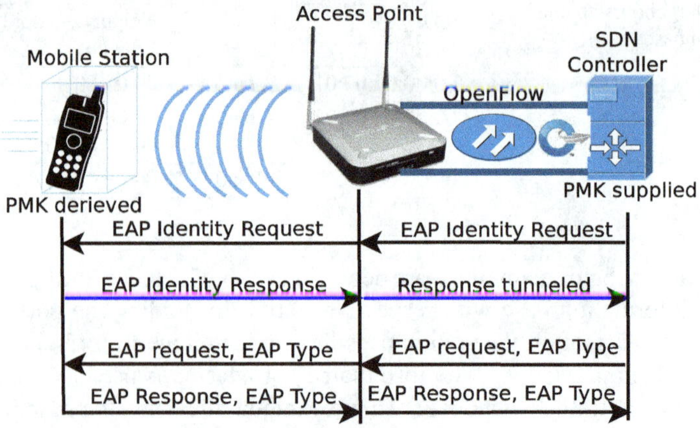

Fig. 2 EAP message exchanged after re-association

unauthorized access. The access of communication port is given on target access point only after success of the four-way handshake. The complete exchange of message are presented in Fig. 2.

The 802.11 X based authentication in our proposed model is done by the SDN controller. The requirement of dedicated radius server for authentication is eliminated. The security-related messages received at the AP are tunnelled to the controller through [15] protocol makes the scheme safer. The validation process is accelerated due to SDN controller is directly connected with access point with low latency link.

4 Emulation Setup

Our work is implemented using Mininet-Wi-Fi [16] which is a extension of Mininet [17]. The data plane devices such as switches and routers are emulated in Mininet-Wi-Fi. The Control plane is implemented using Floodlight [18], which is a Java-based controller used to control and provisioning the traffic of data plane devices. The control plane sends/receives the control messages using OpenFlow protocol as shown in Fig. 3. The topology used in our emulation is shown in Fig. 3. Data sender (video server) is DS as shown in Fig. 3 and data receiver is mobile station (MS) who is moving across the access points. Access point sends special notification for association/disassociation notification to SDN controller. SDN controller receives these notifications through pecket_in event notifier and processes these notifications. The controller communicates the appropriate action/flow to the data plane devices. The special notifications are generated through the scapy tool [19] from data plane. The SDN controller is working with added information base for neighbour information as depicted here in Table 1. This helps the controller for guiding the station for channel selection. This information base is according to our implemented topology shown in

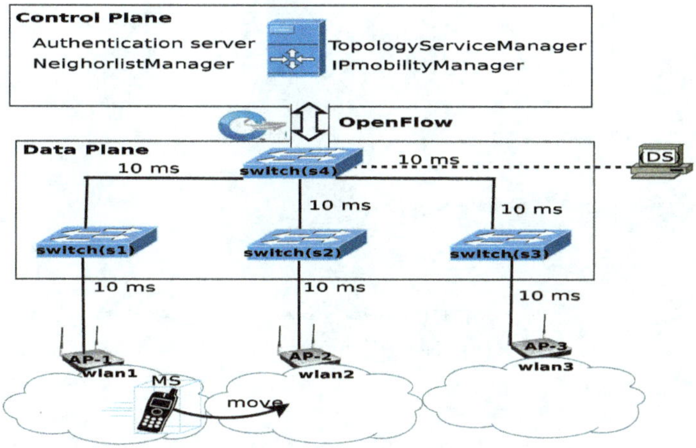

Fig. 3 Implemented topology

Fig. 3. The authentication services are installed at SDN controller, which facilitates the 802.1 X based authentication services.

VLC live streaming server functionality is used at data server (DS) to facilitate live stream of a video from DS. The Network Stream service is used at MS for live streaming the video over network by providing the IP address and port of data server. The MS performs mobility by utilizing the random way point mobility model. MS movement is restricted only in Wi-Fi enabled zone.

4.1 Results and Discussion

Our experiment uses one data server (DS) and one mobile station as shown in Fig. 3. Mobile stations acts as a client and perform video streaming from DS. The wireless channel is assumed to be excellent. We have utilized Wireshark to capture the traffic at station. The captured data traffic is shown here in Fig. 4a, where standard IEEE 802.11 b is used without any optimization for reducing the latency of handover. On the other hand, Fig. 4b presents the SDN-based fast handover traffic, where all proposed optimization of our design is made. We can see that, there is significant break (up to 4 s) in data streaming, due to the handover. On the contrary, we observed that the break in SDN-based fast handover is less than 50 ms presented in Fig. 4b.

In addition to that we have also plotted the graph for performance measurement in terms of throughput. The throughputs are measured with different link delay such as 5, 10 and 15 ms of each link. Figure 5 depicts the comparison of standard 802.11 (b) and SDN-based fast handover in terms of throughput. Table 2 compares the SDN-based fast handover with standard 802.11(b) in terms of latency involved in each phase.

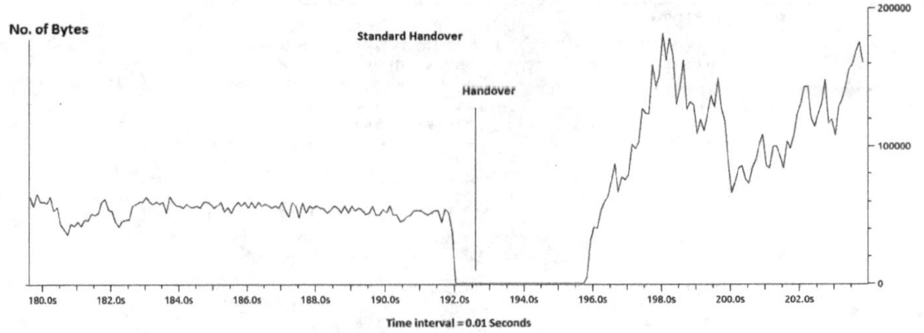

(a) Case I: Standard 802.11

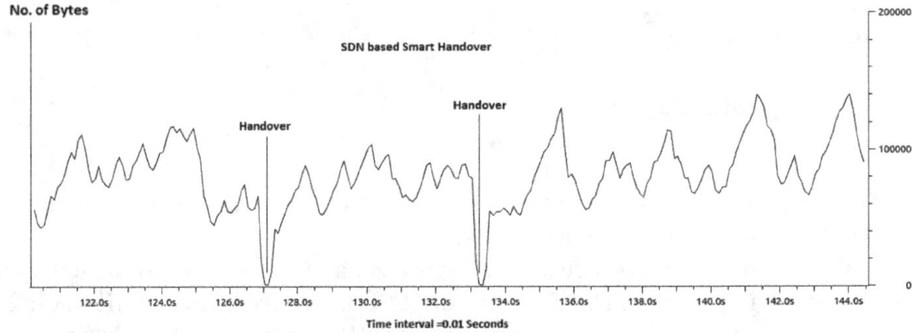

(b) Case II: SDN based

Fig. 4 Captured data traffic at mobile station

Fig. 5 Throughput comparison

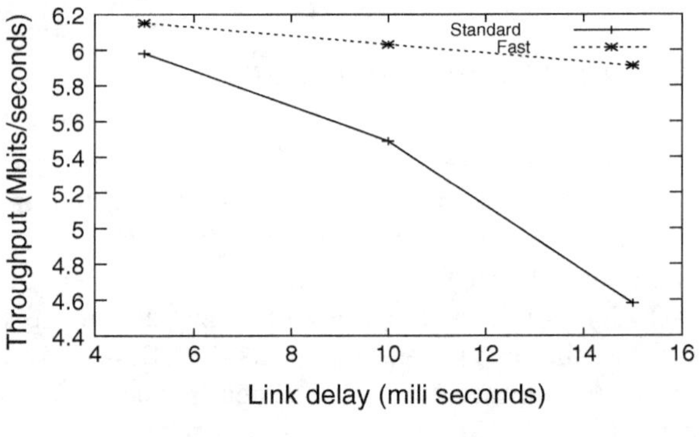

The results presented here clearly depicts that SDN-based fast handover improves the performance by reducing in the handover latency. The solution is not dependent upon specific vendor as it is controlled by SDN. Our proposed solution does not demand any alteration in access points or end hosts. The solution is completely transparent to the DS, and hence, changes induced due to mobility of mobile station are locally handled by SDN controller at edge network.

Table 2 Comparison of latency

Phases of hand over	IEEE 802.11 b	SDN-based smart handover
Scanning	250–400 ms [2, 14]	8–10 ms
Re-authentication	5–8 ms [2, 14]	5–8 ms
Re-association	5–10 ms [2, 14]	5–10 ms
Four-way handshake	400–800 ms (Full 802.1 X)	10–20 ms (802.11 r)
IP configuration	600 ms [20]	0 ms

5 Conclusions

We have proposed a software-defined fast handover solution to provide seamless mobility in Wi-Fi networks. Our approach is able to reduce scan latency to 8–10 ms (from 250–400 ms) and authentication latency which involves four-way handshake to 10–20 ms (from 400–800 ms). We have implemented our approach over Mininet using OpenFlow-enabled switches with Floodlight controller. The gathered results are promising and establish the applicability of our approach for seamless video streaming on the mobile end hosts that perform handover from one AP to another frequently.

References

1. Cisco public (2017) Cisco visual networking index: forecast and methodology, 2016–21. White Paper, June 2017
2. Purushothaman I, Roy S (2010) Fastscan: a handoff scheme for voice over IEEE 802.11 wlans. Wirel Netw 16:2049–2063
3. Powar YA, Apte V (2009) Improving the IEEE 802.11 mac layer handoff latency to support multimedia traffic. In: 2009 IEEE wireless communications and networking conference, pp. 1–6 (April 2009)
4. Chi K-H, Tseng C-C, Tsai Y-H et al (2010) Fast handoff among IEEE 802.11 r mobility domains. J Inf Sci Eng 26(4):1345–1362
5. Singh, KVK, Pandey M (2016) Software-defined mobility in ip based wi-fi networks: design proposal and future directions. In: 2016 IEEE international conference on advanced networks and telecommunications systems (ANTS), pp 1–6 (Nov 2016)
6. Montavont N, Moret AA, Castignani G (2013) On the selection of scanning parameters in IEEE 802.11 networks. In: 2013 IEEE 24th annual international symposium on personal, indoor, and mobile radio communications (PIMRC), pp 2137–2141 (Sept 2013)
7. Shin S, Forte AG, Rawat AS, Schulzrinne H (2004) Reducing mac layer handoff latency in IEEE 802.11 wireless lans. In: Proceedings of the second international workshop on mobility management & wireless access protocols, MobiWac'04, New York, NY, USA, ACM, pp 19–26
8. Nah JW, Chun SM, Wang S, Park JT (2009) Adaptive handover method with application-awareness for multimedia streaming service in wireless lan. In: International conference on information networking, pp 1–7
9. Nunes BAA, Mendonca M, Nguyen XN, Obraczka K, Turletti T (2014) A survey of software-defined networking: past, present, and future of programmable networks. IEEE Commun Surv Tutor 16:1617–1634 (Third 2014)

10. Kobayashi M, Seetharaman S, Parulkar G, Appenzeller G, Little J, van Reijendam J, Weissmann P (2014) Maturing of openflow and software-defined networking through deployments, vol 61, p 151175 (03, 2014)
11. Singh G, Atwal APS, Sohi BS (2008) Effect of background scan on performance of neighbouring channels in 802.11 networks. Int J Commun Netw Distrib Syst 1:19–32
12. Dely P, Kassler A, Chow L, Bambos N, Bayer N, Einsiedler H, Peylo C, Mellado D, Sanchez M (2013) A software-defined networking approach for handover management with real-time video in wlans. J Mod Transp 21:58–65
13. Brik V, Mishra A (2005) Eliminating handoff latencies in 802.11 wlans using multiple radios: applications, experience, and evaluation. In: 5th ACM SIGCOMM conference on internet measurement, pp 27–27
14. Mishra A, Shin M, Arbaugh W (2003) An empirical analysis of the IEEE 802.11 mac layer handoff process. SIGCOMM Comput Commun Rev 33:93–102
15. Openflow switch specification, version 1.4.0 (wire protocol 0x05) 2013
16. Fontes R, Afzal S, Brito S, Santos C (2015) Mininet-WiFi: emulating software-defined wireless networks. In: 2nd international workshop, (ManSDN/NFV 2015), Barcelona, Spain (Nov 2015)
17. O N F promoted, Mininet. http://mininet.org/
18. Floodlight controller. http://www.projectfloodlight.org/floodlight/
19. Scapy tool. https://scapy.net/
20. Kim KY, Ham KG, Cho KS, Choi SG (2011) Seamless handover scheme based on sip in wireless lan. In: 13th international conference (ICACT2011), pp 847–851 (Feb 2011)

Towards DHT-Based P2P Resource Sharing Over Hybrid Infrastructure of Wireless Mesh Network and Mobile Ad hoc Networks

Shabir Ali, Shashwati Banerjea, Mayank Pandey and Neeraj Tyagi

Abstract With the sudden proliferation in the usage of mobile devices, the demand for Internet connectivity is increasing at a fast rate. The public places like the university campuses, railway stations, airports and shopping malls, etc. are facing the challenge of providing uninterrupted Internet connectivity. Wireless mesh network (WMN) presents themself as an elegant alternative for this purpose. These networks exhibit self-organization, self-healing and self-maintenance capabilities. Further, they have less procurement cost and are robust. Also, the mobile devices which are equipped with wireless interfaces can form their own mobile ad hoc network (MANET) which can act as an extension of wireless mesh backbone. This enables to extend the reachability of WMN towards dead zones having no wireless network coverage. In this paper, we propose an approach to combine MANET and WMN. The combination of two types of wireless networks forms the underlay for deploying structured Peer-to-Peer (P2P) applications for resource sharing among the nodes. We have outlined the challenges and propose some possible solutions. Simulation results suggest the applicability of our approach.

Keywords Wireless mesh network · Peer-to-Peer network · Dead zones

---

S. Ali (✉) · S. Banerjea · M. Pandey · N. Tyagi
Computer Science and Engineering Department, Motilal Nehru National Institute of Technology Allahabad, Allahabad, India
e-mail: rcs1501@mnnit.ac.in

S. Banerjea
e-mail: shashwati@mnnit.ac.in

M. Pandey
e-mail: mayankpandey@mnnit.ac.in

N. Tyagi
e-mail: neeraj@mnnit.ac.in

S.-L. Peng et al. (eds.), *Computing and Network Sustainability*, Lecture Notes in Networks and Systems 75, https://doi.org/10.1007/978-981-13-7150-9_15

1 Introduction

The Peer-to-Peer (P2P) technology is envisaged for sharing resources over distributed nodes. These distributed nodes create a logical network over an existing wired or wireless physical underlay network. Due to the rapid proliferation of mobile devices, the physical network is generally wireless. The wireless network can either be infrastructure-based, ad hoc network or hybrid network. The infrastructure-based wireless networks such as WLAN and WMN suffer from the problem of "dead zones". Dead zones are the regions where the signal strength is weak (due to hindrance of walls, trees, doors, windows, etc.) limiting the network connectivity. The electronic gadgets are well equipped with wireless interface. The users in these dead zones can create MANET to share resources.

A MANET is characterized by dynamic self-organized multi-hop wireless network of mobile nodes. The mobile nodes collaboratively forward the data in multi-hop fashion. The routes are made either using reactive [1–3] or proactive [4, 5] routing mechanism. These routing functions are implemented at layer-3 of Internet protocol stack. Further, the MANETs are utilized in the absence of network infrastructure for sharing of resources. The structured P2P applications are popular communication protocols for resource sharing over MANETs. The mobility of nodes may produce isolated disjoint P2P logical overlay network as depicted in Fig. 1. This disjoint logical overlay limits the domain of resource sharing and searching.

To connect the disjoint MANETs, we have utilized infrastructure-based wireless mesh network [6]. The nodes of WMN have low mobility and communicate in multi-hop fashion. WMN uses layer-2 of Internet protocol stack to routing information for transmission of data in multi-hop.

We have designed a DHT-based resource sharing system over a hybrid underlay of WMN and MANET. The challenges involved in the design are as follows.

- The Ieee802.11s draft [7] claims that the mesh portal (MPP)/gateway is used to provide the connectivity to other networks. However, it does not have proper implementation details of MPP. This makes provision of connectivity to other networks difficult.
- In wireless mesh network, the routing is done at layer-2 of Internet protocol stack which means it is based on MAC addresses. While in MANETs, the routing is done at layer-3 of Internet protocol stack using IP addresses of source and destination. The implementation of routing mechanism at different layers creates hindrance in providing communication between MANET and WMN.

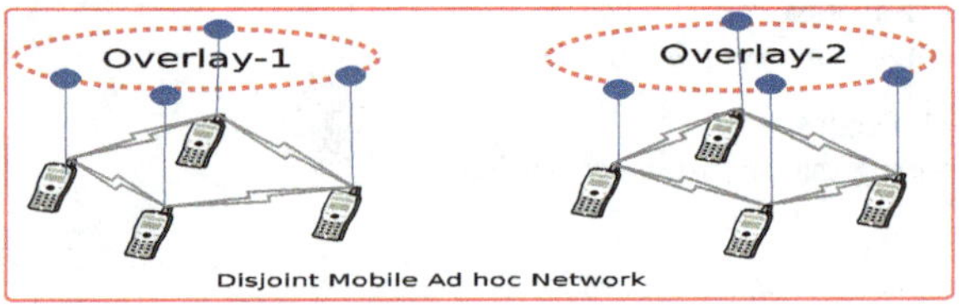

Fig. 1 P2P overlay formation over disjoint MANETs

- Providing the connectivity and reachability of wireless environment in a manner so that it should not result in disjoint DHT-based overlay.

In this paper, we have used an integration of the networks to remove dead zones and to extend the wireless network coverage as needed. We have selected MAP of infrastructure-based WMN as an anchor node for routing between WMN and MANET. Our anchor node has provision of connecting both the networks, i.e. MANET and WMN. We have implemented a table-driven approach at mesh access points. This anchor node provides bridging facility which results into a single logical overlay ring. We have evaluated performance of disjoint overlays in both the scenarios (with/without anchor node). Results suggest that our approach is more stable and scalable.

The remainder of this paper is organized as follows: In Sect. 2, we present works that are closely related to our work. Section 3 explores the formation of P2P overlays over the hybrid wireless network. The simulation set-up is discussed in Sect. 4 and results are discussed in Sect. 5. We have concluded our work in Sect. 6.

2 Related Work

First, we have divided the related work into two parts to accomplish our objective. We explore the works which are focused towards the removal of dead zones [7–12]. Next, we discuss the efforts that have been made for the formation of P2P overlays using hybrid infrastructure of wireless networks [13–16].

In the papers [9, 10], authors have used wireless mesh network to increase the network coverage range of wireless network from few metres to several kilometres in disaster and rural areas. They augmented WMN and geostationary satellite to access the Internet. However, they have not considered the smaller distributed dead zones in scenarios like university campus and smart city. MANET has been prominently used to eliminate dead zones in the paper [8]. Another paper [12] also suggests that the WMN is an elegant solution to extend the network coverage. The integration of networks (WMN and MANET) have been explored in [7, 11] using mesh point portals (MPP). Again both the papers do not provide proper details of implementation.

Several authors [13, 14, 16] have proposed the idea of merging the disjoint cluster of overlays. The authors in [13, 14] have proposed the merging of DHT overlays using cross-layer approach at application layer itself. These works are significant when the mobile nodes are working in same mode (either infrastructure mode or ad hoc mode). Whereas, [16] suggests that dedicated relay node can be placed in close vicinity to Wi-fi which toggles the operating modes. We have also adopted similar method to connect MANETs and WMN using anchor nodes. The major difference in our approach is that we have considered mesh access point as anchor node which has negligible mobility. Whereas, in previous work [16] the relay nodes are the nodes of MANET which have no predictable mobility. Another major problems in selecting mobile nodes as relay node for transmission of data are either mobile nodes have dual wireless network interface card or software level toggling of modes. Both the approaches required customization of mobile nodes which are not feasible solution due to restriction of access.

3 Formation of P2P Overlays Over Hybrid Wireless Network

The P2P resource/message sharing applications can be deployed on smartphones, laptops, PDAs, etc. Each of the mobile devices is equipped with wireless interface. The wireless interface can be utilized to form a MANET. Thus, disjoint MANETs are formed in different dead zones as shown in Fig. 1. These island of MANET have to be connected through infrastructure-based network for exchanging overlay messages. WMN and WLAN are popular infrastructure-based network. WLAN has last hop wireless connectivity, which limits the network coverage. Moreover, the backbone of WLAN is wired which makes the technology costly. On the contrary, WMN is easy to extend.

Figure 2 depicts the scenario, where the disjoint MANETs are connected with infrastructure-based WMN. The data transmission between two disjoint MANETs is conducted in two phases. In first phase, data travel through the first MANET in multi-hop fashion to hand over this data packet to wireless mesh network. In second phase, the data travel through WMN to reach the appropriate node to hand over these data packets again to second MANET. MANET uses layer-3 Internet protocol stack to decide the path to destination. Further, the wireless mesh network uses layer-2 of Internet protocol stack to transmit the messages to destination with in WMN. The differences in routing layers motivate to create a node which transparently handles the routing mechanism for both of the networks, i.e. WMN and MANET. We have created a node called as anchor node which handles the routing job and connects these networks.

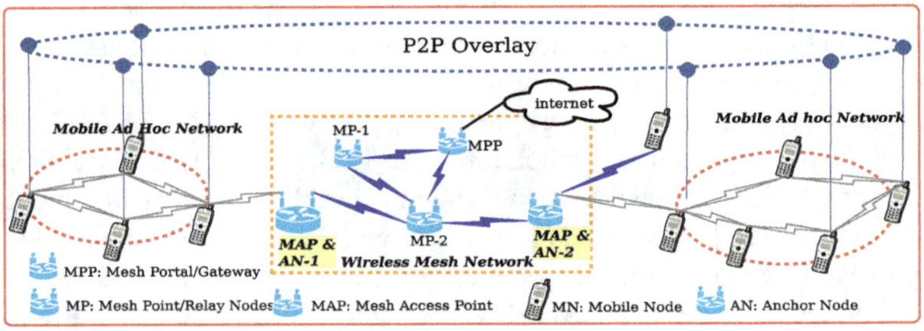

Fig. 2 P2P overlay formation over hybrid network of WMN and MANETs

3.1 Anchor Node: Multi-wireless Interface

The participating nodes of MANET are smartphones, laptops, PDAs, etc. which equipped with wireless network interface card (NIC). This wireless NIC operates in ad hoc mode to form a MANET. To install wireless mesh network, there are four kinds of nodes in WMN: end station (STA), mesh point (MP), mesh point portal (MPP) and mesh access points (MAP). The interconnected MPs form the backbone of the network. MAP is an endpoint of attachment with two wireless network interface card (NIC), one NIC is used to connect end devices such as smartphones, laptops, PDAs, etc. (operating in infrastructure mode with standard 802.11 capability) and, other interface of MAP is used to connect MPs (operating in ad hoc mode with 802.11s mesh capability). Mesh portal (MPP)/gateway nodes facilitate to connect external networks and Internet. We have used a responsible node for routing and data transmission known as "Anchor Node".

The anchor node is mainly designated machine for connecting MANET and WMN. These anchor nodes could either be separately implemented for this purpose or one of the nodes of WMN. We have considered mesh access point of WMN as anchor node because of negligible mobility of this device. As mentioned before, the anchor nodes must operate in dual mode to provide the both kind of connectivity, i.e. infrastructure and ad hoc mode simultaneously. To achieve said target, we attached one additional wireless NIC in mesh access point to provide both kind of connectivity. The anchor node can be placed with in the wireless range of MAP. It can also participate to form overlays. Due to static nature of anchor node, the P2P network produces high stability of overlays.

4 Simulation Setup

In this section, we have described the simulation of peer-to-peer application over hybrid network with and without anchor node as shown in Figs. 1 and 2. Our integrated framework [17] platform comprised of modules of OverSim, INETMANET 3.0 and OMNeT++ 4.6. We have utilized chord (the most popular structured overlay protocol) which builds a ring-based topology of nodes and uses a distributed hash table (DHT) of nodes to perform routing (iterative/recursive) of overlay messages across the overlay ring.

Our integrated framework uses "KbrTestApp" [18] application (based on chord protocol) in overlay nodes to test the reachability between overlay nodes. The "KbrTestApp" performs one-way test by sending PING and PONG messages between the overlay nodes. To measure the performance of chord overlay, we have evaluated "KbrTestApp" application over the underlying MANET. The MANET nodes are forming routing path using a reactive routing protocol DYMO [19]. The WMN underlay uses hybrid wireless mesh routing protocol [20]. To mitigate the mobility model, we have opted random waypoint mobility [21] to control the move-

Table 1 Simulation settings

Simulation parameter	Value
Simulation area	2000 × 2000 m^2
No. of nodes (with and without anchor node)	10, 30, 50, 100, 150, 200, 500
Overlay protocol	Chord
Tier-1 application	Dummy
Tier-2 application	KbrTestApp
Overlay routing	Iterative
Transport layer protocol	UDP
MAC layer	IEEE80211SMAC
Underlay routing protocol	HWMP (layer-2), DYMO (layer-3)
Data rate	54 MB/s
Transmission range	100 m

ment pattern of mobile nodes. We have also utilized no churn model [18] (in which the nodes only join the scenario periodically) to realize the churn dynamics. Further, the stationary mobility model is utilized for simulating the wireless mesh network of fixed nodes. The chord overlay uses key-based routing (KBR) test application [18] as Tier-1 application to test the routing of messages to peers. Table 1 shows the main settings for our simulation set-up.

5 Result and Discussion

This section is intended to discuss the results obtained from our implemented framework. The findings are discussed in terms of mean delivery ratio, one-way hop count and latency of messages chord overlay. The measurement is taken against varying node densities in both the topology as in Figs. 1 and 2. The mean delivery ratio is recorded for one-way test messages sent by the KBR test application running on each overlay terminal. The hop count of overlay messages is measured in terms of mean number of overlay hop count using iterative/recursive routing to reach the destination. We have also recorded mean of latency of overlay messages to reach the destination. We have carried out simulation for two configurations. Every configuration was repeated 10 times in each run.

The performance outcomes are shown in Fig. 3a–c. First set of results was recorded for MANET topology without anchor nodes as in Fig. 1 with random mobility of nodes. Wherein we observed maximum of 42% delivery ratio for node densities of 10, 30, 50, 100, 150, 200 and 500 nodes (Fig. 3a). Another set of results were recorded for MANET topology with WMN and dedicated anchor node (Fig. 3a) with stationary mobility of WMN node (including anchor nodes). In the presence of anchor node, a rapid increase in the performance was seen with increase in node

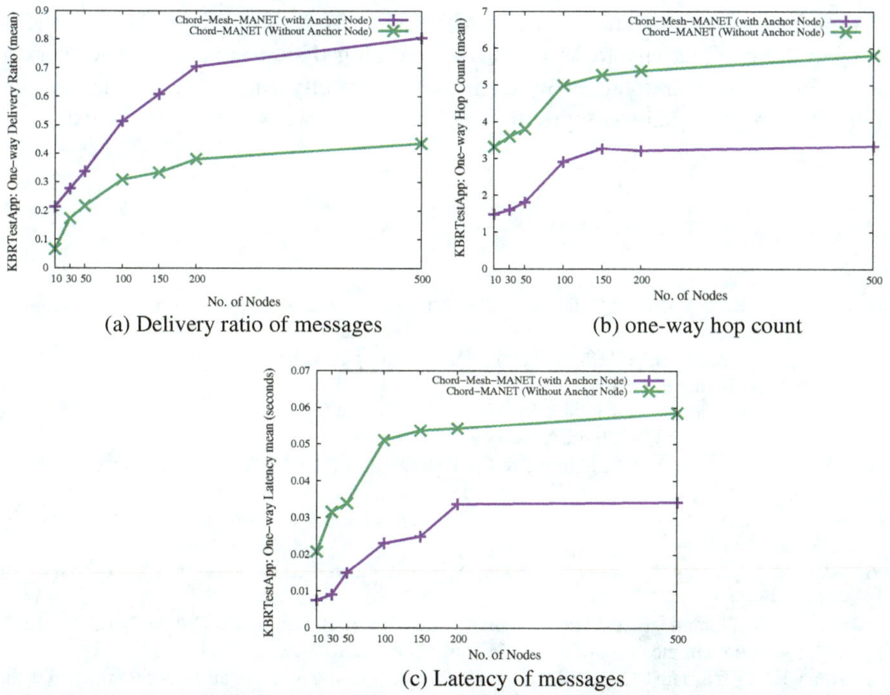

(a) Delivery ratio of messages

(b) one-way hop count

(c) Latency of messages

Fig. 3 **a** Delivery ratio of messages, **b** one-way hop count, **c** latency of messages

densities from 10 to 500 nodes, i.e. delivery ratio increased from 42% to almost 82%. On the other hand, the overlay hop count and latency decreased on increasing the node densities (Fig. 3b) from 5.2 hop count to almost 3 hop count and latency decreased from 6 to 3 ms (Fig. 3c). It is also evident from the plots (Fig. 3a–c) that the performance of chord overlay stabilizes when the node density in the constraint area reaches a certain level.

The results so obtained are in complete accordance with and without anchor nodes which resulting in desired performance of chord overlay over disjoint MANETs. These results are comparable with similar works done in [16, 22].

6 Conclusion

We have proposed an approach for the removal of dead zones by forming MANETs in conjunction with the infrastructure-based WMN. The logical overlay of P2P application running over MANETs in dead zones is split due to unavailability of MANET network coverage. The split of overlay ring results in restrict to search domain. We have proposed WMN-based approach to merge the logical overlay. WMN also poses challenges to provide connectivity to MANETs. We have used an anchor node to provide the connectivity to MANETs for merging disjoint overlays. The work is useful in scenarios like university campuses, smart city, rural area, etc. The results obtained from simulation suggest the applicability of work in similar context.

References

1. Perkins C, Belding-Royer E, Das S (2003) Ad hoc on-demand distance vector (aodv) routing. Technical report
2. Johnson DB, Maltz DA (1996) Dynamic source routing in ad hoc wireless networks. In: Mobile computing. Springer, pp 153–181
3. Neumann A, Aichele C, Lindner M, Wunderlich S (2008) Better approach to mobile ad-hoc networking (batman). IETF draft, October
4. Clausen T, Jacquet P (2003) Optimized link state routing protocol (olsr). Technical report
5. Perkins CE, Bhagwat P (1994) Highly dynamic destination-sequenced distance-vector routing (dsdv) for mobile computers. In: ACM SIGCOMM computer communication review, vol 24. ACM, pp 234–244
6. Akyildiz IF, Wang X, Wang W (2005) Wireless mesh networks: a survey. Comput Netw 47(4):445–487
7. Hiertz GR, Denteneer D, Max S, Taori R, Cardona J, Berlemann L, Walke B (2010) IEEE 802.11s: the wlan mesh standard. IEEE Wirel Commun 17(1):104–111
8. Khan KUR, Zaman RU, Reddy AVG (2008) Integrating mobile ad hoc networks and the internet: challenges and a review of strategies. In: 3rd international conference on communication systems software and middleware and workshops, 2008, COMSWARE 2008. IEEE, pp 536–543
9. Kanchanasut K, Tunpan A, Awal M, Wongsaardsakul T, Das D, Tsuchimoto Y (2007) Building a long-distance multimedia wireless mesh network for collaborative disaster emergency responses. Internet Education and Research Laboratory, Asian Institute of Technology, Thailand
10. Vural S, Wei D, Moessner K (2013) Survey of experimental evaluation studies for wireless mesh network deployments in urban areas towards ubiquitous internet. IEEE Commun Surv Tutor 15(1):223–239
11. Wang J, Xie B, Agrawal DP (2009) Journey from mobile ad hoc networks to wireless mesh networks. In: Guide to wireless mesh networks. Springer, pp 1–30
12. Conti M, Giordano S (2014) Mobile ad hoc networking: milestones, challenges, and new research directions. IEEE Commun Mag 52(1):85–96
13. Shah N, Qian D (2010) Cross-layer design to merge structured p2p networks over manet. In: 2010 IEEE 16th international conference on parallel and distributed systems (ICPADS). IEEE, pp 851–856
14. Shah N, Qian D, Wang R (2015) Merging of p2p overlays over mobile ad hoc network: evaluation of three approaches. Adhoc Sens Wirel Netw 25
15. Shah N, Qian D, Wang R (2012) Manet adaptive structured p2p overlay. Peer-to-Peer Netw Appl 5(2):143–160
16. Sewak A, Pandey M, Gore MM (2016) Forming structured p2p overlays over disjoint manet clusters. In: 2016 IEEE international conference on advanced networks and telecommunications systems (ANTS). IEEE, pp 1–6
17. Ali S, Sewak A, Pandey M, Tyagi N (2017) Simulation of p2p overlays over manets: impediments and proposed solution. In: 2017 9th international conference on communication systems and networks (COMSNETS). IEEE, pp 338–345

18. Baumgart I, Heep B, Krause S (2007) OverSim: a flexible overlay network simulation framework. In: Proceedings of 10th IEEE global internet symposium (GI'07) in conjunction with IEEE INFOCOM 2007, Anchorage, AK, USA, pp 79–84
19. Chakeres I, Perkins C (2010) Dynamic manet on-demand (dymo) routing. draft-ietf-manet-dymo-19.txt
20. Joshi A, Bahr M et al (2006) Hwmp specification. IEEE P802 11(2006)802–11
21. OpenSim Ltd. (2012) INET Framework for OMNeT++ Manual
22. Chowdhury F, Kolberg M (2013) Performance evaluation of structured peer-to-peer overlays for use on mobile networks. In: 2013 sixth international conference on developments in eSystems Engineering (DeSE), pp 57–62

Deep 3D Convolutional Neural Network for Automated Lung Cancer Diagnosis

Sumita Mishra, Naresh Kumar Chaudhary, Pallavi Asthana and Anil Kumar

Abstract Computer-aided diagnosis has emerged as an indispensable technique for validating the opinion of radiologists in CT interpretation. This paper presents a deep 3D convolutional neural network (CNN) architecture for automated CT scan-based lung cancer detection system. It utilizes three-dimensional spatial information to learn highly discriminative three-dimensional features instead of 2D features like texture or geometric shape which need to be generated manually. The proposed deep learning method automatically extracts the 3D features on the basis of spatiotemporal statistics. The developed model is end-to-end and is able to predict malignancy of each voxel for given input scan. Simulation results demonstrate the effectiveness of proposed 3D CNN network for classification of lung nodule inspite of limited computational capabilities.

Keywords CNN · Image processing · Lung cancer · CT scan

1 Introduction

Lung cancer is a prominent cause for cancer-related deaths in India and most of these deaths may be prevented through periodic assessment of an individual's risk of lung cancer. The stages of lung cancer are indicated in the range from initial stage to fourth stage. In early stages, cancer is limited to the lung. Lung cancer spreads to other areas of the body like liver in advanced stages and patient outcome in such cases is not favorable. One of the primary characteristics of lung cancer in early stages is the presence of abnormal tissue growth in lungs known as pulmonary nodule. Screening with the use of low-dose helical computed tomography (CT) may

S. Mishra (✉) · P. Asthana · A. Kumar
Amity School of Engineering and Technology, Amity University
Lucknow Campus, Lucknow, India
e-mail: smishra3@lko.amity.edu

N. K. Chaudhary
Dr. Ram Manohar Lohia Awadh University, Faizabad, UP, India

S.-L. Peng et al. (eds.), *Computing and Network Sustainability*, Lecture Notes in Networks and Systems 75, https://doi.org/10.1007/978-981-13-7150-9_16

improve the mortality rates in high-risk individuals by 20% compared to conventional chest X-ray [1].

In a conventional setup, the radiologist and oncologists play a crucial role in accurate diagnosis of Lung cancer and it also depends upon the experience of specialist. In some cases even for highly trained radiologists, detecting nodules on CT scan and cancer diagnosis become challenging tasks. The recent progress in computer vision and deep learning may be exploited to achieve automatic detection of nodules to provide accurate clinical information, and subsequently, stage of progression of disease may be estimated. Deep neural networks (DNN) are being explored for developing automated tools for lung nodule detection because these algorithms are able to learn features from raw image data. DNN lung nodule identification methods can be classified into 2D deep CNN [2] and 3D deep CNN [3–5].

The 2D CNNs treat each slice of the CT scan as an individual image, and training is done accordingly. However, in actual practice, nodules are dense 3D objects, and in CT scans, they usually appear in several successive CT slices. Therefore, treating each slice as an individual image for nodule detection will result in the loss of highly correlative spatial context information. In the 3D network, the convolutional kernels are three-dimensional and the input data is 3D cubes instead of 2D images leading to increased accuracy, and a major drawback of 3D CNN is increased complexity and requirement of more computational power. The proposed deep 3D convolutional neural network for pulmonary nodule classification is capable of learning good three-dimensional features without manual feature extraction and selection process required in conventional algorithms. This automatic 3D feature selection through the training of the network obtains precise characteristics of pulmonary nodules while retaining relatively better generalization capability of the network.

The paper is arranged in four sections. The next section deals with methods, Sect. 3 deals with results and discussion, and Sect. 4 provides the concluding remark.

2 Methods

The dataset used in the work is Luna which is based on publicly available Lung Image Database Consortium and Image Database Resource Initiative (LIDC/IDRI) data [6]. CT scan images in the database have nodule annotations from leading experts. After data acquisition, image preprocessing is performed to make it suitable for further processing.

2.1 Image Preprocessing

The radius of the average malicious nodule in the dataset is 4.8 mm, with a typical CT volume of 400 mm × 400 mm × 400 mm. First step in the process is data preparation and the dataset provided is in a medical imaging format called MetaImage (mhd/raw) which comprises header data in .mhd files, and multidimensional image

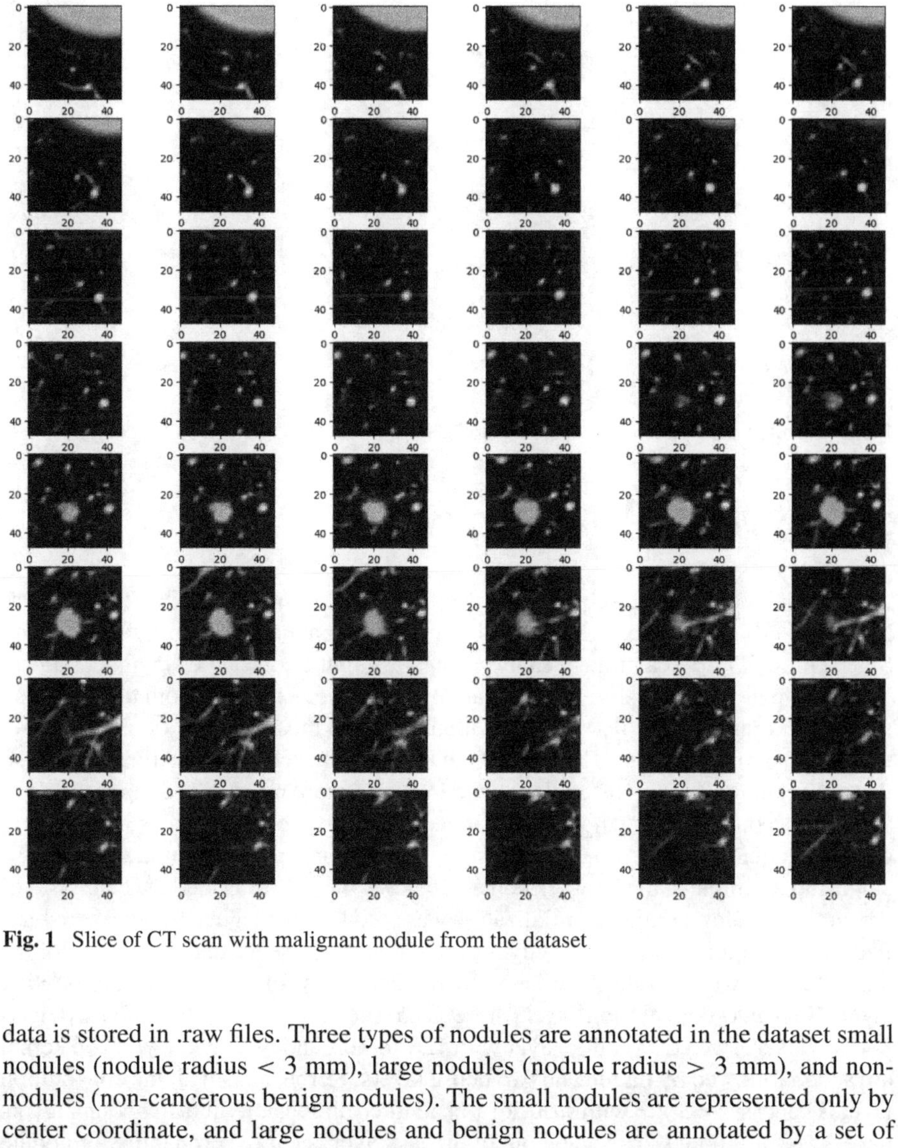

Fig. 1 Slice of CT scan with malignant nodule from the dataset

data is stored in .raw files. Three types of nodules are annotated in the dataset small nodules (nodule radius < 3 mm), large nodules (nodule radius > 3 mm), and non-nodules (non-cancerous benign nodules). The small nodules are represented only by center coordinate, and large nodules and benign nodules are annotated by a set of coordinates representing their edges. Non-nodules are benign nodules and therefore for training the network these are labeled as healthy tissue. After preprocessing, malignant nodules are marked as positive samples, and benign nodules are labeled as negative example. Further, different CT scans have different voxel length for the raw CT scans, so the next step is to convert these different CT scans to the same voxel spacing by resampling. Sample-wide pixel normalization was applied afterward to obtain uniform scaling between samples. These resampled images are then saved in numpy format for further processing. Figure 1 shows a typical input sample having malignant nodule.

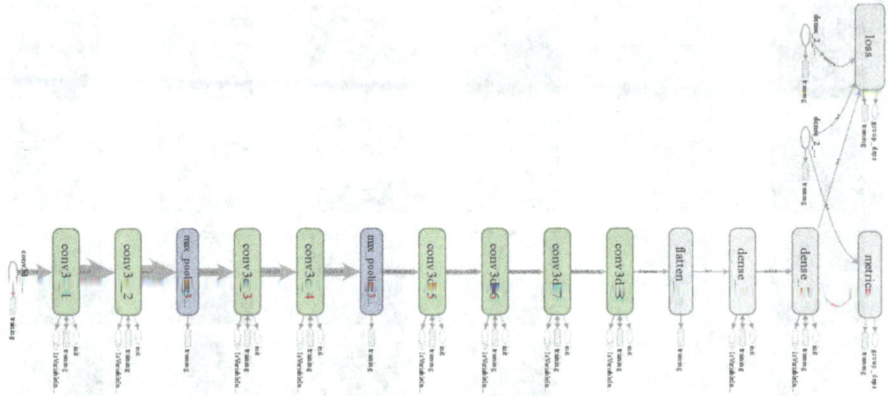

Fig. 2 Detailed architecture of proposed deep 3D CNN

2.2 Architecture of Deep 3D CNN

The detailed architecture of proposed deep 3D CNN is shown in Fig. 2, and it was implemented using Keras API [7] and TensorFlow backend [8] on CPU platform. Since entire 3D CT scan image cannot be fed as input to our 3D CNN model due to the CPU memory constraint; hence, small 3D cubes are extracted from the lung scan volume and input to the network individually. Two kinds of cubes of size 48 × 48 × 48 are randomly selected. First set of the input is selected from positive samples so that they contain at least one malignant nodule. To ensure number of enough negative samples, second set of the inputs are cropped randomly from lung scans with random centers. Latter category of inputs does not contain any malignant nodules. Thus, a single input sample to the network consists of 48 × 48 × 48 cube and the batch size of the input is 2. Gloret uniform initialization was used for all weights [9], with zero bias. Padding was applied around the edges of the input to preserve the dimensions of the data. We use two convolutional layers with 32 kernels followed by a max-pooling layer. Next two convolution layers have 64 kernels. These are followed by max-pooling layer. Two convolution layer with 128 kernels and two layers with 256 kernel are added followed by flatten and two dense layers. Kernel size in all the convolution layer is kept as 3 × 3 × 3 with stride of 1 in all three dimensions and all pooling layers have 2 × 2 × 2 pool size. The output of the last layer is cancer probability of nodule which utilizes sigmoid activation function. The model was trained on 720 scans, 80 scans were used for validation of model and 88 scans were kept for evaluating the model on test data. From each scan, mini-batches were prepared and data was fed to the network in batch size of 2 due to limited memory capability. The structure of each layer in the proposed network including input dimensions is shown in Table 1. Each convolutional layer uses rectified linear unit (ReLU) nonlinearity for efficient gradient propagation. Although ReLU may result in dead neurons but in our case, it led to fast convergence of data. Batch normalization was done to reduce saturation

Table 1 Three-dimensional deep CNN structure including data dimension

Layer (type)	Output shape	Parameters
conv3d1 (Conv3D)	(None, 46, 46, 46, 32)	896
conv3d2 (Conv3D)	(None, 44, 44, 44, 32)	27680
max-pooling3d1 (MaxPooling3)	(None, 22, 22, 22, 32)	0
conv3d3 (Conv3D)	(None, 20, 20, 20, 64)	55360
conv3d4 (Conv3D)	(None, 18, 18, 18, 64)	110656
Max-pooling3d2 (MaxPooling3)	(None, 9, 9, 9, 64)	0
conv3d5 (Conv3D)	(None, 7, 7, 7, 128)	221312
conv3d6 (Conv3D)	(None, 5, 5, 5, 128)	442496
conv3d7 (Conv3D)	(None, 3, 3, 3, 256)	884992
conv3d8 (Conv3D)	(None, 1, 1, 1, 256)	1769728
flatten1 (Flatten)	(None, 256)	0
dense1 (Dense)	(None, 256)	65792
dense2 (Dense)	(None, 1)	257
Total trainable parameters: 3,579,169		

of neurons and improve generalization capabilities [10]. The average error of all pixels in an input sample defines the loss function which is an important metric to quantify the performance of the model on given sample. Overall performance of 3D CNN model is quantified by the average loss over all samples. We have employed binary cross-entropy loss function for training the model. Model is compiled using stochastic gradient descent (SGD) optimizer with Nesterov momentum value of 0.9 [11], and learning rate of 0.003. This set of parameters were found to accelerate the training process by faster convergence. Stochastic gradient descent updates the weight matrix on the basis of negative gradient of the weights but momentum causes the weights to also change in the direction of the previous update [12].

3 Results

The proposed 3D CNN architecture achieved an accuracy of 94.8% on validation data. Figure 3 shows the training accuracy and validation accuracy as the model was trained. It may be observed from the figure that the accuracy is gradually increasing during training. Figure 4 shows that the loss curve of training samples and validation samples.To improve the quality of this model, the validation loss and validation accuracy over epochs was monitored using tensorboard logs and hyper-parameters were appropriately adjusted. The performance with respect to accuracy becomes stable after several iterations. This behavior is correlated with the change in loss as

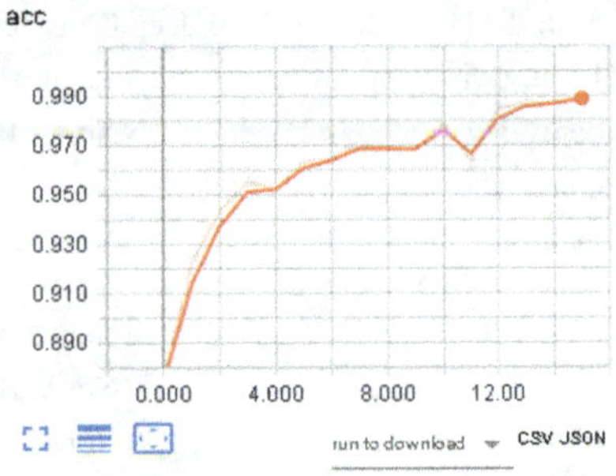

Fig. 3 Training accuracy during training

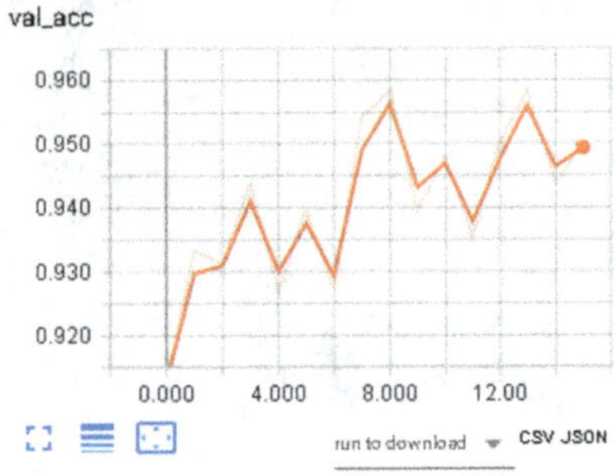

Fig. 4 Validation accuracy during training

shown in Fig. 5 which is decreasing gradually during training, and when the network obtains an optimal point, then the training process becomes stable. Overall, the proposed deep 3D convolutional neural network is effective in classification as shown in Fig. 4 depicting the accuracy during the validation process. In order to demonstrate the performance of our model, a complete CT scan containing annotated malignant nodule was sampled into multiple cubes using sliding window and subsequently fed to the trained 3D CNN network and probabilistic predictions corresponding to each input sample is plotted in two dimensions. Figure 6 shows the predicted probability map from the model in two dimensions, while Fig. 7 shows the regions with probability greater than 90% after removal of noisy predictions. Figure 8 shows 2D slice of corresponding input sample. It may be observed that location of the annotated nodules has very high-probability but a small number of false positives are also predicted as shown in Fig. 6. Evaluation of model on test data yielded a score 0.208 and accuracy of 0.9514. Due to the extensive duration of the training process hyper-parameters initial learning rate, momentum decay was adjusted only once. Further optimization of these parameters is expected to improve the quality of the model and reduce false positives.

Table 2 shows the comparison of our work with similar works and it must be noted that different authors use different resampling and data division methodology for generating test train and validation data. In addition, only some of these methods are end-to-end, thus making it difficult to achieve a fair comparison.

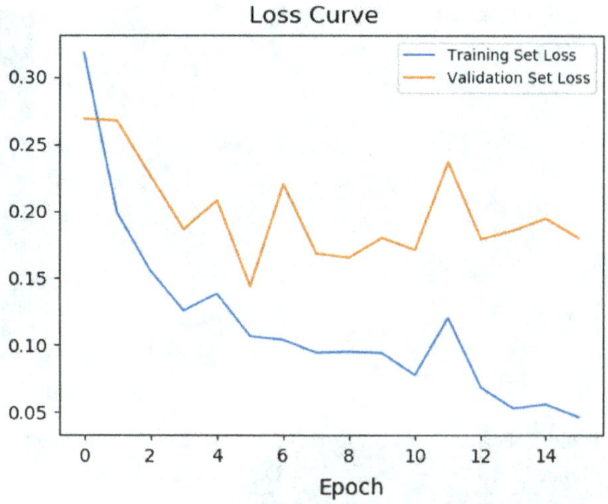

Fig. 5 Loss curve

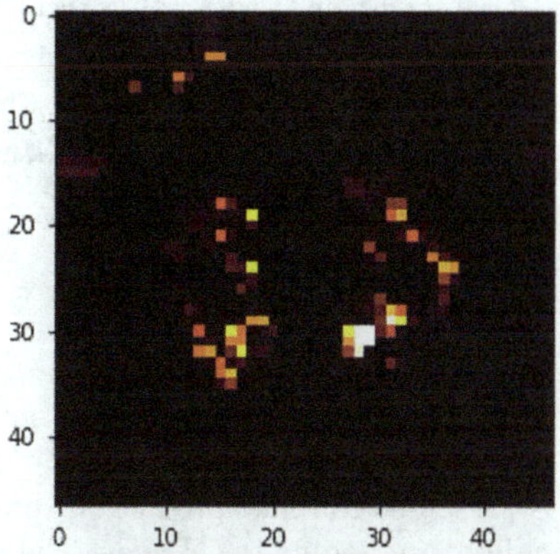

Fig. 6 Prediction of probability map by model

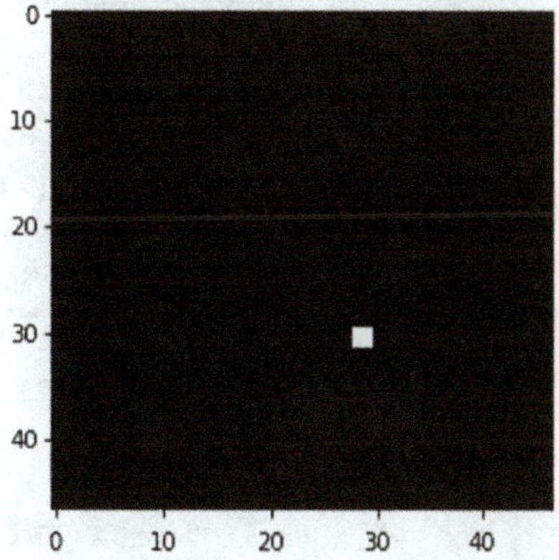

Fig. 7 Predicted sample result

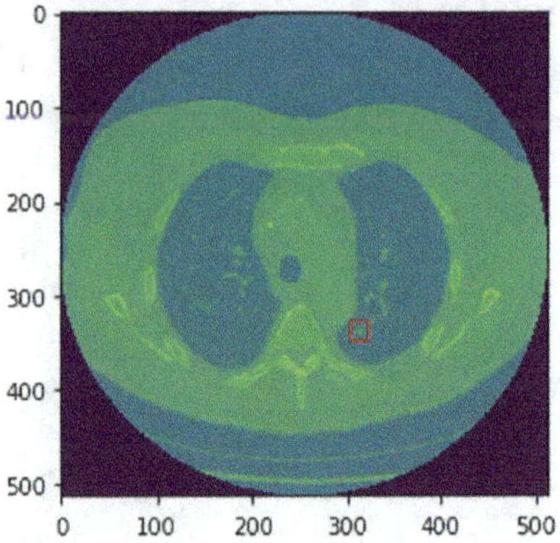

Fig. 8 Malignant nodule

Table 2 Performance comparison of proposed work with other published work

Related work	Dataset characteristics	Performance (%)
Kumar et al. [13]	LDRI, nodule size > 3 mm	Accuracy 75.01
Gruetzemacher and Gupta [14]	LDRI both small and large nodules	Accuracy 86.13
Hamidian et al. [15]	nodule size > 3 mm	Sensitivity 95
Wu and Zhao [16]	NCI, China (nodule size not specified)	Accuracy 77.8
Proposed work	LDRI, nodule size > 3 mm	Accuracy 94.80

4 Conclusion

The chances of successful outcome of lung cancer treatment are improved if it is detected in early stages. In this paper, a deep 3D convolutional neural networks is presented to make predictions regarding the presence or absence of a malignant pulmonary nodule in CT scan. The proposed 3D Deep CNN automatically extracts 3D features for pulmonary nodule classification directly from CT volume. We have demonstrated that our 3D Deep CNN model is able to perform fairly and achieves an accuracy of 94.8% without extensive preprocessing and limited computational capabilities. The performance of the proposed model may be further improved by utilizing suitable data augmentation technique. Although data augmentation using axes swapping was tried in this work but no significant improvement is obtained and over-fitting may be reduced by using dropout layer.

References

1. Siegel RL, Miller KD, Jemal A (2016) Cancer statistics, 2016. CA Cancer J Clin 66:730
2. Yang H, Yu H, Wang G (2016) Deep learning for the classification of lung nodules. arXiv:1611.06651
3. Huang X, Shan J, Vaidya V (2017) Lung nodule detection in CT using 3D convolutional neural networks. In: IEEE international symposium on biomedical imaging. IEEE

4. Zhu W, Liu C, Fan W et al (2017) DeepLung: 3D deep convolutional nets for automated pulmonary nodule detection and classification
5. Valente IRS, Neto EC, Albuquerque VHCD (2016) Automatic 3D pulmonary nodule detection in CT images: a survey. Comput Methods Progr Biomed 124:91–107
6. National Lung Screening Trial Research Team (2011) Reduced lung-cancer mortality with low-dose computed tomographic screening. N Engl J Med 365(5):395–409
7. Chollet F (2015) Keras. http://keras.io
8. Abadi M et al (2016) TensorFlow: large-scale machine learning on heterogeneous distributed systems. arXiv:1603.04467
9. Glorot X, Bengio Y (2010) Understanding the difficulty of training deep feedforward neural networks. Proc Mach Learn Res 9(3):249256
10. Ioffe S, Szegedy C (2015) Batch normalization: accelerating deep network training by reducing internal covariate shift. arXiv:1502.03167
11. Sutskever I, Martens J, Dahl G, Hinton G (2013) On the importance of initialization and momentum in deep learning. In: Proceedings of the 30th international conference on international conference on machine learning, ICML13, pp 1139–1147
12. Ruder S (2016) An overview of gradient descent optimization algorithms. arXiv:1609.04747
13. Kumar D, Wong A, Clausi DA (2015) Lung nodule classification using deep features in CT images. In: 2015 12th conference on computer and robot vision (CRV). IEEE, pp 133–138
14. Gruetzemacher R, Gupta A (2016) Using deep learning for pulmonary nodule detection and diagnosis
15. Hamidian S, Sahiner B, Petrick N, Pezeshk A (2017) 3D convolutional neural network for automatic detection of lung nodules in chest CT. In: Proceedings of SPIE-the international society for optical engineering 10134:1013409. https://doi.org/10.1117/12.2255795
16. Wu Q, Zhao W (2017) Small-cell lung cancer detection using a supervised machine learning algorithm. In: International symposium on computer science and intelligent controls (ISCSIC). Budapest 2017, pp 88–91. https://doi.org/10.1109/ISCSIC.2017.22

Comparative Analysis of Neuro-Fuzzy and Support Vector Approaches for Flood Forecasting: Case Study of Godavari Basin, India

Puneet Misra and Shobhit Shukla

Abstract Forecasting and prediction have been a significant area of study for researchers since very past. Out of various approaches, soft computing data-driven models are very helpful for the purpose of forecasting. Soft computing models are usefully applicable when the relationship between the parameters is very complex to understand. India is a disaster-prone country which requires such major soft computing-based data-driven models to handle disasters like flood, drought and landslide. Flood has a major impact in many regions of India out of which Cauvery, Godavari and Ganges river basins are the most affected regions. The paper attempts to forecast floods by modeling river flow into the area of Godavari river basin of India which has a complicated topography. In this study, two data-driven techniques, adaptive neuro-fuzzy inference system (ANFIS) and support vector machine (SVM), were explored for the purpose of forecasting floods by predicting river flow in Cauvery river sub-basin of southern India.

Keywords Adaptive neuro-fuzzy inference system (ANFIS) · Support vector machine (SVM) · Mean squared error (MSE) · The coefficient of correlation (R) · Nash–Sutcliffe coefficient (NS)

1 Introduction

Forecasting and prediction problems have been researched by numerous researchers since the very past. Forecasting techniques are classified into two primary categories, statistical techniques and data-driven techniques. Statistical techniques are complicated and require advanced statistical and mathematical knowledge and some degree of expertise with those techniques [1], whereas data-driven methods are helpful in forecasting where the relationship between the parameters is very complex to

P. Misra (✉) · S. Shukla
Department of Computer Science, University of Lucknow, Lucknow, UP, India
e-mail: puneetmisra@gmail.com

S. Shukla
e-mail: shobhitshukla89@gmail.com

S.-L. Peng et al. (eds.), *Computing and Network Sustainability*, Lecture Notes in Networks and Systems 75, https://doi.org/10.1007/978-981-13-7150-9_17

understand and model mathematically [3]. River flow prediction is significant for flood prediction as they can lead to loss of life, destruction of infrastructure, etc [6]. In the recent past, two data-driven models that have emerged and became popular among the researchers for solving computationally demanding problems are adaptive neuro-fuzzy inference system (ANFIS) and support vector machine (SVM). In [8], ANFIS has been employed for mapping susceptibility toward a landslide. In [4], the authors give a comparison of ANFIS and evolutionary neural network in the case of predicting floods. In [10], the authors apply ANN and ANFIS methodologies and provide a comparative study for river flow forecasting. The authors of [5] employed ANFIS for hydrological modeling and river flow forecasting of River Great Menderes, located in western Turkey. Tehrany et al. [12] employ SVM methodology for flood c. In [2], used SVM in Bayesian context for long term flow forecasting of Kayacik River, Turkey.

2 Methodology

2.1 *Adaptive Neuro-Fuzzy Inference System (ANFIS)*

Adaptive neuro-fuzzy inference system (ANFIS) is capable of estimating a continuous function on the given dataset to a given degree of precision [9]. The first-order SugenoTakagi system is employed for modeling which is detailed as follows.

The first-order SugenoTakagi system for inputs x, y and an input z can be given as:

Rule 1: If x is A_1 and y is B_1, then $f_1 = p_1\text{x} + q_1\text{y} + r_1$

Rule 2: If x is A_2 and y is B_2, then $f_2 = p_2\text{x} + q_2\text{y} + r_2$

where p_1, q_1, r_1 and p_2, q_2, r_2 are parameters in consequent part. Figure 1 shows the architecture of ANFIS.

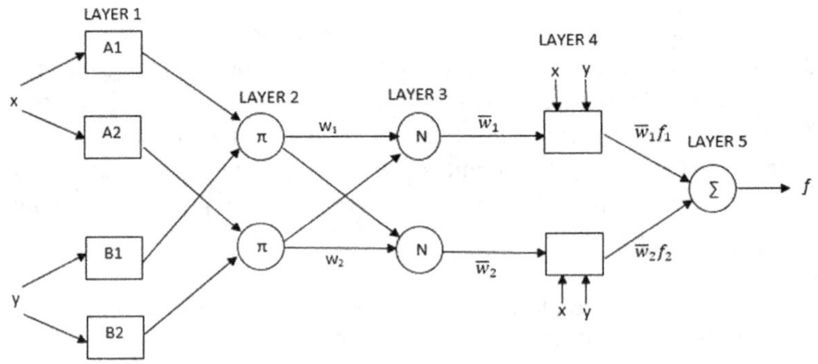

Fig. 1 ANFIS structure for two input systems

Layer 1: Each node in the first layer has the function:

$$O_{1,i} = \begin{cases} \mu_{A_i}(x), & \text{for i} = 1, 2 \\ \mu_{B_{i-2}}(y), & \text{for i} = 3, 4 \end{cases} \tag{1}$$

Here x, y are inputs to node i and A_i and B_{i-2} are labels of the respective nodes. Therefore, $O_{1,i}$ is the membership degree of a set A_1, A_2, B_1 or B_2 characterized by the shape of membership function like triangular, Gaussian, bell or trapezoidal.

Layer 2: This layer is made up of various nodes each of which is labeled Prod and produces the multiplication of all the incoming inputs in it.

$$O_{2,i} = \mu_{A_i}(x) \cdot \mu_{B_i}(y), \quad \text{for i} = 1, 2 \tag{2}$$

Layer 3: Nodes belonging to the third layer are unadaptive nodes called Norm, and the ith node calculates the ratio between the ith rule and the aggregate of the strengths of all the rules.

$$O_{3,i} = \overline{w_i} = \frac{w_i}{w_1 + w_2}, \quad \text{for i} = 1, 2 \tag{3}$$

Layer 4: In this layer, function of the ith node which computes the contribution of the ith rule is:

$$O_{4,i} = \overline{w_i} f_i = \overline{w_i}(p_i x + q_i y + r_i) \tag{4}$$

where w_i is layer 3 strength and $\{p_i, q_i, r_i\}$ are the parameters of the ith node.

Layer 5: This layer contains a single node which calculates the total output of network by aggregating inputs and is given as:

$$O_{5,i} = \sum_i \overline{w_i} f_i = \frac{\sum_i w_i f_i}{\sum_i w_i} \tag{5}$$

ANFIS implemented in this paper is trained using a learning methodology which is composed of least squares technique is used to calculate the parameters in the fourth layer in forward pass and gradient descent technique which propagates error backward and updates parameters.

2.2 Support Vector Machine (SVM)

SVM is a methodology that is formed on the basis of statistical learning. The inputs sustaining the SVM configuration are determined by employing a training procedure. Consider a linear regression which is trained on the dataset $\chi = \{u_i, v_i;$

$i = 1, 2, \ldots, n\}$ with inputs u_i and targets v_i, and then the regression function can be written as:

$$f(u) = w_i \cdot \phi_i(u) + b \tag{6}$$

where w_i is the weight, ϕ_i is a nonlinear activation function which transforms the inputs into a feature space where linear regression can handle nonlinear regression, and b is the bias. Vapnik in [11] gave the subsequent optimization problem with ϵ-insensitive loss for calculating the answer for Eq. 7:

$$Minimize : \frac{1}{2} \cdot \left\| w^2 \right\| + C \left(\sum_{i}^{N} (\xi_i + \xi_i^*) \right) \tag{7}$$

$$\text{Subject to} \begin{cases} w_i \cdot \phi(u_i) + b_i - v_i \le \epsilon + \xi_i^*, & \text{for i} = 1, 2, \ldots, \text{N} \\ v_i - w_i \cdot \phi(u_i) - b_i \le \epsilon + \xi_i, & \text{for i} = 1, 2, \ldots, \text{N} \\ \xi_i, \xi_i^*, & \text{for i} = 1, 2, \ldots, \text{N} \end{cases} \tag{8}$$

where ξ_i and ξ_i^* are slack variables which are brought in for the assessment of the divergence of input–output vectors outside ϵ-insensitive zone and C is constantly regulating the level of penalizing loss when an error occurs during the process of training. Equation 8 is solved by employing the Lagrangian multipliers along with the Karush–Kuhn–Tucker (KKT) optimality condition.

3 Study Area

The suitability of the various mentioned techniques as a time series river flow prediction methodology is studied in this work. To demonstrate ability and validity of these methods for the purpose of river flow forecasting and modeling, the Godavari River, the biggest in southern India, is chosen. The location of Godavari river basin is represented in Fig. 2. Two river-flowing gauge stations, Bhadrachalam and Polavaram on the Godavari River as observed in Fig. 2, are selected for forecasting. As can be observed from the figure, Bhadrachalam is situated upstream of Polavaram. The data records of both these river gauging stations are used for river flow and flood forecast modeling.

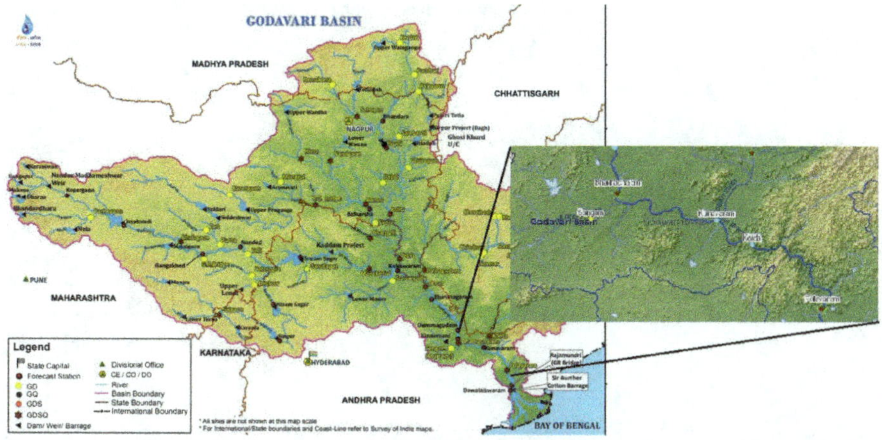

Fig. 2 Location of the Cauvery River and the gauging stations

4 Model Development

4.1 Description of Data

In the paper, the performance of ANFIS and SVM was calculated on daily river flow. For this purpose, 8 years of river flow data from 2007 to 2015 [9] was used. The total number of days for availability of the river flow data was 2630. The data was split into two sets: training dataset composed of years 2007–2014 and a testing dataset of year 2015. The number of lags for river flow forecasting was chosen by employing the partial autocorrelation function (PACF) of daily river flow data of Polavaram gauging station which is given in Fig. 3. It is seen that the first three lags have more effects on M_{t+1}. The cross-correlation of the Polavaram and Bhadrachalam gauging stations presented in Fig. 4 shows a significant correlation for up to three-day lag in the river flow data. Thus, three previous lags of Polavaram and three lags of Bhadrachalam gauging stations were chosen as inputs to the model. Thus, inputs represent previous flow ($t-1$, $t-2$ and $t-3$) and outputs represent river flow at time $t+1$. Thus, structure of the models is given in Table 1 in which the Polavaram gauge river flow data is represented as P and Bhadrachalam gauge river flow data is represented as B.

4.2 Data Preprocessing

For the purpose of obtaining efficient and accurate training of the models, the data is needed to be normalized. In this paper, normalization is performed on all data scaled in the range $0-1$ independently by employing the following equation:

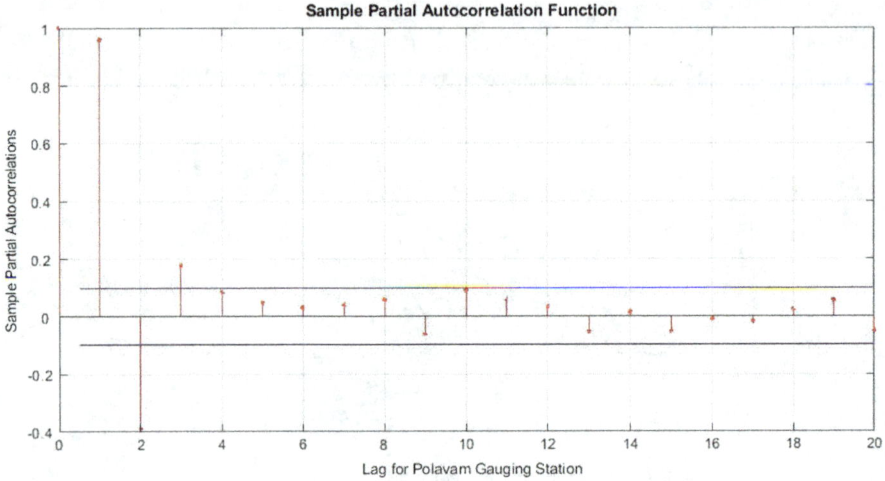

Fig. 3 Partial autocorrelation function of daily river flow data of Polavaram gauge station

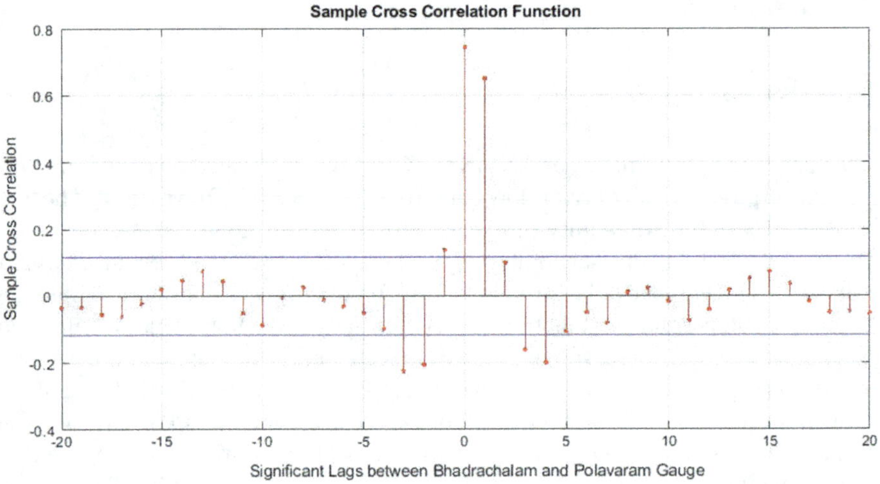

Fig. 4 Cross-correlation function of daily river flow data of Polavaram and Bhadrachalam gauge stations

$$X^{'} = \frac{X - X_{min}}{X_{max} - X_{min}} \tag{9}$$

where $X^{'}$ is the normalized value, X is the original value, X_{min} is the minimum and X_{max} is the maximum.

Table 1 Model structures for forecasting

Model no.	Input structure	No. of variables	Output
M1	P_{t-1}	1	P_{t+1}
M2	$P_{t-1} P_{t-2}$	2	P_{t+1}
M3	$P_{t-1} P_{t-2} P_{t-3}$	3	P_{t+1}
M4	$P_{t-1} B_{t-1}$	2	P_{t+1}
M5	$P_{t-1} P_{t-2} B_{t-1}$	3	P_{t+1}
M6	$P_{t-1} P_{t-2} P_{t-3} B_{t-1}$	4	P_{t+1}
M7	$P_{t-1} B_{t-3}$	2	P_{t+1}
M8	$P_{t-1} P_{t-2} B_{t-3}$	3	P_{t+1}
M9	$P_{t-1} P_{t-2} P_{t-3} B_{t-3}$	4	P_{t+1}
M10	$P_{t-1} B_{t-4}$	2	P_{t+1}
M11	$P_{t-1} P_{t-2} B_{t-4}$	3	P_{t+1}
M12	$P_{t-1} P_{t-2} P_{t-3} B_{t-4}$	4	P_{t+1}
M13	$P_{t-1} B_{t-1} B_{t-3}$	3	P_{t+1}
M14	$P_{t-1} B_{t-1} B_{t-3} B_{t-4}$	4	P_{t+1}
M15	$P_{t-1} P_{t-2} B_{t-1} B_{t-3}$	4	P_{t+1}
M16	$P_{t-1} P_{t-2} B_{t-1} B_{t-3} B_{t-4}$	5	P_{t+1}
M17	$P_{t-1} P_{t-2} P_{t-3} B_{t-1} B_{t-3}$	5	P_{t+1}
M18	$P_{t-1} P_{t-2} P_{t-3} B_{t-1} B_{t-3} B_{t-4}$	6	P_{t+1}

4.3 Model Performance Criteria

Performance of the models generated in the current research was evaluated by employing three standard performance assessment criteria. The indices used were the mean squared error (MSE), coefficient of correlation also known as regression (R) and Nash–Sutcliffe efficiency coefficient (NS). MSE provides the average squared difference between the model output and the actual test outputs. It can be computed as follows:

$$MSE = \frac{1}{n}\sum_{i=1}^{n}(t_i - a_i)^2 \tag{10}$$

where n is the size of dataset , a_i is output of the model and t_i is the corresponding actual output. R can be interpreted as the correlation between actual targets of the dataset and outputs calculated by the model. When the value of R is 1, it means that there is a close relationship, and when the value of R is 0, it means that there is a random relationship between the two. It is calculated by the equation:

$$R = \frac{\sum_{i=1}^{n}(t_i - \bar{t})(a_i - \bar{a})}{\sqrt{\sum_{i=1}^{n}(t_i - \bar{t})^2 \sum_{i=1}^{n}(a_i - \bar{a})^2}} \tag{11}$$

The Nash–Sutcliffe (NS) efficiency coefficient can be measured as:

$$NS = 1 - \frac{\sum_{i=1}^{n}(t_i - a_i)^2}{\sum_{i=1}^{n}(t_i - \bar{t})^2} \tag{12}$$

where n is the size of dataset, a_i is the output of the model and t_i is the corresponding actual output. A model is said to give perfect prediction if the NS value is 1 but it can be regarded as accurate if the NS value is greater than 0.8 as shown in [7].

5 Results and Discussion

In this paper, AVFIS and SVM techniques were applied to the models developed above and the results are described in this section. The implementation and analysis of results of the above-mentioned techniques were performed in MATLAB 2017b.

5.1 Adaptive Neuro-Fuzzy Inference System (ANFIS)

The ANFIS technique was implemented on all eighteen models. The number of membership functions was considered 4. The trapezoidal-shaped membership function was employed for all models. Table 2 lists the performance indices of all eighteen models obtained by employing ANFIS technique.

Table 2 shows that Model 2 composed of the first two antecedent flow data of Polavaram gauging station has the lowest MSE value of 0.00305, highest R value of 0.945 and highest NS value of 0.8969 and is the best fit model for ANFIS technique. The comparison of the observed and the ANFIS computed river flow for Model 2 is shown in Fig. 5.

5.2 Support Vector Machine (SVM)

The SVM technique was applied to all eighteen models, and the performance indices obtained were compared in Table 3. As can be noticed from that Model 1 which consists of both antecedent flow data of Polavaram gauging station and the second antecedent flow data of Bhadrachalam gauging station has the lowest MSE value of 0.00473, highest R value of 0.923 and the maximum NS value of 0.840. The comparison of observed and the SVM computed river flow obtained for Model 1 is shown in Fig. 6. The performances of best fit models of ANN and SVM are shown in Table 4. It can be noticed from Table 4 that ANFIS seems to perform ahead than SVM models as it has minimum MSE and highest R and NS values.

Table 2 Performance indices of ANFIS models

Model	MSE	R	NS
M1	0.00379	0.923	0.8720
M2	0.00305	0.945	0.8969
M3	0.03449	0.937	0.7721
M4	0.05302	0.826	0.6293
M5	0.02828	0.814	0.5998
M6	0.13186	0.867	0.6765
M7	0.03212	0.910	0.5465
M8	0.02960	0.896	0.5096
M9	0.03041	0.901	0.6942
M10	0.17237	0.905	0.7953
M11	0.06121	0.909	0.7072
M12	0.06500	0.915	0.6186
M13	0.02933	0.908	0.8154
M14	0.06560	0.907	0.8205
M15	0.17427	0.877	0.3327
M16	1.23579	0.542	0.5886
M17	0.03221	0.900	0.7737
M18	0.07908	0.899	0.7969

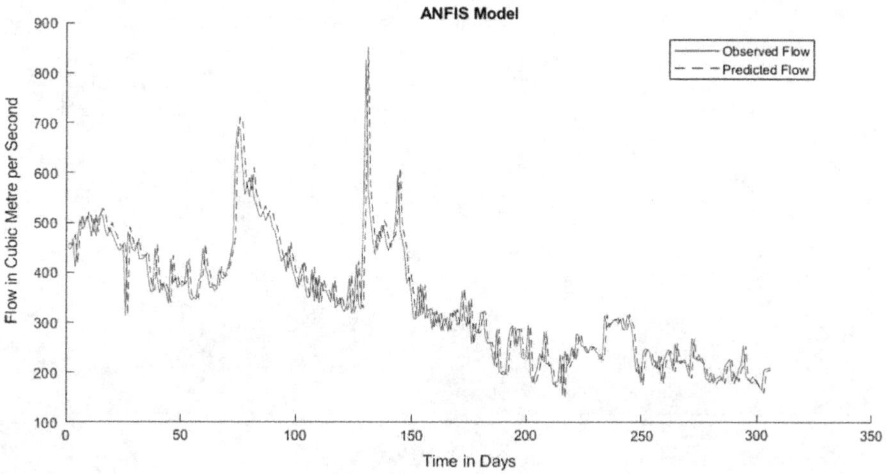

Fig. 5 Actual and ANFIS predicted river flow obtained for Model 2

Table 3 Performance parameters of SVM models

Model	MSE	R	NS
M1	0.00473	0.923	0.8405
M2	0.00569	0.905	0.8083
M3	0.15311	0.805	0.1542
M4	0.01463	0.786	0.5069
M5	0.06104	0.800	0.0558
M6	0.01153	0.819	0.6117
M7	0.20878	0.873	0.0282
M8	0.07367	0.875	0.4802
M9	0.07241	0.900	0.4375
M10	0.04293	0.798	0.4434
M11	0.00757	0.877	0.7453
M12	0.00755	0.874	0.7461
M13	0.14845	0.841	0.3373
M14	0.03271	0.576	0.0998
M15	0.00974	0.840	0.6719
M16	0.00965	0.842	0.6754
M17	0.00870	0.854	0.7070
M18	0.01796	0.707	0.3961

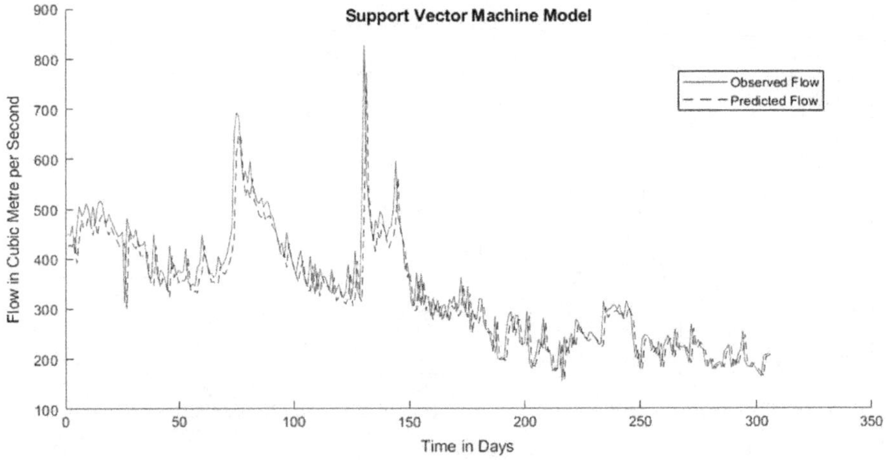

Fig. 6 Actual and SVM predicted river flow obtained for Model 1

Overall, the ANFIS and SVM techniques can provide good forecasting performance and could be used to generate prediction models that provide precise and consistent flood forecasts. The testing results show that the ANFIS model surpassed the SVM model in forecasting.

Table 4 Comparison of the performances of ANN and SVM and tree bagger best fit models

Technique	Model	MSE	R	NS
ANFIS	M2	0.00305	0.945	0.8969
SVM	M1	0.00473	0.923	0.8405

6 Conclusion

In this paper, adaptive neuro-fuzzy inference system and support vector machine techniques were employed to generate models for forecasting of floods based on past values of river flow data. The Polavaram and Bhadrachalam gauging stations located on the Godavari River in southern India were been selected for this purpose. The results obtained during the testing of ANFIS and SVM models were compared and measured based on their testing performance. While comparing the results of these models, it was observed that the MSE values of ANFIS models were lowest among both models. Moreover, the R and NS parameters of ANFIS model were higher than the SVM model. Thus, the ANFIS model enhanced the precision over the SVM model. The testing outcome also showed that ANFIS and SVM models demonstrated good forecast precision for low values of river flow but only ANFIS model was able to sustain its precision for higher values of river flow. Finally, the evaluation done in this paper demonstrates that models generated by employing ANFIS technique were better to the models generated using SVM technique in forecasting floods.

References

1. Aqil M, Kita I, Yano A, Nishiyama S (2007) A comparative study of artificial neural networks and neuro-fuzzy in continuous modeling of the daily and hourly behaviour of runoff. J Hydrol 337(1–2):22–34
2. Firat M, Güngör M (2008) Hydrological time-series modelling using an adaptive neuro-fuzzy inference system. Hydrol Process Int J 22(13):2122–2132
3. Kant A, Suman PK, Giri BK, Tiwari MK, Chatterjee C, Nayak PC, Kumar S (2013) Comparison of multi-objective evolutionary neural network, adaptive neuro-fuzzy inference system and bootstrap-based neural network for flood forecasting. Neural Comput Appl 23(1):231–246
4. Keong KC, Mustafa M, Mohammad AJ, Sulaiman MH, Abdullah NRH (2016) Artificial neural network flood prediction for sungai isap residence. In: IEEE international conference on automatic control and intelligent systems (I2CACIS). IEEE, pp. 236–241
5. Nayak P, Sudheer K, Rangan D, Ramasastri K (2005) Short-term flood forecasting with a neurofuzzy model. Water Resour Res 41(4)
6. Oh HJ, Pradhan B (2011) Application of a neuro-fuzzy model to landslide-susceptibility mapping for shallow landslides in a tropical hilly area. Comput Geosci 37(9):1264–1276
7. Okkan U, Serbes ZA, Samui P (2014) Relevance vector machines approach for long-term flow prediction. Neural Comput Appl 25(6):1393–1405
8. Pramanik N, Panda RK (2009) Application of neural network and adaptive neuro-fuzzy inference systems for river flow prediction. Hydrol Sci J 54(2):247–260

9. Shu C, Ouarda T (2008) Regional flood frequency analysis at ungauged sites using the adaptive neuro-fuzzy inference system. J Hydrol 349(1–2):31–43
10. Tehrany MS, Pradhan B, Jebur MN (2014) Flood susceptibility mapping using a novel ensemble weights-of-evidence and support vector machine models in gis. J Hydrol 512:332–343
11. Vapnik V (2013) The nature of statistical learning theory. Springer Science & Business Media, New York
12. WRIS I (2017) India-wris webgis water resource information system of india. http://www.india-wris.nrsc.gov.in

A Hybrid Feature Reduced Approach for Intrusion Detection System

Lavisha Garg, Akashdeep and Naveen Aggarwal

Abstract Increase in the number of Internet users every day leads to the global concern of information security. Intrusion detection plays an important role in protecting data from any intruders that occur in our networks or system. It detects the problem by observing and analyzing the events occurring in the system. Most of the intrusion detection techniques focus on the feature reduction or selection in order to improve the generalization ability. Feature reduction aims at finding the most compact and informative set of features by removing worthless and redundant data or information from the database. The proposed study uses conventional principal component analysis (PCA) with information gain (IG) and chi-square (CHI) for feature reduction. The reduced features are then used to perform classification using support vector classifier. The validation of the proposed approach has been done on the KDD-99 dataset and promising results have been obtained.

Keywords Feature reduction · PCA · Information gain · Chi-square · SVM

1 Introduction

Intrusion detection system (IDS) is an important tool for protecting computer networks. It is a task that classifies the network activities into normal and abnormal. Traditionally, attackers were able to bypass the first line of defense, i.e., firewalls, passwords, etc. IDS can be termed as the second line of security, which monitors and analyze the events or activities occurring in computer networks by sitting on the side of the network. IDS can be classified into two main strategies: misuse or signature-based system and anomaly-based system. In misuse detection, patterns and

L. Garg (✉) · Akashdeep · N. Aggarwal
University Institute of Engineering and Technology, Panjab University, Chandigarh, India
e-mail: lavisha.1993@gmail.com

Akashdeep
e-mail: akashdeep@pu.ac.in

N. Aggarwal
e-mail: navagg@pu.ac.in

S.-L. Peng et al. (eds.), *Computing and Network Sustainability*, Lecture Notes in Networks and Systems 75, https://doi.org/10.1007/978-981-13-7150-9_18

signatures that are previously present in the database are matched with any new activity occurring in the system. If it matches, then it's a malicious activity. The log files in the database contain the list of know attacks detected from the computer system or networks. In anomaly detection system, the behavior of the activity is identified, that deviates from the normal behavior. It can detect unknown or new attacks but cannot identify the specific type of attacks. The anomaly detection system is adaptive in nature.

Feature reduction is a process for eliminating redundant as well as irrelevant feature. The extra features increase the computation time and affect the accuracy of the classifier. It helps in selecting the small set of the features that forecast the target labels adequately. It also performs additional tasks, i.e., general data reduction, feature set reduction, better data understanding, and also improve performance. Feature reduction as a preprocessing step is essential in building IDS. An optimal feature improves detection rate and increases accuracy. In this paper, we are using three feature subset selection techniques, principal component analysis (PCA), information gain (IG) and chi-square.

PCA is applied to reduce dimensionality and select relevant features. It is concerned with the variance-covariance structure of the set of features. Information gain (IG) is the second method used to find the ranking of the features. IG calculates the entropy of the class label for the whole dataset and then subtracts the conditional entropies for each possible value of that feature [1]. Higher the entropy of the feature, more information it contains. Using IG, we can determine which feature is important for learning purpose to the classifier. The third method used for feature ranking is chi-square (CHI) technique. CHI test determines the correlation between features by calculating the independence of each feature. Correlation is higher, when independence is lower, on the contrary correlation is smaller if the independence is, higher. Important features are useful for better classification of normal and attack class.

SVM plays a necessary role in securing the networks in today's era. It is an important supervised machine learning classifier. Supervised machine learning is targeted at finding the ideal solution within a limited training record. Due to its improved and effective classifications utilization, it is often used in intrusion detection system rather than other classification methods.

The novelty of the proposal is the hybridization of these three different approaches to absorb the effectiveness of all the three. The current study combines the features from the above three approaches and feeds it to support vector machine for classification.

2 Literature Survey

Intrusion detection system suffers from the dimensionality problem. The larger dataset will increase the time complexity of training and testing of IDS. It also consumes resources and may lead to less detection capability. So to make it more accurate, feature reduction or selection came to help. It increases the detection rate and reduces training time by removing unwanted features.

Revathi and Ramesh [2] proposed a dimensionality reduction for feature reduction. They used three dimensionalities like 41 features, 14 features and 7 features which are based on best-first search. The classifiers used to differentiate between normal and attack classes are ID3 and J48. The performances of both the classifiers are compared. WEKA toolkit is used to carry out the experiments. The results show that ID3 is a better classifier. Datti and Verma [3] proposed a technique in which linear discriminant analysis (LDA) is used for feature reduction and used to select only optimal subsets. Experiments were carried out on the NSL-KDD dataset. To detect the intrusions backpropagation algorithm was used as a classifier. The method reduced the training and testing time and was able to detect new attacks. Kayacik et al. [4] did a feature reduction by using information gain as the main reduction technique. It was calculated for each feature and results in information gain for each class. The KDDCUP-99 dataset was used. There work analyzed the usefulness of each feature for a classifier. Eid et al. [5] used linear correlation method for preprocessing. The proposed method includes two layers for reduction. Optimal features were selected among 41 features of NSL-KDD dataset, i.e., 17 features. The method was compared to other feature reduction techniques.

Vasan and Surendiran [6] carried out an experiment on two standard datasets, i.e., KDDCUP and UNB ISCX. The main focus of their work was to reduce features using principal component analysis (PCA) technique. The work determined reduction ratio and ideal no of principal components (PC). Acc. To them, 10 PC is best for classification. Thaseen and Kumar [7] proposed technique they used chi-square as a feature reduction method to find an optimal feature subset without any loss. An enhanced version of KDD-99 dataset, i.e., NSL-KDD dataset is used, also tuning of RBF kernel parameters (C and ɣ) is done for SVM. Chae et al. [8] purposed three algorithms named information gain (IG), correlation-based (CFS) and gain ration (GR) for feature reduction. The NSL-KDD dataset was used. For evaluating the feature reduction method, decision tree classifier was used. Panda et al. [1] purposed the hybrid approach for intrusion detection on NSL-KDD dataset. PCA was used for feature reduction, and different classifiers are used for classification with 10-fold validation. Lakhina et al. [9] proposed a hybrid technique by combining PCA with neural network algorithm for feature reduction and classification. Training and testing were carried on NSL-KDD dataset. It reduced classification time up to some extent and helped in better classification.

Amiri et al. [10]. proposed two feature selection techniques: mutual information and linear correlation coefficient. The KDD-99 dataset was used for feature reduction and other experiments. To improve the IDS system, least square SVM was used as a classifier. Their method was helpful in detecting Probe, R2L attacks efficiently. Farnaaz et al. [11] they built IDS by using random forest classifier. Before this, they did preprocessing on the NSL-KDD dataset and select best feature subset for intrusion detection. Feature selection is classified into three techniques (i) filter, (ii) wrapper, (iii) embedded. Manzoor et al. [12] proposed two algorithms for feature reduction: information gain (IG) and correlation (CR). Three subsets of features were made according to the ranking system after that union and intersection operators were used to find optimal features. The KDD-99 dataset was used for performing experiments. To validate feature subsets, artificial neural networks (ANN) were used as a classifier. This decreases the false alarm rate and helped in increasing the detection rate.

Feature selection is the very first step for the detection. Selecting optimal feature will help in reducing detection time and remove redundant as well as irrelevant feature having no significance. In this paper, feature selection is applied to the dataset using three techniques, i.e., PCA, IG, and CHI. The classifier used in this paper is support-vector machines (SVM). The next section presents the methodology for the proposed system.

3 Proposed Method

The starting step of the proposed work is to use a suitable dataset, i.e., KDD-99 is the conventional dataset for IDS [13]. The KDD-99 dataset contains five classes, one of them is normal class and others are abnormal or attack classes termed as Dos, U2R, R2L, and Probe. The attack classes may have irrelevancy as well as redundancy. Only, 10% of the KDD-99 dataset is taken for the experiment, which contains total 494,020 instances, normal class has 97,277 records; DoS has 391,458 records; Probe has 4107 records; R2L has 1126 records, and U2R has only 52 records. Most of the dataset contains duplicate records. Data preprocessing is done by removing duplicate record manually. The output of preprocessing is the balanced dataset having no redundancies.

The second step is feature ranking, which is carried out by three algorithms namely, principal component analysis, information gain, and chi-square. Principal components (eigenvectors) determines the direction of new feature space were as eigenvalues determine the magnitude. Eigenvalues explain the variance of the data along with new feature axes. Eigenvector which has the lowest eigenvalue bears the least information about the distribution of data. PCA features are split into various subsets as PCA-i, PCA-ii, and PCA-iii. Information gain computes the entropy of each feature if the entropy is higher means it contains more information [12]. Features are then ranked according to their decreasing entropy. These features are also split into subsets as IG-i, IG-ii, and IG-iii. Chi-square finds the correlation between the features, the higher the correlation lowers the independence and vice versa. Features obtained from chi-square are divided into CH-i, CH-ii, and CH-iii. PCA-i, IG-i, and CH-i subsets are built such that first five features are in the span of 1 to 5 are the result of ranking by the corresponding algorithm. PCA-ii, IG-ii, and CH-ii contain features that were ranked in the range of 6–30 and PCA-iii, IG-iii, and CH-iii contain the rest of the features. These subsets were based on the importance of the features and are termed as highly efficient, moderately efficient, and inefficient. Highly efficient features have higher ability and information to differentiate records. Moderately efficient features have higher ability and information to differentiate records than inefficient features.

In the next step, union of PCA-i, IG-i, and CH-i feature subset is performed. Union operator is performed to make sure that all important or strongly useful features can participate in the next level for better detection and classification. Similarly, the intersection is performed among PCA-ii, IG-ii, and CH-ii feature subset and a fresh feature set is obtained. Remaining feature subset, i.e., PCA-iii, IG-iii, and CH-iii are removed as they are useless and their presence will have no difference in intrusion detection. A reduced dataset consists of 22 features is obtained by performing union and intersection operation on different feature subsets.

The third step involves the classification of the dataset into normal and attacks class and SVM has been used for this task. This step involves the training of the classifier. The output obtained from the feature reduction step is given as input to the classifier. Now the training set has 22 attributes or features. The suitable number of samples is taken for training and testing. Training set total consists of 58,625 instances. These instances are divided into five classes, DoS attacks contain 30,020 records, the Normal class contains 25,000 records, Probe contains 2738 records, R2L contains 750 instances and U2R contains 35 records. As the instances of U2R are less than other records, so re-sampling is done to increase the records of U2R. Table 1 shows the dissemination of training datasets. After training of SVM classifier, validation of the proposed method is done using testing datasets.

The test dataset contains 26,526 instances containing five different classes. Out of 26,526, DoS contains 13,657 records, 12,027 are normal instances, 252 are probe records, 473 are R2L records, and 117 are U2R instances after re-sampling is done. Table 2 shows the dissemination of the testing dataset.

Table 1 Dissemination of training dataset

Category of class	No. of instances	Percentage of class occurrence (%)
DoS	30,020	51.20
Normal	25,000	42.64
Probe	2738	4.67
R2L	750	1.27
U2R	35 (3 times)	0.19
Total	58,613	100

Table 2 Dissemination of testing dataset

Category of class	No. of instances	Percentage of class occurrence (%)
DoS	13,657	51.48
Normal	12,027	45.34
Probe	252	0.95
R2L	473	1.78
U2R	35 (3 times)	0.44
Total	26,526	100

4 Experiments and Results

Since the KDD-99 dataset contains a large number of instances, so it is not possible to compute the whole dataset. There may exist redundancy and imbalance of classes. So, first of all, a concise dataset of 41 features is formed by removing redundancies manually. In the next step, feature selection and feature ranking algorithms are applied. Information gain calculates the entropy of the class label for the whole dataset. PCA use principal components and performs linear modification hence move the original dataset to some new feature space with low dimensionality. CHI test determines the correlation between features by calculating the independence of each feature. The feature ranking from the above algorithms is prepared and the results of the ranking are saved.

After calculating the ranking of the features, three subsets of features are formed. The subsets formed are as highly efficient, moderately efficient, and inefficient. Highly efficient features are the strongest features that are highly needed for the intrusion detection. Removal of these features may lead to a decrease in the accuracy of the classifier. Moderately, efficient features are less important features than highly efficient ones but are more useful than inefficient features. These features also contribute to the accuracy of the intrusion detection system. The inefficient features have no significance in the detection system, so these features can be easily removed without hesitation.

The selection process of features is performed after the ranking of features. For that, a union operation is performed on PCA-i, IG-i, and CHI-I feature subsets. Union operation will combine all the highly effective features into one subset. This operation contains 13 features. After that intersection operation is performed for PCA-ii, IG-ii, and CHI-ii. This operation helps us in finding the next efficient feature. Intersection operation contains nine features. The remaining features are discarded as these are useless features.

An SVM classifier is used to check the validness of the proposed method. The classifier is trained with the training set and further experiments are carried out. The performance is observed by certain measures like true positive rate and false positive rate, precision, and recall. The confusion matrix is generated for the measurement of the performance. Table 3 shows the confusion matrix for the test dataset. In the matrix, rows are termed as actual class and columns are termed as a predicted class. Table 4 shows all the statistical parameters for the test dataset.

TP stands for true positive is the condition when attack class is correctly predicted as attack class. TN stands for true negative is the situation when the classifier predicts normal class when actually it is a normal class. FP stands for false positive when classifier predicts the condition, when actually it is absent, i.e., classifier incorrectly predict normal class as attack class. FN stands for false negative is the situation when the classifier incorrectly predicts attack class as a normal class. Figure 1 shows the Accuracy-Precision-Recall graph of the test dataset.

Table 3 Confusion matrix for the test dataset

Predicted class						
		DoS	Normal	Probe	R2L	U2R
True class	DoS	5264	8343	50	0	0
	Normal	148	11,832	23	24	0
	Probe	24	96	132	0	0
	R2L	0	462	1	10	0
	U2R	0	111	0	0	6

Table 4 Statistical parameters of the test dataset

	DoS	Normal	Probe	R2L	U2R
TP	5264	11,832	132	10	6
TN	12,697	5487	26,200	26,029	26,409
FP	172	9012	74	24	0
FN	8393	195	120	463	111
Precision (%)	97	57	64	29	100
Recall (%)	39	98	52	2	5
F1-score	55	52	58	4	10
Accuracy (%)	67	65	99	98	99

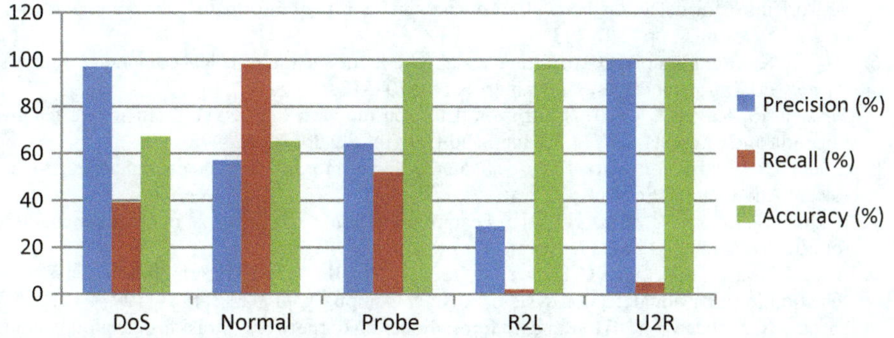

Fig. 1 Accuracy-precision-recall graph for the test dataset

5 Conclusion

This paper proposed a new feature reduction technique for intrusion detection combining stats from three different techniques. The features are selected on the basis of PCA, information gain, and chi-square algorithms. These methods are used to eliminate redundant as well as irrelevant features from the dataset and also help in reducing time complexity. The SVM classifier was designed to classify between normal and the attack classes. The classifier is trained and the proposed method shows that the reduced features influence the accuracy of the system and decrease time complexity.

For further improvement, testing can be performed on the complete KDD dataset. As the preprocessing of the dataset is the important step before classification. The features can be further reduced for selecting more optimal features. It may help in the early convergence of the system for faster and precise detection rates. SVM classifier can also be improved by using some optimization techniques. The accuracy of some classes is still less so these can be improved.

References

1. Panda M, Abraham A, Patra MR (2012) A hybrid intelligent approach for network intrusion detection. Procedia Eng. 30:1–9
2. Revathi M, Ramesh T (2011) Network intrusion detection system using reduced dimensionality. Indian J Comput Sci Eng 2(1):61–67
3. Datti R, Verma B (2010) Feature reduction for intrusion detection using linear discriminant analysis. Int J Eng Sci Technol
4. Kayacik HG, Zincir-Heywood AN, Heywood MI (2005) Selecting features for intrusion detection: a feature relevance analysis on KDD 99 intrusion detection datasets. In: Proceedings of the third annual conference on privacy, security and trust
5. Eid HF, Hassanien AE, Kim T, Banerjee S (2013) Linear correlation-based feature selection for network intrusion detection model. In: Advances in security of information and communication networks. Springer, pp. 240–248
6. Vasan KK, Surendiran B (2016) Dimensionality reduction using principal component analysis for network intrusion detection. Perspect Sci 8:510–512
7. Thaseen IS, Kumar CA (2017) Intrusion detection model using fusion of chi-square feature selection and multi class SVM. J King Saud Univ Inf Sci 29(4):462–472
8. Chae H, Jo B, Choi S-H, Park T (2013) Feature selection for intrusion detection using nsl-kdd. Recent Adv Comput Sci 184–187
9. Lakhina S, Joseph S, Verma B (2010) Feature reduction using principal component analysis for effective anomaly–based intrusion detection on NSL-KDD
10. Amiri F, Yousefi MR, Lucas C, Shakery A, Yazdani N (2011) Mutual information-based feature selection for intrusion detection systems. J Netw Comput Appl 34(4):1184–1199
11. Farnaaz N, Jabbar MA (2016) Random forest modeling for network intrusion detection system. Procedia Comput Sci 89:213–217
12. Manzoor I, Kumar N et al (2017) A feature reduced intrusion detection system using ANN classifier. Expert Syst Appl 88:249–257
13. http://kdd.ics.uci.edu/databases/kddcup99/kddcup99.html

Reservation Policy for Multi-Sharing Resources in Heterogeneous Cloud User

B. K. Dhanalakshmi, G. A. Sumalatha, K. C. Srikantaiah and K. R. Venugopal

Abstract The concept of cloud computing is focused on how to maximize the usage of multi-sharing in heterogeneous cloud user. At present, the cloud service providers are offering outstanding reservation and on-demand plans with restrictions. It restricts sharing of multiple resources with multiple users. There was no provision for user coalition and no option was provided to carry forward the unused resources. In this paper, the improvised model reservation policy for multi-resource sharing model (RPMRS) is proposed. RPMRS takes care of reservation provisioning policy (RPP) and multi-sharing of resource accountability for the end user and it can also carry forward the unused resource. The experimental results show that it is a promising model for carry forward of unused resources (CFUR) for both cloud provider and cloud ultimate user. The results emerged was of 98% efficient by way of resource provisioning with least user collision at a limited cost.

Keywords Cloud computing · Coalitions · Heterogeneous · multi-resource sharing · On-demand · Provisioning plans · Reservation

1 Introduction

Cloud computing has become increasingly in demand, It provides end users with a need-based scalable computing resources. It will be low-cost basis for the purpose of computerizing intensive jobs. The cloud service provider offers outstanding

B. K. Dhanalakshmi (✉) · G. A. Sumalatha · K. C. Srikantaiah
SJB Institute of Technology, Bangalore, Karnataka, India
e-mail: malathi.dl@gmail.com

G. A. Sumalatha
e-mail: sumaanju27@gmail.com

K. C. Srikantaiah
e-mail: srikantaiahkc@gmail.com

K. R. Venugopal
Bangalore University, Bangalore, Karnataka, India
e-mail: venugopalkr@gmail.com

S.-L. Peng et al. (eds.), *Computing and Network Sustainability*, Lecture Notes in Networks and Systems 75, https://doi.org/10.1007/978-981-13-7150-9_19

provisioning plans to cloud user for computing resources on a pay as you go service. The cloud service provider provides two types of plans to users, the first being on-demand and the second being reservation plan. In on-demand, the cost of the resources is expensive because the future use of the resources by the user is uncertain. It is known as optimization cost of resource provisioning (OCRP) model. However, OCRP model has its own downsides in long-term reservation plan and sharing of multiple resources. Over-provisioning problem occurs when resources like (CPU, memory, and bandwidth) are not used exhaustively resulting in accumulation of resources. This would be a big setback for long-term users when the goal is set for optimal computing. It restricts to a single user. The balance of resources left out cannot be carried forward to the next phase of the reservation plan.

Motivation: The Optimization Cost of Resource Provisioning (OCRP) Algorithm has its own downsides in the sharing of multiple resources. It restricts to a single user. The balance of resources left out cannot carry forward to the next phase. Over-provisioning problem occurs when the resources are not used exhaustively resulting in accumulation of resources. This would be a big setback for long-term users when the goal is set for optimal computing. To overcome all these issues, this paper proposes a model called reservation policy multi-resource sharing (RPMRS).

Contribution: RPMRS uses reservation provisioning plans (RPP) and carry forward unused resources (CFUR), a new outlook to reservation policy that meets all the specified properties and are brought up. CFUR is a boon to the heterogeneous cloud users as it facilitates to carry forward the unused resources to the next phase of reservation plan. The goal of RPP is to achieve equality and desirable properties of resource sharing and maximizing utilization of resources and minimizing cost. CFUR would be cost reduction and exchange process which goes to a very great extent to use the unused resources mutually or sharing among multiple users.

The remaining of the paper is structured as follows, related work is described in Sect. 2, background work is described in Sect. 3, Sect. 4 defines the problem definition, the entire structure of the model is described in Sect. 5, experimental results are described in Sect. 6, and Sect. 7 concludes the paper.

2 Related Works

Multi-resource sharing has been identified as a resource utilization issue on cloud computing by several researchers. Xiao et al. [1] stated resource provisioning strategies in distributed cloud systems. Zhou et al. [2] designed an F2C model for cooperative resource management system for infrastructure-as-a-service (IaaS) clouds. In addition, Wei et al. [3] highlight the downsides of a pricing mechanism in a heterogeneous cloud environment which was restricted only for single resources. Qi et al. [4] proposed a multi-resource combinatorial pricing mechanism for optimizing cost. Wang et al. [5] address the issues of heterogeneity-aware dynamic capacity provisioning scheme for cloud data centers.

3 Background

The Optimize Cost Resource Provisioning (OCRP) offers, reservation plan policy for the cloud user. They are of two types of plans, i.e., on-demand and the other is reservation plan. Here, the service provider offers resources in multiple stages: long-term and short-term resources. Long-term duration would avoid the uncertainty of future demand. In this plan, it is limited to a single user. The unused resource cannot be carried to the next phase. The multiple resources (CPU, MEMORY, BANDWIDTH) typically does not allow to share in between the multiple users. However, to overcome this problem, RPMRS is proposed. A novel method to share multiple resources with multiuser for maximizing the resource utilization and minimizing the user collision for using multiple resources and minimizing cost.

4 Problem Definition

Given a set U of n users, $U = \{U_1, U_2, \ldots\ldots U_n\}$ and set R_i of m resources $R_i = \{r_{i1}, r_{i2}, r_{i3}, \ldots, rim\}$, R_i denotes the resource type ith virtual machine in cloud service provider, i.e., CPU, RAM, and memory, these resources are reserved based on reservation policy. Let T denotes the provisioning stages $t \in T$, reserved resources are distributed based on initial share made by user S_i. The user can contribute the unused resource C_i, and he can go for on-demand resource D_i. The unused resources U_i are redistributed to the unsatisfied user. After satisfying the unsatisfied user, any resource is remaining unused, such unused resources are given back to the service provider and service provider add that portion of the resource to the next phase of reserving policy.

Our objective is to design a model called RPMRS for an efficient resource utilization with minimum cost and least collision between users. Assumption: Every machine is of same type. Each machine is allowed to run multiple jobs at a time.

5 RPMRS: System Model

5.1 *Mathematical Model*

Let group of cloud user $U = \{U_1, U_2, \ldots\ldots U_n\}$ requested for cloud provider set of resources R. Where R is set of virtual machines V_m. Each V_m configured with *CPU, MEMORY, BANDWIDTH*. The set of reservation plans has many provisioning stages T_n. $T = \{t_1, t_2 \ldots\ldots, t_n\}$, where n implicates 1:1 relation between U and T. When $t \epsilon T$ if $1 \leq \mathrm{T} \leq 12$.

Once the provision stage is decided by resource reservation plan, let vector be *S(i), C(i), D(i), U(i)*—initial share, contribution, demand, usage of resource, respectively.

The Resource will be equally distributed according to request made by each user. The resource if it is not used by user then he contribute the resources to pool. In case any user wants the resources on demand, then it checks with contribution pool. Always the demanded resources should be less then contribution pool capacity.

$$\vec{\lambda}_t = \vec{s}_n i + \vec{c}_n j + \vec{d}_n k + \vec{u}_n l \tag{1}$$

$$\lambda_1 = \sum_{i=1}^{n} c_i \leq \sum_{i=1}^{n} d_i \tag{2}$$

where c_n is a contribution of n user; in the first stage of reservation plan, share is same as of cost.

$$c_n = s_n - u_n \tag{3}$$

$$R_u = (s_n - u_n) + D_n \tag{4}$$

if $s_n > u_n$ then $\Rightarrow d_n = 0$ else $d_n > 0$

$$S_n^{(r)} = (R_u^r + S_R^r) \tag{5}$$

where r is the stage of reservation plan.

Totalvmneed = Total Share − Unused Virtual Machine Pool.

Algorithm 1 Shows the pseudo-code for collision. Where N number of user and m number of virtual machines is allocated to each user. Each user has k of T_j task. Based on the priority of task, the VM is released. If priority is higher or $task_j < task_{ij}$ then it getVm if condition is true assign VM T_{ij} else return resources to demand function is called. The execution time of T_j is calculated by $Timestamp + System.time + \delta$ where delta is time for assigning VM + current processing time. Then it assigns $t_{ij} = v_{x1}, v_{x2}, v_{x3}$. The allocated VM is locked. If it calls demand function for each VM of x_i, then v_{xy} is unlocked. The V_{xy} execution time is calculated based on (Timestamp $j <$ system.time + α) where alpha is idle time. If V_{xy} is condition is true returns to available. Algorithm 2 CFUR is explained in detail and it checks for reservation plan whether it is day or month. If it is for a day, then at end of each day, unused VM pool x is released to cloud service provider else it is for a month unused VM pool of x release to cloud service provider. For next reservation plan, when user requests cloud service provider for resources, it releases total VM by using this calculation *Totalvmneed = Total Share − Unused Virtual Machine Pool.*

Algorithm 1 The Collision algorithm

```
procedure COLLISION(U, VM, S_i)
    for all U i = 1 to n do
        for each j=1 to k
            if (Priority == HIGH) Or
                T_ij < T_ij.(TimeStamp + System·Time + δ) then
                    if (get_ vm(t_ij))
                        T_ij= {v_x1,v_x2,v_x3}
                        Lock v_x1,v_x2,v_x3
                        return resource of VM x
                    else
                        resource_ t_ij=demand_ virtual_ machine(T_ij)
                    end if
            end if
        end for
    end for
function demand_ virtual _ machine(j)
begin   For each virtual machine x
        For each type of resource y
            V_xy is unlocked
                if V_xy=(Timestamp of j < system.time+α)
                    available=V_xy
                end if
        end for
    end for
    return available
end
end procedure
```

Algorithm 2 CFUR Algorithm

```
procedure CFUR(Used resources, Unused Resources, Type of reservation plan)
    if (type of reservation plan == day)
        for each unused virtual machine pool x
            release x to cloud service provider of complete day
        end for
    else
        for each unused virtual machine pool x
            release x to cloud service provider of complete month
        end for
    end if
    Totalvmneeded = Total Share - Unused Virtual Machine Pool
    Request cloud service provider to release total VM needed for next reservation
    plan
end procedure
```

6 Experimental Results

6.1 Experimental Setup

The model RPMRS has been implemented using 3.3 version cloud simulator tool to assess the resource utilization, thereby reducing the cost. Here, the environment consists of multiusers who are making use of computing resources offered by the service provider for a cost. The setup is made according to amazon cloud in simulator where four data centers are configured with four CPU cores and 16 GB RAM, and virtual machines are created in each data center and each virtual machine is configured with six vCPUs and again VM class is created which requires a different amount of resources.

6.2 Performance Evaluation

We have chosen three different models which are given as input to the OCRP, CORP, and RPMRS. Evaluation of RPMRS is done based on cost and runtime overhead, resource utilization.

Effect of Cost: From the user point of view, the RPMRS highly recommends for resource sharing emphasis is laid upon price saving. In Table 1, detailed explanation is made as to reduction of cost for both with and without reservation plans and also on, on-demand plans by exposing the OCRP and CORP models which would benefit the user to en extent of 98%.

Runtime Overhead: The implementation of RPMRS runtime for different quantum of work can considerably reduced depending upon Reservation Provisioning Policy applied at each node and managed physically. To arrive at the results of 300 VMs on a node coordinated by RPP by decreasing window sizes. Figure 1. RPP causes reasonable CPU load on the host machine even when the window size is restricted to 10 s. To measure the performance overhead of VMs due to resource demand in RPP. It is found that the overhead is minimal in resource sharing. It is proved that RPMRS model takes less time when compared to existing model as shown in Table 2.

Table 1 Comparison of cost with reservation and without reservation and on demand

Model	Cost with reservation ($)	Cost without reservation ($)	On-demand ($)
OCRP	6.24	10.36	14.43
CORP	5.87	9.76	12.53
RPMRS	2.8	3.2	8.2

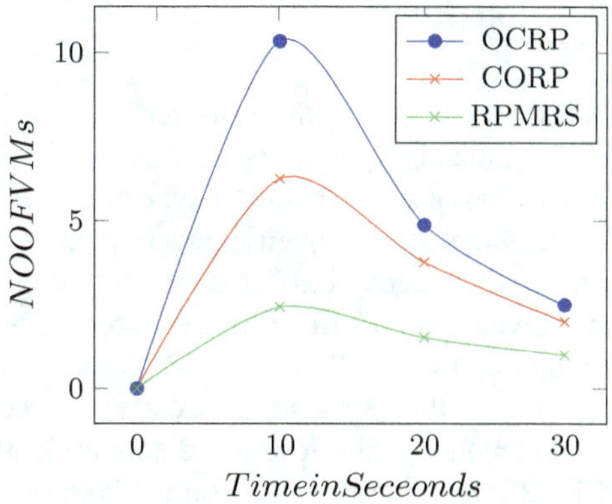

Fig. 1 Comparison of execution time with different models

Table 2 Comparison of execution time with existing model and proposed model

No. of server	OCRP	CORP	RPMRS
30	10.34	6.24	2.43
20	4.87	3.76	1.53
10	3.8	3.2	1.2

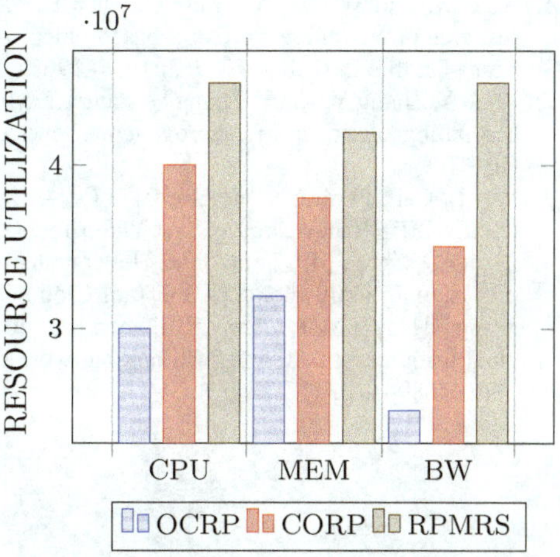

Fig. 2 Parameter comparison of OCRP, CORP, and RPMRS model

Effect of Resource Utilization: Figure 2 shows the utilization of resources, the assigned task was for running in multiple machines with different data size. X-axis represents resource types used by various models and Y-axis represents percentage or resource utilization, and the graph drawn shows the comparison results between existing model and RPMRS model.

7 Conclusion

After careful study, the resource provisioning, multi-resource sharing mechanism is formulated. This is exclusively for heterogeneous cloud user. It overcomes the downsides of single resource type of a single cloud user which poses to underutilized of resources. Consequent upon implementation of this (model) on cloud simulator on an experimental basis, the results emerged from out of this was quite encouraging. It showed that RPMRS is more viable for both cloud users and cloud providers when compared to OCRP and CORP. Cloud user shares a resource with no collision. On account of this, 98% of the cost would be reduced among multiple users. Hence, it is concluded that it is more economically viable if RPMRS including RPP and CFUR models are adopted on the heterogeneous cloud. It definitely would address the economic cost issues and least collision among multiusers while sharing multiple resources.

References

1. Xiao W, Bao W, Zhu X, Wang C, Chen L, Yang LT (2016) Dynamic request redirection and resource provisioning for cloud-based video services under heterogeneous environment. IEEE Trans Parallel Distrib Syst 27(7):1954–1967 (IEEE)
2. Zhou Y, Zhang Y, Xie Y, Zhang H, Yang LT, Min G (2014) TransCom: a virtual diskbased cloud computing platform for heterogeneous services. IEEE Trans Netw Serv Manag 11(1):46–59 (IEEE)
3. Wei L, Cai J, Foh CH, He B (2017) QoS-aware resource allocation for video transcoding in clouds. IEEE Trans Circuits Syst Video Technol 27(1):49–61 (IEEE)
4. Zhang Q, Zhani MF, Boutaba R, Hellerstein JL (2014) Dynamic heterogeneity-aware resource provisioning in the cloud. IEEE Trans Cloud Comput 2(1):14–28 (IEEE)
5. Wang Z, Hayat MM, Ghani N, Shaban KB (2017) Optimizing cloud-service performance: efficient resource provisioning via optimal workload allocation. IEEE Trans Parallel Distrib Syst 28(6):1689–1702 (IEEE)

A Graph-Based Approach for IP Network Analysis

Sirisha Velampalli and Ashutosh Saxena

Abstract With the increased proliferation of the Internet, many of our activities are dependent on its services. Hackers are creating malfunction to particular services. In this work, we present a graph-based approach that analyzes the network data for structural patterns. We collect data about various attributes on computers within a network using *NMap* (Network Mapper) tool and aggregate this information into graphs. Our approach first reports the normative patterns in the data and then discovers any anomalous patterns associated with the previously discovered patterns; i.e., we are able to detect potential vulnerabilities, outdated services, as well as unauthorized ones. Our approach finds a good place when the complete profile of the computer infrastructure is available beforehand; subsequently, we can alert the network administrator to prevent the infrastructure from exploiting potential vulnerabilities by attackers.

Keywords Knowledge · Graph-based knowledge discovery · *NMap* · Intruder · Vulnerabilities · Graph-based anomaly detection

1 Introduction

Due to the evolving nature of networking technology as well as due to the aggressive nature of attackers, it is difficult to handle known and unknown attacks by the intrusion detection systems. Day by day, the Internet users [1] are increasing, and at the same time, vulnerable hosts are also getting increased. When exploring a network for security auditing or administration, one needs to know more than the bare IP addresses of identified machines [2]. Usually, intruders perform network scans for harmful purposes [3]. The intruder scans the target network and systems to identify

S. Velampalli (✉) · A. Saxena
CR Rao Advanced Institute of Mathematics Statistics and Computer Science, University of Hyderabad Campus, Hyderabad, India
e-mail: sirisha.crraoaimscs@gmail.com; sirisha.velampalli@gmail.com

A. Saxena
e-mail: saxenaaj@gmail.com

S.-L. Peng et al. (eds.), *Computing and Network Sustainability*, Lecture Notes in Networks and Systems 75, https://doi.org/10.1007/978-981-13-7150-9_20

active hosts, operating systems, and available services and applications. He then uses this information to exploit potential vulnerabilities. In this work, we believe that if we give a total profile of systems to organization or network engineers beforehand, then we can prevent the systems from exploiting potential vulnerabilities by intruders. Through our approach, one can find normative as well as abnormal patterns of hosts in a network. We structure various properties, attributes, and services from one machine to another machine by means of assigned host address into separate graphs. We decided to use a graph-based approach as it provides a meaningful structural representation to data that can be used for analyzing or discovering interesting patterns in complex relationships. Graph-based approaches can mine patterns from diverse domains such as chemical data, communication networks, traffic networks, and protein interaction networks. In this work, we test graph-based approaches on computer network domain.

The rest of the paper is organized as follows:

We discuss related work in Sect. 2. The proposed methodology and topology are explained in Sect. 3, followed by graph-based anomaly detection (GBAD) system in Sect. 4. We discuss datasets used for our experimentation in Sect. 5, followed by results in Sect. 6.

2 Related Work

Staniford Chen et al. [4] used graph-based scan detection techniques to detect and analyze attacks in individual hosts. However, their approach takes much time for aggregation process into scale to large networks. Kato et al. [5] proposed a method similar to [4] but it attempts to use RST-ACK packet with reduced time for aggregation process. Kim et al. [6] used statistical techniques to detect network port scans. Ertoz et al. [7] developed Minnesota Intrusion Detection System (MINDS) based on anomaly scores. Their system can analyze network traffic and can also detect port scan attacks. Gates et al. [8] used logistic regression to analyze Cisco NetFlow data for port scan attacks. Jung et al. [9] used Random Walk approach to detect port scans using Oracle database that contains the assigned IP addresses. These approaches related to our work use either statistical or data mining approaches to analyze patterns. Unlike those approaches, we use graph-based approach. Graph-based approaches enable one to handle rich contextual data and provide a deeper understanding of data due to the ability to discover patterns in databases that are not using found easily using traditional query or statistical tools. Many traditional data mining algorithms such as classification, clustering, association analysis, and outlier detection have been extended into graph mining. While graph analysis can give us insights on common interaction patterns, it is also a powerful tool for identifying anomalous and often fraudulent behavioral patterns [10] that includes fake user interactions and product ratings to artificially boost popularity, e-mail spam as well as network attacks.

3 Proposed Methodology and Graph Topology

Our proposed methodology to retrieve patterns from network domain is shown in Fig. 1. Data is collected from the required hosts using *NMap* (Network Mapper) tool. *NMap* can determine various network attributes. After obtaining the data using *NMap*, we select attributes, typically host addresses, type of operating system, latency, MAC addresses, service name, state, port, and protocol of corresponding services of our interest and transform them into graphs. Graphs obtained after performing data transformation are fed as input to GBAD tool to find patterns.

There can be multiple graph topologies that one can choose, and in order to address this problem, we used the graph topology shown in Fig. 1. Each machine is differentiated from other machines in the network with host address. We structure various properties and services from one machine to another machine by means of assigned host address.

We use GBAD to discover normative as well as suspicious patterns. One instance of the input file obtained after preprocessing the data is given in Table 1. The input to GBAD is data represented as a graph.

The input to GBAD comprises of a textual representation of a graph. The input file can consist of one or more graphs. Each graph is prefaced (on a line by itself) by an

$$\text{“}XP\# < \# >\text{”} \tag{1}$$

indicating a positive example. The XP is followed by a hash symbol and a unique example ID which must start at 1 and increase by one for each example.
Vertices: Each graph is a sequence of vertices and edges. A vertex is defined as:

$$v < \# >< label > \tag{2}$$

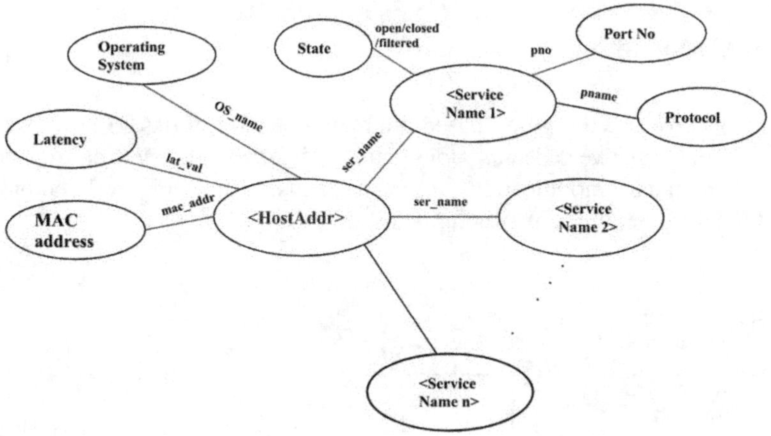

Fig. 1 Proposed graph topology

Table 1 GBAD input file: sample instance

XP # 1
v 1 "10.2.8.1"
v 2 "Linux"
e 1 2 "os_name"
v 3 "0.00094"
e 1 3 "latency_val"
v 4 "00:00:CD:2B:09:5B"
e 1 4 "mac_addr"
v 5 "Telnet"
e 1 5 "ser_name"
v 6 "open"
v 7 "23"
v 8 "TCP"
e 5 6 "state"
e 5 7 "portno"
e 5 8 "protocol"

Edges: An edge is defined as one of the following:

$$e < vertex1\# >< vertex2\# >< label > \quad (3)$$

where <vertex 1#> and <vertex 2#> are the vertex IDs for the source vertex and the target vertex, respectively, and <label> is any string or real number. All labels must be surrounded by double quotes.

4 GBAD System

In this work, we use the graph-based anomaly detection (GBAD) [10] system to discover both normative and anomalous patterns. The GBAD system can detect three structural anomalies: modifications, insertions, and deletions. Figure 2 demonstrates each of the different types of structural changes.

Fig. 2 Example graph showing different types of anomalies

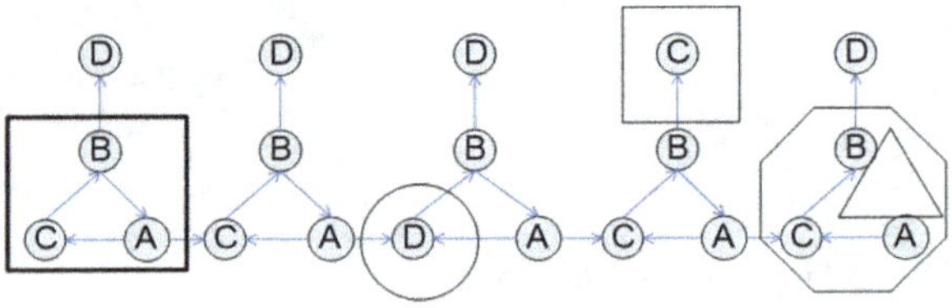

Table 2 Dataset sample

10.2.8.1,00:00:CD:2B:09:5B,Linux,0.00094,Telnet,open,23,TCP
10.2.8.1,00:00:CD:2B:09:5B,Linux,0.00094,HTTP,open,80,TCP
10.2.8.1,00:00:CD:2B:09:5B,Linux,0.00094,HTPS,open,443,TCP
10.2.8.1,00:00:CD:2B:09:5B,Linux,0.00094,Rsync,filtered,873,TCP
10.2.8.1,00:00:CD:2B:09:5B,Linux,0.00094,NFS,filtered,2049,TCP
10.2.8.101,B4:39:D6:8D:AB:00,windows7,0.0013,Telnet,open,23,TCP
10.2.8.101,B4:39:D6:8D:AB:00,windows7,0.0013,HTTP,open,80,TCP
10.2.8.102,F0:62:81:A4:8A:80,windows7,0.0011,Telnet,open,23,TCP
10.2.8.102,F0:62:81:A4:8A:80,windows7,0.0011,HTTP,open,80,TCP
10.2.8.107,F0:62:81:9F:1B:40,windows7,0.0013,HTTP,open,80,TCP

GBAD implements three algorithms GBAD-MDL, GBAD-P, and GBAD-MPS to detect these three types of anomalies. Through these algorithms, GBAD first discovers normative patterns based on the SUBDUE [11] graph-based knowledge discovery system. Each of the three algorithms uses a minimum description length (MDL) [12] heuristic to discover normative pattern.

5 Dataset

Dataset includes various parameters for each machine that include operating system running, latency, MAC address, service running on each host with its corresponding state, port number, and protocol name.

We collect data using *NMap* (Network Mapper) tool. Sample dataset is given in Table 2.

The proposed graph topology is explained in Sect. 3, and we transform the data obtained after preprocessing into a graph as input to GBAD system.

6 Experiments and Results

We collect data about various attributes on computers within a network using *NMap* tool. We then aggregate this information into graphs. Time taken to scan networks using *NMap* in CIDR notation is given in Table 3.

Hardware specifications for all our experiments are as follows:

- Processor Intel(R) Core(TM) i3-5005U CPU @2.00 GHz 2.00 GHz, 2 Core(s), 4 Logical Processor(s).
- RAM 4.00 GB.
- Operating system: xubuntu 16.04.

Table 3 Scan IP addresses using *NMap*

Addresses (CIDR)	Run-times (seconds)
/22	160.71
/24	4.95
/26	1.28
/28	0.59

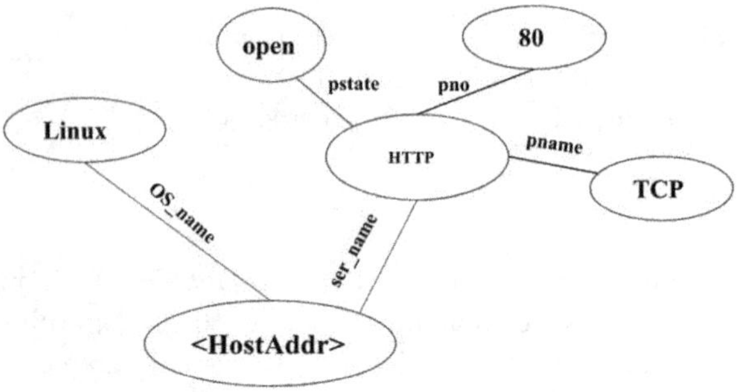

Fig. 3 Normative pattern of OS, protocol, services

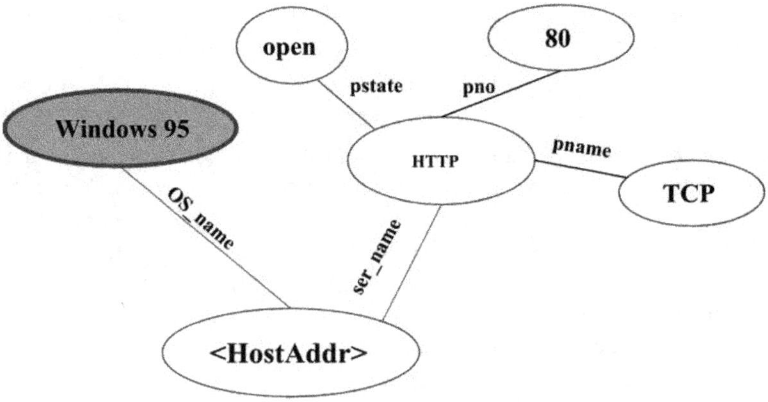

Fig. 4 Anomalous pattern-outdated operating system

Using the graph topology shown in Fig. 1 as input to GBAD, we are able to discover several interesting patterns. GBAD first discovers normative patterns. A normative pattern is one that is most repeatable or common. The top normative pattern discovered by GBAD is shown in Fig. 3. From normative patterns, we can see what are the most common services running, port names, port state, port numbers, and operating system across different machines. After discovering normative patterns, GBAD discovers potential anomalies. One potential anomalous pattern is shown in Fig. 4.

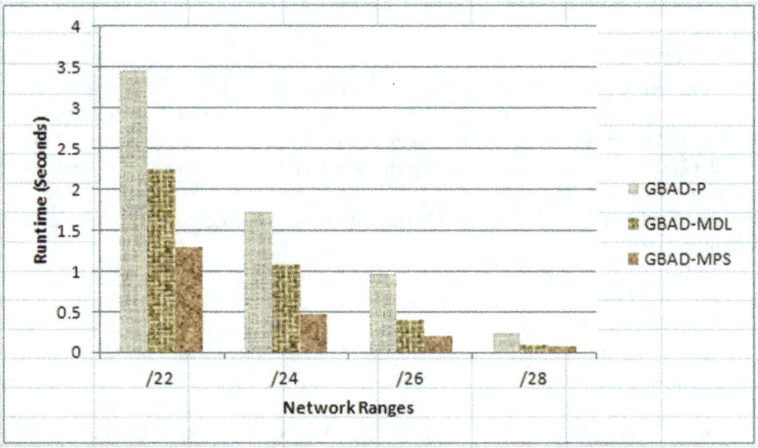

Fig. 5 Discover patterns using GBAD

Anomalous patterns are those that contain outdated operating systems, contains older versions such as SSH, apache, or sendmail or unauthorized services. Runtimes needed to discover patterns using GBAD-MDL are shown in Fig. 5, where we can observe that the run-time needed to discover patterns using GBAD-MPS is comparatively less than GBAD-MDL and GBAD-P because the dataset contains more anomalous modifications and deletions than extensions.

7 Conclusion

In this work, we present how to manifest the data that is represented as a graph and devise an approach to discover normative as well as anomalous patterns. In a typical setup of computers in a network with varied parameters, we are able to detect potential vulnerabilities, outdated services, as well as unauthorized ones. Our approach uses graph-based anomaly detection (GBAD), where we are able to detect anomalous patterns with no false positives. Using our approach, we are able to find the complete profile of the computer infrastructure beforehand, which can alert the network administrator to prevent from exploiting potential vulnerabilities by attackers. In future, we may like to collect data from larger networks, i.e., with wide ranges that contain many hosts. In order to detect vulnerabilities, we may like to explore the use of distributed computing or HPC systems to scan large data networks and detect potential vulnerabilities.

References

1. Shafiq MZ, Farooq M, Khayam SA (2008) A comparative study of fuzzy inference systems, neural networks and adaptive neuro fuzzy inference systems for portscan detection. In: Workshops on applications of evolutionary computation. Springer, Heidelberg, pp 52–61
2. Lyon GF (2009) Nmap network scanning: the official Nmap project guide to network discovery and security scanning. Insecure
3. Orebaugh A, Pinkard B (2011) Nmap in the enterprise: your guide to network scanning. Elsevier, Amsterdam
4. Staniford-Chen S, Cheung S, Crawford R, Dilger M, Frank J, Hoagland J, Levitt K, Wee C, Yip R, Zerkle D (1996) GrIDS-a graph based intrusion detection system for large networks. In: Proceedings of the 19th national information systems security conference, vol 1, pp 361–370
5. Kato N, Nitou H, Ohta K, Mansfield G, Nemoto Y (1999) A real-time intrusion detection system (IDS) for large scale networks and its evaluations. IEICE Trans Commun 82(11):1817–1825
6. Kim H, Kim S, Kouritzin MA, Sun W (2004) Detecting network portscans through anomoly detection. In: Signal processing, sensor fusion, and target recognition XIII. International Society for Optics and Photonics, vol 5429, pp 254–264
7. Ertoz L, Eilertson E, Lazarevic A, Tan P-N, Dokas P, Kumar V, Srivastava J (2003) Detection of novel network attacks using data mining. In: Proceedings of workshop on data mining for computer security
8. Gates C, McNutt JJ, Kadane JB, Kellner MI (2006) Scan detection on very large networks using logistic regression modeling. In: Proceedings of 11th IEEE symposium on computers and communications, ISCC 2006. IEEE, pp 402–408
9. Jung J, Paxson V, Berger AW, Balakrishnan H (2004) Fast portscan detection using sequential hypothesis testing. In: Proceedings of 2004 IEEE symposium on security and privacy. IEEE, pp 211–225
10. Eberle W, Holder L (2007) Anomaly detection in data represented as graphs. Intell Data Anal 11(6):663–689
11. Ketkar NS, Holder LB, Cook DJ (2005) Subdue: compression-based frequent pattern discovery in graph data. In: Proceedings of the 1st international workshop on open source data mining: frequent pattern mining implementations. ACM, pp 71–76
12. Grunwald PD (2007) The minimum description length principle. MIT Press, Cambridge, pp 1–40
13. Depren O, Topallar M, Anarim E, Ciliz MK (2005) An intelligent intrusion detection system (IDS) for anomaly and misuse detection in computer networks. Expert Syst Appl 29(4):713–722

Impact of Ambiguity: Wh-Questions Versus Other Questions in Question Paper Translation

Shweta Vikram and Sanjay K. Dwivedi

Abstract Word Sense Disambiguation (WSD) is a prominent area of research in the field of linguistics. A number of researches have been made to resolve the ambiguity issue in natural sentences. If a sentence has ambiguity or ambiguous word in it, then the meaning of this sentence may differ from context. If the meaning of the sentence is not appropriately inferred from the context, then the WSD algorithms are used to remove the ambiguity. This paper discusses the issue of ambiguity in translation of question paper through various MT tools. In our experiment, we have collected different types of questions for analyzing the impact of ambiguity for wh-questions with respect to other questions (objective, match, fill in the blank and keyword specific). Some machine translators often fail to understand different types of questions and treat them as normal question/sentence. We used five different types of questions in English to translate them using five standard online/offline translators into respective Hindi translation. Our aim is to analyze the impact of translations that arise due to ambiguities. The experiment carried using 150 questions of different types, and the result suggests that most of the translations have performed better in objective questions while the keyword specific questions (such as discuss, explain, etc.) performed poorly.

Keywords Ambiguity · Questions · BLEU score · Machine translation · English language and Hindi language

1 Introduction

This paper discusses the issue which arises wh-question and other questions translation though machine. Ambiguity in question paper translation can be a big issue as the incorrect or inappropriate translation may affect the interpretation of the

S. Vikram (✉) · S. K. Dwivedi
Babasaheb Bhimrao Ambedkar University, Lucknow 226025, India
e-mail: shwetavikram.2009@rediffmail.com

S.-L. Peng et al. (eds.), *Computing and Network Sustainability*, Lecture Notes in Networks and Systems 75, https://doi.org/10.1007/978-981-13-7150-9_21

question. In India, where a number of languages are spoken in different regions, English and Hindi are widely used as preferred languages in most of the competitive examination, questions written as English are often required to be translated in Hindi as well as other regional languages. The manual translation requires linguistics and it often causes delay in the process. Many times when the translation of original questions say, in English, is carried into Hindi and other languages as per the requirement, it is usually instructed that in case of a change in meaning or interpretation of the translated version of the questions, if any, only the original questions will be valid. The issues of multiple translations of the original questions can be effectively handled by suitable MT tools, thus by tool minimizing/eliminating the need of human expert saving time. However, an MT tool may also face problems while translating questions having lexical or structural ambiguity [1].

This paper discusses the experimental analysis of ambiguity in question paper translation using MT and mainly focused on wh-questions versus other types of questions usually asked in the examinations. We took five different types of questions including wh-questions. The other types of questions include objective, matching, fill in the blank, keyword specific. The original questions in English a translated to their respective Hindi version by MT tools.

2 Related Work

A number of researchers have been carried the area of improving machine translation accuracy [1–3]. Handling the different types of question is a bit different process that translating normal, natural sentences researches have been carried in question answering systems [1, 2, 4–8] wherein different questions have been analyzed (such as wh-questions) to understand the meaning in order to produce the correct answer [4–8]. Various MT tools such as Anusaaraka [9], Babelfish [10], Babylon [11], Bing [12], Google [13] produce a quality translation for many language pairs, still, they are not perfect in accurately translating questions every time. Specially when they have complicated structure and ambiguity involved in them.

Numerous word sense disambiguation algorithms have been used by researchers [14, 15] to address the issues in order to improve accuracy of the translators. In our previous work be explored this aspect to understand how the translators perform when the translate questions.

3 Wh-Questions Versus Other Questions

We collected 150 English language questions and broadly classified them as wh and others. Among the other questions, there are five types of questions that are wh-question, objective question, match question, fill in the blank question, and keyword specific question. Categorization of questions we will discuss one by one, firstly we will discuss wh-question [1] after that other type of question.

3.1 Wh-Question

These questions start with wh-word such as "what, why, whom, who, how, where, which," and interrogative symbol (?) comes at the end of the sentence. And most of the time these words are placed at the starting position of the question sentences. Some examples of wh-question for our experimental list are

English Sentence: Why is the study of the origin and growth of sociology important?
Reference Sentence: समाजशात्र के उद्गम और विकास का अधयन क्यों महत्वपूर्ण है?
English Sentence: How does sociology study religion?
Reference Sentence: समाजशास्त्र धर्म का अध्ययन कैसे करता है?
English Sentence: What are the strengths and weaknesses of participant observation as a method?
Reference Sentence: एक पद्धति के रूप में सहभागी प्रेक्षण की क्या-क्या खूबियाँ और कमियाँ हैं?

The reference translations shown in above questions are taken from where the questions are collected [16, 17].

3.2 Other Question

Four types of questions come under this category they are objective, match, fill in the blank, and keyword specific question. Keyword specific question again divide into two categories one is *keyword present within one sentence* (keyword specific-I) and other type of the question has *keyword present outside of the question* (keyword specific-II) as shown in Fig. 1. Some questions from each of the subcategories which have been identified in this work are listed below.

(a) *Objective Question*

English: Swadesh Darshan Scheme launched by Government of India does not include development of which of the following tourist circuit?

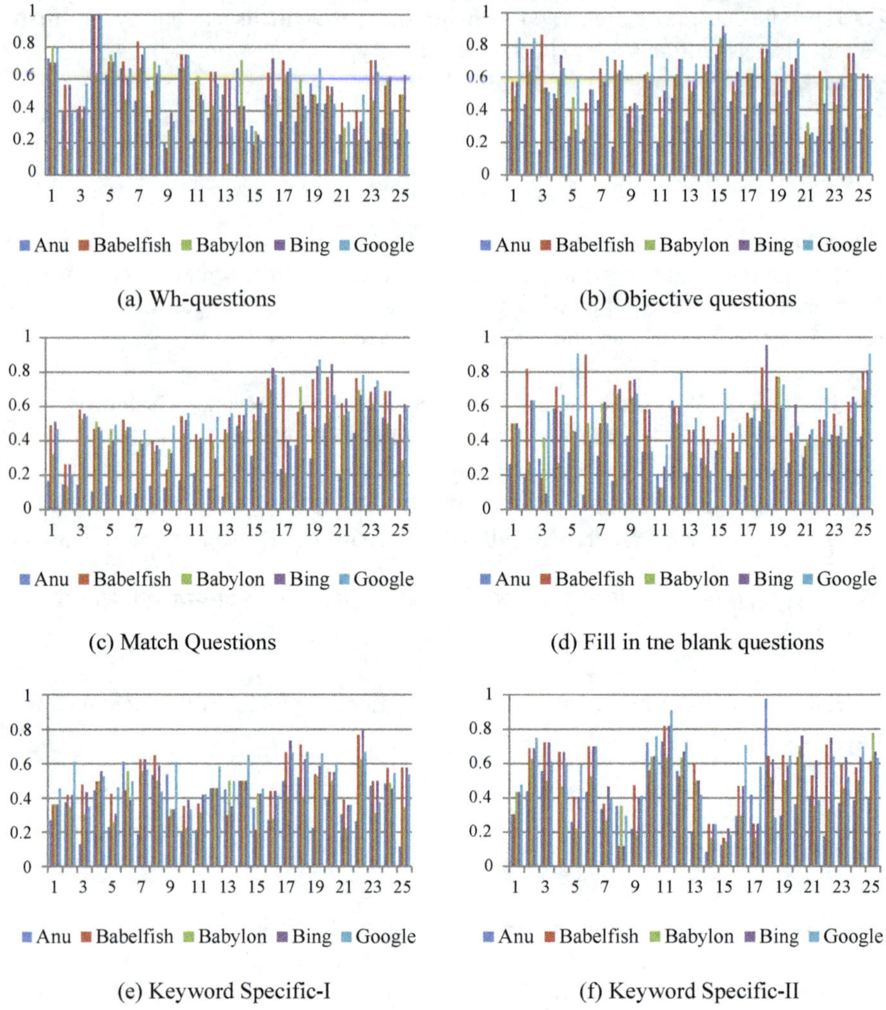

Fig. 1 Performance of Wh-question and other types question

(a) Heritage circuit (b) Sufi circuit (c) Ramayan circuit (d) Coastal circuit

Reference Sentence: भारत सरकार द्वारा प्रारंभ की गई 'स्वदेश दर्शन योजना' में निम्नलिखित में से किस पर्यटन चक्र का विकास सम्मिलित नही है?

(क) विरासत चक्र (ख) सूफी चक्र (ग) रामायण चक्र (घ) तटीय चक्र

(b) *Match Question*

English Sentence: Match list I with List II and select the correct answer from the code given below:

List I [Waterfall]	List II [River]
(a) Dudhsagar	(i) Ghatprabha
(b) Duduma	(ii) Machhkubd
(c) Gokak	(iii) Sharavati
(d) Jog	(iv) Mandavi

Reference Sentence: सूची-। को सूची-।। से सुमेलित कीजिये तथा सूचियों के नीचे दिये गए कूट से सही उत्तर चुनिए|

सूची-। (जल प्रपात)	सूची-।। (नदी)
(क) दूध सागर	(अ) घाटप्रभा
(ख) डुडुमा	(आ) मच्छ्कुं
(ग) गोकक	(इ) शरावती
(घ) जोग	(ई) माण्डवी

(c) *Fill in the Blank Question*

English Sentence: Cotton is a _________ product and cloth is a _________ product. [natural /manufactured]

Reference Sentence: कपास एक...........उत्पाद है और कपड़ा एक .... उत्पाद है? (प्राकृतिक/विनिर्मित)

(d) *Keyword Specific*

In this section two types of keyword specific question discuss, one is keyword present within one sentence and other type of the question has keyword present outside of the question. Keyword specific question has many keys such as explain, discuss, and describe [1].

(i) Keyword Present in Sentence

English: **Explain** how the Uprising of 1857 constitutes an important watershed in the evaluation of British policies toward colonial India.

Reference Sentence: यह <u>**स्पष्ट कीजिये**</u> की 1857 का विप्लव किस प्रकार औपनिवेशिक भारत के प्रति ब्रिटिश नीतियों के विकासक्रम में एक महत्वपूर्ण ऐतिहासिक मोड़ है|

(ii) Keyword Present Outside of the One Sentence

English: Early Buddhist Stupa-art, while depicting folk motifs and narratives successfully expounds Buddhist ideals. <u>**Elucidate**</u>.

Reference Sentence: प्रारम्भिक बौद्ध स्तूप- कला, लोक वर्ण्य-विषयों एवं कथानकों को चित्रित करते हुए बौद्ध आदर्शों की सफलतापुर्वक व्याख्या करती है| <u>**विशदीकरण कीजिये**</u>|

As we see all types of questions do not free from ambiguity, dictionary and transliteration problem.

4 Experiment and Result

In our experiment as already discussed, we took 150 different types of questions which are categories into five different categories. Among these each category has 25 questions, sample questions under each category along with reference translation are shown in the previous second. BLEU matrix has been used as the measurement for finding the accuracy of translations using five popular translators. Table 1 shows the mean BLEU score of different type of questions using five translators. The results show that for all categories of questions Anusaaraka MT has the poorest translation accuracy. Father, among all questions, Anusaaraka has given the poorest translations for match questions that are 0.265 whereas the highest accuracy is achieved for wh-questions. For keyword specific questions all orders translators have shown poor translations compare to other type of questions except Anusaaraka the other translators have shown the best performance for objective type questions among all questions categories. The overall performance by combining all questions as shown in Table 2 indicates that objective-questions perform better than other questions and match questions performed poorly.

Table 1 Mean value of all question types

S. No.	Translator / Question type	Anusaaraka	Babelfish	Babylon	Bing	Google
1	Wh-question	0.440	0.574	0.492	0.567	0.539
2	Objective	0.345	0.607	0.536	0.603	0.685
3	Match	0.265	0.543	0.457	0.542	0.549
4	Fill in the blank	0.311	0.574	0.458	0.537	0.587
5	Keyword specific	0.390	0.541	0.452	0.556	0.576

Table 2 Overall performance by combining all questions

Wh-questions	Objective	Match	Fill in the blank	Keyword specific
0.522	0.555	0.471	0.493	0.503

In Fig. 1, graphs show the comparison among all types of questions. Whereas wh-questions perform better than other questions (except objective-questions) for all MT (except MT Anusaaraka). As our result shows, only MT Google shows wh-questions perform poorly with respect to all questions (except keyword specific questions. Fill in the blank questions slightly better than wh-questions only for MT Babelfish).

5 Conclusion

In this paper, we have tried to analyze the translation accuracy of questions paper through different translators for more robust analysis questions were categories into all possible types as may appear in different examination. The main objective of the work to understand how empty tools behave while translating these categories of questions. Our result indicates that translator behaves differently for different types of questions which clearly indicate that the translation of question sentences is a more challenging task compared to other sentences as they required a high degree of accuracy.

References

1. Dwivedi SK, Vikram S (2017) Word sense ambiguity in question sentence translation: a review. In: International conference on information and communication technology for intelligent systems. Springer, Cham
2. Vikram S, Dwivedi SK (2018) Ambiguity in question paper translation. Int J Mod Educ Comput Sci 10(1):13
3. Hermjakob U, Li Q, Marcu D, May J, Mielke SJ, Pourdamghani N, Murray K (2018) Incident-driven machine translation and name tagging for low-resource languages. Mach Transl 32:59–89
4. Bouziane A et al (2015) Question answering systems: survey and trends. Procedia Comput Sci 73:366–375
5. Pechsiri C, Piriyakul R (2016) Developing a why-how question answering system on community web boards with a causality graph including procedural knowledge. Inf Process Agric 3:36–53
6. Zayaraz G (2015) Concept relation extraction using Naïve Bayes classifier for ontology-based question answering systems. J King Saud Univ-Comput Inf Sci 27:13–24
7. Hartawan A, Suhartono D (2015) Using vector space model in question answering system. Procedia Comput Sci 59:305–311
8. Mishra A, Jain SK (2016) A survey on question answering systems with classification. J King Saud Univ-Comput Inf Sci 28:345–361
9. https://anusaaraka.iiit.ac.in/drupal/node/2

10. https://www.babelfish.com/success
11. http://translation.babylon-software.com/english/to-french
12. https://www.bing.com/translator/help/#AboutMicrosoftTranslator
13. https://translate.google.com
14. Chaplot DS, Salakhutdinov R (2018) Knowledge-based word sense disambiguation using topic models. arXiv preprint arXiv:1801.01900
15. DurgaPrasad P, Sunitha KVN, Rani BP (2018) Context-based word sense disambiguation in Telugu using the statistical techniques. In: Proceedings of the second international conference on computational intelligence and informatics. Springer, Singapore, pp 271–280
16. http://ncert.nic.in/textbook/textbook.htm
17. http://epathshala.nic.in/e-pathshala-4/flipbook
18. Gupta D, Kumari S, Ekbal A, Bhattacharyya P (2018) MMQA: a multi-domain multi-lingual question-answering framework for English and Hindi. LREC, Miyazaki, Japan, pp 7–12
19. Hao T, Wenyin L (2008) Automatic question translation based on semantic pattern. In: Fourth international conference on semantics, knowledge and grid. SKG'08. IEEE, New York
20. Papineni K et al (2002) BLEU: a method for automatic evaluation of machine translation. In: Proceedings of the 40th annual meeting on association for computational linguistics. Association for Computational Linguistics
21. https://www.jagranjosh.com/articles/uppsc-uppcs-prelims-exam-2017-question-paper
22. https://www.jagranjosh.com/articles/uppsc-uppcs-exam-prelims-question-paper-2017
23. https://iasexamportal.com/upsc-mains/papers
24. NCERT: http://ncert.nic.in/NCERTS/textbook/textbook.htm

Impact of ICT Support on e-Governances Services

Praveen Sharma, Sanjay Gaur and Deepak Dashora

Abstract Nowadays the initiative of ICT in India is one of the huge mottos for all Indians. In this mission, e-governance assumes an imperative part. This investigation gives an extensive perspective of the e-governance and its procedure in association with ICT to give better and interfere with governances to the inhabitant. The role of ICT in the foundation of e-governance is forever concern from the services for fundamental needs. In the nation, e-governance services are additionally overseen productively despite the fact that it is one of the regressive and vast ancestral-based pieces of the nation. In this study on the observations of the 300 people from various localities accumulated to investigate the support of ICT in favor of service to the e-governance. Likewise the mapping of the official individual and clients of the governances were taken to know the real usage situation.

Keywords e-governance · Services · ICT · Information · Technology

1 Introduction

In broad spectrum, the information intended by government for the e-governance gives another sight to work with legitimate and reliable information, as it is authentic data of and about any individual of the state and additionally country.

The research concern to these environs is eternally research oriented to give the most immense approach to convey data in total, protected, and sound way. Also

P. Sharma (✉)
Mewar University, Chittorgarh, Rajasthan, India
e-mail: praveenvmou@gmail.com

S. Gaur
Jaipur Engineering College & Research Center, Jaipur, India
e-mail: sanjay.since@gmail.com

D. Dashora
JRN Rajasthan Vidyapeeth University, Udaipur, India
e-mail: deepakdashora786@gmail.com

S.-L. Peng et al. (eds.), *Computing and Network Sustainability*, Lecture Notes in Networks and Systems 75, https://doi.org/10.1007/978-981-13-7150-9_22

the discipline has to follow to operate the operations according to the principle of government.

It is at times monotonous because of not great comprehension of different controls, errands and specialties 'add to the advancement of the combined condition. It is sometimes tedious due to not good understanding of various disciplines, tasks and specialties' contribute to the development of the consolidated environment. Here in this study, we are talking about on the ICT support for e-governance services and the mapping of the perspectives of authority and clients of the services about the real usage situation.

It is well understood that information communication technology is the backbone of the electronic and digital services. So we can say that the development of e-governance is associated with the development of information technology and better management of the data storage.

2 Aim

The aim of the research is to come across the tangible state of affairs and opinion of the official and user subsequent to installation and implementation of e-governance services and also examine the support of IT-enabled infrastructure and communication technology in the real sense. Bec ause it is common factor that e-governance success factor always influence with the information technology. This was carried out by two hypothesis tests.

Hypothesis 1: Customers are satisfied with the ICT-enabled infrastructure and support services to sustain the effectiveness of e-governance citizen-centric services.
Hypothesis 2: e-governance citizen-centric services are significantly managed through the ICT-enabled services.

3 Statistics of Respondent

The inhabitants study section deals in the vivid statistical data; purposes were to summarize data, collected samples of the target population in service center. The primary data were collected through the two questionnaires separately designed: (1) for government official and tech service and (2) for citizen. The total of the government official and tech service respondents are 100 and 200 respondents and are citizen using e-governance services (Fig. 1 and Table 1).

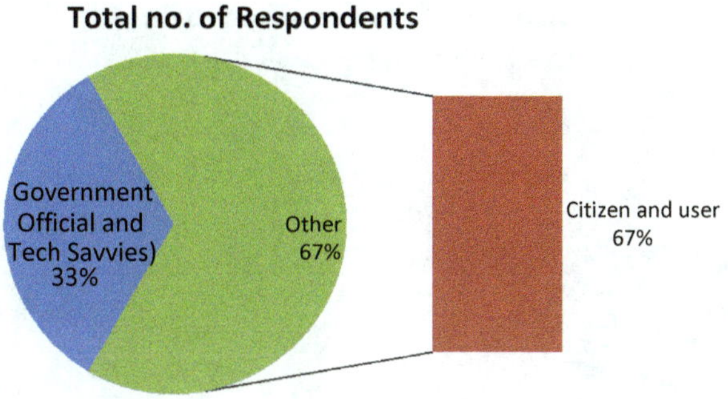

Fig. 1 Information of respondents

Table 1 Population profile of respondents

S. No.	Particular/details of population/samples	Qty	
1	Year of the survey	2018	
2	For government official and tech service (responders)	100	33%
3	Citizen and user (responders)	200	67%
4	Total responders	300	100%

4 Age Distribution

Table 2 describes the respondent's age groups with different mode. Table shows that overall 300 responders are participated in 200 are citizen and 100 are officials. Overall in first group below 18 years, 12 responders participated; between 18 and 40 group, 198 responders participated whereas 90 responders are above 40. So, maximum candidates are between 18 and 40 years group (Fig. 2).

Table 2 Age distribution

S. No.	No. of responder (age group wise)	Govt. officials	Citizen	Overall
1	Below and equal to 18 years	5	7	12
2	Above 18 and equal to 40	63	135	198
3	Above 40 years	32	58	90
	Total	100	200	300

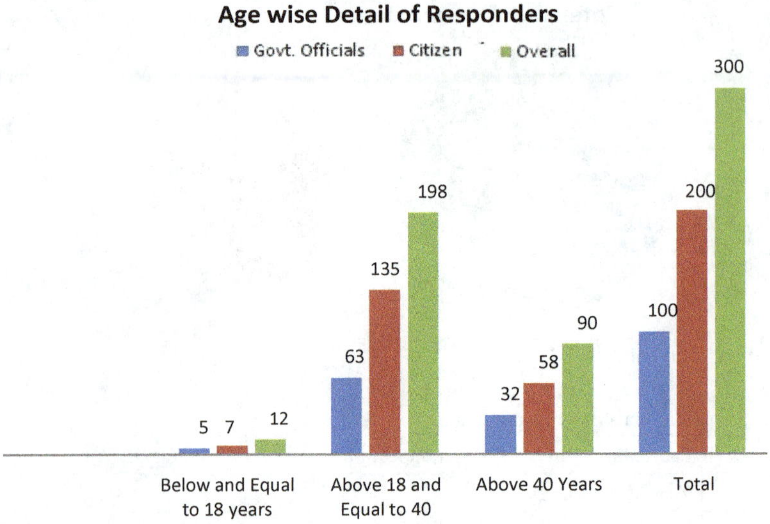

Fig. 2 Showing age distributions service providers and receiver

5 Education Level of Respondents

The level of education shows that overall maximum respondents are graduate 58; and 35.33% respondents are postgraduate and above. The score of matrix is 4.33% and under matrix is 2.33% only. But the most important is that the education of the respondents (service provider) is 52% PG and in citizens 63% responders are graduates so we can say that the responders are well educated (Table 3).

Table 3 Education level of service provider and receiver

S. No.	No. of responder (education level wise)	Govt. officials	Citizen and user	Overall
1	Postgraduate	52	54	106
2	Graduate	48	126	174
3	Matrix	00	13	13
4	Under matrix	00	7	7
	Total	100	200	300

6 Analysis of Data

6.1 Result of Hypothesis 1

Her e, in this part, the opinion of government official staffs with tech savvies associated with and the opinion of the citizen and user operating e-governance were collected and compared separately. The analysis of the opinions of both samples is given in Table 4.

Table 4 gives details of the 'F' test applied on the opinion of the government official and citizen. To test the significance, we are considering the F critical one-tail value on 0.05 level of significance which is 0.74436. The calculated value of 'F' is 0.5139 is the lesser then the one-tail critical value of 'F' at, 0.05 level is 0.74436. Therefore, hypothesis is accepted here in the analysis with test, so it can be said that the customers are satisfied with the ICT-enabled infrastructure and support services to sustain the effectiveness of e-governance citizen-centric services.

6.2 Result of Hypothesis 2

Here, in this part, the assessment of views of government official staff with technological person associated with and the opinion of the citizen including user operating e-governance were composed independently. The analysis of the opinions of both samples is given in Table 5.

Table 5 gives details of the 'F' test applied on the opinion of the government official and citizen. To test the significance, we are considering the F critical one-tail value on 0.05 level of significance which is 1.322. The calculated value of 'F' is 1.19 is the lesser then the one-tail critical value of 'F' at, 0.05 level is 1.322. Therefore, hypothesis is accepted here in the analysis with test, so it can be said

Table 4 'F' test between opinion of the government official and citizen (Hypothesis 1)

Group	N	Mean	SD	F	Result
Government official	100	1.467	0.301	0.5139	S
Citizen	200	1.850	0.421		

F critical one-tail value on 0.05 = 0.74436

Table 5 'F' test between opinion of the government official and citizen (Hypothesis 2)

Group	N	Mean	SD	F	Result
Government official	100	1.50	0.389	1.19	S
Citizen	200	1.59	0.3556		

F critical one-tail value on 0.05 = 1.322

that the e-governance citizen-centric services are significantly managed through the ICT-enabled support services.

7 Conclusion

For the Hypothesis 1, it is found that there is significant similarity among the opinion of the government official staff with technological person associated with and the opinion of the citizen including user operating e-governance in reference of the ICT-enabled infrastructure for e-governance. For the Hypothesis 2, it is found that there is significant similarity among the opinion of the government official staff with technological person associated with and the opinion of the citizen including user operating e-governance which shows that e-governance citizen-centric services are significantly managed through the ICT-enabled support services. From the result of hypothesis analysis that to run the e-governance the support of ICT plays an important role even it is fundamental need of e-governance and digital initiatives' for any nation.

References

1. Alshehri M, Drew S, Alhussain T, Alghamdi R (2012) The effects of website quality on adoption of E-government service. An Empirical Study Applying UTAUT Model Using SEM
2. Azab NA, Kamel S, Dafoulas G (2009) A suggested framework for assessing electronic government readiness in Egypt. Electron J e-Gov 7(1):11–28. Available online at www.ejeg.com
3. Barbosa AF, Pozzebon M, Diniz EH (2013) Rethinking e-Government performance assessment from a citizen perspective. Public Adm 91(3):744–762
4. Bhatnagar SC, Odedra M (eds) (2012) Social implications of information technology in developing countries. Tata McGraw Hill, New Delhi
5. Carter L, Belanger F (2014) Citizen adoption of electronic government initiatives. In: Proceedings of the 47th Hawaii (USA) international conference on system sciences on 6–9 Jan 2014

Challenges Related to Cochlear Implant

Deepti Gupta, Pratistha Mathur and Peeyush Tewari

Abstract Cochlear implant (CI) is a solution to the people who are hearing impaired. With the CI, they can imagine their world with voice. Lots of technical advancement has been done in this area, and still, there are some shortcomings with it. This paper focuses on the challenges which were faced by person equipped with CI in their daily routine life due to technological and different kinds of gap. This paper discusses the basic requirements in the form of hardware and software which are useful and which are difficult for them. In this paper, besides the technical challenges we also include certain social challenge which affects the social life of CI users.

Keywords Cochlear implant (CI) · Speech perception · Language development · Social factor · Electrode insertion

1 Introduction

Cochlear implant (CI) is now a well-known electronic device which is used in the case of sever and profound hearing loss which is also known as sensory-neural hearing loss. It can be used for infants, adults, and old-age people too. There is no age boundary of using it. It is also beneficial for any kind of deafness. CI can be used to persons who became deaf due to loud noises, due to age effect, due to infection, due to hereditary or born deaf. It can also be used with a person having residual hearing [1].

D. Gupta (✉)
Banasthai Vidyapeeth University, Vanasthali, Rajasthan, India
e-mail: 29.deepti@gmail.com

P. Mathur
Manipal University, Jaipur, Rajasthan, India
e-mail: mathurprati@yahoo.com

P. Tewari
Birla Institute of Technology, Noida, India
e-mail: peeyushtewari@gmail.com

S.-L. Peng et al. (eds.), *Computing and Network Sustainability*, Lecture Notes in Networks and Systems 75, https://doi.org/10.1007/978-981-13-7150-9_23

2 Background

There are many challenges related to cochlear implant (CI). Cochlear implant has two parts: (1) internal and (2) external. Few items are related to internal part of the implant, and others are related to outer part of the implant. Some researchers worked on the internal part like changing the number of electrodes, electrode's size and shape. Some worked on external part like changing the coding techniques. However, they were not been able to achieve as much success as required for the best performance by cochlear implant to the users [2].

After the study of many research papers, many aspects have been seen. Here, the discussion has been done on some factors such as electrode shape, size and insertion, communication, training, social interference and technology. Based on these factors, data has been collected, used, and analyzed.

3 Data Collection

The data is collected on some prime factors like electrode size, shape, and insertion problem. Normally, 25 and 30 mm length of electrodes are available in straight shape. The maximum deep insertion of electrode can be up to the 30 mm in the cochlea. There are some other factors like communication and language development problem, performance effect after training, social effect in terms of ignorance and feeling lonely. Last but not least, the point is technology effect as a friend or as a barrier. The feedbacks in the form of comparison with different choices related to all points have been taken and shown in Table 1. Total 15 users have been participated. These numbers of participants have been taken with the suggestion of audiologists and stratified sampling. We have taken samples on the basis of different age group. For correct outcomes in all aspects, we took participants who started using CI from the age of 1–21 years.

1. **The electrode size, shape problem**: There were lots of changes happened related to electrode size and shape. Initially, only four electrodes were present. At present, it has been increased to 8–22 electrodes. The shapes of electrodes were also changed from straight to wave shaped. But these changes also do not have much difference in the speech perception in the form of user's satisfaction. In spite of numerous experiments, none of them is able to give a perfect pitch perception as a normal hearing person can perceive [2].
2. **The electrode insertion problem**: In respect to insertion of electrodes, many theories are present. If we talk about practical work based on theories, they also showed that deep insertion of electrodes gives better result than normal insertion. Deep insertion gives the improved percentage of vowel and consonant recognition. It also reduces electrodes' interaction which minimizes the chances of pitch confusion by the CI users. If user electrode insertion happens at proper place and depth required by the user, then it gives tremendous results. The reason

Table 1 Feedback from the users

Choices	Q1	Q2	Q3	Q4	Q5	Q1%	Q2%	Q3%	Q4%	Q5%
Slightly agree	2	2	2	2	1	13.33	13.33	13.33	13.33	6.67
Agree	4	3	3	2	3	26.67	20.00	20.00	13.33	20.00
Strongly agree	9	6	8	9	11	60.00	40.00	53.33	60.00	73.33
Disagree	0	2	0	1	0	0.00	13.33	0.00	6.67	0.00
Slightly disagree	0	2	2	1	0	0.00	13.33	13.33	6.67	0.00
Strongly disagree	0	0	0	0	0	0.00	0.00	0.00	0.00	0.00
Total	15	15	15	15	15	100.00	99.99	99.99	100.00	100.00

behind it, when insertion happens where user have multiple contacts by which neurons can fire and give the signals to the brain in the form of electrical signal then user perform best. In another case if insertion happens at the place where particular user has less contact to fire, then results degraded drastically. Further wave-shaped electrodes used so that it reduces stiffness and do not push the more as per the requirement. It also reduces trauma and takes care of residual hearing if it is present in the user [3, 4].

3. **Communication and language development**: There are two kinds of persons who are using CI. One is prelingual, and second is post-lingual. Prelingual means who is deaf since birth. Post-lingual means who become deaf or nearby deaf due to some reasons. Basically, speech perception and communication issue find mostly with prelingual CI users. But if they get implanted before language development, then their performance is almost like persons with natural hearing. From the feedback of the persons being consulted, it has been noted that if implantation occurs at young age, then CI users can catch speech perception and speech intelligibility within two years from the date of implantation. If implantation happened at later age, then the development gets delayed to 12 months or more.

In respect to the language clarity point, feedback has been taken from CI users. From the feedbacks, it has been derived that post-lingual users and prelingual users who have received implant at young age have same clarity in language as compared to pre lingual who have received implant at later stage.

If a person is having mixed additional disabilities such as attention-deficit disorder, dyspraxia, learning disability, or cerebral palsy, in these cases also cochlear implant is beneficial. However, results are slow as compared to solving single deafness problem [5].

4. **Performance effect after training**: Consulting the speech therapist and CI users, it has been further noted that CI users need proper training after implantation. It is totally a new and different experience about sound that CI users hear through

CI after implantation. This is the reason users need a proper training to understand electrical signals as a sound. But in most of the cases, users are able to understand the words and sentences which speech therapist teaches users at the time of training because brain gets used to of these electrical signals and sound as words/sentences. But the entire scenario gets changed when users came into real world. The performance degrades drastically as a real world has a number of sounds. The pitch level of the sound varies from person to person which makes it more difficult to understand the speech since these sounds are new to brain.

To accomplish this goal in which performance did not get degrade, a proper training needs to be imparted to the CI users considering the effect of a real environment. Initially, training considering the sounds of real environment needs to be provided in order to make the brain to understand the sound's perception of the environment. Later on time to time, such training needs to be provided to the users so that users become used to of complex and different sounds of real world. As a result, CI users felt more comfortable in the interaction with real world which increases their communication skill with the people. This information can also be ascertained after collecting the feedbacks from the users [6].

The performance of the CI users varies from person to person. In some cases, CI users do very well, and in some cases, they struggle for the basic things of speech. It is very difficult to point out the reason behind it, since all the users are using the same CI and facilities. Some papers show that there are many reasons behind it, like place and depth of insertion of electrodes [4].

5. **Socialization factor**: In most of the cases, CI users felt difficult in social environment. They generally avoid social participation since they found it difficult in interacting with group of people due to low-speech perception level in such environment. They feel more comfortable when to one on one conversation as compared to conservation in group. In a study, it has been found that children using CI had more psychosocial difficulties as compared to normal hearing children. CI users who are good in communication have less psychosocial difficulties as compared to who has a problem in communication. It is generally found that CI users feel lonely in social gathering. From the study, it is clear that at a particular age, all children felt loneliness either they are normal hearing children or children using cochlear implant. From the study, it was also proved that delay in the implantation increases loneliness in CI users. If implantation happens at young age, then CI user's social and communication skill is almost similar to normal hearing. While some researchers said that in respect of quality of life (QOL), all the things are same in normal hearing and cochlear implant users who have implants at early stage of life. However, the user who has implant at later stage, the QOL is comparatively less since they develop their speech and communication skill at slow pace as compared to the users who have implant at early stage. From the feedback of the CI users, it has also been noted that the persons who have implant at early age of life developed the communication skill at much rapid pace as compared to the persons who have implant at later stage.

As a result, early-stage CI users find comparatively easy to learn languages and interaction with the people.

The only difference is that older-age CI user performance is less due to pressure. The pressure may be language learning because at the young age, CI user also gets the same time as a normal hearing child takes to learn language. But if user takes CI at later age, then expectation becomes high to learn language and do all other activities fast like school and social life.

6. **Ignorance from social side**: From the interview of the parents of the CI users, a number of facts were pointed out which belongs to the socialism. Parents had informed that in the school, their child, who is using CI, has no friends to play and chat. The parents of normal hearing children did not invite child who is using CI at their home to play or to sleepover usually. The CI children along with their parents need support from the society. If the people of society are supportive, then CI children show good result with respect to the development of speech perception and communication skills. However, if people neglect them, then children keep themselves away from the society. As a result, their social interaction skill was not developed. In case of group discussion, sometimes repetition is required for CI users. Hence, CI users need support from the society and friend in such situation [7–9].
7. **Technology as a barrier or friend**: In terms of technology, telephone, mobile, Internet, and email are major things nowadays. Both have positive and negative aspects on CI users. Let us talk about mobile; it is very difficult to do conversation on mobile via call for CI users, especially if the other person is new for the CI user. Hence, CI user uses to avoid taking calls. CI users found preferably use text/email as a medium of conversation. However, the only drawback is that the CI user has to wait for the response from other person which can lead to delay in the conversation. As technology advancement, speech-to-text function is also available in most of the mobiles which provide a great relief to the hearing impaired persons in difficult situations.

Internet telephony allows various options for CI users like video calling and audio–video communication. With the help of Internet, there is no boundary in managing long-distance relationship for CI users, and they can make friends via Internet like normal hearing persons. Sound quality and broader frequency bandwidth increase speech perception for CI users. In video calling, video quality is very good now. So, if they feel any difficulty to understand speech, then user can do lip reading too and understand the speech easily. Many things are available via which user can do video calling at very low cost like WhatsApp, Skype. From the feedback of the CI users, it was found that although the above technology does not provide result like normal hearing, however still it provides great help for the CI users in communication [10–12].

The table presents data based on five points. These points are following:

Q1. Communication and language development based on prelingual and postlingual implant.

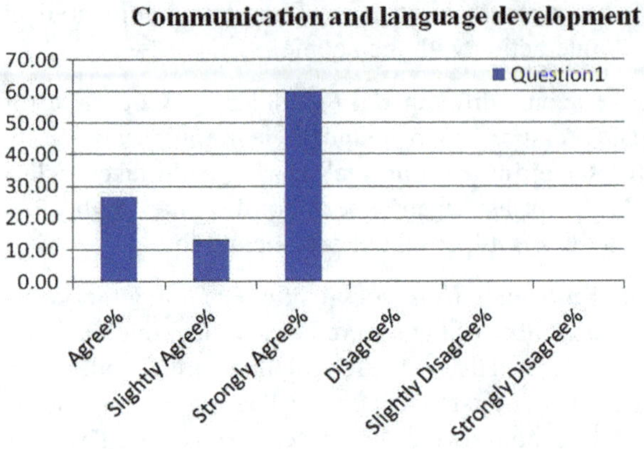

Graph 1 Communication and language development

Q2. Is performance effect different after training with real environment instead of sound-proof room?
Q3. User feels lonely in social gathering.
Q4. Ignorance present from social side for CI users.
Q5. Technology works as a friend.

To answer all questions, different kinds of choices are given to the users. Table 1 shows the counting of choices of user in the form of numbers and percentage.

4 Analysis

On the basis of prime factors, data has been collected and shown in the form of Table 1.

The details analysis can be seen in the form of a graph related to each question.

In Graph 1, data related to communication and language development has been taken in the form of percentage. From Graph 1, it is clear that 60% numbers of users are in strong favor.

The Performance effect after training with real environment instead of sound proof room how to get changed in this regard data has been shown in Graph 2. From Graph 2, it is clear that 40% numbers of users are in strong favor and 20% are in favor. So, we can say most of the users are in favor of training with real environment.

Users feel lonely in social gathering; this statement is strongly accepted by users. Only 12% are not slightly in favor while from Graph 3, it is clear that 70% numbers of users are in strong favor.

The ignorance from the social side affects a lot in the user's life. From the user's feedback, 60% numbers of users are in strong favor while rest are also slightly agreed that can be seen in Graph 4.

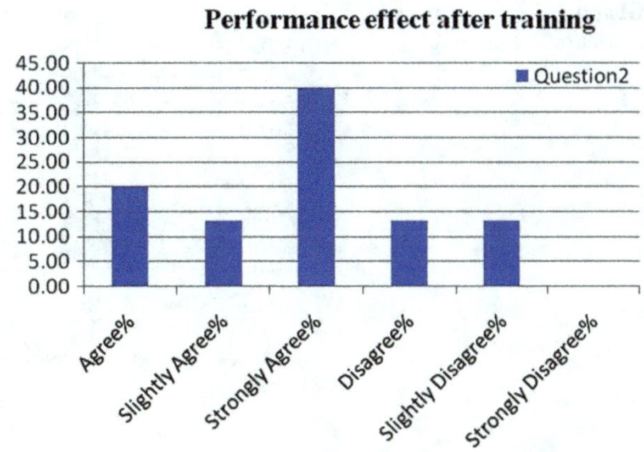

Graph 2 Performance effect after training

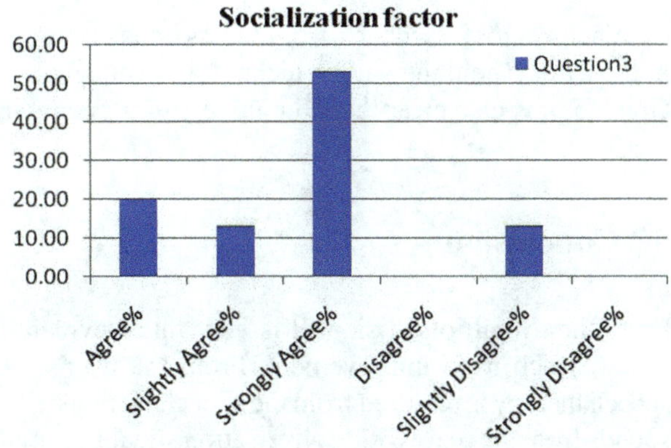

Graph 3 Socialization factor

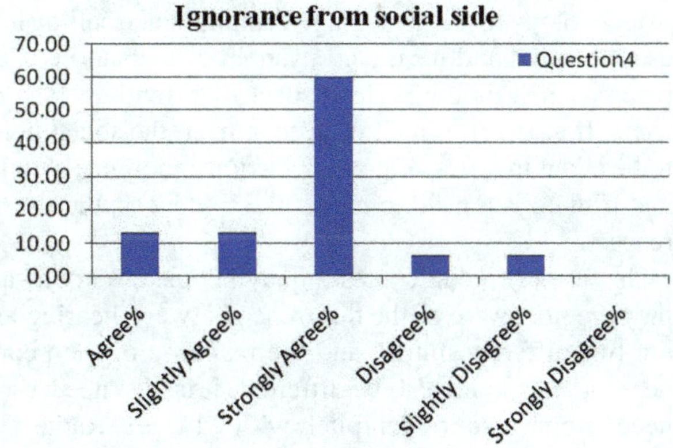

Graph 4 Ignorance from social side

Graph 5 Technology as a barrier or friend

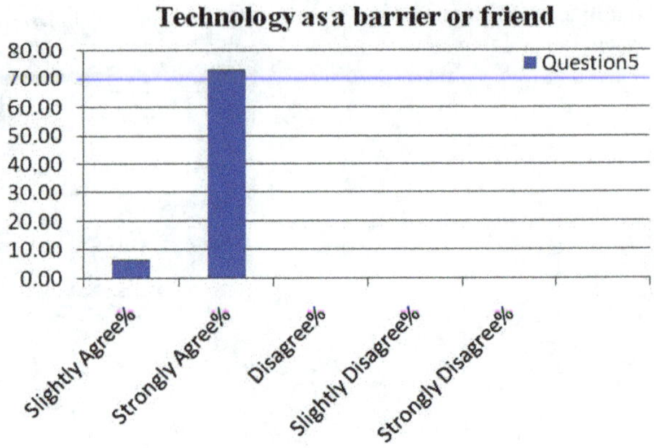

When we took feedback from the users is technology work as a barrier or friend in their life. Then, they agree technology works as a helping hand as friends. From Graph 5, it is also clear that all numbers of users are in strong favor of technology.

5 Conclusion

From the overall discussion, it is very much evident that there are some technical issues which need improvement. From the user's feedbacks, we find that proper speech therapy is required from the speech therapist. During speech therapy, different sounds from the real-world conversation should be used. As a result, CI users found it easy to have conversation in real-time scenario. On the other hand, some advancement in technology like text messages/emails/video calling is also available which provides as a helping hand for the CI users in conversation. We people need to take a social responsibility initiative in respect of providing favorable environment for the CI users. If a person is a CI user, then from the social point of view special care needs to be taken in terms of good behavior. Ignorance should be avoided in the behavior, and if there is a need to repeat the sentences, then it should be done with love and respect.

In case of Indian contest, many people are not aware of CI. As a matter of fact, they are not aware of the difference between hearing aid and CI. CI and hearing aid are two different things, and the response of the people for persons using CI and hearing aid should also be different. It is also needed to be noted that people do not need to converse in high pitch with CI users. Rather, they need to talk a bit slowly to enable CI user to understand conversations more easily. If the CI users can be provided proper behavioral treatment, then they can also lead a good social life just like normal hearing persons.

Acknowledgements There are two audiologists under this committee under whom guidance the data have been collected by the participants, namely a. Shree Apurva and b. Shree Srinivas. Total 15 participants are present in the data collection-based different aspects. I acknowledge the support of all in pursuing my research.

References

1. Wolfe J, Schafer E (2015) Programming cochlear implants, 2nd edn. Plural Publishing Inc, San Diego, United States
2. Zeng F-G (2017) Challenges in improving cochlear implant performance and accessibility. IEEE Trans Biomed Eng 64(8)
3. Gani M, Valentini G, Sigrist A, Kós M-I, Boëx C (2007) Implications of deep electrode insertion on cochlear implant fitting. J Assoc Res Otolaryngol, JARO 8:69–83. https://doi.org/10.1007/s10162-006-0065-4
4. Hochmair, I, Hochmair E, Nopp P, Waller M, Jolly C (2014) Deep electrode insertion and sound coding in cochlear implants. Elsevier B.V. http://dx.doi.org/10.1016/j.heares.2014.10.006
5. Edwards LC (2007) Children with cochlear implants and complex needs: a review of outcome research and psychological practice. Published by Oxford University Press, Advance Access publication on 9 May 2007. https://doi.org/10.1093/deafed/enm007
6. Pisoni DB, Kronenberger WG, Harris MS, Moberly AC (2017) Three challenges for future research on cochlear implants. Chinese Medical Association. Production and hosting by Elsevier B.V. on behalf of KeAi Communications Co., http://creativecommons.org/licenses/by-nc-nd/4.0/
7. Punch R, Hyde M (2011) Social participation of children and adolescents with cochlear implants: a qualitative analysis of parent, teacher, and child interviews. J Deaf Stud Deaf Edu. Advance Access publication on 3 Mar 2011, https://academic.oup.com/jdsde/article-abstract/16/4/474/548933
8. Huttunen K, Välimaa T (2010) Parents' views on changes in their child's communication and linguistic and socioemotional development after cochlear implantation. J Deaf Stud Deaf Edu. Advance Access publication on 2 July 2 2010, https://academic.oup.com/jdsde/article-abstract/15/4/383/335446
9. Hyde M, Punch R, Komesaroff L (2010) Coming to a decision about cochlear implantation: parents making choices for their deaf children. Oxford University, 5 Feb 2010, https://academic.oup.com/jdsde/article-abstract/15/2/162/551818
10. Mantokoudis G, Koller R, Guignard J, Caversaccio M, Kompis M, Senn P (2017) Influence of telecommunication modality, internet transmission quality, and accessories on speech perception in cochlear implant users. J Med Internet Res 19(4); 2017 Apr, PMC5422655. https://doi.org/10.2196/jmir.6954
11. Accessible and User-Friendly Mobile Technology Design White Paper, from http://www.amovil.es/sites/default/files/e-5_1_libro_blanco_ingles.pdf
12. Mantokoudis G, Dähler C, Dubach P, Kompis M, Caversaccio MD, Senn P (2013) Internet video telephony allows speech reading by deaf individuals and improves speech perception by cochlear implant users. PLOS ONE8(1):e54770

An Efficient Hindi Text Classification Model Using SVM

Shalini Puri and Satya Prakash Singh

Abstract In today's world, several digitized Hindi text documents are generated daily at the Government sites, news portals, and public and private sectors, which are required to be classified effectively into various mutually exclusive pre-defined categories. As such, many Hindi text-based processing systems exist in application domains of information retrieval, machine translation, text summarization, simplification, keyword extraction, and other related parsing and linguistic perspectives, but still, there is a wide scope to classify the extracted text of Hindi documents into pre-defined categories using a classifier. In this paper, a Hindi Text Classification model is proposed, which accepts a set of known Hindi documents, preprocesses them at document, sentence and word levels, extracts features, and trains SVM classifier, which further classifies a set of Hindi unknown documents. Such text classification becomes challenging in Hindi due to its large set of available conjuncts and letter combinations, its sentence structure, and multisense words. The experiments have been performed on a set of four Hindi documents of two categories, which have been classified by SVM with 100% accuracy.

Keywords Hindi documents · Text classification · Natural language processing · Feature extraction · SVM

1 Introduction

With the wide availability of the enormous huge-sized text documents and the need to make similar type of document clusters, many efficient text document classification systems using supervised learning paradigm and Natural language processing (NLP) features along with different classifiers have been proposed till date; in which most of the research, contributions are related to the English text classification systems. Such

S. Puri (✉) · S. P. Singh
Birla Institute of Technology, Mesra, Ranchi, Jharkhand, India
e-mail: eng.shalinipuri30@gmail.com

S. P. Singh
e-mail: sp.singh@bitmesra.ac.in

S.-L. Peng et al. (eds.), *Computing and Network Sustainability*, Lecture Notes in Networks and Systems 75, https://doi.org/10.1007/978-981-13-7150-9_24

systems had to face many challenges related to ambiguity removal, document misclassification, loss of important information, curse of dimensionality, dimensionality reduction, etc. Looking toward the linguistic-based perspectives, nowadays, many researchers have been focusing on mono/bi/tri/multi-lingual text document processing and their applicability in real time and distributed environments. In the concern of Indic monolingual systems, the Devanagari scripted Hindi language consists of 33 consonants, 13 vowels, and 10 numerals. It includes a well-defined structured set of grammatical rules and vocabulary and has a large set of conjunctions along with their varying combinations, which make it unique and different from others, while requiring a very high level of efforts and critical processing.

This paper begins with the discussion of Hindi standard encoding schemes and comparison among many existing Hindi-based techniques and applications, which have been developed on diverse factors of identification, detection, extraction, correction, and evaluation of language-based and structure-based processing phenomena. Many researchers have contributed in Hindi-based processing systems [1], in which major applications have been concerned to the language identification and separation in mono/bi/tri/multilingual documents, keyword and key-phrase extraction, such as noun and important multiword names, and idioms identification, named entity recognition, word sense disambiguation, identification and correction of word spellings, and sentence parsing and linguistic-based techniques. Additionally, some others are related to the text summarization, simplification, machine translation, cross-lingual methods, dimensionality reduction, and other concerned methods. Further, the design and algorithm of the efficient Hindi Text Classification using Support Vector Machine (HTC-SVM) are explored, which first accepts Hindi text documents, preprocesses them, extracts their features, trains SVM, and then SVM classifies unknown Hindi text documents.

This paper is organized as follows. Section 2 discusses background of Hindi encoding schemes and an overview and comparison on Hindi-based existing techniques. Section 3 explains proposed HTC-SVM with its design and algorithmic details. The experimental results are shown in Sect. 4. Section 5 includes the conclusion and future extensions.

2 Background

Over two decades, various applications and methods on automated Indic mono/multilingual information retrieval, analysis and processing systems have been implemented successfully with the use of Hindi language essentially. In this, Devanagari-based languages are internally represented either using ISCII or its extension Unicode encoding methods. In this section, the basics of both the encoding standards along with their specific features and complexity issues are discussed. Further, many research contributions toward Hindi text analysis and processing are explored, which includes its automatic NLP techniques in areas of keyword extraction, spell checking, and language dependent/independent-based problem-solving methods.

2.1 Hindi Encoding Standards

The Indian languages internal representation is illustrated with the code assignment of language aksharas, i.e., consonants, vowels, and vowel modifiers. Proposed in the 1980s and revised in the early 1990s, the very first-syllable leveled eight-bit encoding ISCII (Indian Script Code for Information Interchange) is used for its writing system, which contains lower 128 and upper 128 codes for plain ASCII and ISCII-specific, respectively. It is a well-suited robust method for representing Indian language syllables, which includes nukta character, uses halant character once or twice in a conjunct along with soft and invisible halants, follows consonant–vowel combination approach and can work with Romanized Hindi at the cost of multiple varying length representations. On the contrary side, it is a very challenging and complex system, which includes critical parsing of ISCII coded text string, with no preservation of aksharas true sorting order, difficult to check presence of arbitrary syllables, and no error-free transliteration across languages. ISCII has been superseded by Unicode standard as Devanagari (U+0900 to U+097F), Devanagari extended (U+A8E0 to U+A8FF), and Vedic extensions (U+1CD0 to U+1CFF), while preserving the basic ISCII structure. Both standards provide syllables multi-byte representations. Unicode is considered as a stack of planes which is created to effectively handle almost all language alphabets including consonants, vowels, and vowel modifiers, used for thousands of different characters, and used as a standard for multi-lingual documents. On the other side, it has the complexity of rendering syllables, text processing limitations rather than display, and others limitations same as of ISCII.

2.2 An Overview on Existing Hindi Text Processing Techniques

This section elaborates various existing Hindi plain text document processing techniques and methodologies, which were designed to provide many operations, such as language identification and discrimination, important words, keywords, and key phrase extraction, incorrect spelling detection and correction, word sense disambiguation, syntax and semantic analysis, named entity recognition and several other concerned applications. These techniques are discussed and compared with each other as mentioned below.

- **Language Identification and Discrimination**—Language detection and identification are usually performed in mono/multilingual text documents, where they are discriminated and separated from each other. A Fuzzy-SVM-based Romanized plaintexts identification of Indic languages, say, Hindi, Bangla, Manipuri, Urdu and Kashmiri, was proposed in [2], whereas [3] used N-gram-based algorithm to distinguish Hindi and Sanskrit texts, [4] used character N-gram pruning with dictionary to identify Hindi and English code-mixing in social media texts, and a domain in-dependent Hindi language interface to relational database was proposed in [5].

- **Named Entity Recognition (NER)**—NER is used to identify named entities in the document, such as persons, institute names, etc. [6], and is a challenging task in Indic language processing. Hindi NER was implemented in [7] by combining global distributional characteristics and local context information and was presented in [8] by using association rules. [9] used multi-objective optimization to perform feature selection and parameter optimization for NER in Hindi, Bengali, Telugu, and English through four different classifiers, [10] proposed active learning-based NER of Hindi, Bengali, and English through SVM and a combination of SVM-conditional random field, and [11] proposed English NEs mapping with their equivalent representations from different languages recognized independently using Wikipedia.
- **Keyword, Key-phrase, and Stop word Extraction**—The automatic keyword and key phrase extraction from Hindi document were proposed using unsupervised, domain, and corpus independent approach [12]. Sharan and Siddiqi [13] used a supervised approach to distinguish between Hindi keywords and stop-words by using probability distribution functions, whereas [14] evaluated the effect of context window size, stemming, and stop word removal on Hindi word (noun) sense disambiguation.
- **Expressions, Predicate, and Multiword Mining**—The automatic Hindi multiword expressions mining using linguistic knowledge was performed in [15], whereas name mining [16] and complex predicate (meaningful multi-word expression) mining [17] were proposed using parallel Hindi–English corpus. Hindi idioms identification system using rule base was presented in [18] and [19] proposed Hindi temporal expression recognition and classification through rule-based, Conditional Random Field (CRF)-based, SVM-based, and the combination of all three classifiers.
- **Text Parsing**—A parser algorithm was developed for Hindi–English Machine Translation (MT) to resolve grammatical structure issues of root form of the word, category, masculine/feminine/neuter, oblique, direct case, and suffix [20], whereas in [21], self- and co-training were elaborated for Hindi dependency parsing with partial parses.
- **Word Sense Disambiguation (WSD)**—WSD is used to identify the intended correct sense of words in the given context [22]. Sinha [23] proposed Hindi ambiguous proper names recognition using hybridization of rule base and statistical CRF with morphological and context features, whereas [24] disambiguated all open class words (noun, verb, adjective, adverb) and all words present in Hindi sentence simultaneously using graph connectivity measures and Hindi WordNet. The improved Hindi word sense disambiguation methods were proposed by using Lesk algorithm along with dynamic context window [25] and by using graph-based model and Hindi WordNet [26]. Another contribution disambiguated polysemous word meaning by finding context correlation and collocation vector derived from Hindi WordNet [27]. The unsupervised Hindi word sense disambiguation meth-

ods using graph connectivity [28] and using network agglomeration were also proposed [29].

- **Spell Checker and Corrector**—Spelling standardization of varying Hindi spellings using rule-based approach was proposed in [30] to develop Hindi–Punjabi MT, whereas [31] detected and corrected the non-word spelling errors in Hindi.
- **Other Applications-based Methods**—The children story classification in Hindi using keywords and part of speech (POS) density was performed through Naive Bayes (NB), K-Nearest Neighbor (KNN) and SVM [32], and in Hindi and Telugu using the story structure [33]. Several Hindi-based MT systems have been proposed, in which [34] performed human-based and automatic evaluation on some English–Hindi MT systems, whereas [35] improved statistical English–Hindi MT through co-joining parts of verbal constructs. In [36], Hindi text summarization model was proposed using sentence extraction method and optimized it through Genetic Algorithms. [37] compared classifier and baseline approaches for English–Hindi text simplification by human and automatic BLEU, F-measure and Meteor measures, and Hindi text dimensionality reduction using latent semantic analysis was evaluated in [38], whereas [39] provided a corpus-based study on annotating indirect anaphora for Hindi.

An analytical comparison chart of these existing algorithms is shown in Fig. 1, which depicts the % contribution and usage in generating Hindi-based processing systems. It is clearly seen that 21.05% systems have been designed for word sense disambiguation and for other applications separately, whereas keyword extractors, text parsers, and spell checkers have been contributed below than 10% usage. It is also observed that approx. 8% systems used SVM as their classification technique.

Fig. 1 Comparing % contribution and usage of existing Hindi-based processing techniques

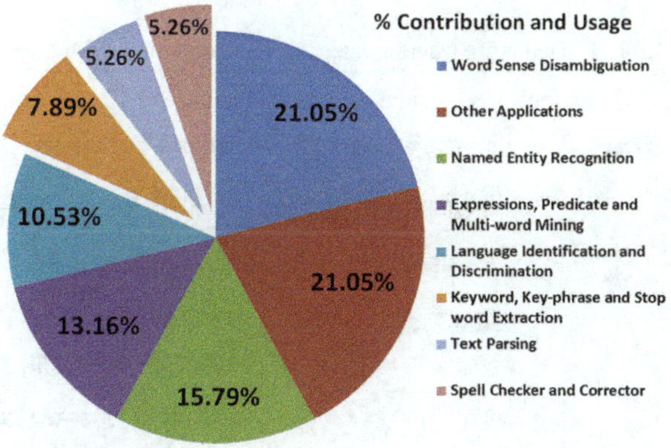

3 Algorithmic Details of HTC–SVM

The Hindi Text Classification using SVM (HTC–SVM) is designed into two phases, Training Processor and Testing Processor, where Training Processor is used to processing the Hindi text document *(HTD$_i$),* where $1 \leq i \leq n$ and *n* represents a number of documents, and converts it into its small parts by using NLP concepts at the sentence, document and integrated corpora levels [40]. Hindi sentences take subject–object–verb (SOV) form as shown in Fig. 2. When the sentences and words are extracted from *HTD$_i$*, they are processed to remove the noise. So, it searches and stores the desired, important, and non-redundant words by discarding stop symbols, invalid and extra words, and then performs word stemming as depicted in Table 1. To discard unimportant words, Feature Frequency Calculator (FFC) is used to calculate the total frequency of each different relevant word occurred in the document, and therefore, all the redundant entries of each word and low-frequency words are removed. At the integrated corpora level, all the resultant words of all the documents are stored, which again keeps only one entry of every different word and updates their frequencies accordingly and thereby making it low dimensional.

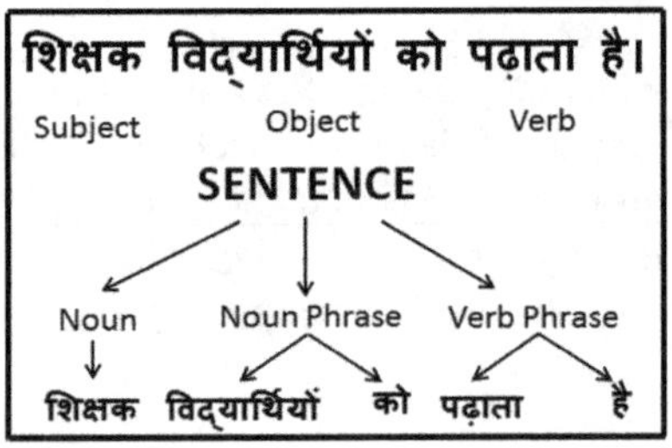

Fig. 2 SOV form and grammar-based representation of a Hindi sentence

Table 1 Depicting a set of category wise keywords with their variations

Category	Keyword	Variations
शिक्षा	शिक्षक/विद्यार्थी	शिक्षकों/विद्यार्थियों
	किसान/कृषक/सिंचाई	किसानों, किसानी/कृषकों/सिंचाईयाँ, सिंचाईय़ों
कृषि	खेत/फसल	खेती, खेतों, खेतिहर, खेतिहरों/फसलें, फसलों, फ़सली
	उत्पाद	उत्पादी, उत्पादों, उत्पादन, उत्पादनों, उत्पादक, उत्पादकी, उत्पादकों, उत्पादकता, उत्पादकताओं
	खेल	खेलें, खेलों, खेलना,खेलनी, खेलने, खेलकर्मी, खेलकर्मियों
खेल	खिलाड़ी/कप्तान	खिलाड़ियों/कप्तानी, कप्तानों
	टीम/मैदान	टीमें, टीमों/मैदानी, मैदानों

At the feature extraction stage, the membership degree of each feature (word) is performed and an analysis is made between every feature of a text document and class. The Pseudo Thesaurus is used which is a database of large set of Hindi words, and is used to check the validity of words, and helps to remove extra words and to perform the word stemming in HTD_i. A class feeder is used which is the set of pre-defined classes and is defined before SVM classification [40]. SVM uses the concept of hyperplanes to identify the suitable category with +1 and −1.

Algorithm Hindi_Text_Classification ( )
STEP 1: Take a set of n known Hindi documents.
STEP 2: Pre-process all n documents.

a. *Extract sentences on the basis of । (purna-viram), . , ?, ! .*
b. *Remove all the stop symbols, dot, semicolon, purna-viram etc.*
c. *Remove all stop words, say, का, है, ने, पर, हम, नहीं, तो, कि, यह, से, में, और etc.*
d. *Perform word stemming.*

STEP 3: For each document i //where $1 \leq i \leq n$
STEP 4: For each sentence j of document i //where $1 \leq j \leq i$
STEP 5: For each word k of sentence j //where $1 \leq k \leq j$
STEP 6: Count the frequency of important words, discard the unimportant ones by using Pseudo Thesaurus and maintain the integrated corpora.
STEP 7: End for. End for. End for.
STEP 9: Extract the features.
STEP 9: Use class feeder and check the matching possibility of each word with the pre-defined classes.
STEP 10: Train the SVM classifier using svmclassify() function and test on unknown documents.

Table 1 depicts various variations of obtained keywords, which are related to the categories 'कृषि', 'शिक्षा', and 'खेल'. Here, stemming is used to extract similar or nearly similar original keyword, i.e., the root node.

4 Experimental Results

The proposed HTC–SVM system is implemented on a small set of four Hindi documents of two classes 'कृषि' and 'खेल', which have been collected from various Hindi blogs and Government sites. These documents contain simple Hindi statements. The implementation is done in MATLAB 2013a and does not use any Hindi–English translator. Figure 3 shows the complete working procedure of the HTC–SVM system.

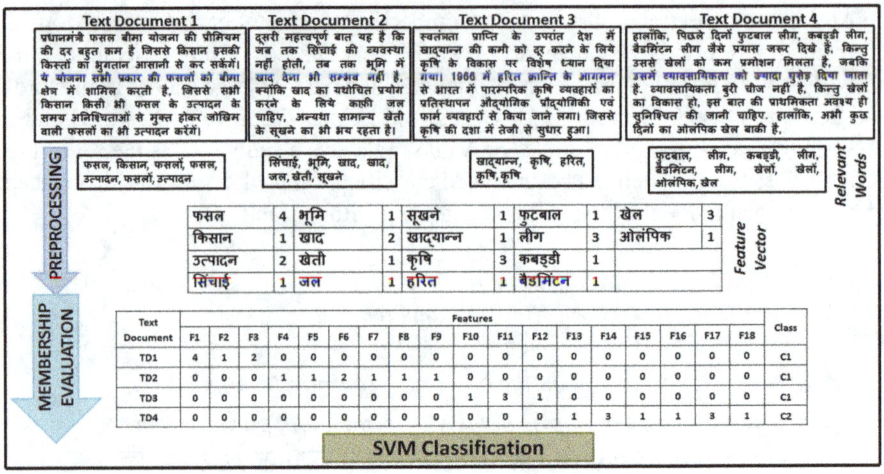

Fig. 3 Detailed Processing of HTC–SVM

5 Conclusion and Future Extensions

The scope of proposed automatic Hindi Text Classification (HTC) system using SVM explores a new and progressively wide range of applicability in NLP and text document classification. Although many Hindi and other Devanagari scripted language-based systems have been successfully implemented and presented with varying types of languages, structures, and contents, yet HTC method is oriented toward accurate text (keyword) identification, extraction and finally its classification with 100% accuracy. As we have seen that no existing system classifies the text into pre-defined categories by accepting Hindi documents, performing a series of operations on them, and classifying the unknown documents using SVM. This work can further be extended into the areas of HTC implementation on a large set of documents, dimensionality reduction, and imaged document classification.

References

1. Sinha RMK (2009) A journey from Indian scripts processing to Indian language processing. IEEE Ann Hist Comput 31(1):8–31. https://doi.org/10.1109/MAHC.2009.1
2. Mishra G, Nitharwal SL, Kaur S (2010) Language identification using Fuzzy-SVM technique. In: 2nd International conference on computing, communication and networking technologies, pp 1–5. IEEE Press. https://doi.org/10.1109/icccnt.2010.5592553
3. Sreejith C, Indu M, Raj PCR (2013) N-gram based algorithm for distinguishing between Hindi and Sanskrit texts. In: 4th International conference on computing, communications and networking technologies. IEEE Press, pp 1–4. https://doi.org/10.1109/icccnt.2013.6726777
4. Kumar R, Singh P (2017) Bilingual code-mixing in Indian social media texts for Hindi and English. In: Singh D, Raman B, Luhach A, Lingras P (eds) Advanced informatics for computing

research. communications in computer and information science, vol 712. Springer, Singapore, pp 121–129. https://doi.org/10.1007/978-981-10-5780-9_11
5. Kumar R, Dua M, Jindal S (2014) D-HIRD Domain-independent Hindi language interface to relational database. In: IEEE international conference on computation of power, energy, information and communication. IEEE Press, pp 81–86. https://doi.org/10.1109/iccpeic.2014.6915344
6. Prasad G, Fousiya KK (2015) Named entity recognition approaches: a study applied to English and Hindi language. In: International conference on circuits, power and computing technologies. IEEE Press (2015), pp 1–4. https://doi.org/10.1109/iccpct.2015.7159443
7. Gupta S, Bhattacharyya P (2010) Think globally, apply locally: using distributional characteristics for Hindi named entity identification. In: Proceedings of the named entities work-shop. Association for Computational Linguistics, Stroudsburg, PA, USA, pp 116–125
8. Jain A, Yadav D, Tayal DK (2014) NER for Hindi language using association rules. In: International conference on data mining and intelligent computing. IEEE Press, pp 1–5. https://doi.org/10.1109/icdmic.2014.6954253
9. Ekbal A, Saha S (2016) Simultaneous feature and parameter selection using multiobjective optimization: application to named entity recognition. Int J Mach Learn Cybernet 7(4):597–611. https://doi.org/10.1007/s13042-014-0268-7
10. Ekbal A, Saha S, Sikdar UK (2016) On active annotation for named entity recognition. Int J Mach Learn Cybernet 7(4):623–640. https://doi.org/10.1007/s13042-014-0275-8
11. Bhagavatula M, GSK S, Varma V (2012) Language-independent named entity identification using Wikipedia. In: Proceedings of the first workshop on multilingual modeling. Association for Computational Linguistics, Stroudsburg, PA, USA, pp 11–17
12. Siddiqi S, Sharan A (2015) Keyword and keyphrase extraction from single Hindi document us-ing statistical approach. In: 2nd International conference on signal processing and integrated networks. IEEE Press, pp. 713–718. https://doi.org/10.1109/spin.2015.7095377
13. Sharan A, Siddiqi S (2014) A supervised approach to distinguish between keywords and stop-words using probability distribution functions. In: International conference on advances in computing, communications and informatics. IEEE Press, pp 1074–1080. https://doi.org/10.1109/icacci.2014.6968206
14. Singh S, Siddiqui TJ (2012) Evaluating effect of context window size, stemming and stop word removal on Hindi word sense disambiguation. In: International conference on information retrieval and knowledge management. IEEE Press, pp 1–5. https://doi.org/10.1109/infrkm.2012.6204972
15. Sinha RMK (2011) Stepwise mining of multi-word expressions in Hindi. In: Proceedings of the ACM workshop on multiword expressions: from parsing and generation to the real world. Association for Computational Linguistics Stroudsburg, PA, USA, pp 110–115
16. Sinha RMK (2009) Automated mining of names using parallel Hindi-English corpus. In: Proceedings of the 7th workshop on Asian language resources. Association for Computational Linguistics Stroudsburg, PA, USA, pp 48–54
17. Sinha RMK (2009) Mining complex predicates in Hindi using a parallel Hindi-English corpus. In: Proceedings of the ACM workshop on multiword expressions: identification, interpretation, disambiguation and applications. Association for Computational Linguistics Stroudsburg, PA, USA, pp 40–46
18. Priyanka Sinha RMK (2014) A system for identification of idioms in Hindi. In: 7th international conference on contemporary computing. IEEE Press, pp 467–472. https://doi.org/10.1109/ic3.2014.6897218
19. Ramrakhiyani N, Majumder P (2015) Approaches to temporal expression recognition in Hindi. ACM Trans Asian Low-Resour Lang Inf Process 14(1):2
20. Mall S, Jaiswal UC (2014) Resolving issues in parsing technique in machine translation from Hindi language to English language. In: International conference on computer and communication technology. IEEE Press, pp 55–58. https://doi.org/10.1109/iccct.2014.7001469
21. Goutam R (2012) Exploring self-training and co-training for Hindi dependency parsing using partial parses. In: International conference on Asian language processing. IEEE Press, pp 37–40. https://doi.org/10.1109/ialp.2012.38

22. Sarika Sharma DK (2015) A comparative analysis of Hindi word sense disambiguation and its approaches. In: International conference on computing, communication and automation. IEEE Press, pp. 314–321. https://doi.org/10.1109/ccaa.2015.7148396
23. Sinha RMK (2011) Learning recognition of ambiguous proper names in Hindi. In: 10th Inter-national conference on machine learning and applications and workshops. IEEE Press, pp. 178–182. https://doi.org/10.1109/icmla.2011.87
24. Jain A, Yadav S, Tayal D (2013) Measuring context-meaning for open class words in Hindi language. In: 6th International conference on contemporary computing. IEEE Press, pp 118–123. https://doi.org/10.1109/ic3.2013.6612174
25. Sawhney R, Kaur A (2014) A modified technique for word sense disambiguation using Lesk algorithm in Hindi language. In: International conference on advances in computing, communications and informatics. IEEE Press, pp 2745–2749. https://doi.org/10.1109/icacci.2014.6968470
26. Jain A, Lobiyal DK (2014) A new method for updating word senses in Hindi WordNet. In: International conference on issues and challenges in intelligent computing techniques. IEEE Press, pp 666–671. https://doi.org/10.1109/icicict.2014.6781359
27. Agarwal M, Bajpai J (2014) Correlation based word sense disambiguation. In: 7th International conference on contemporary computing. IEEE Press, pp 382–386. https://doi.org/10.1109/ic3.2014.6897204
28. Nandanwar L (2015) Graph connectivity for unsupervised word sense disambiguation for Hindi language. In: International conference on innovations in information, embedded and communication systems. IEEE Press, pp 1–4. https://doi.org/10.1109/iciiecs.2015.7193083
29. Jain A, Lobiyal DK (2015) Unsupervised Hindi word sense disambiguation based on network agglomeration. In: 2nd International conference on computing for sustainable global development. IEEE Press, pp 195–200
30. Goyal V, Lehal GS (2010) Automatic standardization of spelling variations of Hindi text. In: International conference on computer and communication technology. IEEE Press, pp 764–767. https://doi.org/10.1109/iccct.2010.5640441
31. Jain A, Jain M (2014) Detection and correction of non word spelling errors in Hindi language. In: International conference on data mining and intelligent computing. IEEE Press, pp 1–5. https://doi.org/10.1109/icdmic.2014.6954235
32. Harikrishna DM, Rao KS (2015) Classification of children stories in Hindi using keywords and POS density. In: International conference on computer, communication and control. IEEE Press, pp 1–5. https://doi.org/10.1109/ic4.2015.7375666
33. Harikrishna DM, Rao KS (2015) Children story classification based on structure of the story. In: International conference on advances in computing, communications and informatics. IEEE Press, pp 1485–1490. https://doi.org/10.1109/icacci.2015.7275822
34. Joshi N, Mathur I, Darbar H, Kumar A, Jain P (2014) Evaluation of some English-Hindi MT systems. In: International conference on advances in computing, communications and informatics. IEEE Press, pp 1751–1758. https://doi.org/10.1109/icacci.2014.6968570
35. Arora KK, Sinha RMK (2012) Improving statistical machine translation through co-joining parts of verbal constructs in English-Hindi translation. In: ACM proceedings of the 6th workshop on syntax, semantics and structure in statistical translation. Association for Computational Linguistics Stroudsburg, PA, USA, pp 95–101
36. Thaokar C, Malik L (2013) Test model for summarizing Hindi text using extraction method. In: Proceedings of conference on information and communication technologies. IEEE Press, pp 1138–1143. https://doi.org/10.1109/cict.2013.6558271
37. Tyagi S, Chopra D, Mathur I, Joshi N (2015) Comparison of classifier based approach with baseline approach for English-Hindi text simplification. In: International conference on computing, communication and automation. IEEE Press, pp 290–293. https://doi.org/10.1109/ccaa.2015.7148390
38. Krishnamurthi K, Sudi RK, Panuganti VR, Bulusu VV (2013) An empirical evaluation of dimensionality reduction using latent semantic analysis on Hindi text. In: International conference on Asian language processing. IEEE Press, pp 21–24. https://doi.org/10.1109/ialp.2013.11

39. Singh P, Dutta K (2014) Annotating indirect anaphora for Hindi: a corpus based study. In: 6th international conference on computational intelligence and communication networks. IEEE Press, pp 525–529. https://doi.org/10.1109/cicn.2014.120
40. Puri S, Kaushik S (2012) An enhanced fuzzy similarity based concept mining model for text classification using feature clustering. In: Students' conference on engineering and systems. IEEE Press, pp 1–6. https://doi.org/10.1109/sces.2012.6199126

A Novel Design and Implementation of 32-Bit Hybrid ALU

Suhas B. Shirol, S. Ramakrishna and Rajashekar B. Shettar

Abstract Arithmetic logic unit (ALU) is a building block of many digital calculating systems like calculator, cell phone, computer, processors, etc. In present electronic market as technology is upgrading day by day, fast-growing technologies with portable devices demand for low-power VLSI design. Demand for less delay and low area is increasing. Reversible logic circuit helps in reduction of power dissipation. ALU is designed with both reversible and irreversible logic gates to reduce power and delay. This type of design can be named as hybrid ALU. In digital adder, time taken to propagate carry is decreased by using carry select adder (CSA) and Kogge–Stone adder (KSA), and area is reduced by using binary-to-excess-one converter (BEC) instead of ripple carry adder (RCA). This adder design is used in Vedic multiplier to add partial products, which also reduce delay in digital multiplier.

Keywords Arithmetic logic unit (ALU) · Carry select adder (CSA) · Kogge–Stone adder (KSA) · Binary-to-excess-one converter (BEC) · Ripple carry adder (RCA) · Hybrid ALU

1 Introduction

ALU plays vital role in the digital calculations. It is part of calculator, computer, and processer. ALU performs arithmetic, logical, and shift operations. ALU consists of digital adders, subtractor, and multiplier [1]. In the present electronic market as technology is upgrading day by day, the demand for less delay, low power, and low area is increasing and plays crucial role. A design engineer needs to keep accept in

S. B. Shirol (✉) · S. Ramakrishna · R. B. Shettar
Department of Electronics and Communication Engineering, B V Bhoomaraddi College of Engineering and Technology, Hubballi 580031, Karnataka, India
e-mail: suhasshirol@gmail.com

S. Ramakrishna
e-mail: ramakrishna@bvb.edu

R. B. Shettar
e-mail: raj@bvb.edu

S.-L. Peng et al. (eds.), *Computing and Network Sustainability*, Lecture Notes in Networks and Systems 75, https://doi.org/10.1007/978-981-13-7150-9_25

mind before starting with any design. In digital adder, the time taken to propagate carry is reduced by using carry select adder; area is reduced by using binary-to-excess-one converter (BEC) instead of RCA ripple carry adder. These adders are used in Vedic multiplier to add partial products which also reduce delay in digital multiplier [2]. Fast-growing technologies with portable devices demand for low-power VLSI design. Problem with low-power VLSI design is power dissipation. Reversible logic provides solution for this problem [3]. As reversible logic provides solution for this problem, reversible logic circuit helps in reduction of power dissipation. ALU is designed with both reversible and irreversible logic gates to reduce power and delay. We can name this type of design as hybrid ALU. ALU performs digital calculations and forms important part of computer all forms, smart phones, tablets, calculator. Since delay and power have crucial role in today's digital market, designing to reduce delay and power is essential.

2 Literature Survey

Adder is subcircuit of ALU. Delay in adder is mainly contributed by carry propagation. In order to decrease delay and area, modified carry select adder (CSA), i.e., binary-to-excess-one converter (BEC) is used instead of using ripple carry adder (RCA) with carry input as one. CSA reduces delay, and BEC reduces area [4–6].

Multipliers find huge importance in digital computing electronic circuits, like ALU, digital processing units, image processing, security system. In this paper [2, 7], multipliers are designed according to Vedic sutra called Urdhva Tiryagbhyam, and Kogge–Stone (K-stone) adder is used to add the partial products. Since K-stone is one of the fastest adders, this reduces delay and Vedic sutra improves functional efficiency of multiplier.

Adders are used in Vedic multiplier to add partial products. In this paper [3, 7–9], modified CSA with BEC is designed to add partial products. Comparison of booth multiplier and Vedic multiplier is done. Result resembles that Vedic multiplier with modified CSA is area efficient and delay efficient.

ALU is prominent subsystem of many systems like calculator, cell phone, computer, processors, etc. Reversible logic gates have gained large importance in low-power VLSI industry as it has ability to reduce power dissipation. Circuits with irreversible logic gates consume more compared to that of reversible logic gates [1].

Kogge–Stone adders are one of the fastest adders and are known for their high-speed performance. In this paper [5, 10–12], K-stone adder is designed using reversible logic gate, to have reduced delay and power dissipation.

There is no loss of data as it is one-to-one mapping between outputs and inputs of reversible logic gate. This results in less power dissipation; hence, it fulfills the demand of low power dissipation in low-power VLSI. Code-converting circuits play important role in enhancing security system reducing complexity in arithmetic circuits. This results in reduction of hardware and increases speed and power saving [11].

3 Design Methodology

ALU is digital electronic circuit that operates on binary integers; it performs arithmetic operations, logical operations, and bit-wise operations. ALU is building block of many digital computing systems like computer, calculator, cell phone, and tablet. Figure 1 represents the block diagram of ALU.

3.1 Design and Implementation

Figure 2 shows the design and implementation of the 32-bit ALU using 4:1 multiplex.

3.2 Reversible Logic Gates

Reversible logic circuits are the circuits where inputs can be recovered from outputs. The number of inputs and the number outputs are equal; there is one-to-one mapping between output and input. There is no loss of information, and hence, power dissipation is reduced. Outputs are computed from inputs in the same way inputs can be computed from output, at any point of operation we stop and can compute

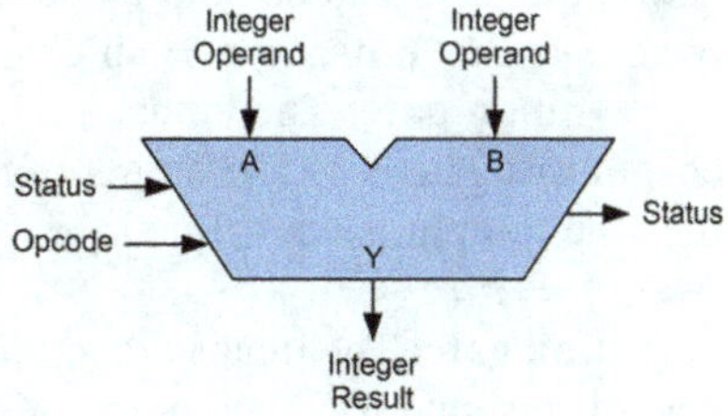

Fig. 1 Block diagram of ALU

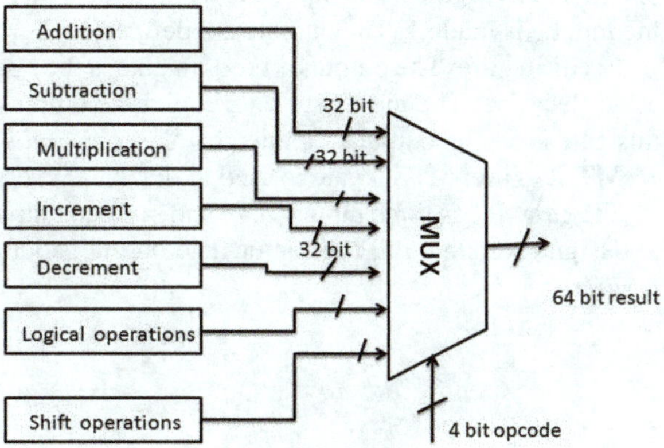

Fig. 2 ALU implementation

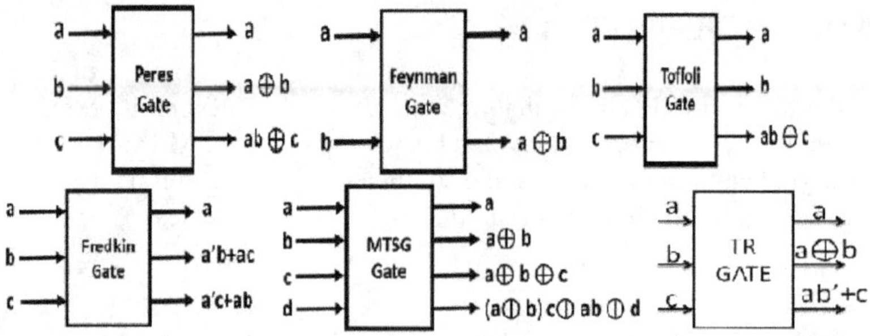

Fig. 3 Reversible logic gates

back the inputs. There is less power dissipation; this provides solution for low-power applications. In computing system when there is change of bits from zero to one and one to zero, energy gets dissipated in the form of heat. In irreversible logic circuits when bits change from one to zero, there is change in voltage levels; this change in voltage level gives energy in the form of heat. However, reversible logic gate moves charge from one node to another node. To achieve less power dissipation in ALU, in this project, we have used many reversible logic gates to design adder, MUX, BEC, subtractor, logical operations like XOR, OR, AND, and many more which are shown in Fig. 3. Quantum cost of any reversible logic gate is how many basic irreversible logic gates are required to design that particular gate.

Peres gate: Half adder is implemented using Peres gate by making input c as zero. Full adder can be built by cascading two half adders. The outputs are defined by $X =$ a, $Y =$ a^b, and $Z =$ ab^c.

Feynman gate: The inputs are *a* and *b*, and the outputs are *X* and *Y*. Quantum cost of this gate is 1. The output of Feynman gate is XOR operation of a and b, and the other output is maintained for the reversibility. The outputs are defined by $X = a$ and $Y = a\hat{}b$.

Toffoli gate: The inputs are *a*, *b*, and *c*, and the outputs are *X*, *Y*, and *Z*. Quantum cost of this gate is 5. The output *Y* of Toffoli gate can be used for two operations: one as an AND gate if the input c is made to be 0 and another as an XOR gate when the input *c* is made 1. The outputs are defined by $X = a$, $Y = b$, and $Z = ab\hat{}c$.

Fredkin gate: The output of Fredkin gate is the operation of MUX with the input as a select line. The inputs are *a*, *b*, *c*, and the outputs are *X*, *Y*, *Z*. Quantum cost of this gate is 5. The output is defined by $X = a$, $Y = ab\hat{}ac$, and $Z = ac\hat{}ab$.

MTSG gate: MTSG gate is used to design a reversible full adder circuit.

TR gate: It has three inputs: *a*, *b*, and *c*; three outputs *X*, *Y*, and *Z*. TR gate is used to design reversible full subtractor. The output is defined as $X = a$, $Y = a\hat{}b$, and $Z = ab\hat{}c$.

4 Arithmetic Operations Using Hybrid Design

4.1 Adder

Adders are combinational circuits, which add two binary integers. Adders find huge application in digital electronic circuits, to add numbers, in multipliers to add partial products in many arithmetic operations. On the other hand, today's technology demands for less delay and minimum power dissipation. Therefore, adder with minimum delay and minimum power dissipation is designed using reversible logic gates. There are many types of adders: carry select adder, ripple carry adder, parallel prefix adder, carry skip adder, etc.

4.1.1 Ripple Carry Adder (RCA)

Ripple carry adder is formed by cascading full adder, where carry out of full adder acts as carry in for next full adder. RCA is functionally efficient but delay increases as we go higher bit addition, because of time taken for carry generation and propagation.

4.1.2 Carry Select Adder (CSA)

CSA is combinational circuit; it comprises RCA and MUX. Here delay is reduced because carry propagation and generation is not included, as availability of precalculated sum for both values of carry either one or zero. MUX is used to select the sum depending on the value carry. For summation circuit, any adder circuit like carry-lookahead adder, prefix adder (Kogge–Stone adder), carry skip adder, etc., can replace RCA. Carry select adder reduces delay. Disadvantage of carry select adder is complexity of the adder increases because of presence of MUX, additional RCA with carry as 1, increases area of the circuit.

4.1.3 Kogge–Stone Adder (KSA)

KSA is known for its speed performance and finds various applications in industry. KSA requires long wires to connect cells of adder; this does not affect the area of the circuit but increases power consumption.

4.1.4 Binary-to-Excess-One Converter (BEC)

Code converters are used to reduce the complexity of arithmetic circuits, which therefore helps in reduction of area and delay. Hence, it improves performance of the circuit 4-bit RCA adder and uses 8 AND gates, 8 XOR gates, and 4 OR gates, where

5-bit BEC requires four XOR gates, three AND gates and one inverter. Comparison concludes that BEC has less area. Hence, in carry select adder we can use BEC instead of adder with carry equal to one, which helps in reduction area.

4.1.5 Modified Carry Select Adder

CSA is modified to minimize delay and area. Figure represents architecture of modified CSA. Figure 4 represents architecture of modified 32-bit CSA. Modified 32-bit CSA comprises two 16-bit KSA and 17-bit BEC.

4.1.6 Reversible Modified CSA

Modified CSA is designed to achieve minimum delay and area. In order, reduce power dissipation modified CSA is implemented using reversible logic gate. Figure 5 represents architecture BEC using reversible logic gate. Peres gate is cascaded to build binary-to-excess-one converter (BEC).

Fig. 4 32-bit CSA adder with K-Stone

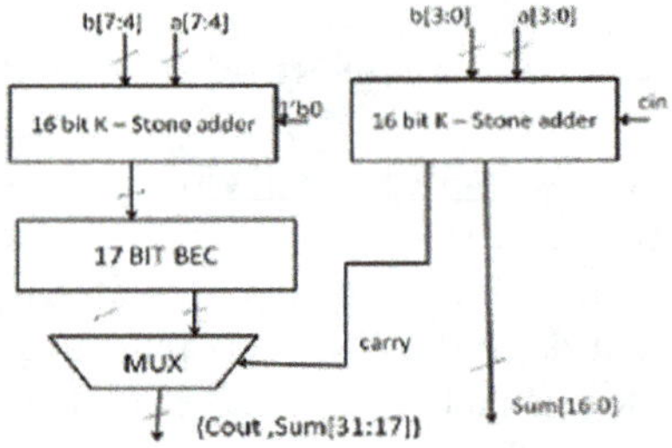

Fig. 5 BEC using reversible logic

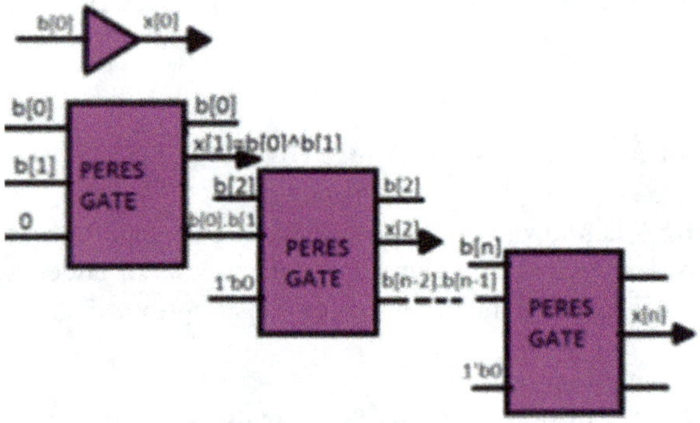

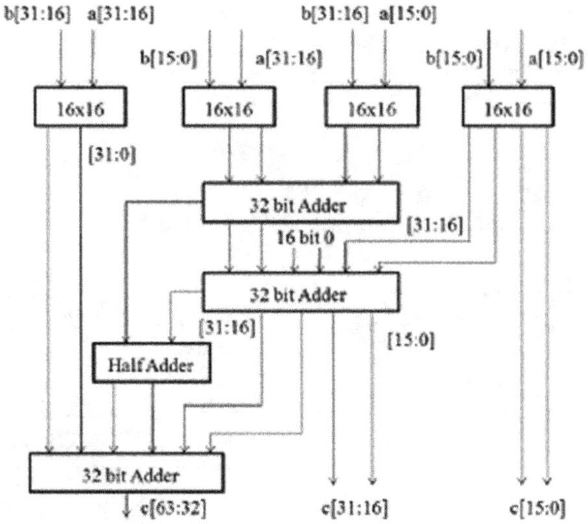

Fig. 6 32 × 32 Vedic multiplier

5 Multiplier

Multipliers are combinational logic circuit, which multiplies binary integers. Multipliers are building blocks of digital processing units, ALU, convolutions, matrix multiplication. There are many types of multipliers: Vedic multipliers, booth and array multipliers. If speed of multipliers is increased, it is possible to decrease the processing time required by digital processing units and ALU. In ancient Vedic mathematics, various sutras were used for multiplication. Among those sutras, Urdhva Tiryagbhyam sutra gives more efficient multiplier. In Urdhva Tiryagbhyam, adders are used to add the partial products, speed and these are also decides the speed of multiplier and area of multiplier. Therefore, to have minimum delay in our project we have used modified CSA adder to add the partial products.

Figure 6 represents architecture of 32 × 32 Vedic multiplier. From architecture, it is found that in addition to partial products, addition of carry is not needed so modified CSAs area and delay can be reduced by replacing KSA with RCA and half adders. Figure represents design of modified CSA with RCA and half adders.

6 Results and Analysis

The coding of design is done by using Verilog HDL code. The code is simulated and synthesized by expending Xilinx 14.2 ISE, using Verilog language. 32-bit adders, multipliers, and subtractors are designed using both reversible and reversible logic gates using Verilog. Delay, area, and power are compared. 32-bit hybrid ALU implemented using ChipScope Pro on Spartan 6 FPGA. The results are represented in

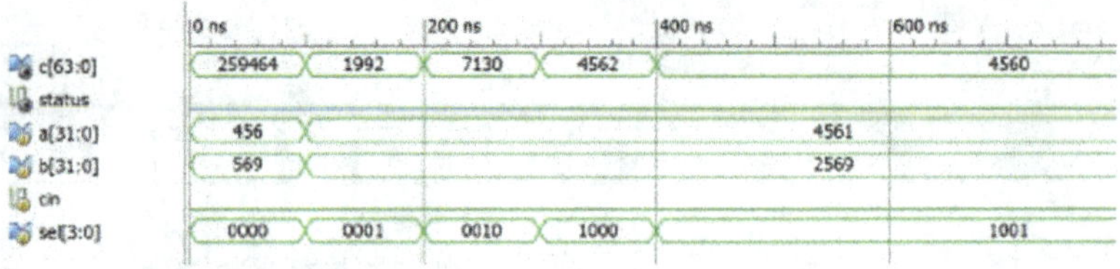

Fig. 7 Arithmetic operations of 32-bit hybrid ALU

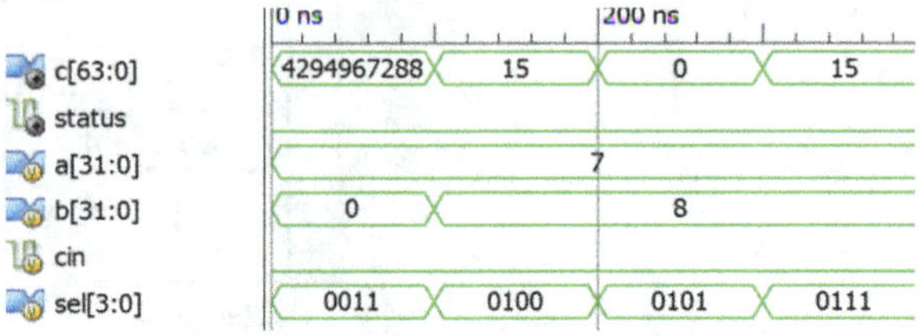

Fig. 8 Logical operations of 32-bit hybrid ALU

below sections. Figure 7 represents simulation result of arithmetic operations 32-bit hybrid ALU.

Arithmetic operations

1. Opcode 0000: 0000 codes for multiplication. *a* and *b* are the operands. *a* is forced to value 456, and *b* is forced to value 569. Sel is forced to 0000. Hence, result *c* is 259464.
2. Opcode 0001: 0001 codes for subtraction. *a* and *b* are the operands. *a* is forced to value 4561, and *b* is forced to value 2569. Sel is forced to 0001. Hence, result c is 1992.
3. Opcode 0010: 0010 codes for addition *a* and *b* are the operands. *a* is forced to value 4561, and *b* is forced to value 2569. Sel is forced to 0010. Hence, result *c* is 7130.
4. Opcode 1000: 1000 codes for increment. *a* and *b* are the operands. *a* is forced to value 4561, and *b* is forced to value 0. Sel is forced to 1000. Hence, result *c* is 4562.
5. Opcode 1001: 1001 codes for decrement *a* and *b* are the operands. *a* is forced to value 4561, and *b* is forced to value 0. Sel is forced to 1001. Hence, result *c* is 4560.

Logical operations of ALU

The results are represented in below sections. Figure 8 represents simulation result of logical operations of 32-bit hybrid ALU.

1. Opcode 0011: 0011 codes for NOT. *a* and *b* are the operands. *a* is forced to value 7, and *b* is forced to value 0. Sel is forced to 0011. Hence, result *c* is 4294967288.
2. Opcode 0100: 0100 codes for OR. *a* and *b* are the operands. *a* is forced to value 7, and *b* is forced to value 8. Sel is forced to 0100. Hence, result *c* is 15.

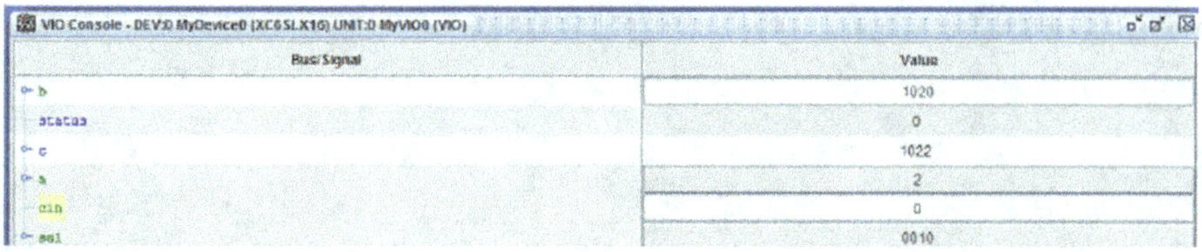

Fig. 9 Implementation of hybrid ALU using ChipScope

Fig. 10 16-bit adder design

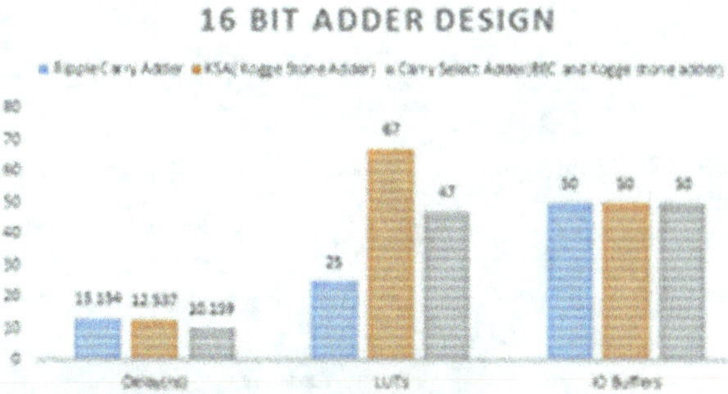

3. Opcode 0101: 0101 codes for AND. *a* and *b* are the operands. *a* is forced to value 7, and *b* is forced to value 8. Sel is forced to 0101. Hence, result *c* is 0.
4. Opcode 0111: 0111 codes for increment. *a* and *b* are the operands. *a* is forced to value 7, and *b* is forced to value 8. Sel is forced to 0111. Hence, result *c* is 15.

Implementation using ChipScope

Figure 9 represents result of 32-bit hybrid ALU using ChipScope, where input *a* is forced to 1020, *b* is forced to 2, c is to 0, and sel is set to 1010. It selects addition as operation, and sum is stored in *c* is 1022.

Analysis of Results

16-bit Adder design Comparison, 32-bit Adder design comparison: Irreversible and Reversible Logic, 32-bit Vedic Multiplier design comparison: Irreversible and Reversible Logic, 32-bit ALU design comparison: Irreversible and Hybrid Logic, Maximum obtained path delay, LUTs and IO Buffers comparison of different 16-bit adders is shown in Figs. 10, 11, 12 and 13 that shows the less power dissipation.

Conclusion

The ALU designed here makes use of reversible logic gates, which give minimum delay and power dissipation, modified CSA with reversible logic gates for addition, which contributes for minimization of delay, area, and power dissipation. Vedic multiplier is designed by using modified CSA using BEC, which reduces the area and time. Subtractor is designed using reversible logic gates. The used adder model, multiplier model, and subtractor model reduce the time involved in the operation, and thus increase the speed and reduces power dissipation. The design is implemented on Spartan 6 FPGA using ChipScope Pro. The whole design is coded by using Verilog

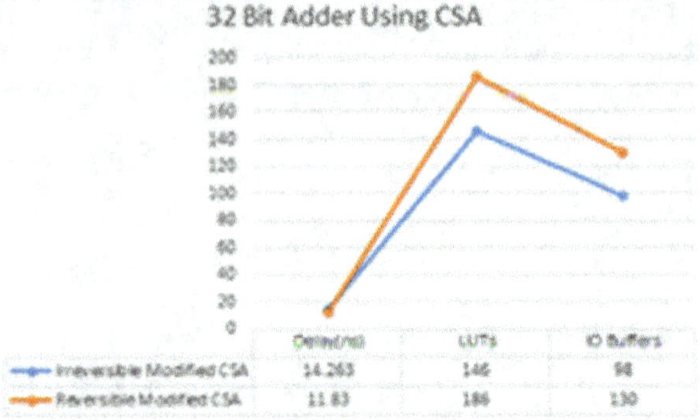

Fig. 11 32-bit adder using CSA

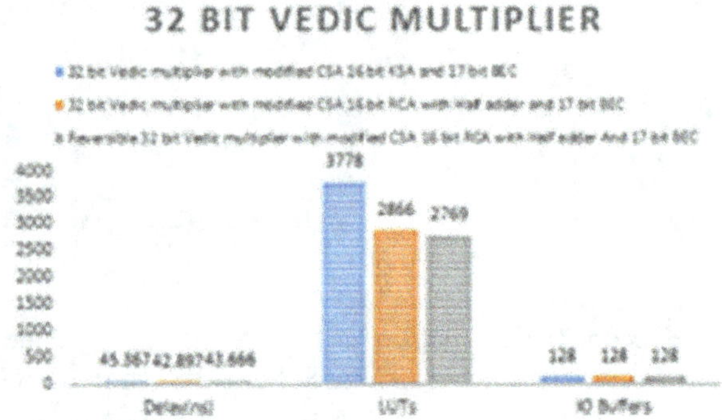

Fig. 12 32-bit Vedic multiplier

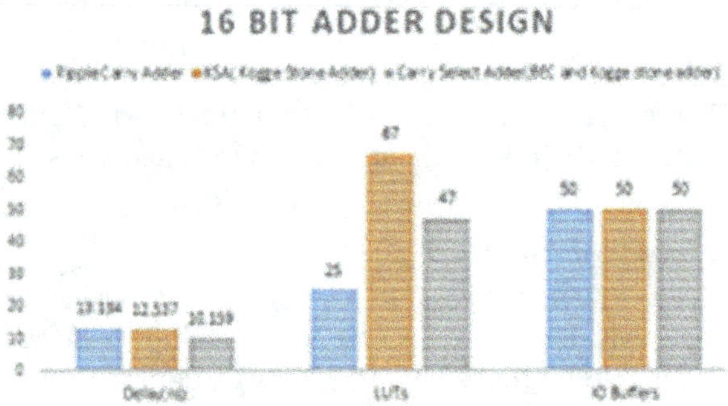

Fig. 13 32-bit ALU (irreversible and reversible logic)

HDL. The design is simulated and synthesized using Xilinx 14.2 ISE simulator using Verilog language.

References

1. Swamynathan S, Banumathi V (2017) Design and analysis of FPGA based 32 bit ALU using reversible gates. In 2017 IEEE international conference on electrical, instrumentation and communication engineering (ICEICE). IEEE, pp 1–4
2. Sudeep M, Vucha M et al. (2014) Design and FPGA implementation of high speed vedic multiplier. Int J Comput Appl 90(16): 2014
3. Gokhale G, Gokhale S (2015) Design of area and delay efficient vedic multiplier using carry select adder. In: 2015 International Conference on Information Processing (ICIP). IEEE, pp 295–300
4. Nautiyal P, Madduri P, Negi S (2015) Implementation of an alu using modified carry select adder for low power and area efficient applications. In: 2015 International conference on computer and computational sciences (ICCCS). IEEE, pp 22–25
5. Elangadi S, Shirol S (2014) Design and characterization of high speed carry select adder. Int J Ethics in Eng Manag Edu 1(6)
6. Shirol S, Ramakrishna S, Shettar S (2019) Design and implementation of adders and multiplier in FPGA using Chipscope: a performance improvement. Inf Commun Technol Comp Str, lecture notes in networks and systems. https://doi.org/10.1007/978-981-13-0586-3_2
7. Saha P, Banerjee A, Bhattacharyya P, Dandapat A (2011) In: High speed ASIC design of complex multiplier using vedic mathematics. IEEE 2011
8. Jais A, Palsodkar P (2016) SM-IEEE, Design and implementation of 64 Bit multiplier using vedic algorithm. Apr 6–8, 2016, IEEE
9. Ramkumar B, Kittur HM (2012) Low-power and area-efficient carry select adder. IEEE Trans Very Large Scale Integr (VLSI) Syst 20(2):371
10. Morrison M, Lewandowski M, Meana R, Ranganathan N (2011) Design of a novel reversible alu using an enhanced carry look-ahead adder. In: 2011 11th IEEE conference on nanotechnology (IEEENANO). IEEE, pp 1436–1440
11. Pattnaik SK, Nanda U, Nayak D, Mohapatra SR, Nayak AB, Mallick A (2017) Design and implementation of different types of full adders in alu and leakage minimization. In: 2017 international conference on trends in electronics and informatics (ICEI). IEEE, pp 924–927
12. Banerjee A, Das DK (2016) A new alu architecture design using reversible logic. In 2016 sixth international symposium on embedded computing and system design (ISED). IEEE, pp 187–191

A Novel Approach for Security of Data in IoT Environment

Priyanka Anurag Urla, Girish Mohan, Sourabh Tyagi and Smitha N. Pai

Abstract The Internet of Things (IoT) is an emerging technology in this era. The luxury of technology comes with security, which is a noteworthy issue to be addressed. The paper's soul purpose is to provide security to the data in the IoT environment. Data collected from multiple IoT devices have been provided multilevel security using elliptical curve cryptography (ECC) and fully homomorphic encryption(FHE), mitigating the generalized cryptographic attacks. The data are collected at the roof and on customization sent to the cloud. The proposed method enhances the security by preventing modification of data from any brute-force attack using a short key with less computation power.

Keywords IoT · Fully homomorphic encryption · ECC · Roof · Encryption standards

1 Introduction

The Internet has been around for a while, largely for the betterment of the human lives. The data, images, recordings, games, books and commerce have been created for the people, by the people and about the people. The Internet is one of the most important and transformative technologies ever invented. The Internet is not just about connecting people, it is about connecting things (devices), called as Internet of Things (IoT). A network of things connected to Internet, capable of data collection and data

P. A. Urla (✉) · G. Mohan · S. Tyagi · S. N. Pai
Department of Information and Communication Technology, Manipal Institute of Technology, Manipal Academy of Higher Education, Manipal 576104, Karnataka, India
e-mail: priyankaraoau432@gmail.com

G. Mohan
e-mail: Mohangiri1999@gmail.com

S. Tyagi
e-mail: Tyagi.sourabh77@gmail.com

S. N. Pai
e-mail: smitha.pai@manipal.edu

S.-L. Peng et al. (eds.), *Computing and Network Sustainability*, Lecture Notes in Networks and Systems 75, https://doi.org/10.1007/978-981-13-7150-9_26

exchange operating with the aid of sophisticated embedded sensors, together constitute IoT. IoT makes the environment speak for itself and to understand the factual world better for people and machines. Real-time context as well as patterns which can be observed over the period of time are built with the help of the data continuously generated by IoT systems. New emerging technologies, also the advancement of current existing technologies has resulted in giving the user, a friendly experience leaving to deal with certain aspects pertaining to privacy, scalability, functionality and security. Among the listed issues to be addressed, security is the most vital and critical aspect. The data from the IoT environment can be subjected to attacks like brute force, masquerade, man in the middle, etc. There are many techniques to avoid each of these attacks. In the current work, a multilevel encryption and decryption process is addressed. In order to secure the environment from the listed attacks, the algorithm which are put into effect are: lightweight encryption technique, the secured hash function, elliptical curve cryptography and fully homomorphic encryption.

2 Literature Review

Security and efficiency go hand in hand. They define each other in a very prominent manner. Security issues are highly disturbing and suggestive solutions have to be obtained to protect data in IoT environment from various sources like sensors and application [1, 2]. The architecture of IoT being a very crucial aspect and the study of the communication protocols [3, 4] existing in the IoT environment give us an open platform to take measures in the security area and also to render new solutions to mitigate security threats [5, 6].

Gentry's methods are employed to convert a plain 'somewhat homomorphic' encryption algorithm utilising elementary modular arithmetic [7]. The main aim of the methodology is that it is conceptually simple. A new modulus technique of switching for the Dijk, Gentry, Halevi and Vaikuntanathan's (DGHV) methodology is described along with the foremost objective which is a compression technique where there is a reduction in the size of public key of DGHV's fully homomorphic scheme [8, 9]. As an alternative to traditional cryptographic, namely RSA, the use of elliptical curve cryptography has an enormous amount of advantage in terms of performance in the wireless framework, whereas the voluminous and vigilant study of the software administration on the workstations of the elliptical curves over binary fields is carried out and recommended by NIST [10, 11]. For the reduction of the key size and to improve the efficiency, a new strategy for the compression of public key size is advocated for fully homomorphic encryption(FHE) having its origin from quadratic parameters along with correction. To operate on real numbers, an advanced FHE scheme is proposed. This methodology gives us the advantage of accuracy along with efficiency [12, 13]. An encryption scheme is formulated where it differentiates the credentials of private and public keys. End-to-end secure connectivity can be obtained with protocols and apt architecture [14, 15].

The work discussed in the related field deals with encryption in different domains where data are fragile. The current paper addresses the issues of security in the IoT environment providing enhanced security using multiple levels.

3 Methodology

Data acquisition is currently carried out by interfacing with Raspberry Pi 3/Arduino board. The data measured are temperature, humidity and soil moisture using various sensors. The flow chart of the project is given in Fig. 1.

Encryption of data is carried out using the multilayered approach, using three security algorithms, i.e. light-weighted encryption technique, secured hash function, elliptical curve cryptography(ECC) and fully homomorphic encryption(FHE). The encryption involves the computations as shown in Fig. 2.

Fig. 1 Flow chart

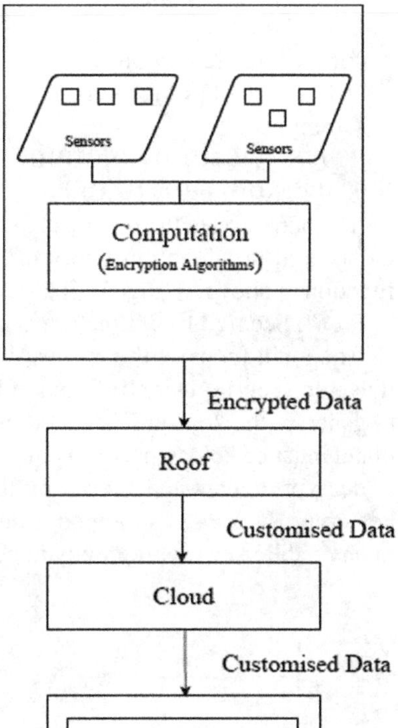

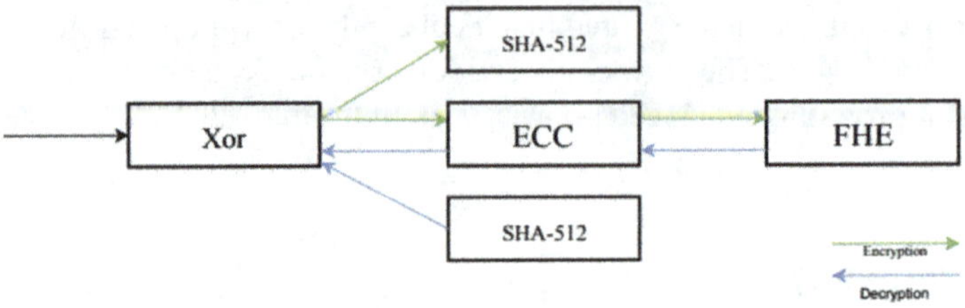

Fig. 2 Computational flow

At the first level, for light-weighted encryption, XOR is used to secure the data from brute-force attacks. In XOR, the output is true if two inputs are distinct with respect to each other.

Pseudocode for the generation of the key

Plaintext = P, Ciphertext = C, Output = r, input = s, Key = k and KeyLength = b.

for each bit of plain text P

C ← P XOR k.

for i ← 0 to length of(s)

r[i] ← s[i] XOR k[i mod b]

Example:

10011100101101010011101O XOR 0101101000011011110110OO = 110001101010111011100010

The output from the first level of security is the input to the second level. This uses the secured hash function algorithm. The hash functions are a structured mathematical function as shown in Fig. 3. The frame format of the secured hash function is shown in Fig. 4. Secured hash function is used to ensure that the data are not modified.

The main focus of this project lies on two important algorithms providing multilayered security using ECC and FHE. The properties of elliptical curves are used to generate the keys in ECC. The most desirable feature of multilevel encryption is that it must be able to perform computations on encrypted data without the necessity to decrypt it. This is achieved by the using FHE. ECC is an asymmetric or public key cryptosystems. Its method is similar to RSA or Diffie–Hellman. Just like RSA or any public key cryptography in general, it relies on trap door function. In terms of

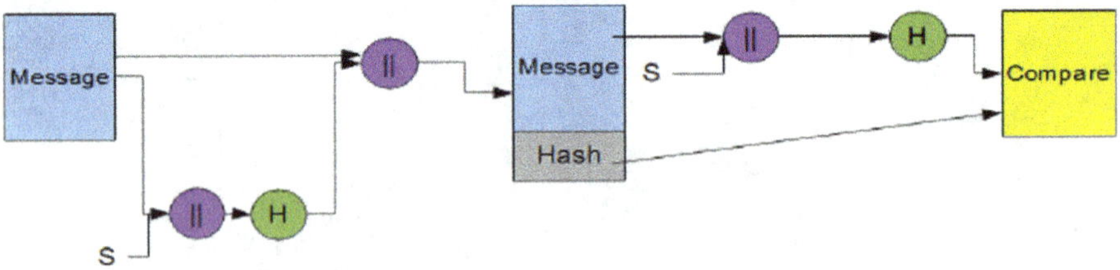

Fig. 3 Flow of data information using message digest [16]

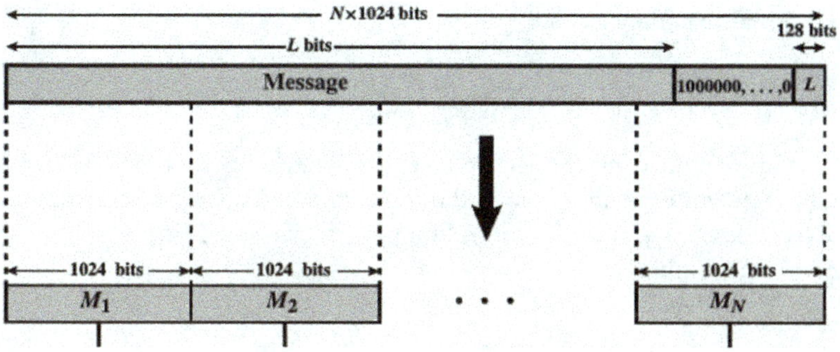

Fig. 4 Frame format [17]

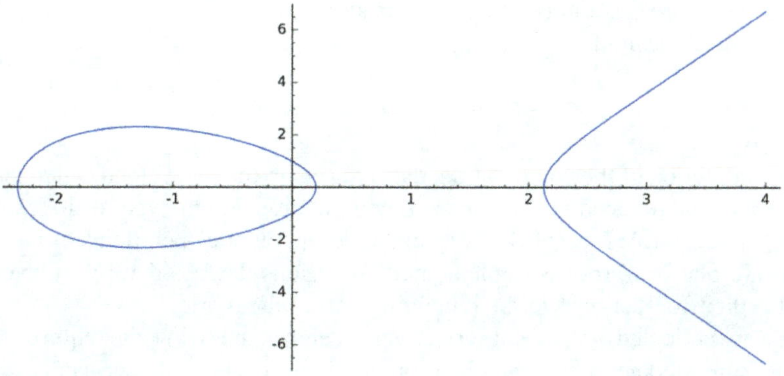

Fig. 5 Elliptic curve of ECC

application, it is similar to RSA and Diffie–Hellman but more complex mathematics is used. The curves which are used in this context are of the form $y^2 = x^3 + Ax + B$ as shown in Fig. 5.

ECC has following two characteristics: With horizontal symmetry, at any point on the curve, it gets be reflected over the x-axis and it will still be on the curve and the other characteristic is that any non-vertical line that is drawn on the curve will intersect this curve at the max three times. Using these characteristics, the key generation, encryption and decryption are done using the formulas from Eqs. (1, 2 and 3).

Key generation:

$$Q = d * p \tag{1}$$

where Q is public key, P is the point on curve and d is the private key, respectively.

Encryption formula:

$$C1 = K * p,\ C2 = M + K * Q \tag{2}$$

Decryption formula:

$$M = C2 - d * C1 \tag{3}$$

FHE is an operation performed on a set of cipher texts such that decrypting the result of the operation is the same as the result of some operation performed on the plain texts. The encryption and decryption processes use the Eqs. (4–5)

Encryption formula:

Cipher text for $m \in \{0,1\}$

$$C = q.p + 2r + m \tag{4}$$

where r and q are random integers, p is the secret key.

Decryption formula:

$$(C \bmod p) \bmod 2 = m \tag{5}$$

On completion of the computation, data are sent to the local cloud called the roof. Roof is an amalgamated networking and computational archetype for IoT, which is invariably accessible for real-time context building, on-site operation facilitation. As a software platform, roof is implemented on various devices that proxy the things and also their IoT services to the remaining part of the world.

Data are collected at the local server. Values above and below the required threshold are being updated on to the cloud on encryption. The end-user can access this data through an end device as and when needed (continuously or at specific time interval, or averaged data for within some interval). When the end-user accesses the data, it is decrypted in the same sequence as it was encrypted before.

4 Implementation

Data are read and collected from sensors interfaced with Raspberry Pi 3/Arduino board and stored in the system in the form of a file. Application-specific sensors can be used. Other devices which can send text, image, etc. can also be taken into account. In the current scenario, humidity, temperature and soil moisture content detector are used. The baud rate chosen in this example is 3600baud. Figure 6a, b illustrate the set-up of sensors with Arduino. The data collected are shown in Figs. 7 and 8. Experiments are carried out using Raspberry Pi 3 board.

Four levels of encryption are applied to the collected data. As an example, the following is the input, Fig. 9.

At the first level, a simple lightweight encryption standard is used whose output is as shown in Fig. 10. The input to the file is the humidity and temperature information.

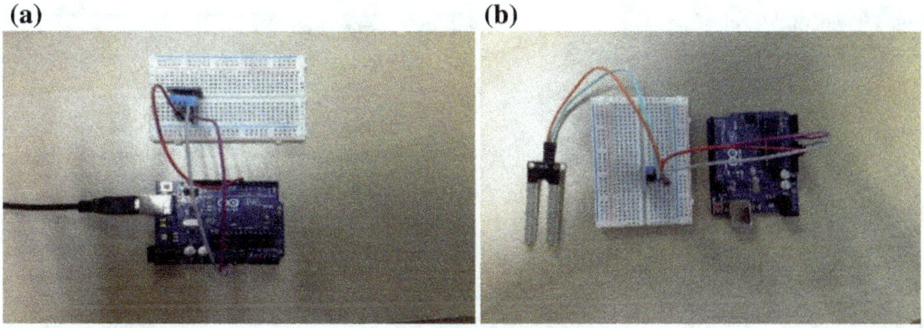

Fig. 6 A.Temperature and humidity sensor connectivity with Arduino. B. Moisture sensor connectivity with Arduino

Fig. 7 Measured temperature and humidity sensor data

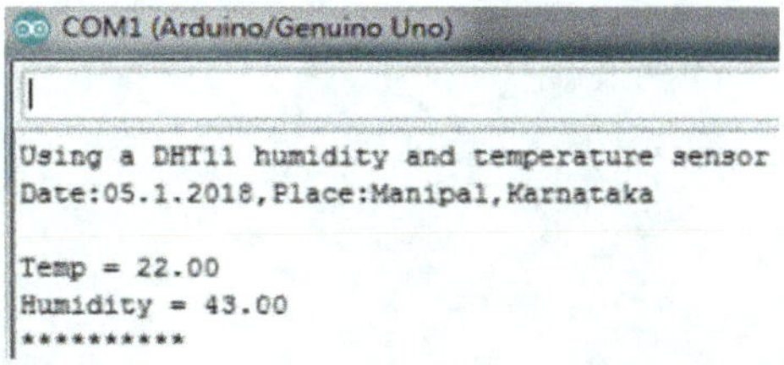

Fig. 8 Moisture sensor data

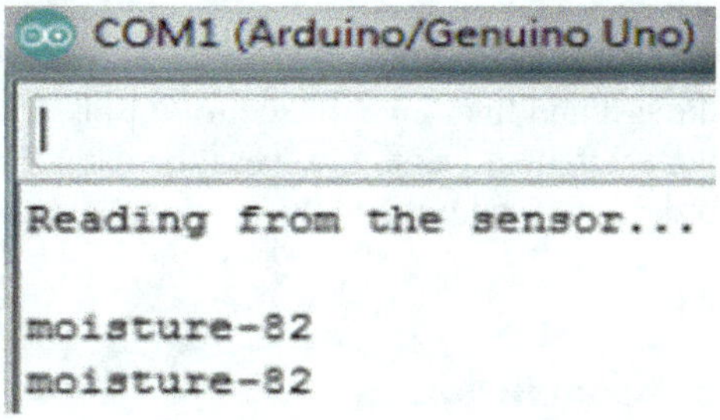

Fig. 9 Input to the algorithms

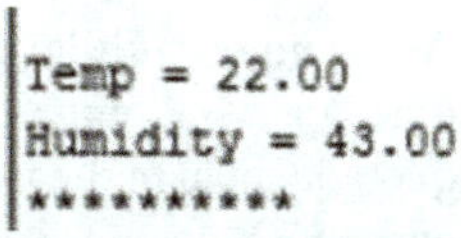

The security at the next level is obtained using the secure hash function whose input is as from Fig. 10. The output generated by the SHA algorithm is as shown in Fig. 11

Fig. 10 Level 1 encrypted data

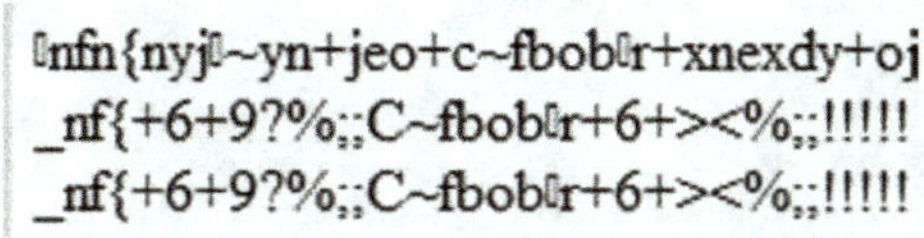

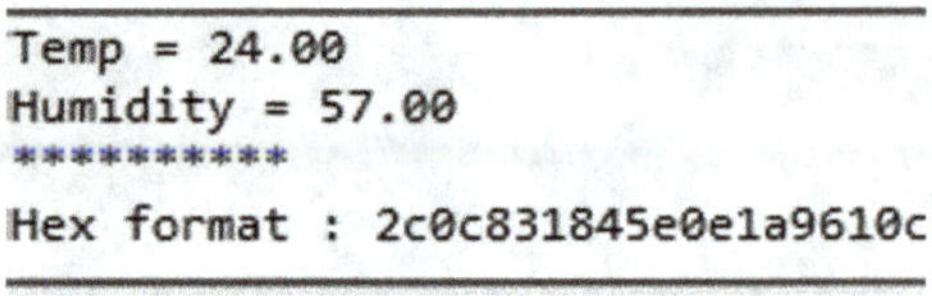

Fig. 11 Second level of encryption

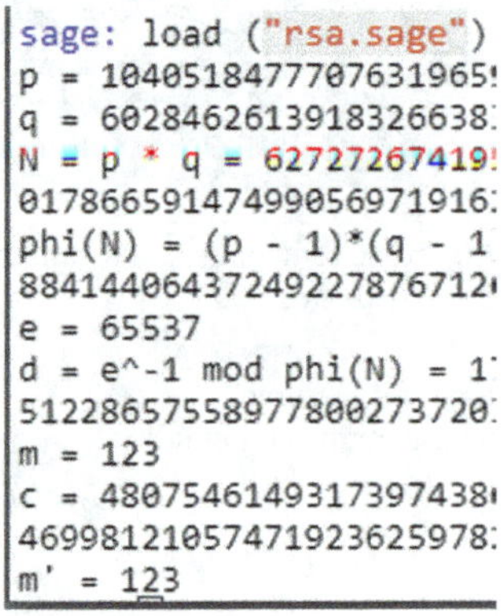

Fig. 12 ECC output

The output generated by the hash function is verified for modification and the output of the lightweight encryption is fed to elliptical curve cryptosystem. The elliptical curve cryptosystem output is as shown in Fig. 12 and is fed as the input to the last and final level of security, fully homomorphic encryption algorithm. When the user/client wants to access the data, the decryption process in the same reverse order is carried out and the original data are retrieved.

5 Conclusion

A system of multilevel security to the data collected from IoT devices is proposed in this paper. In doing so, the security level is enhanced mitigating the attacks, i.e. masquerade, man in the middle, data modification, denial of service, etc. The first level of encryption ensures security from brute-force attacks. The second level of encryption prevents data being modified. The third level of encryption increases security using public key for encryption operation. The key generation grows exponentially. Fourth level ensures that other attacks are not possible as it employs relatively a shorter key. The short key is faster, entails less computation of power and ensures transfer of the data faster.

References

1. Suo H, Wan J, Zou C, Liu J (2012) Security in the internet of things: a review. In: Proceedings—2012 international conference on computer science and electronics engineering, ICCSEE 2012 (2012)
2. Gan G, Lu Z, Jiang J (2011) Internet of things security analysis. In: 2011 international conference on internet technology and applications (2011)
3. Yaqoob I, Ahmed E, Hashem IAT, Ahmed AIA, Gani A, Imran M, Guizani M (2017) Internet of things architecture: recent advances, taxonomy, requirements, and open challenges. IEEE Wirel Commun 24:10–16
4. Pan Y, Cheng X, Xia C (2017) World Symposium on Mechanical and Control Engineering (WSMCE) Research and Design of Lightweight Encryption for MQTT Protocol 1:143–145 (2017)
5. Zhao K, Ge L (2013) A survey on the internet of things security. In: Proceedings of the—9th International Conference Computational Intelligence in Security CIS 2013, pp 663–667 (2013)
6. Granjal J, Monteiro E, Sa Silva J (2015) Security for the internet of things: a survey of existing protocols and open research issues. IEEE Commun Surv Tutorials 17:1294–1312
7. Van Dijk M, Gentry C (2010) Fully homomorphic encryption over the integers. Adv Cryptology—... 1–28 (2010)
8. Coron J-S, Naccache D, Tibouchi M Public Key Compression and Modulus Switching for Fully Homomorphic Encryption over the Integers
9. Yi X, Paulet R, Bertino E (2014) Homomorphic encryption and applications. (2014)
10. Lauter K (2004) The advantages of elliptic curve cryptography for wireless security. IEEE Wirel Commun 11:62–67
11. Hankerson D, Hernandez JL, Menezes A (2000) Software implementation of elliptic curve cryptography over binary fields. In: Lecture notes in computer science (including subseries lecture notes in artificial intelligence and lecture notes in bioinformatics) (2000)
12. Chen L, Lim M, Fan Z (2017) A public key compression scheme for fully homomorphic encryption based on quadratic parameters with correction. IEEE Access 5:17692–17700
13. Gai K, Qiu M, Li Y, Liu X-Y (2017) Advanced fully homomorphic encryption scheme over real numbers. In: 2017 IEEE 4th international conference on cyber security and cloud computing (CSCloud) (2017)
14. Sha P, Zhu Z (2016) The Modification of RSA Algorithm To Adapt Fully Homomorphic Encryption Algorithm in Cloud Computing
15. Chahid Y, Benabdellah M, Azizi A (2017) Internet of things security. In: 2017 international conference on wireless technologies, embedded and intelligent systems, WITS 2017 (2017)
16. https://www.slideshare.net/karteek10/secure-hashin
17. https://crypto.stackexchange.com/questions/22956/what-is-the-length-field-in-sha-512-padding

A Proposed Device for Sending Sms Automatically to Doctor When Patient Body Temperature Goes Above Certain Temperature Based on IoTs

Vijay Kumar and Anal K. Jha

Abstract Hereby, we are proposing a device, which uses temperature to voltage sensor, Arduino Uno board, gsm 900 board, fitted on arm of patient to send the sms when the body temperature of patient increases above certain level. This is an idea regarding sending sms automatically. By using this device, there will be no requirement to manually check the temperature on regular time basis. The future work will be to send multiple messages at different temperature with their temperature.

Keywords Sms · IoTs · Temperature to voltage sensor · Arduino Uno board · Gsm 900 board

1 Introduction

We are proposing an IoT-based system for sending sms automatically to doctor when body temperature of a patient goes above certain limit. For this, we require temperature to voltage sensor, gsm 900 board, sim, and Arduino Uno board with proper programming. For this, thin copper plate may be tightly bound to the one temperature sensing body of temperature to voltage sensor. Voltage sensor has three legs: 1Vcc leg 2; +ve voltage leg 3; and Gnd voltage leg.

The +ve voltage leg is connected to analog port pin 3 of Arduino Uno board. Gnd voltage leg is connected to the ground pin of Arduino Uno board. Similarly, the gsm 900 board is fitted with sim and connected to Arduino Uno board.

The analog port pin 3 can sense a voltage and can send the message with gsm 900 board.

V. Kumar (✉) · A. K. Jha
Aryabhatta Knowledge University, Patna, India
e-mail: Vijay898627@gmail.com

S.-L. Peng et al. (eds.), *Computing and Network Sustainability*, Lecture Notes in Networks and Systems 75, https://doi.org/10.1007/978-981-13-7150-9_27

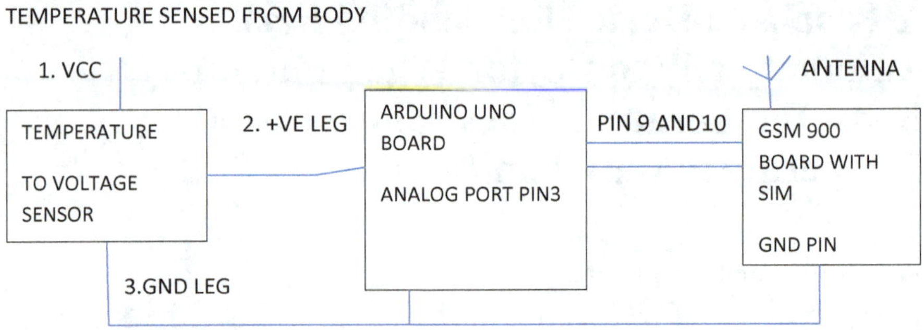

Fig. 1 Block diagram showing procedure

2 Procedure and Method to Accomplish the Task

The procedure and method to perform the task are shown via block diagram and flowchart. We can use LM35 temperature to voltage sensor for converting temperature into voltage. The LM 35 is IC which linearly converts the temperature to voltage. It has ±0.25 °C accuracy at room temperature and ±0.75 °C at temperature range −55–150 °C. The conversion rate is +10 mV/degree celcius. The temperature is measured from the body of device.

After measuring the temperature, the device converts it to the voltage which is provided through the second leg to the Arduino Uno board's analog port pin 3. Third pin of LM35 is connected to the ground pin. Arduino Uno board is properly programmed to sense the voltage at which message is to be sent. For this, the gsm 900 board is fitted with sim and its Tx and Rx pins are connected with pin 9 and pin 10 of Arduino Uno board, respectively. The maximum operating frequency of Arduino Uno is 16 MHz (Fig. 1).

3 Calculations and Other Data

In this paper, we will take the crucial body temperature to be 101°F which is 38.333 °C. So the voltage generated will be 383.33 mV. The Arduino Uno board will be sensing this much voltage, i.e., when analog port 3 of Arduino Uno board will sense 383 mV, it will send the message through gsm 900 board and proper programming of the Arduino Uno board.

As the maximum operating frequency of the Arduino Uno board is 16 MHz so the sensing frequency of the Arduino Uno board will be in mega hertz which time is more than gradual increase in temperature of human body (Fig. 2).

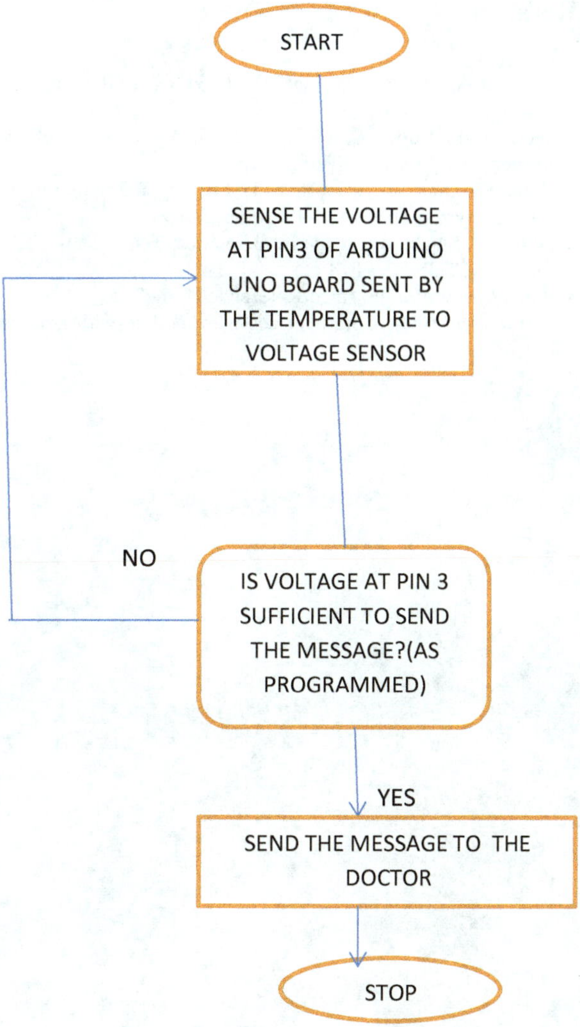

Fig. 2 Flowchart of procedure

4 Conclusion and Future Work

The above method can be used for monitoring the body temperature of a patient. When the body temperature exceeds certain limit, this system sends the message to the doctor. Presently, I done the message sending part and temperature to voltage sensing part is under investigation. It can be a success.

Future work will be to send temperatures in routine format with sms with different temperature.

References

1. Sharma A, Bhuriya D, Singh U (2015) Secure data transmission on MANET by hybrid cryptography technique. IEEE, 2015
2. Verma UK, Kumar S, Sinha D (2016) A secure and efficient certificate based authentication protocol for MANET. IEEE, 2016
3. Jathe SS, Dhamdhere V (2015) Hybrid cryptography for malicious behavior detection and prevention system for MANETs. IEEE, 2015
4. Vijay M, Sujatha R (2014) Intrusion detection system to detect malicious misbehaviour nodes in MANET. IEEE, 2014
5. https://www.arduino.cc/en/Reference/HomePage
6. www.circuitstoday.com/interface-gsm-module-with-arduino

Automated Community Feedback and Monitoring Assistant

T. S. Aswin

Abstract There is a need for social media platforms to provide a streamlined and seamless experience to share one's comments on any event or content. The need for community moderation is ever increasing, as well as the need to make the best use of any and all of the huge data generated, in an ethical and responsible manner. The proposed system allows users to mark their responses either by text or by a speech-to-text conversion tool. Thereon, the system classifies the data and initiates corresponding action workflows. Once the data is found not to be platform abuse, the content is reviewed for any sentiment analysis information that could be extracted. In real time, users can express their opinions freely, responsibly, adhering to community standards of information sharing, as well as allow concerned advertising groups to monitor trends and make informed decisions with the user generated data. The system enforces community moderation standards with least manual effort and processes the content uploaded instantly, to provide valuable insight and add commercial value.

Keywords Automated community moderation · Real-time text analysis · Real-time sentiment analysis · Social media moderation · Platform abuse filtering · Spam filtering

1 Introduction

With the ever-increasing online presence on social media, huge volumes of data are generated and stored online each day. Due to scale and variety of data, monitoring the content for spam, abuse, etc., has proven to be very challenging. It is the need of the hour to understand how best to ensure that the platform is not being misused, as well as to make the best use of the generated data, subject to the rules and stipulated policies. The monitoring assistant program is intended to be applied to public groups and web pages, rather than individuals' accounts. Social media is intended to be a platform for individuals to express their opinion freely, without fear of persecution, apart from

T. S. Aswin (✉)
Data Science Group, 247.ai, Bengaluru, India
e-mail: aswin.tekur@gmail.com

S.-L. Peng et al. (eds.), *Computing and Network Sustainability*, Lecture Notes in Networks and Systems 75, https://doi.org/10.1007/978-981-13-7150-9_28

sharing information about their lives, content including pictures, micro-blog content, etc. There is need for a robust and efficient system to monitor the content to ensure the platform is not being misused by spam, abusive, or intentionally misleading content. At the same time, the system must not be intrusive or in any way interfere with the core goal of an impartial and open system. The generated data may be responsibly used by entities to improve their service, collect valuable information about public opinion, and frame their programs or policies to best fulfill the market requirements.

2 Background Information

Within the past two decades, the trend of social media has been steadily growing and has been predicted to be a permanent medium of interaction among the masses, as well as a way for eminent personalities to interact directly with their followers. Industries, corporations, and even governments have set up affiliated groups and accounts to openly share information about their working, to ensure transparency. A huge challenge for the social media administrators is to combat platform abuse, which could include bullying, spam (e.g., uploading irrelevant content that could be intended for marketing a product at a public web page which was set up only for personal opinion and content) or fake news (intentionally spreading wrong news) to stir up tensions, aggravate groups, cause unrest, etc. Social media companies are exploring options to identify such content and deter such action [1–5]. Engaging personnel to manually identify platform abuse content may not be a scalable solution, given the ever-increasing huge volume of data uploaded on a daily basis [6–9]. There is also a consensus that large corporations need to understand public opinion on a wide range of issues, and framing policies, understanding the needs of the people, all of which require the opinion of the masses. The data obtained by mining the users' entered data would help extract information on a wide range of issues, and applying the best classification techniques would help in building an automated system.

3 System Design

See (Fig. 1).

By using a speech-to-text conversion module, users can enter their oral comments in a user convenient manner. Manual entering of text content is also supported. The content is then passed on to a text classifier built to identify the category of the content. If the content is found to be a case of platform abuse (spam or online bullying or fake news), a user-designed workflow can be followed (e.g., the content may be deleted and the user could receive a warning or account suspension, based on the severity of the offense). On the other hand, if the content is not a case of platform abuse, the content follows a different user-designed workflow. In this case, the content is passed

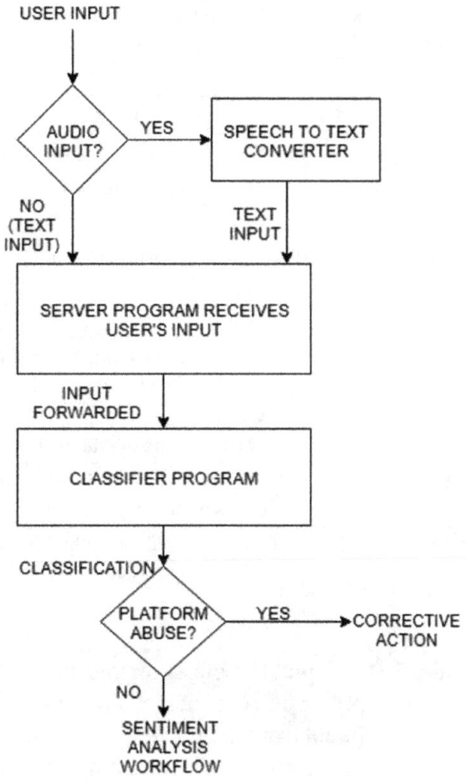

Fig. 1 System workflow

through a sentiment analysis block. The content is examined to identify if the user has left a positive or negative (or neutral) response.

The resulting data can be examined in real time and the course of action followed, with least amount of manual intervention.

4 System Working

Step1: The trained data set is used to build an intent classifier (with the application of NLP preprocessing techniques on the data). The classifier is stored as a pickled file on a remote machine. The pickled file is the result of the training of the model, and any modifications made to build a new classifier require rebuilding the pickled file too. The classifier is tested thoroughly to ensure highest accuracy score-based intent is returned [10–12]. The classifier for platform abuse has been trained using the data set containing the related data (few sample lines listed above in Table 2) [13]. The sentiment analysis data set lines have been listed above (refer Table 1) [14].

Table 1 Sample lines from the sentiment analysis training data set of Twitter content for a commercial airliner

Tweet ID	Text content	Sentiment
570010571707256000	Wow this just blew my mind	Positive
570038941497192000	When can I book my flight to Hawaii??	Neutral
569967019958730000	Everything was fine until you lost my bag	Negative

Table 2 Sample lines from the 'Russian troll tweets' data set on Kaggle

External author ID	Text	Label
1513801268	Slavery in the United States was never abolished. It evolved. #13TH	Platform abuse
1513801268	Racist white trolls in my mentions might be some kin to Gollum. Just a thought	Platform abuse
1513801268	The Election BOYCOTT is the first step to liberation, and will prove that citizens do not select presidents	Platform abuse
1513801268	Black defendants, white jurors: Does race make a difference in the courtroom?	Platform abuse

Step 2: The pickled classifier file is stored in a remote system, integrated with a program. A request to this remote program with a given input can fetch the intent from the pickle file and yield the intent classified. A server application is hosted on this remote machine so that any input to the application is fed as a request to the pickle file.

Step 3: When the application receives a request with some text, the text is passed to the classifier as input. The classifier identifies the intent, and returns the standard result, and/or performs the stipulated action. In this case, the first classifier identifies whether the given input (transcript of audio input, or typed text content) is 'platform abuse' or not. If the text is classified as 'platform abuse', the user defined workflow is carried out. Action could be taken against the user based on the severity of the offensive content posted (if any). If the content is not found to be under the 'platform abuse' category, the text is passed to the next module, where the content is passed to the next classifier to identify the class of the sentiment analysis (positive, negative or neutral, neutral implying no significance in data). Statistics of the sentiment analysis would be of significant value to various fields such as advertisers and researchers. The model is required to be accurate to ensure the classification of the category is accurate.

The system uses a trained intent prediction classifier using a support vector machine (SVM) algorithm (code in Fig. 2). SVM has been chosen to build the classifiers, rather than unsupervised methods because the accuracy of supervised algorithms conventionally outperforms unsupervised approaches [15, 16]. The open

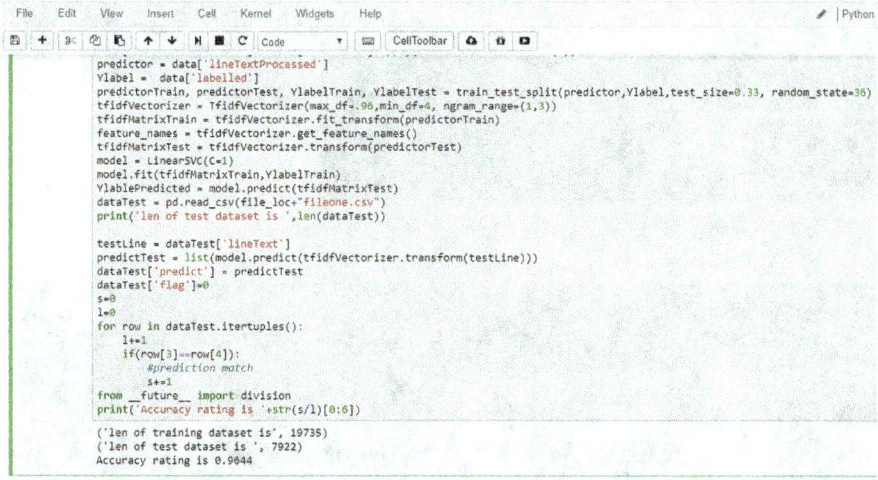

```
predictor = data['lineTextProcessed']
Ylabel = data['labelled']
predictorTrain, predictorTest, YlabelTrain, YlabelTest = train_test_split(predictor,Ylabel,test_size=0.33, random_state=36)
tfidfVectorizer = TfidfVectorizer(max_df=.96,min_df=4, ngram_range=(1,3))
tfidfMatrixTrain = tfidfVectorizer.fit_transform(predictorTrain)
feature_names = tfidfVectorizer.get_feature_names()
tfidfMatrixTest = tfidfVectorizer.transform(predictorTest)
model = LinearSVC(C=1)
model.fit(tfidfMatrixTrain,YlabelTrain)
YlablePredicted = model.predict(tfidfMatrixTest)
dataTest = pd.read_csv(file_loc+"fileone.csv")
print('len of test dataset is ',len(dataTest))

testLine = dataTest['lineText']
predictTest = list(model.predict(tfidfVectorizer.transform(testLine)))
dataTest['predict'] = predictTest
dataTest['flag']=0
s=0
l=0
for row in dataTest.itertuples():
    l+=1
    if(row[3]==row[4]):
        #prediction match
        s+=1
from __future__ import division
print('Accuracy rating is '+str(s/l)[0:6])

('len of training dataset is', 19735)
('len of test dataset is ', 7922)
Accuracy rating is 0.9644
```

Fig. 2 Code snippet for the text classifier program

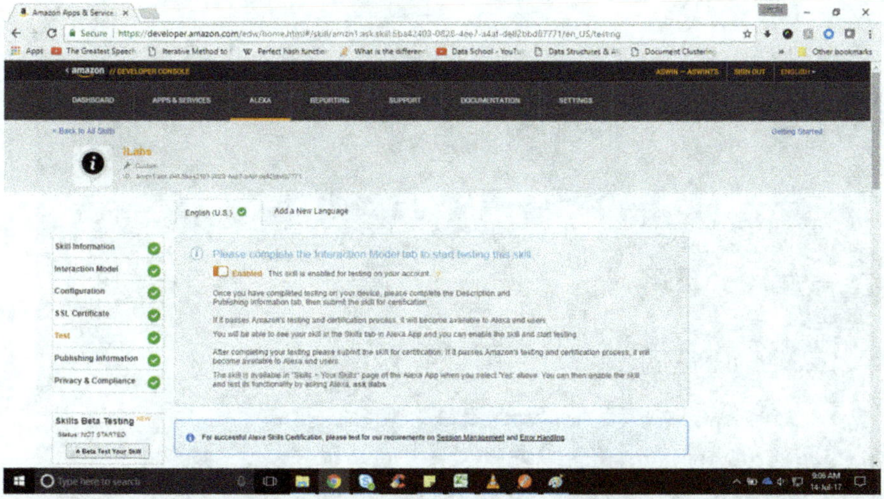

Fig. 3 Interface for the customized speech-to-text conversion program

available Alexa speech-to-text conversion tool has been used to demonstrate the speech-to-text conversion functionality (see Fig. 3).

Once the server application is set up, the application is tested by simulating a client request. The system must be checked by confirming that the system is responding to basic invocation requests (Fig. 4).

Once the remote server application is hosted, a request is sent to the server and the program is run multiple times to confirm that the intent is accurately classified for different inputs (see Fig. 5).

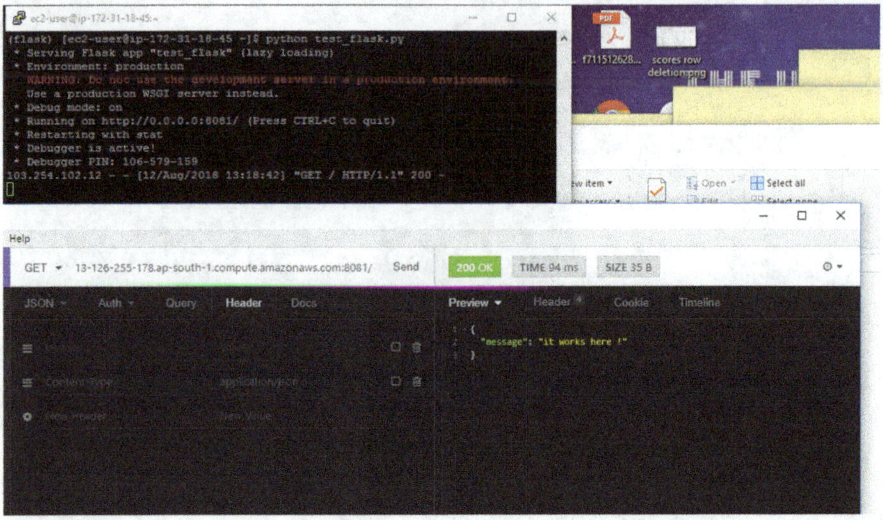

Fig. 4 Testing of remote server application

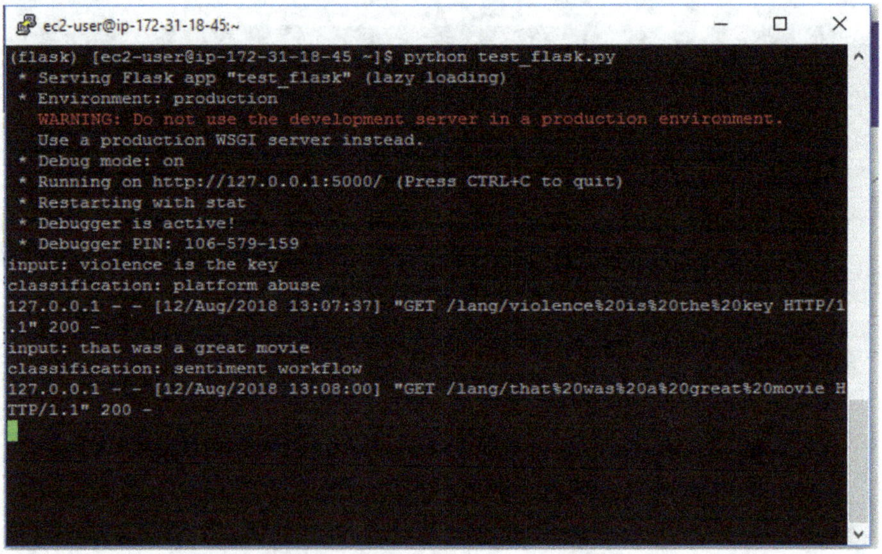

Fig. 5 Testing the remote application for accurate intent prediction

5 Limitations

1. It is presumed that the speech-to-text conversion module is capable of recognizing users' input content.
2. It is presumed that the libraries used in building the program are sufficiently capable of dealing with the languages that the program is designed to support.
3. The program may support other languages as per user's requirement, subject to the availability of NLP tools to process other languages.
4. It is presumed that the system can handle large number of input requests concurrently and can handle with the requests in near real time.
5. The limitations of the NLP tools and techniques may impact accurate recognition of input from the user and thereon the identification of the intent accurately.
6. The system is designed for non-personal media accounts. Rather than mine the content of personal accounts' it would be more beneficial to monitor the content posted on media accounts of large corporations, public bodies, institutions, eminent personalities, etc., so as to monitor the platform for any unwanted content, as well as receive information about the public opinion and response to newly launched programs, products, etc.

6 Challenges

1. It is essential that the intent predicted is the correct one, as the workflows and path taken up may differ based on the intent predicted.
2. From a user convenience perspective, the ideally optimized system would ensure the entire prediction life cycle is completed in a few seconds.
3. The nature of trained data will shape the system's functionality and application. It is highly crucial that newer data is appended to the training data set and newer models are trained so as to incorporate new changes and trends in the data to ensure the classifier models are relevant, accurate, and efficient.

7 Future Scope

1. The workflows can be customized, to extend the system application to trigger warnings or alerts, based on the criticality of the action.
2. The system could be used as a real-time reporting tool. Adding some customized functionality could ensure reports are generated on user-specified parameters.
3. Furthermore, the sentiment analysis content could be ranked to yield more information about the sentiment of content than just indicate about positive, negative, or neutral. Moreover, consider a rating of 1–5 (e.g.) along with the sentiment of positive, negative, or neutral. The additional information insight would add

more value, if considered as a reporting tool. In the scenario that the workflow is considered spam or abusive content also, the content could be checked for severity and a score is assigned to indicate a higher severe offense, action is to be initiated based on the score assigned.
4. Newer techniques for pattern matching have risen, requiring large computational frameworks and expensive hardware for processing large quantities of data and building models quickly. These 'deep learning' technologies may advance accuracy scores of prediction, even with huge training data sets [17].
5. Many different options are present to make best use of available data to build highly efficient classifiers [18]. Care must be taken not to overfit the classifiers on the training data sets [19].

8 Conclusion

Thus, a community moderation tool has been built to operate on social media account data (operating in real time) to identify cases of platform abuse and identify the severity of the same, if any, and follow the workflow for custom built action. Data that is not considered platform abuse is used to generate sentiment analysis statistics, thus generating valuable user feedback.

Acknowledgements The Data Science Group of 247.ai India is thanked for all the assistance that made this project possible. Sincere thanks are also conveyed to Kaggle and their associated contributors for the data sets sought that were essential to the building of the classifiers.

References

1. The Guardian (2018) Facebook offers plan to tackle fake news ahead of US midterms. https://www.theguardian.com/technology/2018/mar/29/facebook-fake-news-political-ad-security-us-midterms-2018. Retrieved from 29 March 2018
2. Business Insider (2018) Facebook is finally launching a new feature to combat fake news, after six months of testing—here's how it works. https://www.businessinsider.in/Facebook-is-finally-launching-a-new-feature-to-combat-fake-news-after-six-months-of-testing-heres-how-it-works/articleshow/63603421.cms. Retrieved from 4 Apr 2018
3. Business Standard (2018) Researchers develop new algorithm to detect fake accounts on FB and Twitter. https://www.business-standard.com/article/technology/researchers-develop-new-algorithm-to-detect-fake-accounts-on-fb-and-twitter-118041800601_1.html. Retrieved from 18 Apr 2018
4. Business Standard (2018) Facebook, Google warn Singapore against new laws to combat fake news. https://www.business-standard.com/article/international/facebook-google-warn-singapore-against-new-laws-to-combat-fake-news-118032201346_1.html. Retrieved from 22 March 2018
5. Business Standard (2018) Facebook begins 'fact-checking' photos, videos to reduce false news stories. https://www.business-standard.com/article/international/facebook-begins-fact-checking-photos-videos-to-reduce-false-news-stories-118033000778_1.html. Retrieved from 30 March 2018

6. The Washington Post (2018) As Facebook confronts data misuse, foreign governments might force real change. https://www.washingtonpost.com/news/worldviews/wp/2018/04/05/as-facebook-confronts-tough-questions-on-data-misuse-europe-might-force-real-change/?noredirect=on&utm_term=.0a2e0dfd6d74. Retrieved from 4 Apr 2018
7. The Washington Post (2018) Facebook is now in the data-privacy spotlight. Could Google be next? https://www.washingtonpost.com/news/the-switch/wp/2018/04/11/facebook-is-now-in-the-data-privacy-spotlight-could-google-be-next/?noredirect=on&utm_term=.ebdd054539bc. Retrieved from 11 Apr 2018
8. BBC News (2018) Social media firms criticised for lack of cyber-bullying action. https://www.bbc.com/news/av/uk-politics-44197761/social-media-firms-criticised-for-lack-of-cyber-bullying-action. Retrieved from 21 May 2018
9. Bloomberg (2018) U.K. Seeks penalties for tech giants to stop cyber bullying. https://www.bloomberg.com/news/articles/2018-05-20/u-k-seeks-penalties-for-tech-giants-to-stop-cyber-bullying. Retrieved from 20 May 2018
10. Yu B (2008) An evaluation of text classification methods for literary study. Lit Linguist Comput 23(3):327–343
11. Tong S, Koller D (2001) Support vector machine active learning with applications to text classification. J Mach Learn Res 2:45–66
12. Ikonomakis M, Kotsiantis S, Tampakas V (2005) Text classification using machine learning techniques. WSEAS Trans Comput 4(8):966–974
13. Kaggle, NBC News. The Russian Troll tweets. Retrieved from https://www.kaggle.com/vikasg/russian-troll-tweets
14. Kaggle, Twitter US Airline Sentiment. Retrieved from https://www.kaggle.com/crowdflower/twitter-airline-sentiment/home
15. Dalianis H (2018) Computational methods for text analysis and text classification. Clinical text mining. Springer, Cham., pp 83–96
16. Deng X, Li Y, Weng J, Zhang J (2018) Feature selection for text classification: a review. Multimed Tools Appl pp 1–20
17. Singh J, Singh G, Singh R, Singh P (2017) Optimizing accuracy of sentiment analysis using deep learning based classification technique. In International conference on recent developments in science, engineering and technology, Springer, Singapore, Oct 2017, pp 516–532
18. Wang YW, Feng LZ (2018) A new feature selection method for handling redundant information in text classification. Front Inf Technol Electron Eng 19(2):221–234
19. Sun X, He J (2018) A novel approach to generate a large scale of supervised data for short text sentiment analysis. Multimed Tools Appl, pp 1–21

Power-Efficient Approach to Optimize SHA-256 Bit Using Reversible Logic

S. H. Rohini, G. Nischal and Rajashekar B. Shettar

Abstract With the growing network to connect people all over the world, network security has received major attention in this field. Network security provides a secured way of communication eliminating unauthorized access and misuse of important data. Cryptography is the major entity of network security and involves conversion of readable data to apparent nonsense. Secured hash algorithm (SHA) is the family of cryptographic hash functions which are one-way faster with good security and consumes low power but they have attacks like brute-force attack, side-channel attack(particularly differential power analysis) and the main reason behind these power attack is power consumption. This paper is intended to implement SHA-2 design using reversible logic. Reversible logic has received great attention in low-power VLSI design, nanotechnology and quantum computing, etc. SHA-2 contains different length output hash functions which are 224, 256, 384 and 512 bits and are used in significant applications like digital signatures, authentication and message integrity checks. A novel approach in design of SHA-2 to achieve more security, less hardware and power consumption is minimized by using reversible logic. Proposed SHA-2 algorithm of 256 bits has been simulated using Xilinx through reversible logic. Proposed SHA-2 is power efficient and having less hardware complexity in comparison with existing work. Performance analysis of reversible implementation of SHA-2 with respect to gates used, garbage and quantum cost is done. Synthesis analysis of our work shows that total time delay is 626 ns and the final result shows 12% reduction in hardware requirements.

Keywords SHA · Reversible · Cryptography · VLSI

S. H. Rohini (✉) · G. Nischal · R. B. Shettar
B V Bhoomaraddi College of Engineering and Technology, Hubli 580031, Karnataka, India
e-mail: rohini_sh@bvb.edu

G. Nischal
e-mail: nischal.giriyan@gmail.com

R. B. Shettar
e-mail: raj@bvb.edu

S.-L. Peng et al. (eds.), *Computing and Network Sustainability*, Lecture Notes in Networks and Systems 75, https://doi.org/10.1007/978-981-13-7150-9_29

1 Introduction

In recent years with the growing network, communication between people has become very easy but along with that, it also matters how secure the communication is, because of all the important and confidential data that are being exchanged over the network [1]. Cryptography is the practice and study of techniques for secured communication in presence of the third parties called adversaries. One more important term used in cryptography is cryptanalysis means the study of how to crack encryption algorithms. Modern cryptography is broadly classified into three categories. (1) symmetric cryptography, (2) asymmetric cryptography and (3) hash functions . Symmetric-key cryptography refers to encryption methods in which both the sender and receiver will have the same key [2]. Asymmetric cryptography is a type of cryptography in which two different keys(private key and public key) but related by their generation are used. [2]. In a public key encryption system, the public key is used for encryption, while the private or secret key is used for decryption. Hash functions are the algorithms which do not use any key for encryption but a fixed-length hash value is computed based upon the plain text that makes it impossible to recover data from it. They are used to provide the digital fingerprint of a file's content. They have very special properties like pre-image resistance, collision resistance and second pre-image resistance which make them harder to crack even by means of brute-force attack [3]. The MD5 algorithm is a widely used hash function producing a 128-bit hash value. Message digest (MD5) was designed by Ronald Rivest in 1991 to replace an earlier hash function MD4 [3, 4]. Secure hash algorithms (SHA) are one of the best examples of cryptographic hash function with major applications in the field of security. SHA-1 is the hash function which takes an input and produces 160-bit hash value as output. It uses similar structure as that of MD4 and MD5 [5]. SHA-2 includes significant changes compared to SHA-1. It consists of a family of six hash functions that are with different output bit lengths and they are 224, 256, 384, 512, 512/224 and 512/256 bit. SHA-256 is computed with 32-bit word and SHA-512 is computed with 64-bit word. They both have a similar structure with only difference in shift amounts, additive constants and number of rounds. SHA-224 and SHA-384 are simply truncated versions of the first two, respectively, computed with different initial values. SHA-256 based on MD5 structure with input length less than 264 bit is the widely used hash function as it has enough high output length for most of the practical applications. SHA-256 algorithm has major steps like padding, word expansion and main loop which will run for 64 times with initial hash values as input. It involves 64 constants each of 32 bit derived from the fractional parts of the cube roots of the first sixty-four prime numbers [6]. Each time after main loop runs, there is swapping of registers involved in the round and after the completion of last round, it is added with initial hash values to get the final output hash. SHA-2 hash functions are widely used in protocols like TLS, SSL and PGP. Currently, the best public attacks break pre-image resistance for 52 out of 64 rounds of SHA-256 or 57 out of 80 rounds of SHA-512, and collision resistance for 46 out of 64 rounds of SHA-256 [7]. SHA-3 is the new member of the SHA family released by NIST in 2015. SHA-3 is based on the cryptographic family KECCAK, which is based on a novel approach called sponge construction [8]. In the recent years, due to their ability to reduce the power dissipation, reversible logic has received attention in the field of VLSI and research. Reversible logic supports the unique way of generation of input from the output which is one-to-one mapping. The actual idea behind this logic is energy dissipation can be reduced if the computation becomes information lossless [9]. Reversible logic has found a major application in cryptography because this field needs a method to address and reduce the power consumed while carrying out such a large number of operations in loop.

2 Proposed Idea

The algorithm SHA-256 has initial hash values fixed (32 bits) which are derived from the square root of the first eight prime numbers and there are no specific reasons for the selection of such fixed values. So in order to make it more secured instead of directly taking these fixed values, we propose the idea of making those hash values dependent on the input text message. Our proposed algorithm can be added as module to the initial pre-processing stage of the current SHA-256 implementation leaving less changes in the rest of the rounds. Keeping hardware, power requirements and cost of implementation in mind, we have modified initial hash values based on the incoming message and reduced further rounds (if standard implementation has *r* rounds) in main loop to (*r*-16) to maintain the complexity of the hashing and security level as that of SHA standard implementation. Flow chart of the proposed algorithm is given in Fig. 1.

2.1 Proposed Algorithm

(1) Initial Hash Value Modification:
The initial hash value $H^{(0)}$ is the following sequence of 32-bit words as in the standard implementation by NIST [6]:

$H_1 = \text{6a09e667}$; $H_2 = \text{bb67ae85}$; $H_3 = \text{3c6ef372}$; $H_4 = \text{a54ff53a}$
$H_5^{(0)} = \text{510e527f}$; $\mathrm{H}_6^{(0)} = \text{9b05688c}$; $H_7^{(0)} = \text{1f83d9ab}$; $H_8^{(0)} = \text{5be0cd19}$

Considering the input message (represented each character in ASCII representation). Add first two bytes, middle byte and last two bytes of the input message, represent this number in binary 32-bit format (m) and calculate,

1. A = (Middle byte of message) mod 32 2. B = (Total sum of the message) mod 32

To update initial hash values ($H_1^{(0)}$–$H_8^{(0)}$) considering two conditions there are four cases,

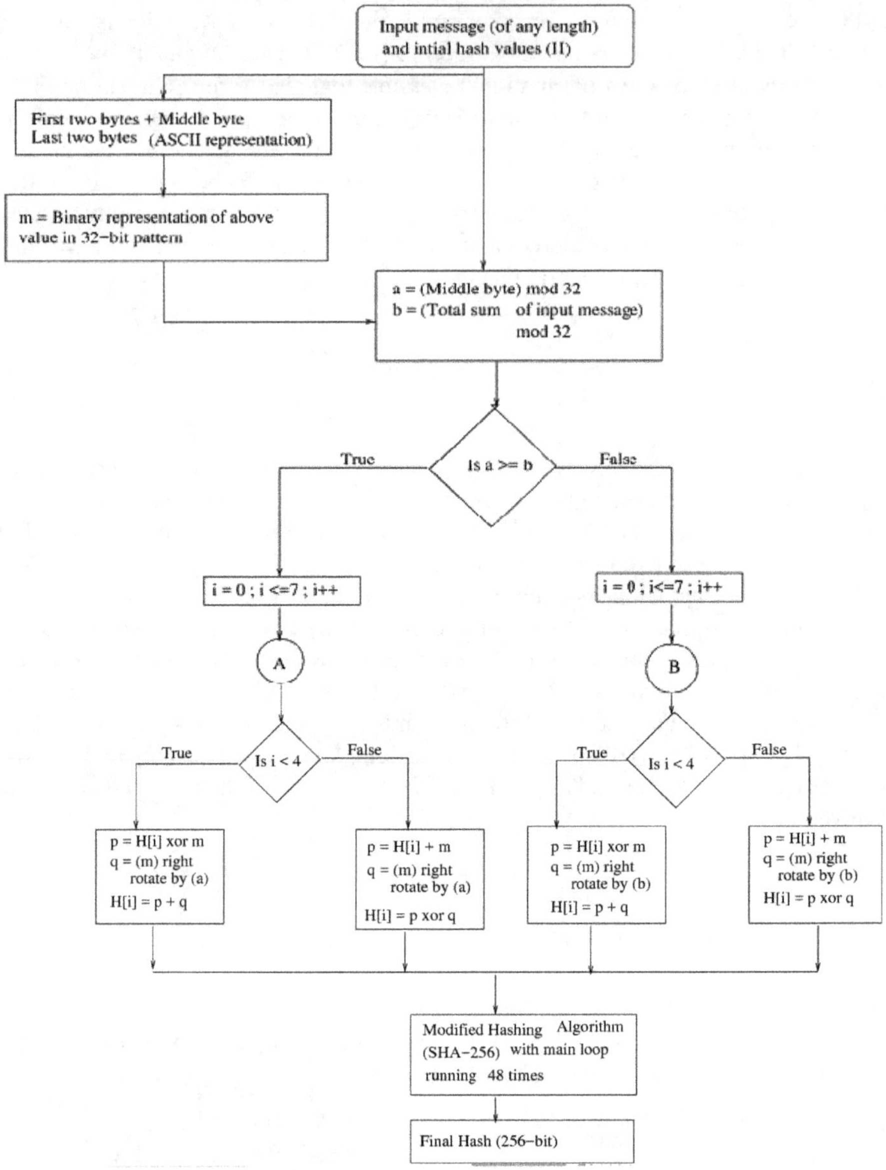

Fig. 1 Flowchart of proposed algorithm

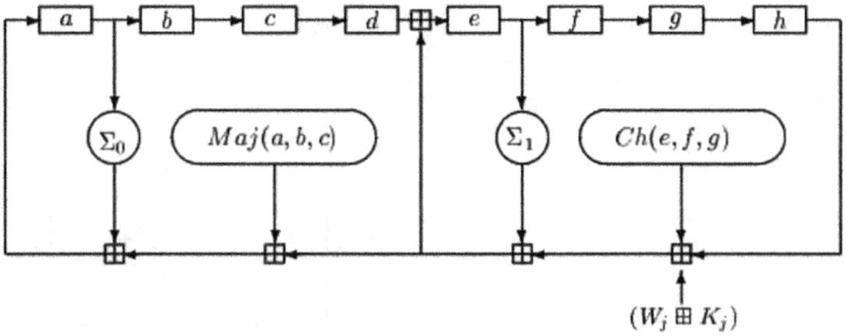

Fig. 2 Compression function

(a) If (A ≥ B) and (i < 5) then, $H_i^{(0)} = ((H_i^{(0)}$ XOR m) + (m rotate right by A))
(b) If (A ≥ B) and (i ≥ 5) then, $H_i^{(0)} = ((H_i^{(0)}$ + m) XOR (m rotate right by A))
(c) If (A < B) and (i < 5) then, $H^{(0)} = ((H^{(0)}$ XOR m) + (m rotate right by B))
(d) If (A < B) and (i ≥ 5) then, $H_i^{(0)} = ((H_i^{(0)}$ + m) XOR (m rotate right by B))
Now all these eight hash values are fed to the main loop of SHA algorithm [6] and the number of rounds in the main loop is reduced to (*r*-16), i.e., for SHA-256 (64-16), 48 rounds. These changes will make this proposed work more hard to crack by means of brute-force attack as the initial fixed hash values are incoming text dependent.

(2) Word Expansion:
This module involves indeed multiple steps to arrive at final expanded word list. SHA-256 algorithm is flexible enough to operate on a wide range of message length till <2^{64} bit. Incoming message is padded with the zero's and message length in the multiples of 512 bit and then divide this padded block into message blocks $M^{(1)}$, $M^{(2)}$, …, $M(N)$ each of 512 bits. Each message block is divided into 16 words each of 32 bit. These are fed to the word expansion block which will run for 32 times to generate 32 unique words. Word expansion equation is same as mentioned in the standard implementation of SHA-256 [6].
(3) Compression Function:
SHA-256 compression function consists of main loop where actual hashing of input message takes place. This module takes expanded words, pre-defined 48 key constants and modified initial hash values as input. Initially modified hash values are copied into temporary registers and then the loop will run on these registers for 48 times. The values of the registers in the final round will be added with initially modified hash values in a specified manner to get the final 256-bit hash. Single run of main loop is shown in Fig. 2. Six functions are used in SHA-256, each of which operating on 32-bit word, which internally they uses XOR, Right rotate, Right shift and Modulo 2^{32} addition operations and their definitions are kept as mentioned in standard implementation of SHA-256 by NIST [6].

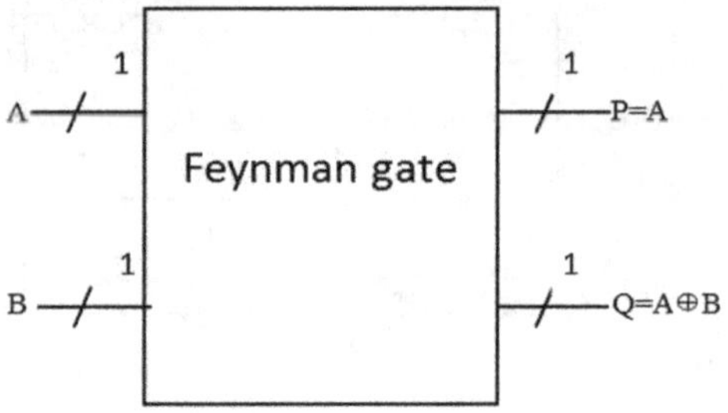

Fig. 3 XOR using Feynman gate

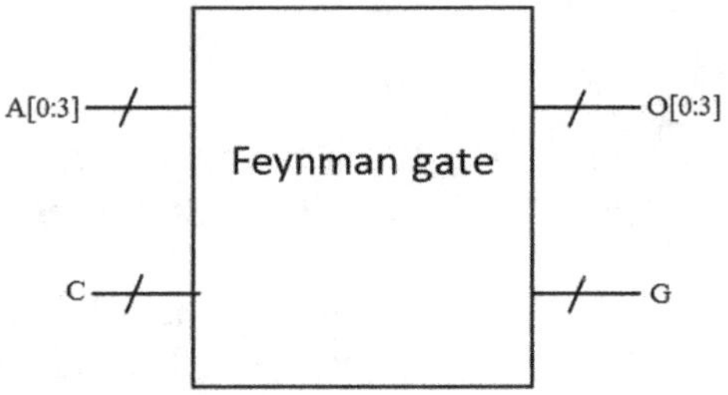

Fig. 4 Right rotate using Feynman gate

2.2 *Reversible Logic Implementation*

During the implementation of proposed algorithm with reversible logic, Feynman gate is used to replace all the XOR gates and the block diagram of same is given in Fig. 3. The outputs are defined by $P = A$ and $Q = A \oplus B$. Quantum cost of a Feynman gate is 1. In implementation of proposed algorithm, the output P is garbage when used in replacement of XOR gate and output Q is garbage when used in replacement of Right shift or Right rotate. So in general, whenever this gate is used in implementation, it produces one garbage. Block diagram of Right rotate operation using Feynman gate is shown in Fig. 4 where A, B, C and D are of 8 bits, so together making a 32-bit word [10, 11].

2.3 *Example of Proposed Algorithm*

With the help of one example, proposed algorithm is explained to generate hashing values in this section.

2.3.1 Example 1

Consider an example of input message abc and ASCII equivalent representation is,

$$a\ b\ c\ 97\ 98\ 99$$

Hexadecimal representation of these ASCII values are 61, 62 and 63, respectively, and the equivalent binary representation is: 011000010110001001100011. Padding is done as explained below: Step (1) Add 1 to the end and put the numbers together: 0110000101100010011000111

Step (2) Append 0's to the end as explained before in the algorithm zero's should be added to the end until the message chunk is multiple of 512 bits and message length is 24 bit. 01100001011000100110001110000…0000000000 $\Rightarrow$ 48 bit

Step (3) Append original message length. You will now add the 64-bit representation of the original message length at the end of the current message. 01100001011000100110001110000…0000000000 000….00011000 $\Longrightarrow$ 512 bit. The message length should now be an exact multiple of 512.

Step (4) Chunk the padded message: Break the padded message into chunks each of 512 bit and break the chunk into words.

Step (5) Change the initial hash values depending on the message. According to the proposed algorithm, add first two bytes, middle byte and last two bytes of the message. This is possible when the message length is at least 40 bit, but here the message length is 24 bit. Consider the message: abc

a	b	c
First byte	Middle byte	Last byte

Due to the absence of second and last but one byte only other three bytes are considered for the addition. Sum is 126 (in Hex) m = 00000000000000000000000100100110 And A = (Middle byte) mod 32 = 98 mod 32 (in Decimal) = 62 mod 20 (in Hex)

A = 02, B = (Total sum) mod 32 = 294 mod 32 (in Dec) = 126 mod 20 (in Hex) = 06

As (A< B) according to algorithm, rotation amount is B. To update $H_1{}^{(0)}$ to $H_4{}^{(0)}$ the operations carried out according to the technique explained in algorithm and hash values are

$H_1^{(0)}$ = 11101010000010011110011110001010 (0209E745)
$H_2^{(0)}$ = 01010011011001111010111110100111 (5367AFA7)
$H_3^{(0)}$ = 11010100011011101111001001011000 (D46EF258)
$H_4^{(0)}$ = 00111101010011111111010000100000 (3D4FF420)

According to proposed algorithm, to update $H_5^{(0)}$–$H_8^{(0)}$ the operations carried out is

$H_i^{(0)}$ = (($\mathrm{H}_\mathrm{i}^{(0)}$ + m) XOR (m right rotate by B)) and hash values are
$H_5^{(0)}$ = 11001001000011100101001110100001 (C90E53A1)
$H_6^{(0)}$ = 00000011000001010110100110110110 (030569B6)
$H_7^{(0)}$ = 10000111100000111101101011010101 (8783DAD5)
$H_8^{(0)}$ = 11000011111000001100111000111011 (C3E0CE3B)

Fig. 5 xilinx simulation result for example 1

Table 1 Analysis of proposed algorithm

Rounds	Gates used	Garbage	Quantum cost	No. of gates
Modifying initial hash	Feynman	256	32	32
Word expansion	Feynman	128	32	32
Main loop	Feynman	224	32	32

Step (6) So these values can be fed to the main loop of SHA-256 to carry on further steps. As explained earlier in the proposed algorithm, the main loop will run for 48 rounds instead of 64 rounds, and final hash values will be calculated and will be appended in the order H_8–H_1 to get the final 256-bit hash.

Final hash value of input 'abc' is:

F8DD01AAB491CE0FD868965049BCE4755605CF988D72308B9FB5BCB7295 DCF0A.

3 Discussion of Results

All the examples mentioned above are theoretically verified with respect to the proposed algorithm and proven to be efficient compared to the standard implementation in regard to security and other parameters like hardware usage and time consumption in xilinx simulation. The simulation results of proposed algorithm are shown in Fig. 5 for input 'abc'. Performance analysis of proposed algorithm is given in Table 1. The parameters analyzed are rounds used, gates used, garbage, quantum cost and number of gates used.

4 Summary and Conclusions

Novel approach for SHA-2 algorithm is proposed by modifying initial hash values and reducing number of rounds in main loop. In order to overcome differential attacks, the reversible design of proposed SHA-256 is proposed and verified using xilinx tool. Our work shows efficient in terms of power, hardware and computation time with same security level as that of existing work. Synthesis analysis of our work shows that total time delay is 626 ns and the final result shows 12% reduction in hardware requirements.

References

1. Mahajan P, Sachdeva A (2013) A study of encryption algorithms AES, DES and RSA for security. Glob J Comput Sci Technol Netw, Web Secur 13(15):7–8
2. Bhanot R, Hans R (2015) A review and comparative analysis of various encryption algorithms. Int J Sec Appl 9(4):289–306
3. Al-Kuwari S, DavenPort JH, Bradford RJ (2011) Cryptographic hash functions: recent design trends and security notions 2011, pp 565
4. Rivest R (1992) The MD5 message-digest algorithm. RFC Editor. https://doi.org/10.17487/rfc1321
5. FIPS 180-1.Secure hash standard (SHS) (1995) National institute of standards and technology, Apr 1995
6. FIPS 180-2.Secure hash standard (SHS) (2002) National institute of standards and technology, Aug 2002
7. IlyaMironov (2005) Hash functions—theory, attacks and applications. Microsoft Research, Silicon Valley Campus, 14 Nov 2005
8. FIPS 202. SHA-3 standard (2015) permutation-based hash and extendable-output functions, National Institute of Standards and Technology, Aug 2015
9. Yelekar PR, Chiwande SS (2011) Introduction to reversible logic gates and its application. In: 2nd national conference on information and communication technology (NCICT)
10. Rohini H, Rajashekar S, Kumar P (2016) Design of basic sequential circuits using reversible logic. In: International conference on electrical, electronics, and optimization techniques(ICEEOT), March 2016, pp 2110–2115
11. Rohini H, Rajashekar S (2016) Design of reversible logic based combinational circuits. Commun Appl Electr (CAE) 5:38–43

A Review of the Medical Image Segmentation Algorithms

J. E. Anusha Linda Kostka

Abstract The discovery of X-ray in 1895 initiated the era of medical imaging diagnostics. Since then, medical imaging systems have realized unprecedented advancements. These systems have also turned out to be invaluable tools in the practice of diagnostic medicine. However, despite the significant development in medical imaging technologies, processing medical images still pose a substantial challenge especially when it comes to image segmentation. That problem is gradually being alleviated by the implementation of digital medical image processing, especially in the diagnosis and treatment of brain tumors. But the capability of most of the contemporary image delineating algorithms remains limited. Therefore, there is a need to come up with the new medical image segmentation programs to fully utilize the power of digital image processing. In light that, this article reviews some of the contemporary algorithmic protocols for brain tumor delineation systems and how effective they are.

Keywords Algorithm · Brain tumor · Digital imaging processing · Image segmentation

1 Introduction

Image processing is "the manipulation and analysis of information contained in images" (Maintz 2005, p. 10). On the other hand, the digital medical image processing means the delivery of digital images processing for medicine [1]. Image processing includes various core stages. These are the image creation, visualization, analysis, management, and enhancement phases. At the creation stage, an image is captured and then rendered into a digital image matrix by the use of suitable sensors [1, 2]. At visualization leg, the model formed is manipulated to output an optimized image.

During the analysis point, the image is quantitatively measured and abstractedly interpreted [1]. This stage requires prior knowledge and a precise set of algorithms to

J. E. Anusha Linda Kostka (✉)
Noorul Islam Centre for Higher Education, Kumaracoil, India
e-mail: anuday222@yahoo.com

S.-L. Peng et al. (eds.), *Computing and Network Sustainability*, Lecture Notes in Networks and Systems 75, https://doi.org/10.1007/978-981-13-7150-9_30

ensure that whatever the image represents is discerned correctly [1]. The management part of image processing involves efficient storage, communication, and transmission, archiving, and retrieval of the image [1]. The last phase, image enhancement, is a low-level processing step. Unlike analysis, it requires little knowledge and can either be manual or automatic [1].

2 The Challenges of Image Segmentation

The most significant problems doctors encounter in medical image processing are low quality, the varying nature of image captured even with standardized image-creating protocols, problematic delineation of objects, and complicated algorithms [1]. As a result of these challenges, image segmentation is one of the most crucial parts of image processing since it ensures the accuracy of an interpretation and ultimately that of diagnosis. Segmentation is a technique for separating objects from the background [3]. This process is one of the most challenging procedures in medical imaging processing. The reason for this is because most of the modalities used for capturing medical images are harmful to the body. Consequently, they are required to be used in small doses and at lower energy [1]. Therefore, the outputs usually have poor signal-to-noise ratio. Since manual segmentation needs a trained radiologist and takes a lot of time and energy, [3] argue that it is crucial to automate this process to expedite the process and improve on the accuracy and dependability of diagnosis.

3 Proposed Segmentation Algorithms: Sobel Operator, Dependent Thresholding, and Close-Contour Methods

With the aim to create a perfect algorithm for brain tumor segmentation, Aslam et al. [3] propose a modified model combining Sobel operator and automatic dependent thresholding methods for the extraction of tumor edges with the aim of extending to object segmentation (p. 431). Then, closed-contour is applied on those edges to locate closed areas of an image. Finally, the tumor is extracted from the MRI image. The projected algorithm has four stages. These are (1) looking for gradient image by Sobel operator, (2) calculating image-dependent threshold repeatedly, (3) use closed-contour algorithm, and (4) separating the object based on the pixel concentration inside the closed contour [3]. Figures 1 and 2 denote the modified algorithm compared to the conventional one.

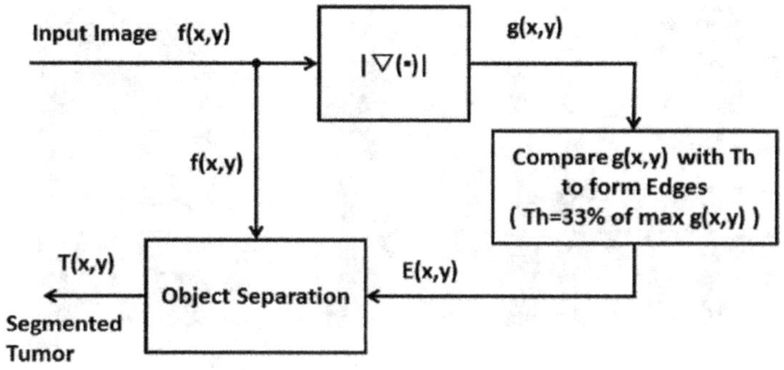

Fig. 1 Conventional algorithm

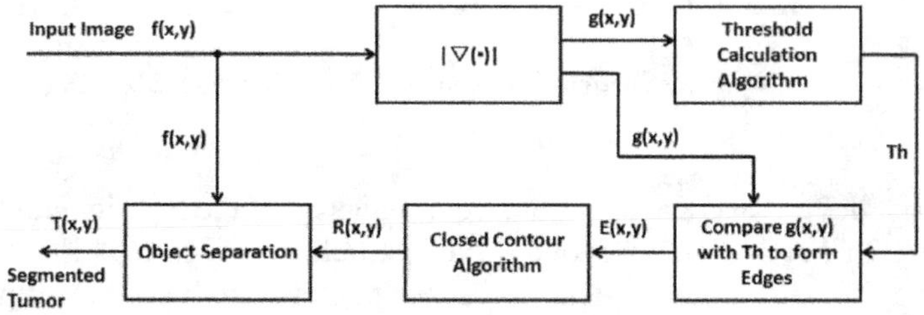

Fig. 2 Proposed algorithm

4 New Triple Modality: MRI-Photoacoustic-Raman-Nanoparticle

Kircher et al. [4] propose the use of a novel triple protocol that combines magnetic resonance imaging (MRI), photoacoustic imaging, surface-enhanced Raman scattering (SERS), and injections of nanoparticles (MPRs) to the tumor to map its margins accurately (p. 3) that allows for the complete removal of a tumor during surgery which is essential for the treatment of tumors [4]. The results of an experiment conducted on mouse glioblastoma models indicate that this technique performance is superior to most of the usual methods in use today. The triple-modality method depends on injecting MPRs, which have higher permeability properties and are only absorbed by the tumor. The MPRs are also retained for a more extended period by the malignant cells [4]. Once the three modalities are applied, the entire tumor location is mapped in more precise details. Also, given that the injected MPR stays for a protracted period in the tumor, both the pre-operational evaluations and the operational processes since the radiologist and neurosurgeon see the same probe preoperatively and during surgery. The photoacoustic imaging allows for high spatial resolution, three-dimensional imaging while the Raman imaging offers high sensitivity and high-resolution surface imaging. The Raman imaging properties also allow for post-operation analysis, which provides for explicit confirmation of margins [4].

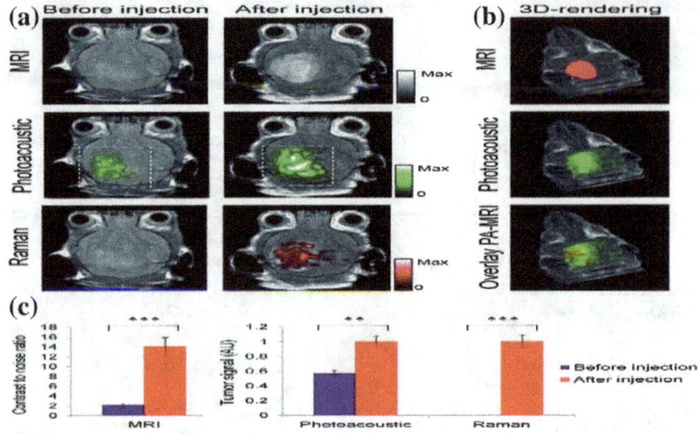

Fig. 3 Kircher et al.'s [4] representation of pre- and post-triple-modality trials

The tests' definitive conclusions are a consequence of the high longitudinal relativity of MPRs that is the highest of any nanoparticle ever reported [4]. The photoacoustic imaging allows the imaging of deeper tissues with high three-dimensional resolution. With the MPR's excellent optical absorbance and the 3D capability of the photoacoustic, the process allows even for the tumor hidden behind normal brain cells to be detected [4]. Raman imaging, which has super sensitivity, is then applied to map the tiniest of the tumor masses [4]. The property of the MPRs permits repeated imaging without the need for more injections [4]. Overall, the triple-modality method of delineating brain tumor tissues improves the localization of the tumor and significantly improves the signal-to-noise ratio as shown in Fig. 3.

5 Conclusion

The technology for medical imaging has taken big strides since the invention of the X-ray in 1895. Despite the unparalleled technological advancement in the photography field, medical imaging still suffers from many shortcomings. One of the biggest issues that hinders medical imaging is low-quality images. Since most of the modalities used for imaging in the healthcare sphere are harmful to the body, they are applied in low doses and operated at low-energy levels. As a result, medical images are afflicted by poor signal-to-noise ratios that make it difficult for medical professional to evaluate a diagnosis, especially when it comes to segmentation of brain tumor. But recent studies propose new models and protocols for image alienation that aims at improving this process. One of the proposed processes involves the combination of Sobel operator, automatic dependent thresholding, and close-contour methods with the final stage being the extraction of the tumor image from the MRI. The other method includes the use of a triple-modality system that utilizes magnetic resonance imaging (MRI), photoacoustic imaging, surface-enhanced Raman scattering (SERS), and injections of nanoparticles (MPRs) to the tumor to enhance the mapping of the tumor margins accurately. Both of these proposed protocols have shown great potential as shown by the sample images above.

References

1. Deserno T (2009) Medical image processing. In: Optipedia. SPIE Press, Bellingham, WA
2. Dougherty G (2009) Digital image processing for medical applications. Cambridge University Press, Cambridge, UK
3. Aslam A, Khan E, Beg MS (2015) Improved edge detection algorithm for brain tumor segmentation. Proc Comput Sci 58:430–437. https://doi.org/10.1016/j.procs.2015.08.057
4. Kircher MF, De la Zerda A, Jokerst JV, Zavaleta CL, Kempen PJ, Mittra E, Gambhir SS (2012) A brain tumor molecular imaging strategy using a new triple-modality MRI-photoacoustic-Raman nanoparticle. Nat Med 18(5):829–834. https://doi.org/10.1038/nm.2721
5. Cheng Y, Morshed RA, Auffinger B, Tobias AL, Lesniak MS (2014) Multifunctional nanoparticles for brain tumor imaging and therapy. Adv Drug Deliv Rev 66:42–57. https://doi.org/10.1016/j.addr.2013.09.006
6. Despotović I, Goossens B, Philips W (2015) MRI segmentation of the human brain: challenges, methods, and applications. Comput Math Methods Med 2015:1–23. https://doi.org/10.1155/2015/450341
7. Luo Y, Liu L, Huang Q, Li X (2017) A novel segmentation approach combining region- and edge-based information for ultrasound images. Biomed Res Int 2017:1–18. https://doi.org/10.1155/2017/9157341
8. Meyer-Baese A, Plant C, Gorriz Saez JM (2014) Advanced computer vision approaches in biomedical image analysis. Comput Math Methods Med 2014:1–2. https://doi.org/10.1155/2014/347265
9. Nuster R, Slezak P, Paltauf G (2014) High resolution three-dimensional photoacoustic tomography with CCD-camera based ultrasound detection. Biomed Opt Express 5(8):2635. https://doi.org/10.1364/boe.5.002635
10. Sharma N, Aggarwal L (2010) Automated medical image segmentation techniques. J Med Phys 35(1):3–14. Retrieved from http://www.jmp.org.in/text.asp?2010/35/1/3/58777
11. Walter T, Shattuck DW, Baldock R, Bastin ME, Carpenter AE, Duce S, Hériché J (2010) Visualization of image data from cells to organisms. Nat Methods 7(3):S26–S41. https://doi.org/10.1038/nmeth.1431
12. Wang D, Wu Y, Xia J (2016) Review on photoacoustic imaging of the brain using nanoprobes. Neurophotonics 3(1):010901. https://doi.org/10.1117/1.nph.3.1.010901

Deep Learning Neural Network for Identification of Bird Species

Sofia K. Pillai, M. M. Raghuwanshi and Urmila Shrawankar

Abstract The species information is fundamental for ensuring biodiversity. The distinguishing proof of birds by customary keys is intricate, tedious, and because of the unavailability of information about exact name, it is difficult to identify and often baffling for non-specialists. This makes a difficult to conquer leap for tenderfoots intrigued by procuring species information. Today, there is an expanding enthusiasm for computerizing the procedure of species distinguishing proof. The accessibility and pervasiveness of important advancements, such as, computerized cameras and cell phones, the remote access to databases, new strategies in picture preparing and design acknowledgment let computerized species recognizable proof move toward becoming reality. This paper presents a deep learning neural network technique for identifying bird species. Tensor flow framework is used for the implementation.

Keywords Machine learning · Deep learning neural networks

1 Introduction

Biodiversity is declining relentlessly all through the world. The present rate of termination is to a great extent the consequence of immediate and roundabout human exercises Building precise information of the personality and the geographic dissemination of birds is fundamental for future biodiversity preservation. In this way, quick and exact plant distinguishing proof is essential for viable investigation and

S. K. Pillai (✉) · M. M. Raghuwanshi · U. Shrawankar
G.H. Raisoni College of Engineering, Nagpur, Maharashtra, India
e-mail: sofia.pillai@raisoni.net

M. M. Raghuwanshi
e-mail: m_raghuwanshi@rediffmail.com

U. Shrawankar
e-mail: urmila@ieee.org

M. M. Raghuwanshi
YCCE, Nagpur, Maharashtra, India

S.-L. Peng et al. (eds.), *Computing and Network Sustainability*, Lecture Notes in Networks and Systems 75, https://doi.org/10.1007/978-981-13-7150-9_31

administration of biodiversity. In a manual distinguishing proof process, botanist utilizes variant bird attributes such as shape, size, color, texture, color of beak, and wing color to recognize bird species.

2 Image Recognition

Our brains impact vision to give off an impression of being straightforward. It does not require effort to individuals to recognize any bird, object, or even a human face. Human brain trains itself for years to associate an image with a text. For example, we are able to identify the name of a person just by looking at him. Of course, this is possible only after a particular time of training by the brain. However, these are actually troublesome cases to enlighten with a substantial framework; they simply have all the earmarks of being basic because our brains are unfathomably incredible at understanding pictures. Since past years, the field in machine learning has increased huge ground on different applications. In particular, a model called a profound convolutional neural framework can be used for much complex applications. Researchers have shown reliable progress in computer vision by affirming their contention with ImageNet. Dynamic models continue demonstrating improvements, each time achieving another best in class result: QuocNet, AlexNet, Inception (GoogLeNet), BN-Inception-v2. However, these techniques are not easy to implement for real-time applications.

3 Softmax Classifier

The classifier must assign a label to a class of images. The drawback of kNN is that the classifier must recollect the greater part of the preparation information and reserve it for any examinations with the sample data. This is often area wasteful on the grounds that datasets may effectively be high in estimate.

The identification of an image is really difficult because it has to be associated with its training images. The proposed approach is more compelling approach to manage pictures gathering that will at last regularly connect with entire NN and CNN. The identification of a bird species will require storing all the training datasets and then comparing the unknown image to identify the correct species. This will both give true predictions as well as false prediction, but it should highly be noted that the total prediction should range from 0 to 1 and false prediction should be as low as 0.01–0.005 and true prediction to be in the range of greater than 0.5. To achieve the above, a score function and a loss function are required for belonging to the training data and for measuring the false prediction. The foremost piece in the proposed method focuses on is to identify the exact species from true prediction and false prediction. As previously, how about we expect a preparation dataset of pictures $xi \in$ RD, each related with a name yi that maps the crude image pixels to class functions.

The other prevalent decision is to use Softmax classifier, compralternate loss function. Unlike logistic regression previously, the Softmax classifier is its speculation of different classes. The SVM gives $f(xi, W)$ as (potentially hard for translating) and providing the exact prediction from a set of given class of bird species.

$$Li = -\log(efyi\Sigma jefj) \text{ or equivalently } Li = -fyi + \log \Sigma jefj \quad (1)$$

where we are utilizing the documentation *fj* until the mean of *j*th part of the vector of sophistication scores *f*. The capacity

$$fj(z) = ezj\Sigma kezk \quad (2)$$

is known as the *sm* function: A vector of self-assertive real reputable values (in z) is taken and places it to a vector of qualities within the locality of zero and unity that whole to 1. The total cross-entropy function which includes the Softmax capability could seem alarming. The cross entropy between a "genuine" appropriation p furthermore and an expected circulation q is characterized as:

$$H(p, q) = -\Sigma xp(x) \log q(x) \quad (3)$$

In short, it can be explained as:

$$h\theta(x) = 1/\exp(-\theta \top x), \quad (4)$$

where θ is will be repeatedly used for minimizing the cost function.

$$\underset{i=1}{\overset{m}{J(\theta)}} = -\left[\Sigma y(i) \log h\theta(x(i)) + (1 - y(i)) \log(1 - h\theta(x(i)))\right] \quad (5)$$

In bird species identification, we require a multi-class classification, for 12 species $K = 12$ different species. To evaluate a cost function, we have:

$$\overset{mK}{J(\theta)} = -\left[\Sigma\Sigma 1\{y(i) = k\} \log \exp(\theta(k) \top x(i))/\Sigma Kji = 1k = 1j = 1 \exp(\theta(j) \top x(i))\right] \quad (6)$$

4 Dataset—The Caltech-UCSD birds 200-2011 and CIFAR-10

For image identification of bird species, it is required to have a strong dataset on which the system can be trained, tested, and validated. For example, the Caltech bird dataset consists of around 200 bird species of more than 30 images in each class. Offspring 200-2011 is an expanded form of CUB-200, a testing dataset of 200 flying creature species. The broadened form generally duplicates the quantity of pictures per class and includes new part limitation comments. All pictures are clarified with bouncing boxes, part areas, and property names. Pictures and comments were sifted by various clients of Mechanical Turk.

The dataset contains 11,788 pictures of 200 winged animal species. Flying creatures 200 has various special properties which are important to the examination group: Subordinate class acknowledgment: Methods that are generally prominent on datasets, for example, Caltech-101 [1].

The CIFAR-10 dataset comprises of 60,000 32 $\times$ 32 shading pictures in 10 classes, with 6000 pictures for each class. There are 50,000 preparing pictures and 10,000 test pictures.

The dataset is separated into five preparing clumps and one test group, each with 10,000 pictures. The test bunch contains precisely 1000 haphazardly chose pictures from each class. The preparation clumps contain the rest of the pictures in irregular request, yet some preparation groups may contain a greater number of pictures from one class than another. Between them, the preparation groups contain precisely 5000 pictures from each class.

5 Results and Conclusion

The system is tested for 12 species of birds taken from Caltech dataset. The training is done for the dataset images as given by the TI which represents the total images of that class. True prediction is obtained from one of the images from training data itself. Unknown image which is not present in the training dataset is taken next to check the true prediction from unknown image. As we can see the result, the TP(D) as well as TP(U) is always above 0.5. In many cases, both the values are above 0.9. This proves the accuracy of the algorithm.

Similarly, the incorrect predicted values are labeled as false prediction which ranges from 0.01 to 0.2 and not more than that. Thus, it proves the accuracy of the algorithm (Table 1).

As we look at the train accuracy, it starts from 29% and goes till 100% for dataset. The cross entropy and validation accuracy are inversely proportional (Fig. 1).

Table 1 Accuracy table for 4000 iteration

Step	Train accuracy (%)	Cross entropy	Validation accuracy (%)
0	29.00	2.432535	17.00
100	87.00	1.200826	89.00
500	96.00	0.447151	89.00
1000	99.00	0.235821	96.00
1500	100.00	0.17149	96.00
2000	100.00	0.116124	96.00
2500	100.00	0.096606	95.00
2800	100.00	0.068634	96.00
3999	100.00	0.060213	96.00

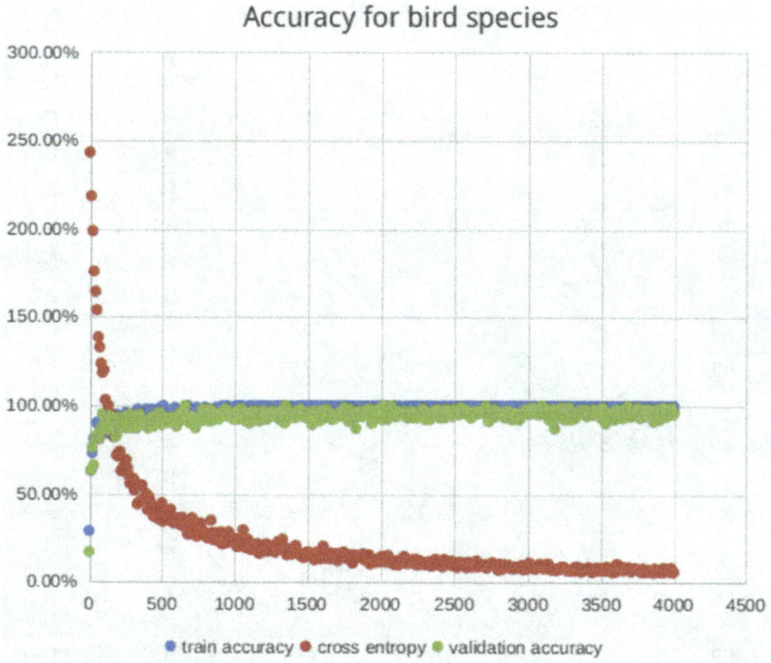

Fig. 1 Accuracy of bird species

S. No.	BS	TI	TP(D)	TP(U)	FP(D)	FP(U)
1	B1	36	0.86	0.58	B3−0.11	B3−0.28
					B2−0.02	B9−0.03
					B6−0.01	B110.02
					B5−0.01	B5−0.02
2	B2	30	0.96	0.96	B1−0.01	B1−0.02
3	B3	30	0.71	0.5	B2−0.21	B2−0.25
					B1−0.04	B1−0.21
					B7−0.01	B5−0.01
					B6−0.01	B7−0.01
4	B4	35	0.99	0.92	B1−0.01	B9−0.03
					B1−0.01	B110.02
					B9−0.01	B100.01
					B1−0.01	B5−0.01
5	B5	31	0.91	0.95	B7−0.05	B7−0.02
					B8−0.01	B11−0.01
					B6−0.01	B8−0.01
6	B6	30	0.67	0.89	B4−0.13	B5−0.05
					B7−0.07	B11−0.02
					B5−0.05	B7−0.02
					B8−0.02	B9−0.01
7	B7	33	0.87	0.79	B6−0.08	B10−0.09
					B5−0.02	B7−0.065
					B10−0.01	B11−0.02
					B11−0.01	B9−0.01
8	B8	32	0.63	0.95	B11−0.15	B5−0.01
					B5−0.07	B11−0.01
					B10−0.05	B10−0.01
					B9−0.03	B6−0.005
9	B9	30	0.97	0.87	B11−0.02	B11−0.12
					B10−0.01	B10−0.01
					B8−0.001	B5−0.001
					B12−0.001	B6−0.001
10	B10	30	0.65	0.97	B9−0.287	B11−0.01

(continued)

(continued)

S. No.	BS	TI	TP(*D*)	TP(*U*)	FP(*D*)	FP(*U*)
					B6−0.026	B5−0.010
					B11−0.018	B9−0.004
					B4−0.005	B12−0.01
11	B11	30	0.81	0.96	B6−0.089	B9−0.036
					B10−0.042	B10−0.01
					B9−0.028	B5−0.001
					B8−0.01	B12−0.01
12	B12	30	0.99	0.99	B10−0.01	B10−0.01
					B11−0.01	B11−0.01
					B4−0.01	B9−0.01
					B9−0.01	B7−0.01

where S. No.—serial number
BS—bird species
TI—total no. of images
TP(*D*)—true prediction (training dataset)
TP(*U*)—true prediction (unknown image)
FP(*D*)—false prediction (dataset)
FP(*U*)—false prediction (unknown)

S. No.	Name of species	No.
1	Black-footed albatross	B1
2	Layson albatross	B2
3	Sooty albatross	B3
4	Grooved-bill ani	B4
5	Crested auklet	B5
6	Least auklet	B6
7	Parakeet auklet	B7
8	Rhinoceros auklet	B8
9	Brewer's blackbird	B9
10	Red-winged blackbird	B10
11	Rusty blackbird	B11
12	Yellow-headed blackbird	B12

References

1. Wah C, Branson S, Welinder P (2011) California Institute of Technology, The Caltech-UCSD Birds-200-2011 Dataset WahCUB_200_2011, CNS-TR-2011-001
2. Li J, Zhang L, Yan B (2014) Research and application of bird species identification algorithm based on image features. In: International symposium on computer, consumer and control. IEEE Computer Society. https://doi.org/10.1109/is3c.2014.47
3. Ertam F, Aydan G (2017) Data classification with deep learning using tensorflow. In: IEEE 2nd international conference on computer science and engineering
4. Chan T-H, Jia K, Gao S (2015) Effective microorganisms: deep residual learning for image recognition. In: IEEE transactions on image processing, vol 24, no 12, Dec 2015
5. Branson S (2012) Interactive learning and prediction algorithms for computer vision applications. UC San Diego Electronic Theses and Dissertations
6. Russakovsky O, Deng J, Su H, Krause J, Satheesh S, Ma S, Huang Z, Karpathy A, Khosla A, Bernstein M et al (2014) ImageNet large scale visual recognition challenge. arXiv:1409.0575
7. Saxe AM, McClelland JL, Ganguli S (2013) Exact solutions to the nonlinear dynamics of learning in deep linear neural networks. arXiv:1312.6120
8. Ciresan DC, Meier U, Masci J, Gambardella LM, Schmidhuber J (2011) High-performance neural networks for visual object classification. Arxiv preprint. arXiv:1102.0183
9. Deng J, Dong W, Socher R, Li L-J, Li K, Fei-Fei L (2009) ImageNet: a large-scale hierarchical image database. In: CVPR09
10. Deng J, Berg A, Satheesh S, Su H, Khosla A, Fei-Fei L (2012) ILSVRC-2012

DITFEC: Drift Identification in Traffic-Flow Streams for DDoS Attack Defense Through Ensemble Classifier

K. Munivara Prasad, V. Samba Siva, P. Krishna Kishore and M. Sreenivasulu

Abstract The overwhelming of the request flow beyond the target server capacity leads to the service denial to the legitimate users. Because of the server's oversized potential, the act of flooding requests beyond the server capacity is carried by the malicious attackers from distributed environment called distributed denial-of-service attack. Hence, applying the knowledge gained from the findings of previous request distributions research works seems to be the suitable strategy to cease the DDOS attacks. This strategy indispensable limitation is skipping to detect the new patterns of request flooding dug by the attacker at the server from the previous knowledge on earlier attack distribution patterns. Therefore, this paper endeavors to contribute on how to handle the limitation by proposing a novel-trained ensemble classifier with new features which reflects in the traffic-flow properties, so that the traffic-flow tuple shows distribution diversity from each other which is considered and attached to individual classifier. With the application of KS test, the proposed model tries to find the distribution diversity among the traffic-flow tuples using the features set. Later, the similar policy is used to discover the distribution resemblance amid the renewed tuple as well as the tuples involved to the multiple classifiers in the ensemble classification model. The experiment worked out on the voluminous traffic flow with visible distribution variety.

Keywords DDoS attack · Ensembles approach · KS test · Application-layer DDoS attacks

1 Introduction

As the Internet technology boom, almost everything is carried out on the Internet. Online shopping, learning trading, and teleconferences are much more common in nowadays. The Internet has become an integral part of people's lives in the network

K. Munivara Prasad (✉) · V. Samba Siva · P. Krishna Kishore · M. Sreenivasulu
Chadalawada Ramanamma Engineering College, Tirupati, India
e-mail: prasadkmv27@gmail.com

S.-L. Peng et al. (eds.), *Computing and Network Sustainability*, Lecture Notes in Networks and Systems 75, https://doi.org/10.1007/978-981-13-7150-9_32

information times. Therefore, the security of the Internet has been an important issue of whole society. It is vital to ensure the network service available and defense the DDoS attack. Distributed denial of service (DDoS) is a consuming type of attacks which can consume the bandwidth and resource of the Internet and even stop the normal service of the network [1]. According to the report of DDoS attack of Arbor Networks [2], the largest reported DDoS attack traffic reached 500 Gbps in 2015, and attacks reached 450, 425, and 337 Gbps have also been found in other reports. In the survey, the largest attack scale has been more than 60 times ten years ago. Besides, the types of the DDoS attack are updated with the Internet developing. More complex DDoS attacks have already appeared in these years such as lots of low-speed attacks launched by many compromised hosts at the same times rather than high-speed attacks by one host. The huge size and high frequency of the attacks put pressure on detection the DDoS attacks timely and accurately. To overcome the problem, data mining method is introduced to the DDoS detection. An ensemble learning framework is proposed using the data mining method to improve the accuracy of DDoS attack detection effectively.

This methodology is used for modeling the defense system for DDoS attacks, built with a drift detection capacity on an ensemble classifier at level of the streaming service request. This model works on aspects such as incorporating service request streaming features defining process, enabling drift recognition through definite service request stream characteristics, clustering the sessions using learning corpus, which depends on service request stream features' diversity, detecting drift through the diversified clusters as source, and using session groups for training ensemble classifiers. An experimental study by establishing a setup utilizing synthesized service request stream; using statistical metrics: positive prognostic value, accuracy, true negative rate the results are tested.

2 Related Work

The exponential growth of network traffic as well as multifarious Internet services ushered new challenges to find the behavior of the network attacks. In order to overcome these challenges, distributed denial-of-service (DDoS) attacks are existent and voluminous peril to the availability of Internet services and networks. To mitigate massive DDoS attacks, service global-scale distribution solution does effective work, however, many cases like Telco cloud demands the protection of the network or service at local stratum.

Though it performs effectively on the familiar attack patterns, this method is used prominently as a defense mechanism to identify and distinguish the normal patterns and attack traffic forms to sieve the attack-time traffic. Nevertheless, conventional methods and techniques with limitations fail to detect the diverse and intricate attacks traffic (DDoS attacks) efficiently in high-speed network. In [3], Palmieri et al.

explained that malicious attack squander energy using the computing resources and upsurge costs. Yan et al. [4] states that the DDoS attack's rate increases significantly when the network migrates to the cloud computing environments.

These various DDoS attacks affect the network in different ways; for instance, Smurf attack pushes voluminous Internet organized message protocol packets to the authentic users; DDoS is R-U-Dead-Yet (RUDY), explained in [5] just consumes never-ending Web application sessions, viz block the Web service responding to new request received from new users; according to the work [6], most popular HTTP POST/GET [BIFAD] attackers push the legitimate posted messages at 1 byte/240 s speed into the Web server which hosts the Web application that leads to temporary crashing of the server and weakening the quality of the Web service.

SQL Injection Dos (SIDDoS), a modern DDoS attack, which is discussed in [7], here, the attackers send a string malicious SQL statement into the data base of the websites database like an equation and access the resources or save the data on servers illegally.

Alkasassbeh et al. [8] claims that many of the open-access datasets contain replicated as well as redundant information leading to DDoS detection and classification impractical. According to the contribution of the work [9], the NSL-KDD datasets have new DDoS kinds like HTTP flood, SIDDOS. Therefore, for the present research, the scholar gathered a fresh dataset in a controlled environment. This dataset contains four types of DDoS attacks, (1) UDP flood, (2) Smurf, (3) HTTP flood, and (4) SIDDOS [10].

3 Drift Identification in Traffic-Flow Streams for DDOS Attack Defense

The proposed attack detection process has two phases: (1) training phase and (2) detection phase. In the training phase of the process, the given corpus containing either attack-prone or salubrious labeled sessions is partitioned into two sets: attack-prone and salubrious sessions. Then, the training model implements sessions' clustering process, notices the clusters with distribution resemblance, and selects the clusters as cause to notice drift in request. The next section provides a detailed explanation of KS test assessment of the distribution similarity, process of clustering, process of ranking cluster based on distribution similarity, and using the chosen clusters as source to detect drift.

Table 1 Following inputs of the distribution similarity demonstrate the session cluster similarity

Begin intervals of the session	The feature shows the previous time between the both sessions starting time in sequence
Completion intervals of the session	The feature represents the past time between both sessions end time in sequence
Begin intervals of page access	The feature denotes the past time between the page access requests onset time in sequence
Completion interval of page access	The feature exhibits the past time between the pair of pages accessed termination time of in arrangement
Bandwidth ingesting at session level	The feature shows each session consumption of the bandwidth over the number of requests

3.1 KS Test

Ghasemi and Zahediasl [10] presents the potential ability of Kolmogorov–Smirnov test (KS test), i.e., concluding the distribution miscellany amid two numeric vectors. Model has potential to detect the distribution diversity without knowing the data distribution type of target vectors, and the size of the vector may vary. Therefore, the present work explores the model distance metric attracts. The process of KS test execution is described here.

3.2 Characteristics Used

With the distribution similarity of subsequent inputs, the similarity of the session groups is depicted as shown in Table 1.

3.3 Clustering the Sessions

Cluster groups are formed from the sessions of the specified traffic flow because the sessions with overlapping time intervals come under the same cluster. The process of clustering process is shown here. Individual attack prone and salubrious traffic flow are taken separately, and the clustering of the given sessions of the target traffic flow is explained below. Until the ordered session list turns out to be nil, this iterating of the process continues. The sessions' clustering algorithm formulation is represented as follows.

Algorithm 1: Session Grouping Algorithm

```
Loop 1: while(|TSL|> 0) Begin // whereas TSL is not vacant
  c_j = {s_1 ∃s_1 ∈ TSL} // take the order depicted the first session in list TSL as
  centroid of the j^th cluster scl_j
  scl_j ← c_j
  nc_j ← c_j
  Loop 2: ∀_{i=1}^{|TSL|} {s_i ∃s_i ∈ TSL ∧ s_i ≠ c_j} begin
      if (b(s_i) < e(c_j)) begin
          scl_j ← s_i
          if (e(s_i) > e(c_j)) begin
              nc_j ← s_i
          End
      End
  End
  if (c_j ≠ nc_j) begin
      c_j = nc_j // the cluster scl_j new centroid
      scl_j = null // the cluster becomes nil
      Go to loop 2
  End
  else if (c_j ≡ nc_j) begin
      TSL = {TSL \ scl_j} // the cluster scl_j entries are detached from the
      sorted list TSL
      j = j+1
  End
End
```

3.4 *Analysis of Cluster Similarity*

The procedure of this model commences with the organization of the resulting clusters in the descendent order based on each cluster's amount of sessions. Clusters are grouped if these clusters have one as distribution similarity value, because the distribution similarity imitates the equal symmetric property; for example, if $a = b$, then $b = a$ as well as equal transitive property; viz, if $a = b, b = c$, then $c = a$. This is done with KS test.

3.5 Source Cluster Selection for Drift Detection

Once the completion of cluster grouping process based on the distribution resemblance, the proposed model selects the cluster with the peak sessions amid clusters groups projecting the equality transitive property. And then, in the request streams, as drift detection source, these chosen clusters are utilized. In addition, the process of the proposal model utilizes the cluster groups which are selected as source to detect drift for training the ensemble classification classifiers. Thus, the drift detection depends on the traffic flow of ensemble classification. Therefore, the suggested model is appropriate to train any classifier which is suitable to ensemble model.

4 Experimental Study

Here, creating the dataset strategic approach generation, setting up the experiment background, and comparing the results to assess the performance are discussed in the section.

4.1 DDoS Attack Dataset

In the duration of assessing against DDOS attack log datasets like NSL-KDD, the attack defense models are prone to false issue because of iteration of the requests in the respective datasets [11]. So, the proposed model conducts the assessment on dynamic traffic flow which includes different kinds of DDOS attacks like Smurf attacks, SIDDOS attacks, HTTP flooding attacks, UDP flooding attacks. Therefore, the critical issue for argument is that assessment of IDS becomes important challenge, as the datasets of DDoS lack appropriate level. In public domain, the three DARPA/Lincoln Laboratories, KDD [12], CAIDA [13, 14] datasets are accessible to detect intrusions. The datasets of DAPRA have several weeks' network activity information which is based on simulated network of air force. The available data is not artificial and do not possess any kinds of modern attacks. As per the work [15], DARPA has disparaged data. The CAIDA dataset is accessible by just dropping a user request. This CAIDA has DDoS attack dataset of the year 2007.

Therefore, as mentioned in the works [16–18], the experiments were worked out on DDOSIM, Tor's Hammer, and JMeter and generated request streams. The core potential of the DDOSIM is to generate traffic flow on target server through zombie hosts, viz random and dynamic IP addresses underneath the comprehensive TCP communication.

The transaction statistics gathered here are further utilized in the experimental study results presented in Table 2.

Table 2 Statistics of the traffic flow produced for DITFEC

	Particulars of escalated load given as flood	Normal load
No. of sessions	36,152	32,373
Bandwidth consumed	56,010 MB	31,096 MB
No. of requests	614,584	615,087
No. of sources	521	302

Table 3 Statistics of source cluster selection process and the cluster formation

	No. of sessions used for training	No. of cluster groups	Usage of classifiers in ensemble method	No. of session clusters	Amount of sessions in source clusters
From DDOS attack transactions	25,306	8	8	37	In range of 684–1026
From normal transaction	22,661	5	5	52	In range of 436–654

4.2 Experimental Results

Powers [19] contributes on the experiment carried out using cross-justification metrics like true negative rate (TNR), true positive rate (TPR), and positive predict value. The observation of the cross-validation metrics are: TPR depicts the existing model sensitivity toward the recognition of the attack-prone sessions; TNR projects the corresponding model sensitivity to identification of the session resistant to attacks; the positive predictive value demonstrates the true positives' probability, i.e., sessions prone truly to attack from the sessions prone to attacks; i.e., the positives are found by the other model. Thus, the accuracy of the metrics evinces the final ratio of the sessions prone to truly attacks and the real original sessions of the comprehensive sessions prone to attack and the sessions that are normal identified by the other model. And the corresponding model identified the normal sessions ratio as prone to attacks are shown through false alarming rate indicates.

Furthermore, the section explains the proposed model results and is represented in Tables 3 and 4. In [20], the prominent classifier, AdaBoost, is utilized for performing the process of ensemble classification which is used for the model. Table 3 represents the obtained results obtained of clustering and grouping clusters from the labeled session based on the distribution similarity and statistics of source clustering. Table 4 consisting of performance metrics is calculated for proposed model and NFBoost and the proposed model showing the better performance than the NFBoost.

Table 4 Performance metrics and comparison between NFBoost and DITFEC

Performance metric	NFBoost	DITFEC
For training the number of attack-prone sessions	25,306	25,306
For training the number of salubrious sessions	22,661	22,661
Groups formed from salubrious sessions given for training	52	52
Groups formed from attack-prone sessions given for training	37	37
Negatives for testing	9712	9712
Positives for testing	10,846	10,846
Clusters formed from salubrious sessions given for testing (total negatives)	22	22
Groups formed from attack-prone sessions given for testing (total positives)	25	25
Groups represented incorrectly attack-prone (false positives)	5	2
Groups represented truly attack-prone (true positives)	20	23
Groups represented incorrectly negative (false negatives)	4	2
Groups represented correctly salubrious (true negatives)	18	20
Accuracy	0.8826	0.914894
TPR	0.873	0.92
TNR	0.8745	0.909091
FAR	0.8785	0.08
Positive predict value	0.1174	0.92

5 Conclusion

This paper endeavors to contribute on how to handle the limitation by proposing a novel-trained ensemble classifier with new features which reflects in the traffic-flow properties, so that the traffic-flow tuple shows distribution diversity from each other which is considered and attached to individual classifier. With the application of KS test, the proposed model finds the distribution diversity among the traffic-flow tuples using the features' set. Later, the similar policy is used to discover the distribution resemblance amid the renewed tuple as well as the tuples involved to the multiple classifiers in the ensemble classification model. The experiment was worked out on the voluminous traffic flow with visible distribution variety. The main findings of the experimental study are that even the ensemble model NFBoost depicted the inferior performance on the accuracy of drift and attack detection because this ensemble model is not proficient on the contrary traffic-flow possessions but proficient on the packet flow forms and source features. But the NFBoost represents the lower recognition precision rate when compared to the suggested model, as the DITFEC is trained on the voluminous traffic flow that comprises noticeable distribution variety.

References

1. Bhuyan MH, Bhattacharyya DK, Kalita JK (2015) An empirical evaluation of information metrics for low-rate and high-rate DDoS attack detection. Pattern Recogn Lett 51(C):1–7
2. https://www.arbornetworks.com/
3. Palmieri F, Ricciardi S, Fiore U, Ficco M, Castiglione A (2015) Energy-oriented denial of service attacks: an emerging menace for large cloud infrastructures. J Supercomput 71(5):1620–1641
4. Yan Q, Yu FR, Gong Q, Li J (2016) Software-defined networking (SDN) and distributed denial of service (DDoS) attacks in cloud computing environments: a survey, some research issues, and challenges. IEEE Commun Surv Tutorials 18(1):602–622
5. Najafabadi MM, Khoshgoftaar TM, Napolitano A, Wheelus C (2016) RUDY attack: detection at the network level and its important features. In: FLAIRS conference, 30 Mar 2016, pp 288–293
6. Prasad KM, Reddy AR, Rao KV (2017) BIFAD: bio-inspired anomaly-based HTTP-flood attack detection. Wirel Pers Commun 97(1):281–308
7. Vivin Sandar S, Shenai S (2012) Economic denial of sustainability (edos) in cloud services using http and xml based ddos attacks. Int J Comput Appl 41(20)
8. Alkasassbeh M, Al-Naymat G, Hassanat AB, Almseidin M (2016) Detecting distributed denial of service attacks using data mining techniques. Int J Adv Comput Sci Appl 7(1)
9. Revathi S, Malathi A (2013) A detailed analysis on NSL-KDD dataset using various machine learning techniques for intrusion detection. Int J Eng Res Technol (ESRSA Publications)
10. Ghasemi A, Zahediasl S (2012) Normality tests for statistical analysis: a guide for non-statisticians. Int J Endocrinol Metab 10(2):486
11. Bai Y, Kobayashi H (2003) Intrusion detection systems: technology and development. In: 17th international conference on advanced information networking and applications 2003. AINA 2003, 27 Mar 2003. IEEE, pp 710–715
12. KDDdataset (1999). http://kdd.ics.uci.edu/databases/kddcup99/kddcup99.html
13. The CAIDA (2007) DDoS attack 2007, Dataset Paul Hick, Emile Aben, KC Claffy, Josh Polterock. Available from http://www.caida.org/data/passive/ddos20070804dataset.xml
14. CAIDA UCSD Network Telescope, Three days of conficker—Nov 2008, Paul Hick, Emile Aben, Dan Andersen, KC Claffy. Available from http://www.caida.org/data/passive/telescope-3days-conficker_dataset.xml
15. Behal S, Kumar K (2017) Characterization and comparison of DDoS attack tools and traffic generators: a review. IJ Netw Secur 19(3):383–393
16. Sommer R, Paxson V (2010) Outside the closed world: on using machine learning for network intrusion detection. In: 2010 IEEE symposium on Security and privacy (SP), 16 May 2010. IEEE, pp 305–316
17. Badve OP, Gupta BB (2016) Taxonomy of recent DDoS attack prevention, detection, and response schemes in cloud environment. In: Proceedings of the international conference on recent cognizance in wireless communication & image processing 2016. Springer, New Delhi, pp 683–693
18. Kiran S, Mohapatra A, Swamy R (2015) Experiences in performance testing of web applications with unified authentication platform using Jmeter. In: 2015 international symposium on technology management and emerging technologies (ISTMET), 25 Aug 2015. IEEE, pp 74–78
19. Powers DM (2011) Evaluation: from precision, recall and F-measure to ROC, informedness, markedness and correlation
20. An TK, Kim MH (2010) A new diverse AdaBoost classifier. In: 2010 international conference on artificial intelligence and computational intelligence (AICI), 23 Oct 2010, vol 1. IEEE, pp 359–363

Comparative Analysis of Supervised Machine Learning Algorithms for GIS-Based Crop Selection Prediction Model

Preetam Tamsekar, Nilesh Deshmukh, Parag Bhalchandra, Govind Kulkarni, Kailas Hambarde and Shaikh Husen

Abstract Regardless of the technological progress, the agricultural sector remains unorganized in India. With respect to present geographic, social and economic trends, the Indian agricultural sector needs to change and adopt artificial intelligence by combining the practical knowledge derived from generations and the scientific basis. By performing the computations on the historical data and using AI, we can predict the crop for cultivation. This article deals with the comparative analysis of supervised machine learning algorithms for GIS-based crop selection prediction model (CSPM).

Keywords GIS · AI · CSPM

1 Introduction

This article work deals with the study and analysis of various techniques for crop selection using supervised machine learning algorithms and GIS. It is an interdisciplinary work which draws expertise from machine learning, GIS and DSS. ML techniques are the subset of artificial intelligence which empowers software application to become more accurate in prediction without being explicitly programmed. ML

P. Tamsekar (✉) · N. Deshmukh · P. Bhalchandra · G. Kulkarni · K. Hambarde · S. Husen
S.C.S, Swami Ramanand Teerth Marathwada University, Nanded 431606, Maharashtra, India
e-mail: pritamtamsekar@gmail.com

N. Deshmukh
e-mail: nileshkd@yahoo.com

P. Bhalchandra
e-mail: srtmunparag@gmail.com

G. Kulkarni
e-mail: govindcoolkarni@gmail.com

K. Hambarde
e-mail: kailassrt@gmail.com

S. Husen
e-mail: husen09@gmail.com

S.-L. Peng et al. (eds.), *Computing and Network Sustainability*, Lecture Notes in Networks and Systems 75, https://doi.org/10.1007/978-981-13-7150-9_33

algorithms are classified as supervised and unsupervised classification techniques. The current research work deals with supervised machine learning algorithms. GIS is a location-based system which plays a vital role in spatial analysis to make firm decisions. DSS is defined as human–computer systems which collect information, process information and provide information based on computer systems [1].

In India, the foremost challenge in the field of agriculture is feeding a swelling human population (i.e. 1.25 billion) with the declining weather conditions and soil quality is making it a tough task. The restricted scope to increase the natural resources area under agriculture makes the challenge tougher to serve the world's second most populous country which is 18% population of the world along with the 15% of livestock only with the 2.2% of the world geographical area, where only 46% of the land is cultivable [2]. To tackle the increasing appetite need precision farming is the key, and correct crop selection is important.

2 Related Work

In [3], Zhu et al. have introduced the GIS-based agriculture technically aid scheme to provide technology-based system for strong intelligent decision support system.

In [4], Kaklauskas reveals that integration of genetic algorithm with a decision support system can be useful to resolve the optimization, search and other problems.

In [5], Rahimi et al. used AHP and GA techniques for geospatial tool to investigate the features/attributes which affects the artificial groundwater recharge process.

In [6], Mokarram et al. reveal that the use of AI and ML techniques for the classification of land suitability classification automation. Rapidly, growing techniques such as multiple classifier systems provided precise and robust than an excellent single classifier in many fields.

3 Design and Implementation

The below systematic representation shows the workflow of the comparative analysis of supervised machine learning algorithms for GIS-based crop selection prediction model (Fig. 1).

1. **Identification of Parameters**
   The first and foremost thing to build the machine learning model is to identify the important input parameters to get the optimal result. To build the CSPM, the essential fields, i.e. biological and chemical details of soil, irrigation details and record of past 10 years crop cultivated.
2. **Data Collection**
   The data for the identified parameters are collected from the Soil Health Card and crop cultivation records obtained from the authorized government agency.

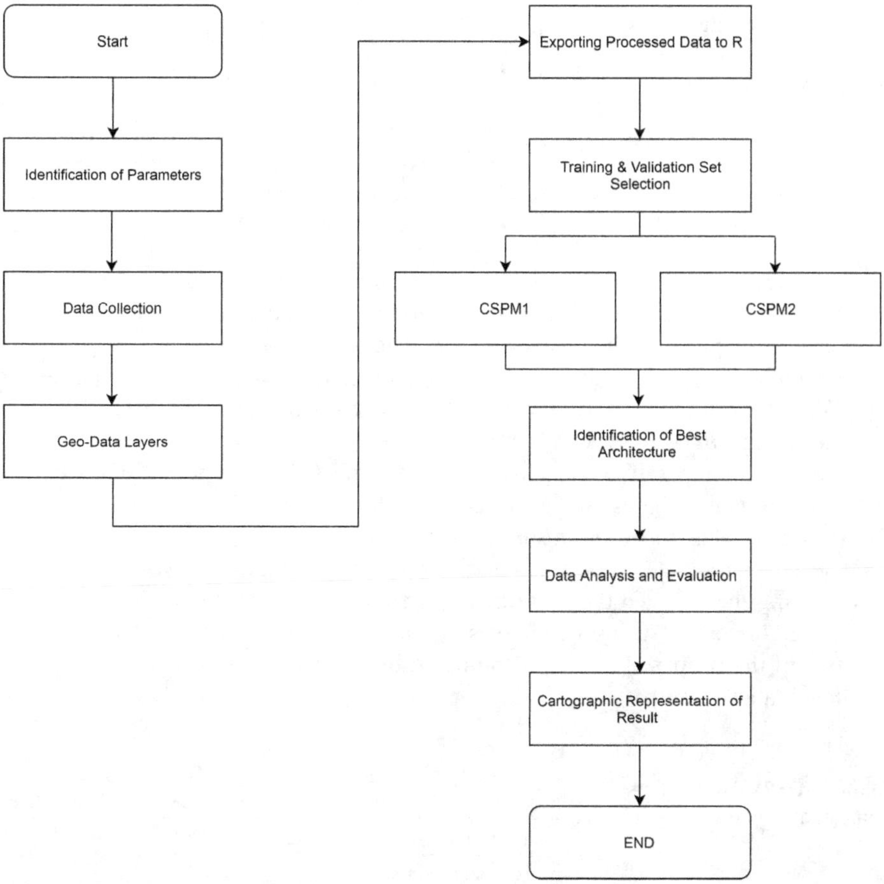

Fig. 1 Workflow of CSPM

3. **Geo-Data Layers**
   The collected data is stored in the geo-database of GIS software and geo-data layers of this data are prepared.
4. **Exporting Processed Data to R**
   The processed data is exported in R software for implementing machine learning algorithm.
5. **Training and Validation Set Selection**
   The processed data consist of two data sets from which one data set with the expert advice is selected for training the model and the other data set is used for validating the model.
6. **CSPM1 and CSPM2**
   From the training set selected in CSPM1, the supervised machine learning classification algorithms, namely CART, support vector machine, *k*-nearest neigh-

Table 1 CSPM1 algorithms comparative analysis

Algorithm	Accuracy
Classification and regression tree	0.53846154
K-nearest neighbours	0.1538462
Random forest	0.1538
Support vector machine	0.6154

bours and random forest, are implemented on the data set and the result obtained is shown in Table 1 and Fig. 1, further to improve the accuracy PCA algorithm has been applied on the data set and then again all the classification algorithms are applied and the results are shown in Table 2 and Fig. 2 where we can see the SVM algorithm has shown the highest accuracy which is 0.7692%.

7. **Identification of Best Architecture**
   From the comparative analysis in CSPM1 and CSPM2, the SVM's accuracy 0.7692% is the highest and so the SVM is the best algorithm.
8. **Data Analysis and Evaluation**
   Once the anticipated supervised machine learning techniques have been executed by using the selected data set, the required values are computed. The model is trained on testing data set and assessed on the validation data set. The effectiveness of the proposed model is computed by investigating the accuracy level of the obtained results.

Table 2 CSPM2 algorithms comparative analysis

Algorithm	Accuracy
Classification and regression tree	0.6923
K-nearest neighbours	0.4615
Random forest	0.4615
Support vector machine	0.7692

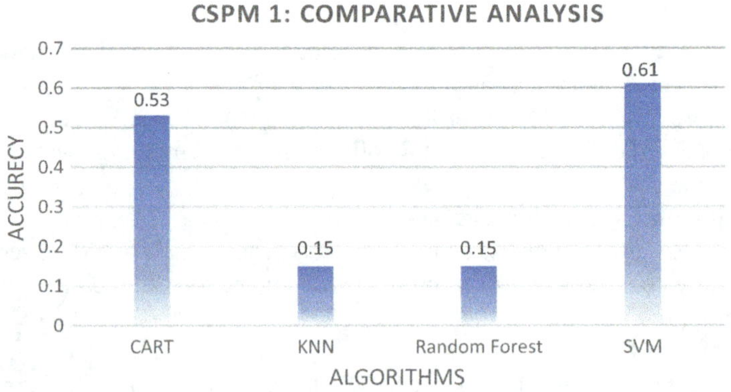

Fig. 2 CSPM1 algorithms comparative analysis

4 Result and Discussion

This experiment uses an integrated methodology on machine learning algorithm with GIS to assess parameters to build a crop selection prediction model (CSPM). An inventory of 130 soil health card collected which includes 117 from Hadgaon ruler area farm plot which was used to build the machine learning model and 13 collected from Belgaon area farm plot were used for validation of the model. The data used for training the model was soil parameters from the soil health card and irrigation details. Two CSPMs have been built, namely CSPM1 and CSPM2. In CSPM1 directly, the data set has been used and in CSPM2, PCA has been applied to the data set. Both CSPM1 and CSPM2 are trained using supervised machine learning algorithms classification and regression tree, support vector machine, *k*-nearest neighbours, random forest and the model was validated on the Belgaon area data set, the accuracy results are shown in Table 1 for CSPM1 and Table 2 for CSPM2, which shows that classification and regression tree has the highest accuracy in CSPM2. So we can conclude that CSPM2 with classification and regression tree is the best model for crop selection prediction.

5 Conclusion

The present investigation reveals the use of supervised machine learning algorithms and GIS technologies in the agricultural sector. Our experimental outcome shows that (Fig. 3 and Table 2) SVM algorithm with the PCA algorithm on the data set has given the highest accuracy, and hence, the CSPM2 is the best model for predicting the crop for cultivation. Further, this study is also useful to bridge the knowledge vacuity of farmers and improve decision-making in the selection of respective crops.

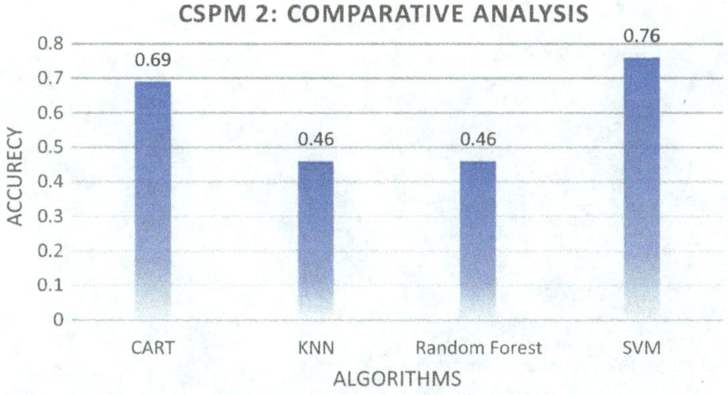

Fig. 3 CSPM2 algorithms comparative analysis

References

1. Sivanandam SN et al (2007) Introduction to genetic algorithms. Springer, Berlin Heidelberg New York. ISBN 978-3-540-73189-4
2. Srivastava P et al (2016) An urgent need for sustainable thinking in agriculture—an Indian scenario. In: Ecological indicators, vol 67, pp 611–622. ISSN 1470-160X. https://doi.org/10.1016/j.ecolind.2016.03.015
3. Zhu Z, Zhang R, Sun J (2009) Research on GIS-based agriculture expert system. In: 2009 WRI world congress on software engineering, Xiamen, pp 252–255
4. Kaklauskas A Intelligent decision support systems. In: Biometric and intelligent decision making support, intelligent systems reference library 81. https://doi.org/10.1007/978-3-319-13659-2_2
5. Rahimi S et al (2014) Using combined AHP–genetic algorithm in artificial groundwater recharge site selection of Gareh Bygone Plain, Iran. Springer-Verlag, Berlin, Heidelberg
6. Mokarram M et al (2010) Land suitability evaluation for wheat cultivation by fuzzy theory approach as compared with parametric method. In: The international archives of the photogrammetry, remote sensing and spatial information sciences, vol 38, Part II
7. Tamsekar P et al (2017) Architectural outline of decision support system for crop selection using GIS and DM techniques. In: Vishwakarma H, Akashe S (eds) Computing and network sustainability. Lecture Notes in Networks and Systems, vol 12. Springer, Singapore
8. Tamsekar P et al (2018) Architectural outline of GIS-based decision support system for crop selection. In: Satapathy S, Bhateja V, Das S (eds) Smart computing and informatics. Smart innovation, systems and technologies, vol 77. Springer, Singapore

Text Summarization Using Adaptive Neuro-Fuzzy Inference System

Pratiksha D. Warule, S. D. Sawarkar and Archana Gulati

Abstract Nowadays, data present on the World Wide Web is growing exponentially. People use search engines like Google, Bing, and Yahoo for retrieving the required information. But as the information present on the Web is huge, it is necessary for user to make the summary of this information. User can easily understand the large volume of data with the help of summary and does not require spending so many for analyzing the collected information. Text summarization is the process of transforming the large text into short and meaningful text. While summarizing the text, one should preserve its data matter and general message. It is laborious for individual person to summarize the large documents as it takes much time. It is a very difficult and time-consuming process for humans to summarize large documents. Different methods are used for summarizing the text till now like neural network, fuzzy logic, genetic algorithm, and many more. The proposed system is a hybrid system of neural network and fuzzy logic, which is known as adaptive neuro-fuzzy inference system. So it will overcome the drawbacks of both neural network and fuzzy logic. This proposed system takes the learning ability of neural network and uncertainty data handling of fuzzy system.

Keywords Neural network · Fuzzy logic · Adaptive neuro-fuzzy inference system

---

P. D. Warule (✉) · S. D. Sawarkar · A. Gulati
Computer Engineering, Datta Meghe College of Engineering, Airoli, India
e-mail: pnsonawane128@gmail.com

S. D. Sawarkar
e-mail: sudhir_sawarkar@yahoo.com

A. Gulati
e-mail: gulatiarchana81@gmail.com

S.-L. Peng et al. (eds.), *Computing and Network Sustainability*, Lecture Notes in Networks and Systems 75, https://doi.org/10.1007/978-981-13-7150-9_34

1 Introduction

1.1 Text Summarization Approaches

The main idea behind text summarization is to find the important content from the text, understanding its meaning clearly, and generate the short text from it. Here, understanding of text is natural language processing problem. Linguistic approach uses linguistic knowledge to study the text semantically, and according to that, it selects the sentences for summary by considering the verb, subject, noun, etc. This method finds the term relationship between the documents by using part of speech tagging, grammar analysis, and thesaurus usage and selects the relevant sentences. Statistical approach summarizes the text without understanding the meaning of the words. It finds statistical distribution of some features. The sentences which have these features are considered as summary sentences. In this method, the sentences are selected based on the word frequency and indicator phrases without considering the meaning of the words. Hybrid approach uses the combination of statistical and linguistic method for generating meaningful and short summary.

2 Text Summarization Process

2.1 Preprocessing

In preprocessing step, text document is prepared for finding the text features. This step involves the following phases. In sentence segmentation phase, the given text is divided into sentences with its word count. In Tokenization phase system finds commas, special symbols, and spaces present between the words so that the sentences gets decomposed into words. In next phase, stop words are removed from the text by removing the words with less important meanings and which do not add any major information to the job. Stop words are common words that give less relevant information than other words. In case-folding step, text gets converted either in uppercase or lowercase [1]. Lemmatization process uses vocabulary and morphological analysis of words. In this, it removes inflectional endings only and gives the root of a word which is known as lemma [2].

2.2 Text Features

Title Feature. If the words which are present in the document title are also present in the sentence, it indicates that the sentence is having high weight and should be added in summary. This feature can be calculated by finding the number of title words which are present in sentence and total number of words in the document title [3].

$$f_1 = \frac{\text{Total no. of title words in sentence}}{\text{Total no. of word in document title}} \quad (1)$$

Term Weight. By using term weight feature, one can find weight or importance of every word present in the document. The tf.idf method is used for calculating the weight WT_i of word i. The proposed system uses this method as tf.isf (term frequency, Inverse sentence frequency):

$$\mathrm{WT}_i = tf_i \times isf_i = tf_i \times \log Nn_i$$

where tf_i gives the word i's term frequency in the document. Here, the number of sentences present in document is given by N and the sentences in which word i is present is given by n_i. The following equation is used for finding the term weight for particular sentence.

$$f_2 = \frac{\sum_{i=1}^{M} \mathrm{WT}_i(S)}{\mathrm{Max}\sum_{i=1}^{M} \mathrm{WT}_i(S)} \quad (2)$$

where $\mathrm{WT}_i(S)$ is the term weight of word i in sentence S and M is the words count in sentence S.

Sentence Length. This feature finds the length of the sentence. For generating precise summary, it is important to find the sentence length. Too short or long sentences should not be considered as summary sentences.

$$\mathrm{SL} = 0 \text{ if } \mathrm{LN} < \mathrm{Min} \text{ or } \mathrm{LN} > \text{Otherwise}$$
$$f_3 = \mathrm{SL} = \mathrm{Sin}((\mathrm{LN} - \mathrm{Min}) \times ((\mathrm{Mx}\ \Theta - \mathrm{Mn}\ \Theta)/(\mathrm{Max} - \mathrm{Min}))) \quad (3)$$

where

LN Sentence length
Min Minimum sentence length
Max Maximum sentence length
Mn Θ Minimum angle (0)
Mx Θ Maximum angle (180) (Fig. 1).

Sentence Position. Position of the sentence in the document is important while generating the summary. Sentences which are present at the start state the subject of the document, whereas sentences at the end arrive at the judgment [4]. Threshold value states the number of sentences in the beginning and at the end will be kept in summary whose weight is given here, SP = 1. The sentence position value for remaining sentences is calculated as follows:

$$f_4 = \mathrm{SP} = \mathrm{Cos}((P - \mathrm{MinVal}) \times ((\mathrm{Max}\ \Theta - \mathrm{Min}\ \Theta)/(\mathrm{MaxVal} - \mathrm{MinVal}))) \quad (4)$$

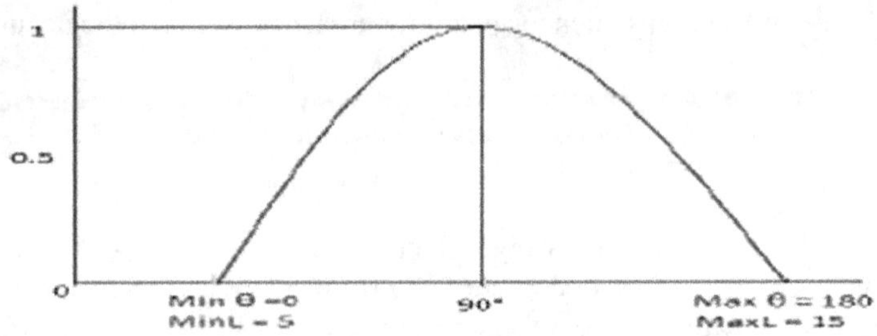

Fig. 1 Sentence length

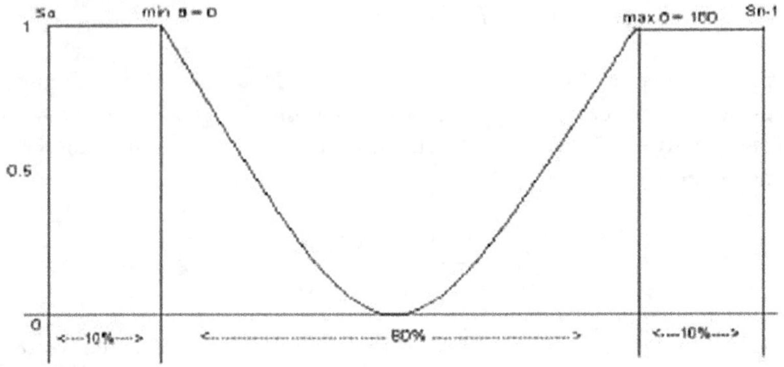

Fig. 2 Sentence position

where TR = threshold value

MinVal	N * TR (minimum sentence length)
MaxVal	N * (1 − TR) (maximum sentence length)
N	number of sentences in document.
Min Θ	minimum angle (0)
Max Θ	maximum angle (180)
P	position of sentence in document (Fig. 2).

Numerical Data. The sentences which have numbers get more weightage and are considered to be more important, so they should be included in summary [3].

$$f_5 = \text{ND} = \begin{matrix} 1, & \text{Numbers exist in the sentence} \\ 0, & \text{Numbers does not exist in the sentence} \end{matrix} \tag{5}$$

Sentence to Sentence Similarity. It finds the closeness between one sentence with every other sentence in the given text. The proposed system uses token-matching method for finding the closeness between the sentences. The two-dimensional matrix is formed of the size [N] [N], where N is sentence count in the document. In this matrix, diagonal elements should be assigned 0 values as sentence should not get compared with itself [5].

$$\mathrm{Sim}(S_i, S_j) = \mathrm{TM}\left[(t_i)^n{}_1, (t_j)^m{}_1\right]$$

where TM is token-matching method. The score for sentence to sentence similarity is calculated as ratio of summary of closeness of sentence S with every other sentence over the maximum summary.

$$f_6 = \sum\left[\mathrm{Sim}(S_i, (S_j))_1^N\right]/\mathrm{Max}\sum\left[\mathrm{Sim}(S_i, (S_j))\right]\left[\mathrm{Sim}(S_i, (S_j))_1^N\right]\left[\mathrm{Sim}(S_i, (S_j))_i^N = 1\right] \tag{6}$$

SVO Qualification. English language has SVO sentence structure. In SVO structure, the first subject comes before verb and then object comes. For this, tagging algorithm is used in which by considering the tags assigned the first noun word will be assumed as the subject of the sentence. Then, complete sentence will be analyzed till its end, and if the last word is object, then sentence will be considered as SVO qualified sentence. Only those sentences which are marked as SVO structure will be taken for further processing of text summarization.

$$f_7 = \mathrm{SVO}\ (S_i) = \begin{matrix} 1, & \text{SVO Qualification successful} \\ 0, & \text{SVO Qualification unsuccessful} \end{matrix} \tag{7}$$

Subject Similarity. Here, system finds whether the subject of both the sentence and the title is same or not.

$$f_8 = \mathrm{Sub}\ (S_i) = \begin{matrix} 1, & \text{if the subject of title and the subject of sentence is same} \\ 0, & \text{if both the subjects are different} \end{matrix} \tag{8}$$

3 Proposed Method

The proposed system uses extractive multi-document text summarization using adaptive neuro-fuzzy inference system which uses classification. The proposed system is based on [6]; in addition, it uses lemmatization for finding the root of word and case folding in preprocessing stage, and it changed the features based on which sentence score will be calculated. This model divides the text into summary sentence and non-summary sentence. ANFIS merges neural network and fuzzy logic so that it can explain the relationship between input and output data and learns the knowledge [2]. So by using adaptive neuro-fuzzy inference system, a better system is formed which can identify the summary sentences precisely (Fig. 3).

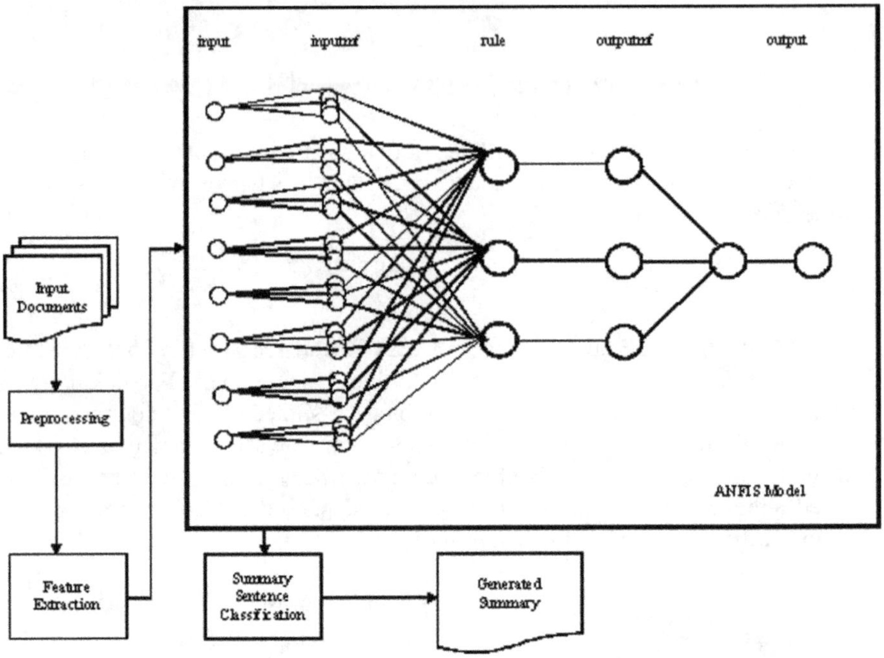

Fig. 3 Architecture of the proposed system [6, 8]

3.1 Fuzzifications

In fuzzification process, crisp values get changed into fuzzy values. The system uses trapezoidal membership function. According to sentence features and rules present in the knowledge base, weight for each sentence is calculated which will range from 0 to 1. The output of membership function will depict the degree of importance of sentence in the given text. The proposed system uses trapezoidal membership function. In this function, lower limit is defined by a, an upper limit is defined by d, a lower support limit is defined b, and an upper support limit is defined by c, where $a < b < c < d$.

$$f(x, a, b, c, d) = \begin{cases} 0, & (x < a) \text{ or } (x > d) \\ \frac{x-a}{b-a}, & a \le x \le b \\ 1, & b \le x \le c \\ \frac{d-x}{d-c}, & c \le x \le d \end{cases}$$

Or more precisely, it can be stated as

$$F(x, a, b, c, d) = \max\left(\min\left(\frac{x-a}{b-a}, 1, \frac{d-x}{d-c}\right), 0\right) \tag{9}$$

The parameters a and d will find the base of the trapezoid, whereas parameters b and c find shoulders of trapezoid.

Fuzzy Rules. The human knowledge is represented by forming fuzzy rules. Here, fuzzy rules are nothing but natural language expressions.

IF antecedent, THEN consequent

The above expression states the general IF–THEN rule-based form. The example of fuzzy rule which is used in the proposed system is as follows:

IF (Titlefeature is VH) and (TW is VH) and (SL is H) and (SP is H) and (ND is VH) and (SentencetoSentenceSimilarity is VL) and (SVOQualification is VH) and (SubjectSimilarity is VH) THEN (Sentence has more weight).

3.2 ANFIS Learning Method

As shown in Fig. 3, feature score for every sentence will act as the input to ANFIS model. After the training of ANFIS model, the proposed system will able to estimate the score of the new document's sentences. The trained model will state whether the sentence is summary sentence or non-summary sentence. The eight features which system has extracted will be the input to ANFIS model. Then, with the help of trapezoidal membership function, crisp values will get converted into fuzzy values. These fuzzy values will used for calculating the firing strength of the fuzzy rules. The linear combination of input variables will be combined with the output of each rule. Then, all the incoming signals are aggregated to get the final score of the sentence [7].

In the proposed system, subtractive clustering is used for automatically generating fuzzy rules. Subtractive clustering algorithm maps the input–output data and predicts the cluster number and its center. It treats each instance as a possible cluster center and adds the instances in the first cluster which have the value in the limit of the first cluster. If not, then it will create a new cluster. This task repeats itself till all the instances are added in the clusters [6, 8].

Least square estimation along with back-propagation gradient descent method is used for training ANFIS model. Least square estimation (LSE) is used for reducing the squared error between the real output which system gets and the desired output. The arguments of the membership function changed with the help of back-propagation along with LSE. In back-propagation network, every input weight has its own learning rate which changes after each iteration [6, 8]. In back-propagation network, forward pass is used for finding the error, whereas in backward pass, error rates are transmitted from output end to input end. According to these error rates, all parameters like learning rate are updated. By using hybrid combination of neural network along with fuzzy logic parameters of the membership functions are changed directly from the output data.

3.3 Sentence Classification and Summary Generation

The output of the ANFIS model is used for classifying the sentences as summary sentences and non-summary sentences. The proposed system uses binary classifier for classifying the sentences. The sentences which have sentence score 1 are considered as summary sentences, and they are added in class 1, whereas the sentences which have sentence score 0 are considered as non-summary sentences and are added in class 0. Here, the proposed system uses threshold value for classifying the sentences. After classifying the sentences in two classes as class 0 and class 1 respectively, the proposed system will take the class 1 sentences and show them as the summary of text [6, 8].

4 Results Analysis and Future Scope

To judge the functioning of the proposed system, precision, recall, and *F* score are used. Precision score will be given by dividing the total number of true positive means correct summary sentences by number of all the sentences extracted by the proposed system. Recall score is nothing but the fraction of relevant sentences that are successfully retrieved by the system. *F* score will be calculated by harmonic mean of precision and recall scores. This value is roughly equal to the average of precision and recall when they are close. Extracted sentences are extracted by the proposed system, and relevant sentences are sentences which are included in human-generated summary. Human-generated summaries are considered as standard summaries because humans can understand the text and find the main information from the text. The system calculates precision, recall, and *F* measure for the proposed system-generated summaries with human-generated summaries for DUC 2002 datasets. The results show that the proposed method gives better average precision, recall, and *f* score (Fig. 4).

Table 1 shows the comparison for the average precision, recall, and *F* score between ANFIS method and human-generated summary for the five datasets, which is taken from standard dataset DUC 2002, namely d061j, d062j, d063j, d064j, d065j. There are in total of 34 documents. Then, system has taken three documents which are based on same topic. After comparing the obtained results with the precision,

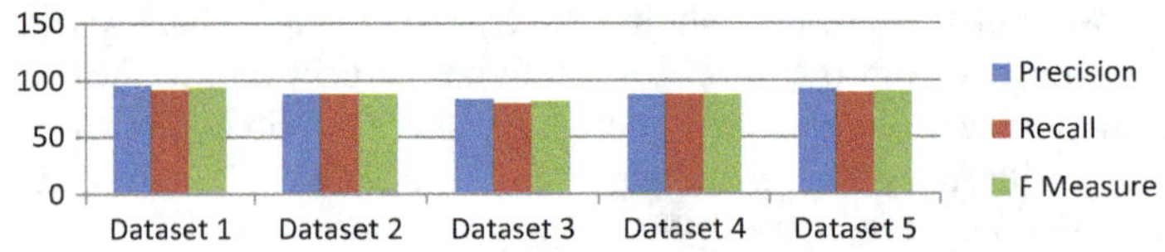

Fig. 4 Graph showing precision, recall, and *F* score

Table 1 Comparisons of machine-generated summary and human-generated summary

Dataset	Precision	Recall	*F* score
Dataset 1	95.56	91.49	93.48
Dataset 2	88	88	88
Dataset 3	83.33	78.95	81.08
Dataset 4	87.50	87.50	87.50
Dataset 5	92.59	89.29	90.91
Avg	89.396	87.046	88.19

Table 2 Comparison of different algorithms

Summarization method	Precision	Recall	*F* score
Fuzzy logic with GA-GP	0.831	0.650	0.728
BM25 and neural network-based method	0.94	0.932	0.933
Sentence fusion using fuzzy logic	0.47589	0.46660	0.47019
Fuzzy genetic semantic-based method	0.49958	0.45191	0.47048
Fuzzy inference method	0.58	0.6	0.59
Fuzzy logic and WordNet method	0.46824	0.43411	0.44829
Clustered genetic sematic graph method	0.67	0.395	0.2806
ANFIS-based text summarization	89.396	87.046	88.19

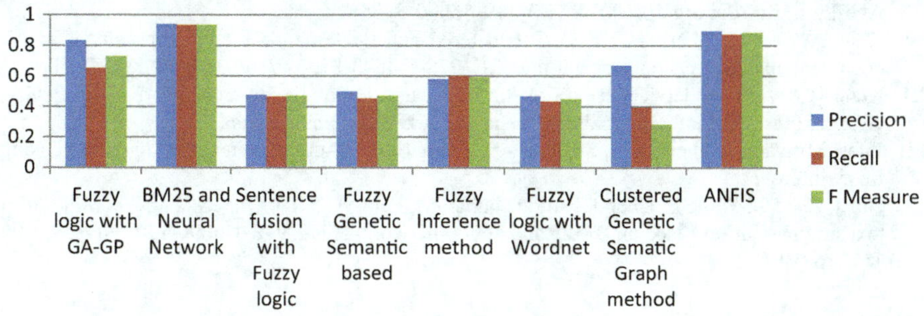

Fig. 5 Comparison of the proposed system with other methods

recall, and F score of the other methods [1, 4, 5, 9–12], the following findings have been listed (Table 2 and Fig. 5).

From the above comparison, we can observe that fuzzy logic with neural network produces better results than other methods. Only BM25 with neural network gives better results than the proposed system, but that method is used only for summarizing the short text. The proposed system can summarize long documents precisely. This proposed method can overcome the drawbacks of neural network and fuzzy logic by using ANFIS. In the future, one can add more features for calculating the sentence score and rank them accordingly. Also, one can combine ANFIS with other methods like genetic algorithm. In the future, one can use classification algorithm for classifying the sentences instead of threshold value for getting better accuracy in text summarization. The system is used only for English documents; in the future, one can use this system for other languages too.

References

1. Sunmali L, Binwahlan MS, Salim N (2009) Sentence features fusion of text summarization using fuzzy logic. 978-0-7695-3745-0/09 $25.00© 2009 IEEE
2. Alparslan E, Karahoca A, Bahsi H (2011) Classification of confidential documents by using adaptive neuro fuzzy inference systems. 1877-0509. Elsevier. https://doi.org/10.1016/j.procs.2011.01.023
3. Dixit RS, Apte SS (2012) Improvement of text summarization using fuzzy logic based method. IOSR J Comput Eng
4. Yadav J, Meena YK (2016) Use of fuzzy logic and wordNet for improving performance of extractive automatic text summarization. In: ICAACCI, 21–24 Sept 2016
5. Suanmali L, Salim N, Binwahlan MS (2011) Fuzzy genetic semantic based text summarization. IEEE. 978-0-7695-4612-4/11 $26.00 © 2011 IEEE. https://doi.org/10.1109/dasc.2011.192
6. Azhari M, Kumar YJ (2017) Improving text summarization using neuro-fuzzy approach, vol 1, no 4. Taylor & Francis, pp 367–379
7. Aik LE (2008) A study of neuro–fuzzy system in approximation based problems. In: Neuro—fuzzy system ANFIS: adaptive neuro–fuzzy inference system, vol 24, no 2, pp 113–130
8. Kumar YJ, Kang FJ, Goh OS, Khan A (2017) Text summarization based on classification using ANFIS. https://doi.org/10.1007/978-3-319-56660-3_35
9. Khan A, Salim N, Farman H (2016) Clustered genetic semantic graph approach for multi-document abstractive summarization. IEEE. 978-1-4673-8753-8/16/$31.00 ©2016 IEEE
10. Kiani-B A, Akbarzadeh-T MR (2006) Automatic text summarization using: hybrid fuzzy GA-GP. In: 2006 IEEE international conference on fuzzy systems, Canada
11. Jafari M, Wang J, Gheisari M, Shahabi AS, Tao X (2016) Automatic text summarization using fuzzy inference
12. Niu J, Zhao Q, Wang L, Chen H, Atiquzzaman M, Peng F (2016) OnSeS: a novel online short text summarization based on BM25 and neural network. IEEE. 978-1-5090-1328-9/16/$31.00 © 2016 IEEE

Autonomous Traffic Light Control System for Smart Cities

S. Suresh Kumar, M. Rajesh Babu, R. Vineeth, S. Varun, A. N. Sahil and S. Sharanraj

Abstract Nowadays, in the current problems of the smart city implementations, traffic management and control is the most imperative. The conventional traffic control systems lack in the capability of tackling the growth of the road traffic. A Sustainable Urban mobility (Arena et al., AEIT IEEE annual conference 2013, [1]) plan should be put in place by the urban planning authorities in order to efficiently address the traffic-related problems in the smart cities having a focus on the long-term scope of its implementation. This paper focuses on implementing smart traffic control system using intelligent algorithms in an Internet of Things (IoT) (Bhasin et al., IEEE international conference on big data analytics and computational intelligence (ICBDAC), 2017, [2], Kim et al., IEEE transaction on intelligent transportation systems, vol 17, no 8, 2016, [3], Alam and Pandey, IOSR J Comput Eng 16(3):3644, 2014, [4]) based environment. This system follows a decentralized approach which will, in turn, tend to optimize the road traffic by controlling the mobility of the vehicles in the traffic signals (Arena et al., AEIT IEEE annual conference 2013, [1], Uddin et al., Electrical engineering and information communication technology (ICEEICT) 2015 international conference, [5]). This system will overcome the flaws of the current traffic control system by accurately managing all traffic scenarios. This paper primarily presents a comprehensive study of the methodology used by the autonomous traffic light control (ATLC) system, then it focuses on the algorithms used by the system and finally, the paper concludes by stating the advantages and the future work of the ATLC System.

Keywords Urban planning · ATLC · IoT · Zigbee · Tensorflow · Commuters

---

S. Suresh Kumar (✉) · M. Rajesh Babu · R. Vineeth · S. Varun · A. N. Sahil · S. Sharanraj
Sri Ramakrishna Engineering College, Coimbatore, Tamil Nadu, India
e-mail: sureshkumar.s@srec.ac.in

S.-L. Peng et al. (eds.), *Computing and Network Sustainability*, Lecture Notes in Networks and Systems 75, https://doi.org/10.1007/978-981-13-7150-9_35

1 Introduction

The conventional traffic control system in India is inefficient due to a change in the traffic density on the roads. Failure of signals and lack of intelligence for traffic signals has led to traffic congestion. A survey by Ford motors has revealed that nearly 49% of the respondents [6] in India spend at least 12 h a day driving their cars. Conventional traffic signals have preprogrammed timing for all the traffic density patterns. But as traffic patterns evolve, the timer used in conventional traffic lights becomes ineffective. Due to this, the vehicles spend a very long time even there is no traffic congestion [7] or less traffic density patterns. This results in the increased emission of carbon dioxide and other vehicular pollutants. If the traffic signal is programmed to analyze and calculate the signal time for varying traffic density patterns, the traffic congestion can be greatly reduced. This proposed system is fully decentralized. So, each signal makes its own timing decisions, making it a truly smart signal.

2 Methodology

This proposed system uses object detection and classification techniques to manipulate the signal timer according to the traffic density. It uses high-resolution cameras to sense the traffic patterns [5] and process the results to the control system. This model uses TensorFlow for vehicle detection and classification and Raspberry Pi GPIO [8] for image processing and control. At present, the traffic signals in India lack intelligence and acts in an open-loop system, with no feedback or sensing network. The objective is to improve the traffic system in terms of both efficiency and intelligence. The Raspberry Pi [8] computer collects the images of all the roads from the camera and processes them to produce output, which is the number of vehicles on all the roads. The output from the image is calculated and compared and they are transmitted to the traffic light timer control [9]. This acts as a control system. So, the control system uses the developed timer algorithm to predict the time for all the signals and the information is transmitted to the timer display wirelessly by using Zigbee [10] or other transmitters. The control system sets the timer according to the traffic density and the camera senses for every change in the signal timer. In other words, the imaging device snap images after one complete cycle of traffic signal manipulation. The information from the control system is transmitted through wireless means of communication [11]. The control system is executed with a developed timer algorithm to vary the time according to the traffic density, making it an autonomous system [4]. So, this reduces delay in traffic.

The autonomous traffic control system consists of the following entities:

- High-resolution imaging device.
- Raspberry Pi GPIO [8, 12] for image classification (using Tensorflow framework).
- Microcontroller [11] based traffic light timer control.

Fig. 1 A sample traffic road scenario

- Wireless transmission using UART principles.

2.1 Vehicle Detection and Classification

TensorFlow is an open-source machine learning framework [12] used for object detection and classification. The object classification proceeds in the following way. Assuming daytime analysis, four snapshots were obtained from the four roads of the intersection. The TensorFlow uses coco dataset to detect and classify the vehicles. The TensorFlow object detection algorithm detects and classifies the number and the type of vehicle and sends the information to the processing tool. One advantage of TensorFlow is that it can detect and classify objects thereby ensuring accuracy. The Raspberry Pi [8] is a tiny computer in which the TensorFlow framework is installed. The Raspberry Pi is the main component which detects objects and compares the values for optimal time prediction [5]. The image from the camera [11] is sent to this computer where the algorithm is used to detect vehicles on all the roads of the intersection and send the output to the microcontroller device (Figs. 1 and 2).

2.2 Traffic Light Timer Control

The traffic light timer control is used to provide control and coordination in traffic management and ensure safety to the commuters. This device uses LED for indication purpose [3, 11] and uses a microcontroller to change the signal at a specific interval of time automatically. The number of vehicles for all the roads in an intersection is

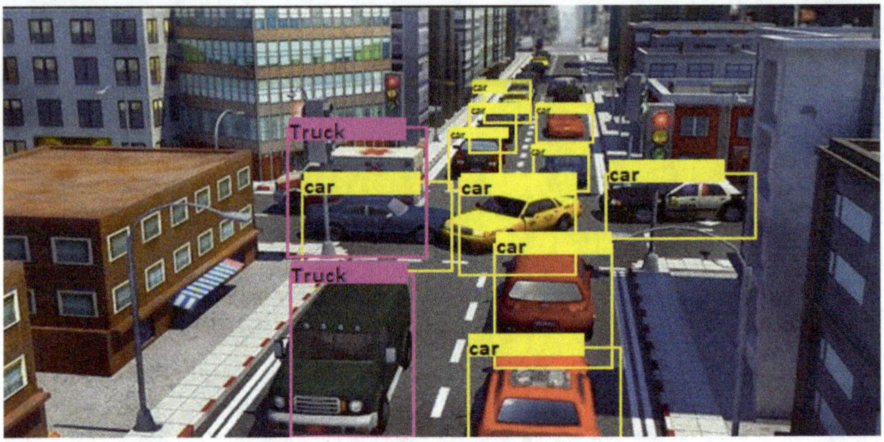

Fig. 2 Traffic road after using TensorFlow for object classification

compared [4] and then the output of the comparison goes to the control unit. Based on the developed timer algorithm, the control unit predicts the time for all the roads based on the traffic density and sends information to the timer display.

2.3 Wireless Transmission

Zigbee is a communication device used for transferring data between the computers like Raspberry Pi and others anything with a serial port. As it works with low power consumption, the transmission distance is limited to 10–100 m line-of-sight. Here we use Zigbee with Raspberry Pi3 for proper wireless communication. Raspberry Pi3 has got four USB ports, so it is better to provide Zigbee interface [10] for the transmission. The wireless communication can be done using the two connected Zigbee dongles.

3 Process Involved

See Fig. 3.

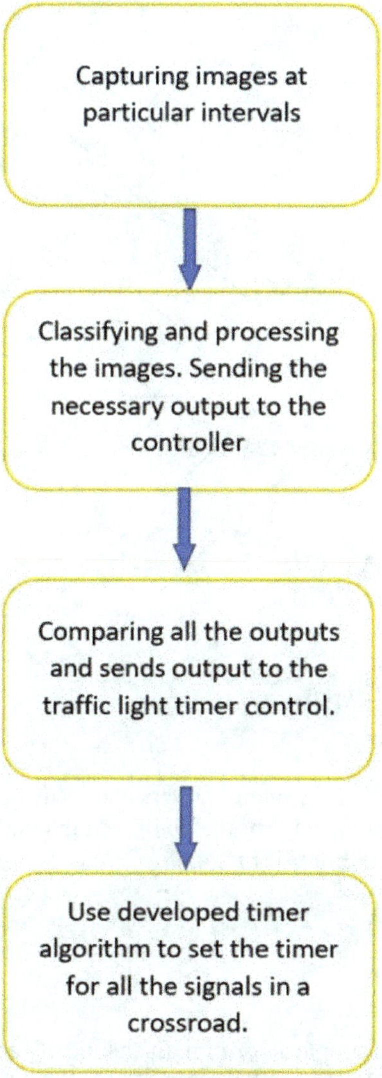

Fig. 3 Process involved in the ATLC system

4 Simple Architecture of the Proposed System

Here the imaging device detects vehicles for a limited distance in order to reduce wrong predictions. So, for a limited distance, it analyzes the number of vehicles. The timer works only for the number of vehicles identified within the distance. So as soon as the one which has the more traffic density is allowed to go, the other three in a crossroad is being stopped. In case the two of the roads are allowed to move

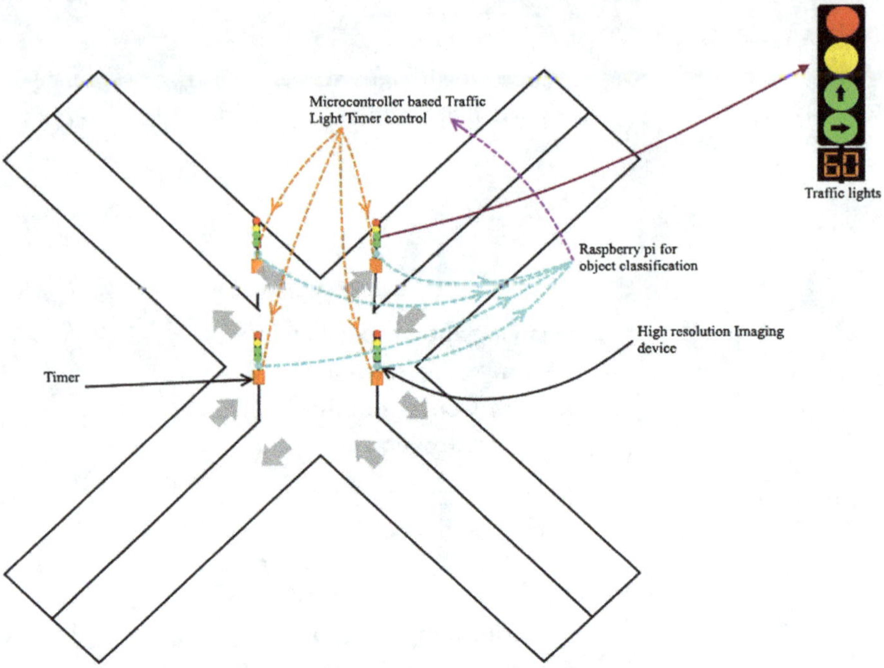

Fig. 4 Simple architecture of ATLCS

further, there will be confusion among drivers and leads to traffic congestion. Only one road is allowed to move only for the optimized interval of time predicted using the developed timer algorithm [9] (Fig. 4).

5 Timer Algorithm

The timer makes use of a simple algorithm. Each kind of vehicle is allotted a specific count and correspondingly, the total counts on a road will be calculated. Each count is allotted an average of 5 s of time. The counts for vehicles are as follows:

- Two wheelers—1/8 Counts.
- Four Wheelers—1/2 Counts.
- Heavy vehicles—1 Count.

To calculate the moving time for a road,

$$\text{Signal green} = (\text{count} * 5)/2$$

Table 1 Real-time traffic signal scenario

Road	No. of counts	Priority
Road 1	5	1
Road 2	3	3
Road 3	4	2
Road 4	2	4

By calculating the total count of vehicles on a road using the corresponding count values, the time for them to cross the signal will be calculated. The road with the highest number of counts will be given the first priority followed by the second highest and so on. To be fair, each road will get 50% of their aggregate time to cross the signal. For example, a road with 10 counts will be allocated a pass time of 25 s rather than their aggregate of 50 s.

5.1 Illustration of Timer Algorithm

For a better understanding, a real-time example is illustrated here. Consider a four roads junction with respective counts of vehicles. The data will be as follows (Table 1).

5.1.1 Example Calculations

Count of vehicles in road 1 = 5
Priority for road 1 = 1
Green signal time for road 1 = (count * 5)/2 = (5 * 5)/2 = 13
Red signal time for road 1 = 0
Priority for road 3 = 2
Green signal time for road 3 = (count * 5)/2 = (4 * 5)/2 = 10
Red signal time for road 3 = green signal time for road 1 = 13
Priority of road 2 = 3
Green signal time for road 2 = (count * 5)/2 = (3 * 5)/2 = 8
Red signal time for road 2 = green signal time for road 3 + red signal time for road 2 = 23
Priority of road 4 = 4
Green signal time for road 4 = (count * 5)/2 = (2 * 5)/2 = 5
Red signal time for road 4 = green signal time for road 2 + red signal time for road 2 = 31.

5.2 Comparison with Other Algorithms

So, from the previous example scenario, we will be going to compare and calculate the total cycle length of the traffic signal in a four-way intersection (Tables 2, 3, and 4).

Once a signal passes from green to red in a cycle, there will be no timer displayed during the red signal until the full cycle is completed [13]. It means the first signal which turns green will have a timer showing the moving time left, and after the timer becomes zero, the signal passes to red but showing no timer. When the last signal in an intersection turns green to red, the imaging device senses the other three signals and uses the same timer algorithm and again the timer for every signal is set.

As soon as the green signal for the last road starts, the camera will start detecting for other three roads. The highest of the three will be allocated first and for a moment, the last road will be considered the last. As the now highest road's green signal goes on, the camera will recognize the count of the fourth road and will move it accordingly and the timers will be reset. So, this makes drivers less boring.

Table 2 Using single-phase timer algorithm (conventional)

Cycle	Road 1	Road 2	Road 3	Road 4	Cycle length
1st cycle	20	20	40	60	20
2nd cycle	60	20	20	40	20
3rd cycle	40	60	20	20	20
4th cycle	20	40	60	20	20
Total cycle length					80

Table 3 Using two-phase timer algorithm

Cycle	Road 1	Road 2	Road 3	Road 4	Cycle length
1st cycle	30	30	30	30	30
2nd cycle	0	0	0	0	0
3rd cycle	30	30	30	30	30
4th cycle	0	0	0	0	0
Total cycle length					60

Table 4 Implementation of the proposed timer algorithm

Cycle	Road 1	Road 2	Road 3	Road 4	Cycle length
1st cycle	13 Nil	23 Nil	13 Nil	31 Nil	13
2nd cycle	Nil Nil	10 Nil	10 Nil	18 Nil	10
3rd cycle	Nil Nil	8 Nil	Nil Nil	8 Nil	8
4th cycle	Nil Nil	Nil Nil	Nil Nil	5 Nil	5
Total cycle length					36

This algorithm is more effective than the conventional traffic networks as it reduces the delay time significantly. The major advantage of this system is its ability to predict signal timings accurately than the other smart traffic networks. It also ensures safety and efficiency [13] as only one road operates with a quick response and the other roads will be operated one by one at a lower interval of time.

6 Results

The cycle length of the traffic signal in a four-way intersection was calculated using both conventional and the proposed algorithm from the previous example and the results are displayed as follows (Fig. 5).

From this chart, it is proved that the proposed algorithm is much efficient than the conventional algorithms thereby having lesser cycle lengths and much less total cycle length than others. The proposed algorithm has resulted in the following characteristics

- Ensures safety since only one road is allowed to move while the others at rest.
- Analyze the varying traffic density patterns and set the timer accordingly. In other words, it operates with intelligence to decide the signal time and reduce the traffic congestion.
- More number of phases with minimum cycle lengths to confirm quick transport [3, 7] and clearing traffic congestions.
- Waiting time for commuters will be reduced due to less cycle lengths and quick phase completion.

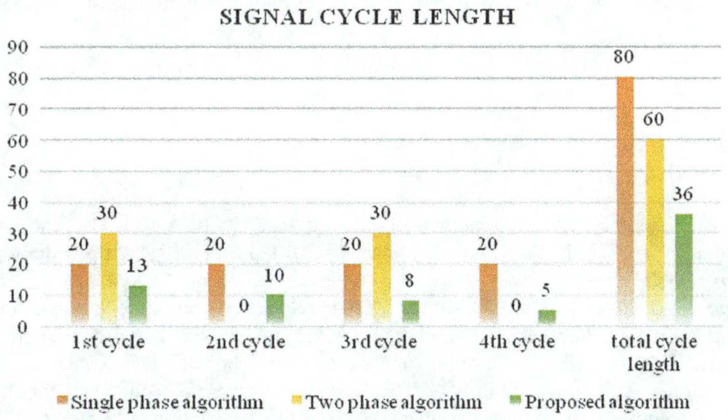

Fig. 5 Comparison of cycle lengths using conventional and proposed algorithms

7 Conclusion

From this system, every road of an intersection is active at a frequent interval of time, thereby making the average delay to reduce at a significant level. This model not only identifies the traffic density but also the number of vehicles and the type of each vehicle. The proposed system reduces the cycle length and increases the number of cycles thereby reducing the waiting time for the commuters in the traffic congestion [7]. Depending upon that, it predicts the time which improves efficiency.

8 Future Work

The major advantage of the proposed system is its robustness and efficiency in reducing traffic congestions [3, 7]. With the integration of techniques such as display boards and use of solar power sources, high-resolution night vision imaging devices, the efficiency, and accuracy of the system can be greatly improved. In India, a system has to be accurate enough to parked vehicles on the road, for which adaptive image cropping techniques will be required. However, the proposed ATLC system is that it is not programmed for detecting authorized emergency vehicles and accidents, for which the concepts of digital signal processing along with RFID tags may be needed. The problem can also be solved by using improved object detection and classification techniques and algorithms. The limitation of the system is that it could detect the pedestrians but could not adapt the traffic control for them.

Acknowledgements The authors would like to thank the Management, Director (Academics), Principal and Head of the Department of Computer Science and Engineering of Sri Ramakrishna Engineering College, Coimbatore, Tamil Nadu, India for their motivation and the support rendered to carry out this work.

References

1. Arena M, Zaninelli FCD, Zaninelli D, Lamedica R, Piccolo A (2013) Smart mobility for sustainability. In: AEIT IEEE annual conference 2013, 11 Dec 2013. https://doi.org/10.1109/aeit.2013.6666811
2. Bhasin S, Choudhry T, Gupta SC, Kumar P (2017) Smart city implementation model based on IoT. In: 2017 IEEE international conference on big data analytics and computational intelligence (ICBDAC), 19 Oct 2017. https://doi.org/10.1109/icbdaci.2017.8070836
3. Kim C, Ong YS, Cheong T, Tan PS (2016) Solving the dynamic vehicle routing problem for traffic congestion. In: IEEE transaction on intelligent transportation systems, vol 17, no 8, Aug 2016
4. Alam J, Pandey MK (2014) Development of intelligent traffic light system based on congestion estimation using fuzzy logic. IOSR J Comput Eng 16(3):3644
5. Uddin MS, Das AK, Abu Taleb M (2015) Real-time area based traffic density estimation by image processing for traffic signal control system: Bangladesh perspective. Presented at the

electrical engineering and information communication technology (ICEEICT) 2015 international conference
6. https://timesofindia.indiatimes.com/india/Indians-spend-more-time-behind-the-wheel-than-Chinese-Aussies-Survey/articleshow/46337734.cms
7. Lopez-Garcia P, Onieva E, Osaba E, Masegosa AD, Perallos A (2016) A hybrid method for short-term traffic congestion forecasting using genetic algorithms and cross entropy. In: IEEE transactions on intelligent transportation systems, vol 17, no 2, Feb 2016
8. Azni MN, Vellasami L, Zianal AH, Mohammed FA, Daud NM, Vejasegaran R, Basharudin NW, Jusoh M, Azir K, Eh Kan PL (2016) Home automation system with android application. In: 2016 IEEE 3rd international conference on electronic design (ICED), 05 Jan 2017. https://doi.org/10.1109/iced.2016.7804656
9. Hsieh P-C, Chen Y-R, Wu W-H, Hsiung P-A (2014) Timing optimization and control for smart traffic. In: 2014 IEEE international conference on internet of things (iThings), and IEEE green computing and communications (GreenCom) and IEEE cyber, physical and social computing (CPSCom), 16 Mar 2015. https://doi.org/10.1109/ithings.2014.11
10. Firdaus, Nugroho E, Sahroni A (2014) ZigBee and Wifi network interface on wireless sensor networks. In: 2014 Makassar IEEE international conference on electrical engineering and informatics (MICEEI), 26 Mar 2015. https://doi.org/10.1109/miceei.2014.7067310
11. Pang J (2015) Review of microcontroller based intelligent traffic light control. In: IEEE 12th international conference & expo on emerging technologies for a smarter world (CEWIT), 03 Dec 2015. https://doi.org/10.1109/cewit.2015.7338166
12. Yi H, Jung H, Bae S (2017) Deep neural networks for traffic flow prediction. In: IEEE international conference on big data and smart computing (BigComp), 20 Mar 2017. https://doi.org/10.1109/bigcomp.2017.7881687
13. Brasil RH, Machado AMC (2017) Automatic detection of red light running using vehicular cameras. IEEE Lat Am Trans 15(1)

A Flow-Based Network Intrusion Detection System for High-Speed Networks Using Meta-heuristic Scale

V. Jyothsna, D. Mukesh and A. N. Sreedhar

Abstract Internet users are drastically increasing, along with new network services which are leading to more serious network attacks and threats. A network intrusion detection system (NIDS) is designed mainly to detect such attacks. While some NIDSs work by inspecting the complete traffic called as packet-based NIDSs, other NIDSs inspect cumulative information related to the packets in the form of flow called as flow-based NIDSs. Even though packet-based NIDSs generate high detection accuracy with low false alarms, they are hard, or even impossible, to achieve at the speed. In the flow-based IDSs, the amount of information to be processed is lesser, so it is the logic suitable for high-speed networks. But the problem is with flow-based high false alarm rate and low accuracy. In this paper, a flow-based intrusion detection system (IDS) for high-speed networks using meta-heuristic scale is proposed. Initially, a flow-based approach is applied on request stream to define feature metrics. Next, these feature metrics are used to define the scale, which is further used to define whether the flow is normal or malicious. In order to confirm the effectiveness of the model proposed, it has been experimented on NSL-KDD. The experimental results exhibit that the designed models provide improved accuracy with less computational time and minimal false alarm rate.

Keywords Network IDS · Packet-based · Flow-based · High-speed networks · False alarm · Performance · Accuracy

---

V. Jyothsna (✉)
Sree Vidyanikethan Engineering College, Tirupati, India
e-mail: jyothsna1684@gmail.com

D. Mukesh · A. N. Sreedhar
Chadalawada Ramanamma Engineering College, Tirupati, India
e-mail: dhandumukesh.crec@gmail.com

A. N. Sreedhar
e-mail: ansreedhar01@gmail.com

S.-L. Peng et al. (eds.), *Computing and Network Sustainability*, Lecture Notes in Networks and Systems 75, https://doi.org/10.1007/978-981-13-7150-9_36

1 Introduction

In the early days, only a small group of users were using the Internet. So, ensuring the security was an easy task. With the evolution to its current massively larger form, this no longer holds good. The Internet is a critical business factor, and data is worth more than ever before which enables security as a core issue. Most important challenge in today's world is maintaining and keeping the computer network safe from malicious activities. Research has provided a broad range of contributions to improve network security, in the form of hardware devices (firewalls), software (anti-virus software), and also principles (standards) [1]. Firewalls are mostly of static nature and operate on the lower levels (network layer). They are not able to act upon complex attack scenarios. A network IDS is one that is designed to detect such attacks. While some NIDS operate by inspection of the complete traffic (packet level), for known attack signatures (byte patterns) and fully state-aware systems allow for more enhanced detection techniques [2]. Still, the majority of the systems may potentially misclassify attacks. With the growing speed of broadband Internet, the information in the backbone networks increases steadily. Monitoring of links by NIDS may need more computational resources than available. This problem of resource demand is commonly solved by engineering dedicated hardware that can process detection tasks more efficiently. Flow-based mechanism is proposed as a solution to identify attacks in IDS.

The organization of the paper is as follows: Section 2 specifies the literature of flow-based approaches. Section 3 presents feature metrics to identify the request behavior. Section 4 illustrates the design of heuristic scale. Empirical analysis of the proposed model is presented in Sect. 5. Finally, Sect. 6 includes conclusion and present and future research activities.

2 Related Work

A flow-driven framework is designed and implemented by Song et al. [4] which derives flow-level viewpoint. By using these viewpoints, fine-grained controllability can be achieved. A scheme with two threshold based on ideal exit point (ATEP) is used to reduce the overhead during flow and also used to provide efficient crash detection mechanism. The experimental result indicates that this technique can achieve higher throughput and less packet loss. Alaidaros [5] studied the influence of new technologies on network intrusion detection systems in terms of performance and accuracy. It also describes the performance consequences in payload-based and flow-based NIDSs for high-speed networks. Packet-based NIDSs process every packet and produce low false alarms. But the time taken to process is high which is hard, or even difficult, to perform this approach on high-speed networks of the capacity, multiple gigabits per second (Gbps). However, flow-based IDSs are suitable for high-speed networks as it process lower amount of data. But, it produces high false alarms. Therefore, a hybrid or a mixture NIDS model for high-speed network environment may guarantee a higher capability to respond on the wider scope of attacks. Schaffrath [6] designed a hybrid network-based IDS framework integrating flow-based and packet-based. It aimed to detect malicious activities by inspecting the network traffic. Over the last years, network speed increases which leads to unspecified growth of the traffic volume, and payload-based intrusion detection system became difficult to analyze which leads to the development of alternative technique called flow-based analysis. Though each approach solitary suffers from weaknesses, hybrid approaches show the potential in improving the performance of the system in less time. Abuadlla [7] proposed a two-stage neural network for flow-based anomaly intrusion detection system. The significant variations in the traffic that leads to potential attacks are detected in the first stage, while the classification of attack once the known patterns are identified is performed in the second stage. Li [8] proposed, implemented, and evaluated a DoS-resilient high-speed flow-level IDS (HiFIND).

3 Flow-Based Metrics for Intrusion Detection System

In this section, a collection of metrics are proposed for calculating length of request chain, to know the ratio of packets in each flow, to evaluate the type of packets, distance between the packets as ratio of request intervals, and propagation of requests as request chain context" were formulated which converts the flow into these attributes, which is used for "early and fast detection" of the collective requests [9] . The proposed method identifies sessions as input to find the attacks in the networks.

3.1 Calculating Time Frame Length

For each cluster, the difference between the end times of maximum session and begin time of least session is called as time frame [10].

The time frame ($\mathrm{tf}(c_i)$) of the cluster c_i is

$$\mathrm{tf}(c_i) = \sqrt{(\mathrm{se}_1 - \mathrm{sb}_1)^2} \quad (1)$$

For all clusters, the average time frame length is defined as

$$\langle \mathrm{tf}(C) \rangle = \frac{\sum_{i=1}^{K} \mathrm{tf}(c_i)}{K} \quad (2)$$

Absolute deviation (tfAD) is defined as

$$\mathrm{tfAD} = \frac{\sqrt{\sum_{i=1}^{K} (\langle \mathrm{tf}(C) \rangle - \mathrm{tf}(c_i))^2}}{K} \quad (3)$$

At the end, time frame (tf) is defined as

$$\mathrm{tf} = \langle \mathrm{tf}(C) \rangle + \mathrm{tfAD} \tag{4}$$

3.2 Empirical Metrics

3.2.1 Request Chain Length (RCL)

For each time frame, length tf is calculated as (refer Sect. 3.1).

$$\mathrm{clAD} = \frac{\sqrt{\sum_{i=1}^{|\mathrm{TS}(\mathrm{CS}_N)|} (\langle \mathrm{TS}(\mathrm{CS}_N) \rangle - (|\mathrm{ts}_i \,|\, \exists\, \mathrm{ts}_i \in \mathrm{TS}(\mathrm{CS}_N))^2}}{|\mathrm{TS}(\mathrm{CS}_N)|} \tag{6}$$

$$\mathrm{rcl}(\mathrm{CS}_N) = \langle \mathrm{TS}(\mathrm{CS}_N) \rangle + \mathrm{clAD} \tag{7}$$

3.2.2 Ratio of Packet Count

The ratio of packets in traffic for every time frame $\{\mathrm{ts}_i \,\exists\, \mathrm{ts}_i \in \mathrm{TS}(\mathrm{CS}_N) \wedge i = 1, 2, \ldots |\mathrm{TS}(\mathrm{CS}_N)|\}$ observed is calculated as

$$\mathrm{rp}(\mathrm{ts}_i) = \sum_{i=1}^{|\mathrm{TS}(\mathrm{CS}_N)|} \frac{|P(\mathrm{ts}_i)|}{\sum_{k=1}^{|\mathrm{TS}(\mathrm{CS}_N)|} |P(\mathrm{ts}_k)|} \tag{8}$$

The absolute deviation (tflpsAD) for ratio of packets is calculated

$$\mathrm{tflpsAD} = \frac{\sqrt{\sum_{j=1}^{|\mathrm{TS}(\mathrm{CS}_N)|} (1 - \mathrm{rp}(\mathrm{ts}_j))^2}}{|\mathrm{TS}(\mathrm{CS}_N)|} \tag{9}$$

At the end, the ratio of packet count (RPC) of CS_N is calculated as

$$\mathrm{rpc}(\mathrm{TS}(\mathrm{CS}_N)) = \frac{\sum_{i=1}^{|\mathrm{TS}(\mathrm{CS}_N)|} \mathrm{rp}(\mathrm{ts}_i)}{|\mathrm{TS}(\mathrm{CS}_N)|} + \mathrm{tflpsAD} \tag{10}$$

3.2.3 Ratio of Intervals Between Flow Requests

Requests between the intervals are defined as

$$\mathrm{Im}(\mathrm{ts}_k) = \frac{\sum_{j=1}^{|I(\mathrm{ts}_k)|} i_j}{|I(\mathrm{ts}_k)|} \tag{11}$$

The global mean $\mathrm{gm}(\mathrm{CS}_N)$ for the intervals is defined as

$$\mathrm{gm}(\mathrm{CS}_N) = \frac{\sum_{j=1}^{|I(\mathrm{CS}_N)|} i_j}{|I(\mathrm{CS}_N)|} \tag{12}$$

For all time frames, the deviation (iAD) in the dataset is calculated as

$$\mathrm{iAD} = \frac{\sqrt{\sum_{j=1}^{|\mathrm{TS}(\mathrm{CS}_N)|} (\mathrm{gm}(\mathrm{CS}_N) - \mathrm{lm}(\mathrm{ts}_j))^2}}{|\mathrm{TS}(\mathrm{CS}_N)|} \tag{13}$$

At the end,

$$\mathrm{ri}(\mathrm{CS}_N) = \frac{\sum_{i=1}^{|\mathrm{TS}(\mathrm{CS}_N)|} \mathrm{lm}(ts_i)}{|\mathrm{TS}(\mathrm{CS}_N)|} + \mathrm{iAD} \tag{14}$$

3.2.4 Ratio of Packet Types in Fixed Time Allotment

For every time frame ts_i $\{\mathrm{ts}_i \exists \mathrm{ts}_i \in \mathrm{TS}(\mathrm{CS}_N)\}$ and for every packet type pt_j $\{\mathrm{pt}_j \exists \mathrm{pt}_j \in \mathrm{PT} \wedge j = 1, 2, \ldots, |\mathrm{PT}|\}$ of tf_i, the no. of happening of every dissimilar pt_j is calculated as

$$c_{\mathrm{ts}_i}(\mathrm{pt}_j) = \sum_{k=1}^{|P(\mathrm{ts}_i)|} \{1 \exists t(p_k) \equiv \mathrm{pt}_j\} \tag{15}$$

The local support of packet types tf_i is measured as

$$\mathrm{ls}_{\mathrm{ts}_i}(\mathrm{pt}_j) = \frac{c_{\mathrm{ts}_i}(\mathrm{pt}_j)}{|P(\mathrm{ts}_i)|} \tag{16}$$

The global support of deviating packets is measured as

$$\mathrm{gs}(\mathrm{pt}_j) = \frac{c(\mathrm{pt}_j)}{\sum_{k=1}^{|\mathrm{TS}(\mathrm{CS}_N)|} |P(\mathrm{ts}_k)|} \tag{17}$$

The absolute deviation (ptsAD) is calculated as

$$\mathrm{ptsAD}(\mathrm{pt}_j) = \frac{\sqrt{\sum_{i=1}^{|\mathrm{TS}(\mathrm{CS}_N)|} \left(\mathrm{gs}(\mathrm{pt}_j) - \mathrm{ls}_{\mathrm{ts}_i}(\mathrm{pt}_j)\right)^2}}{|\mathrm{TS}(\mathrm{CS}_N)|} \tag{18}$$

Finally,

$$\text{rpt}(\text{pt}_j) = \frac{\sum_{i=1}^{|\text{TS}(\text{CS}_N)|} \text{ls}_{\text{ts}_i}(\text{pt}_j)}{|\text{TS}(\text{CS}_N)|} + \text{ptsAD}(\text{pt}_j) \tag{19}$$

3.2.5 Request Chain of Context

The absolute deviation (rpsAD) of request pair support for the local supports of pair $p_i \{p_i \exists p_i \in \text{rps}_N \wedge i = 1, 2, \ldots, |\text{rps}_N|\}$ from global support is described as

$$\text{rpsAD} = \sqrt{\frac{\sum_{j=1}^{|\text{TS}(\text{CS}_N)|} \left(\text{gs}(p_i) - \text{ls}_{ts_j}(p_i)\right)^2}{|\text{TS}(\text{CS}_N)|}} \tag{20}$$

The request chain context $\text{rcc}(p_i)$ of all pairs received in all time frames of training set CS_N is defined as

$$\text{rcc}(p_i) = \frac{\sum_{j=1}^{|\text{TS}(\text{CS}_N)|} \text{ls}_{ts_j}(p_i)}{|\text{TS}(\text{CS}_N)|} + \text{rpsAD} \tag{21}$$

4 Feature Association Impact Scale for IDS

In this section, a meta-heuristic statistical scale called feature association impact scale (FAIS) is proposed to estimate whether the transaction is safe, suspicious, or attack [11]. The feature association metric (fas) considers the above-specified metrics as record set of network transaction.

4.1 Assumptions

Let $\{f1, f2, f3, \ldots, fn \, \forall \, f_i = \{f_i v_1, f_i v_2, \ldots, f_i v_m\}\}$ be set of features which are categorical values used to form the transactions T [12].

4.2 Process

Step 1: Weight of edge between the features $\text{val}(f_1)$ and $\text{val}(f_2)$ is estimated as:

$$w(\text{val}(f_1) \leftrightarrow \text{val}(f_2)) = \frac{\text{ctvs}}{|\text{STVS}|} \tag{22}$$

Step 2: The edge weight among transaction value sets and its categorical values is measured as

$$E = \{(\mathrm{tvs}_i, \mathrm{val}_j) : \mathrm{val}_j \in \mathrm{tvs}_i, \mathrm{tvs}_i \in \mathrm{STVS}, \mathrm{val}_j \in v\} \tag{23}$$

Step 3: The pivots and feature categorical values are measured.
Step 3.1: Consider matrix u, which denotes pivot initial value as 1
Step 3.2: Transpose the matrix A as A'
Step 3.3: Calculate prerogative weights by multiplying A' with u
Step 3.4: Calculate original weights using multiplication between A and V
Step 4: Calculate the feature categorical value fas of $f_i v_j$ as

$$\mathrm{fas}(f_i v_j) = \frac{\sum_{k=1}^{|\mathrm{STVS}|} \sum \{u(\mathrm{tvs}_k) : (f_i v_j \rightarrow \mathrm{tvs}_k) \neq 0\}}{\sum_{k=1}^{|\mathrm{STVS}|} u(\mathrm{tvs}_k)} \tag{24}$$

Step 5: The fais for every transaction value set tvs_i is estimated as

$$\mathrm{fais}(\mathrm{tvs}_i) = 1 - \frac{\sum_{j=1}^{m} \{\mathrm{fas}(\{\mathrm{val}_j \exists\, \mathrm{val}_j \in V\}) : (\mathrm{val}_j \subset \mathrm{tvs}_i)\}}{|\mathrm{tvs}_i|} \tag{25}$$

Step 6: The feature association impact scale threshold (faist) can be measured as:

$$\mathrm{faist} = \frac{\sum_{i=1}^{|\mathrm{STVS}|} \mathrm{fais}(tvs_i)}{|\mathrm{STVS}|} \tag{26}$$

Step 7: Calculate the standard deviation as:

$$\mathrm{sdv}_{\mathrm{faist}} = \sqrt{\frac{\left(\sum_{i=1}^{|\mathrm{STVS}|} \mathrm{fais}(\mathrm{tvs}_i) - (\mathrm{faist})^2\right)}{(|\mathrm{STVS}| - 1)}} \tag{27}$$

Step 8: The FAIS range is defined in the following steps.
Step 8.1: Calculate lower threshold value of faist as $\mathrm{faist}_l = \mathrm{faist} - \mathrm{sdv}_{\mathrm{faist}}$
Step 8.2: Calculate higher threshold value of faist as $\mathrm{faist}_h = \mathrm{faist} + \mathrm{sdv}_{\mathrm{faist}}$

Further, the conditions used to estimate the network record (nt) are normal, suspicious, or attack as follows:

(i) If $\mathrm{fais}(nt) < \mathrm{faist}_l$, then the network record (nt) is said to be safe.
(ii) If $\mathrm{fais}(nt) \geq \mathrm{faist}_l$ && $\mathrm{fais}(nt) < \mathrm{faist}_h$, then the network record (nt) can be defined as suspicious.
(iii) If $\mathrm{fais}(nt) \geq \mathrm{faist}_h$, then the network transaction (nt) can be conformed as attack.

5 Empirical Analysis of the Proposed Model

The reliability of the projected model is explored network transactions dataset NSL-KDD [13]. Every feature is calculated independently with the dataset CS consisting of N (normal) as CS_N and D (intrusion) as CS_D, and its accuracy is evaluated. According to the results given in Table 2, the proposed model is 95% accurate. The failure percentage is approximately 5%, which is considerable and occurred due to categorical values of the features. The rate of successful intrusion detection is represented by the metric called sensitivity, which is identified as around 0.945535536. The rate of successful fair transaction detection is represented by the metric called specificity that is identified as 0.925100789, which is not significant in these experiments due to the negligible count of fair transactions given as test input (Tables 1 and 3, Fig. 1).

Table 1 Metric calculation for normal and intrusion transactions

The total no. of transactions (CS): 247,150			
		Training (60%)	Testing (40%)
N (normal) CS_N	Transactions	79,250	55,500
	Sessions	2872	1948
	Clusters	287	194
	Length of time frame (tf)	610	605
	No. of time frames	240	151
	Ratio of packet count (see Sect. 3.2.2)	0.678	0.612
	Request chain length (RCL) (see Sect. 3.2.1)	22.356	21.789
	Ratio of intervals between requests (see Sect. 3.2.3)	0.265	0.252
D (attack) CS_D	Transactions	65,440	46,960
	Sessions	2548	1732
	Clusters	254	173
	Length of time frame (tf)	730	763
	No. of time frames	250	172
	Ratio of packet count	0.445	0.413
	Request chain length (RCL)	28.456	27.568
	Ratio of intervals between requests	0.132	0.119

Table 2 Performance values of the empirical analysis

Total no. of records for both training and testing	24,7150
The total no of records considered for training	134,750
Total no. of records for testing	112,400
Feature association impact scale threshold found	0.8021
Feature association impact scale threshold upper bound	0.8625
Feature association impact scale threshold lower bound	0.7416
True positive (tp)	37,631
False positive (fp)	4354
True negative (tn)	46,244
False negative (fn)	642
Precision (*P*)	0.91
Recall/sensitivity (*R*)	0.98
Specificity (*S*)	0.924
Accuracy (*A*)	0.955
F-measure (*F*)	0.945

Table 3 Evaluation of the proposed approach with FAIS [11] and ARTP [9]

	BIHS	ARTP	FAIS
P	0.91	0.895	0.889
R	0.98	0.985	0.935
S	0.924	0.914	0.485
A	0.955	0.944	0.851
F	0.945	0.938	0.911

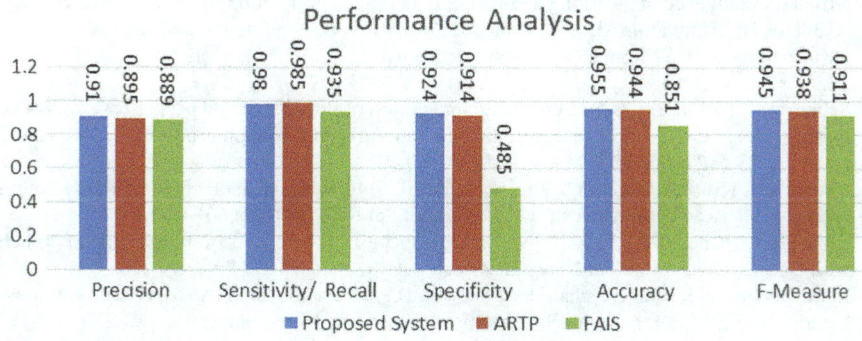

Fig. 1 Comparison of the proposed method with FAIS [11] and ARTP [9]

6 Conclusion

In this paper, NIDS is evaluated at flow level instead of considering the traffic as request stream. Many machine learning features are defined to represent the flow of request, and each metric is validated individually. The experimental result shows that the proposed model is 95% accurate when compared to FAIS and ARTP. The failure percentage is approximately 5%, which is considerable and occurred due to categorical values of the features. The rate of successful intrusion detection is represented by the metric called sensitivity, which is identified as around 0.945535536. The rate of successful fair transaction detection is represented by the metric called specificity that identified as 0.925100789, which is not significant in this experiments due to the negligible count of fair transactions given as test input. An ensemble base incremental model is designed with more accuracy that can be used to detect and classify anomalies with less time for training.

References

1. Adnan M, Abdulazeez B, Adel SI (2011) Intrusion detection and attack classifier based on three techniques. Comp Study Eng Technol J 29(2):233–254
2. Gao M, Zhang K, Lu J (2006) Efficient packet matching for gigabit network intrusion detection using TCAMs. In: Proceedings of 20th international conference on advanced information networking and applications (AINA'06), pp 249–254
3. Myung S, Hun K, Seong H, Seung C, James H (2004) A flow-based method for abnormal network traffic detection. In: IEEE/IFIP network operations and management symposium, pp 599–612
4. Song B, Yang W, Chen M, Zhao X, Fan J (2010) Achieving flow-level controllability in network intrusion detection system. In: 11th ACIS international conference on software engineering, artificial intelligence, networking and parallel/distributed computing, IEEE computer Society
5. Alaidaros H, Mahmuddin M, Al Mazari A (2011) An overview of flow-based and packet-based intrusion detection performance in high speed networks. In: The international arab conference on information technology (ACIT'2011)
6. Schaffrath G, Stiller B (2008) Conceptual integration of flow-based and packet-based network intrusion detection. In: IFIP International Federation for Information Processing 2008, AIMS 2008. LNCS 5127, pp 190–194
7. Abuadlla Y, Kvascev G, Gajin S, Jovanovic Z (2016) Flow-based anomaly intrusion detection system using two neural network stages. Comput Sci Inf Syst 11(2):601–622
8. Li Z, Gao Y, Chen Y (2009) HiFIND: a high-speed flow-level intrusion detection approach with DoS resiliency. Comput Netw (Elsevier)
9. Munivara Prasad K, Rama Mohan Reddy A, Venu Gopal Rao K (2018) An experiential metrics-based machine learning approach for anomaly based real time prevention (ARTP) of App-DDoS attacks on web. In: Artificial intelligence and evolutionary computations in engineering systems. Springer, pp 99–112
10. Munivara Prasad K, Rama Mohan Reddy A, Venu Gopal Rao K (2017) BIFAD: bio-inspired anomaly based HTTP-flood attack detection. Wirel Pers Commun 97(1):281–308
11. Jyothsna V, Rama Prasad VV (2016) Anomaly based network intrusion detection through assessing feature association impact scale. INDERSCIENCE-Int J Inf Comput Secur (IJICS) 8(3):241–257

12. Jyothsna V, Rama Prasad VV (2016) FCAAIS: anomaly based network intrusion detection through feature correlation analysis and association impact scale. J Inf Commun Technol Express 2(3):103–116 (Elsevier)
13. Dhanabal L, Shantharajah, SP (2015) A study on NSL-KDD dataset for intrusion detection system based on classification algorithms. Int J Adv Res Comput Commun Eng 4(6)

Predictive Digital Forensic Model to Track Antisocial Behavior Based on Dermatoglyphics

M. D. Vasan and B. R. Thakar

Abstract Recent trend of antisocial activity monitoring needs advance method to track and reduce relevant crime. Fingerprint used as biometric print for unique access ID and evidence marking in forensic investigation and in financial transaction authentication. Dermatoglyphics markers characteristics can be used to develop forensic model for antisocial activity tracking. Every human contains fingerprints basically categorized in Whorl, Arch, and Loop, which creates unique biometric pattern. Antisocial behavior appeared mainly by impulsivity which is caused by social isolation, depressed mentality, negative thought process, and lack of emotion. Impulsive behavior can be analyzed by Barratt Impulsiveness Scale, Version 11 (BIS-11) and Impulsive/Premeditated Aggression Scale (IPAS) method. Correlative study of finger ridges and impulsive behavior tests can be used to develop advanced precognition technique which can predict antisocial or criminal intent. Digital forensics algorithm can pre-cure potential antisocial issues from dermatoglyphics (fingerprint) and IPAS, BIS-11 study results. This model can also be used in employment procedure, anti-national threat identification, suicidal attempt precautions, advanced criminal investigation, and during court testimony of identity forensics as psychology index.

Keywords Forensic · Predictive digital forensic model · Fingerprint · Dermatoglyphics · Criminal identity · Impulsive behavior · BIS-11 · IPAS · Advanced forensics · Antisocial behavior · Digital forensics

1 Introduction

Recent criminal research shows that antisocial behavior primely caused impulsivity, which includes aggregation, periodic emotional isolation, potential revenge thinking, unstable lifestyle, insulting incidents, and egoistic communication approach. Human

M. D. Vasan (✉) · B. R. Thakar
Raksha Shakti University, Ahmedabad, India
e-mail: margivasan@outlook.com

B. R. Thakar
e-mail: br.thakar@outlook.com

S.-L. Peng et al. (eds.), *Computing and Network Sustainability*, Lecture Notes in Networks and Systems 75, https://doi.org/10.1007/978-981-13-7150-9_37

behavior indicates the common approach to any situation related to his/her comfort thinking, moral valuing, social connections, and mental goals. Detailed forensic and psychological study identified that usually motives of antisocial behavior are based on impulsive action concerned to person. Fingerprint counted as the presence of person at crime place and could be helpful to identify potential attacker in investigation and testimony procedure. Person, who has persistently shown violent actions and intentional negative thinking, may cause damage to individual or society, and also he/she can commit crime lead by impulsivity. Impulsivity control can be observed by two universally accepted tests IPAS and BIS-11. Fingerprint comparison database consists of different human grouped by gender, age, profession, intelligent quotient and psychology collected and distributed as per tests mentioned above which can give us effective identification method of potential antisocial activity performed by any person. Digital modeling by dermatoglyphic prints and statistical analysis of impulsive fundamentals can give us predictive forensic approach for precognition of such antisocial personality.

2 Methodology

Impulsivity describes instant action reasoning, social disconnections, abusive communication arrogant reaction, and negative mental values, which can consequented into unstable decision making, harsh insulting, and merciless thinking. Above discussed symptoms can be commonly observed in person by poor work performance, directionless thinking, delayed and confused decision, messy task results, and failure sticking. Impulsivity observation methods discussed below are vital information to predict the criminal intent personality.

2.1 Impulsive/Premeditated Aggression Scale (IPAS)

Aggression behavior observation results based on violent/abusive activity incorporating mental conditions with scale of

- Impulsive Aggressive (IA)
- Premeditated aggressive (PA)

The IPAS questionnaire was introduced and developed by Dr. Matthew Stanford, which is designed to evaluate individual impression of aggressive acts. Aggressive participant classification (i.e., impulsive and premeditated) done by categorical scoring mechanism based on answered of an individual participant with thirty questions, observed in 5-Likert Scale (1 = Strongly Agree, 2 = Agree, 3 = Neutral, 4 = Disagree, 5 = Strongly Disagree) further aggregates to total scores. The percentage of positive items for each aggression scale is then calculated on ten questions to determine impulsive aggressive (IA) characteristics, eight questions for premeditated aggressive (PM) characteristics, and remaining twelve questions are not scored for this model, but each question is mandatory to be answered.

2.2 *BIS-11*

BIS-11 was created by Dr. Ernest S. Barratt at the Department of Psychology, Texas Christian University, Galveston, foremost used for track behavioral/personality based on impulsivity. Thirty question response record of an individual unveil the nature and types of impulsivity. Concise detailed report based on four-point calculation (Rarely/Never [1], Occasionally [2], Often [3], Almost Always/Always [4]) rating information. This distinguished scale unit estimates impulsivity in various categories like cognitive instability, complexity, perseverance, cognitive complexity, self-control, motor, and attention for six first-order factors. The Barratt Impulsiveness Scale, Version 11 (BIS-11) questioner explained in thirty basic items values to acquire multi factor-resultant stats of impulsiveness. This test also acquires second-order symptoms described as attentional impulsiveness (attention and cognitive instability), motor impulsiveness (motor and perseverance), non-planning, and impulsiveness (self-control and cognitive complexity).

2.3 *Fingerprint Pattern*

- Whorl—plain, oblong plain, spiral, double spiral
- Arch—exceptional, tented, simple
- Loop—plain pocket, central pocket, lateral pocket, ulnar, radial, twinned, double spiral (Fig. 1).

Two-dimensional impression of friction ridges sophistically acquired by fingerprint digital scanning device specially designed for forensics instead of old paper and ink approach, which is essential for the predictive digital forensics modeling.

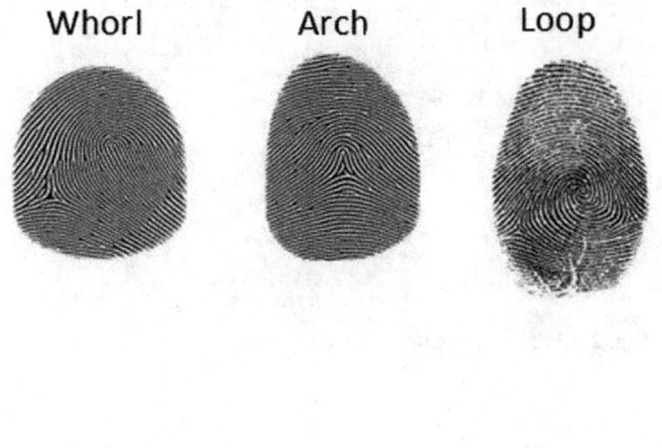

Fig. 1 Basic dermatoglyphics marker prints (fingerprints)

The administration and data collection done by two questionnaires (IPAS/BIS-11) have to be filled under systematic, simple, and in familiar language instruction set. A self-reporting inventory made of total sixty questions and fingerprint scan followed by permission page signed as self-declaration given to each probe candidate. The information (fingerprints, test results) acquired from probe candidates is preserved, and individuals are assured of non-disclosure of identity.

2.4 Digital Modeling

Modeling requirement descriptions

- Fingerprint types (Whorl, Arch, and Loop)
- Match algorithm (custom application developed in MATLAB)
- Individual IPAS report outcomes
- Individual IPAS report outcomes
- Correlative reporting from fingerprint and both test results (Fig. 2).

Primary experimental approach starts with fingerprint data collection of volunteers for research study program from different background. Three hundred candidate dermatoglyphics data collected and impulsivity tests (IPAS, BIS-11) have been performed as a pilot project which can help to analyze statistical analysis). Example study is explained with thirty probe candidates (Tables 1 and 2, Fig. 3).

- PC: Probe candidate
- R: Right palm finger (R1–R5)

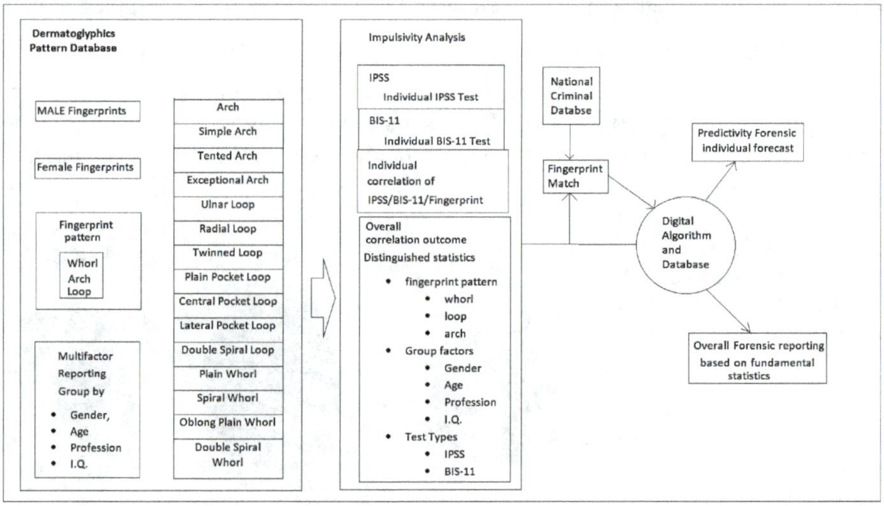

Fig. 2 Predictive digital forensic model

Table 1 Statistical fingerprint data on IPAS, BIS-11

PC	L1	L2	L3	L4	L5	R1	R2	R3	R4	R5	BIS-11	IPAS
PC1	PL	PL	UL	UL	UL	PL	CPL	UL	UL	UL	Highly impulsive	PM
PC2	DSL	UL	UL	CPL	UL	PL	DSL	UL	UL	UL	Normal	PM
PC3	DSL	SA	SA	UL	UL	UL	SA	UL	UL	UL	Normal	IA
PC4	SL	SL	SL	SL	UL	SL	SL	SL	SL	UL	Highly impulsive	IA
PC5	UL	UL	UL	SL	UL	UL	UL	UL	SL	DSL	Highly impulsive	IA
PC6	UL	RL	SL	DSL	UL	UL	RL	UL	SL	UL	Normal	PM
PC7	UL	UL	UL	SL	UL	TL	UL	UL	SL	UL	Normal	PM
PC8	TL	TL	UL	TL	DSL	TL	UL	UL	TL	DSL	Normal	PM
PC9	UL	PL	PL	PL	CPL	CPL	UL	PL	PL	PL	Normal	IA
PC10	UL	UL	UL	UL	UL	UL	PL	UL	PL	UL	Normal	IA
PC11	SL	SL	DSL	SL	UL	DSL	PL	UL	SL	UL	Normal	IA
PC12	UL	RL	UL	UL	UL	UL	UL	UL	UL	UL	Highly impulsive	PM
PC13	TL	RL	UL	CPL	UL	UL	UL	UL	UL	UL	Highly impulsive	PM
PC14	UL	CPL	SW	CPL	UL	UL	SW	RL	UL	UL	Normal	PM
PC15	UL	UL	UL	UL	UL	SL	UL	UL	SW	UL	Normal	PM
PC16	UL	UL	UL	UL	UL	UL	UL	UL	CPL	UL	Highly impulsive	PM
PC17	PL	PL	PL	PL	PL	TL	PL	PL	PL	PL	Highly impulsive	PM
PC18	UL	UL	UL	PL	UL	UL	UL	UL	PL	UL	Highly impulsive	PM
PC19	UL	UL	UL	UL	UL	UL	UL	UL	UL	UL	Normal	Normal
PC20	PL	EW	EW	EW	UL	PL	EW	UL	EW	UL	Normal	IA
PC21	UL	UL	UL	UL	UL	UL	SA	UL	UL	UL	Normal	IA
PC22	TL	EW	EW	EW	UL	UL	PL	EW	EW	UL	Normal	PM
PC23	UL	UL	EW	EW	UL	UL	UL	UL	EW	UL	Normal	PM
PC24	TL	EA	UL	UL	UL	EA	SA	UL	UL	UL	Highly impulsive	IA
PC25	UL	UL	UL	UL	UL	UL	UL	UL	UL	UL	Normal	PM
PC26	UL	PL	TL	EW	UL	TL	PL	TL	PL	UL	Normal	IA
PC27	UL	UL	UL	UL	UL	UL	UL	UL	UL	UL	Highly impulsive	IA
PC28	UL	UL	UL	CPL	UL	UL	UL	UL	UL	UL	Normal	PM
PC29	TL	TL	UL	PL	UL	TL	TL	UL	SL	SL	Highly impulsive	IA
PC30	UL	PL	PL	PL	UL	TL	PL	UL	UL	UL	Normal	IA

Table 2 Total probe count as per IPAS and IBS-11

Count of probe candidate	IBS-11			
IPAS	Highly impulsive	Highly impulsive	Normal	Grand total
IA	3	2	8	13
Normal			1	1
PM	4	1	8	13
PM	1		2	3
Grand total	8	3	19	30

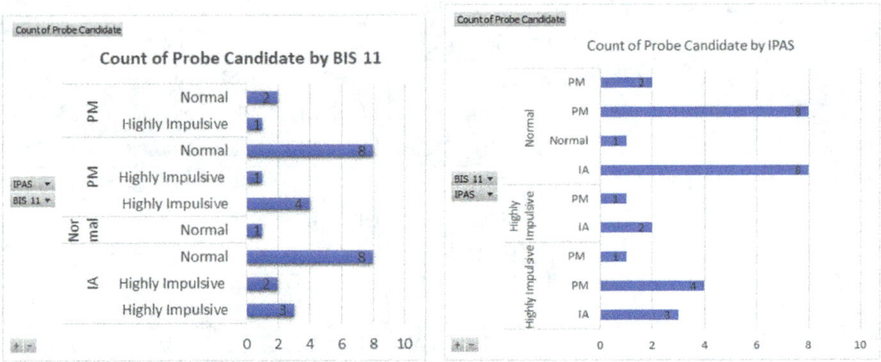

Fig. 3 Probe count distribution according to BIS-11 and IPAS

- L: Left palm finger (L1–L5)
- IPAS: Impulsive/Premeditated Aggression Scale
- BIS-11: The Barratt Impulsiveness Scale, Version 11
- IA: Impulsive aggressive
- PM: Premeditated aggressive (PA)

Thus, statistical study of impulsive behavior correlation with fingerprints can give us high precision forensics model that can be analyzed from existing digital criminal data to prevent antisocial activity.

3 Discussion and Future Scope

Recent biological and medical research assessment presented connection between mental health condition and fingerprint pattern. Scientific study of psychological disorders stats relation of fingerprint and somatic nervous system/psychic characteristics. Also, dermatoglyphics and brain development stages contain similarity for human personality development. Schizophrenia, bipolar disorder, Alzheimer,

Down's syndrome, and all mentioned mental diseases have common factors associated with fingerprint markers. Enormous collection of fingerprint and IPAS—BIS-11 test result analysis can acquire precise confirmation of personal impulsive change in order to maintain social balance. To create more accurate, forensic digital modeling of fingerprint datasets analyzed with poor executive function test (WCST) and individual finger marker statistical analysis. A well-designed inventory can be developed for predictive forensics modeling which consists of existing criminal records (fingerprints), criminal psychological history digitally available from nationwide law enforcement agenesis. Also, IPAS and BIS-11 test can be conducted in manner of online forms to prison centrally to national forensics investigation laboratory. The forensic data acquisition can be distributed in jails on different geolocations allover country. In this digital age of the Internet and online communication, fingerprint is used as promising authentication (e.g., ADHAR—India), and in order to correct data, modeling needs only two Google form of thirty questions (IPAS and BIS-11) which can obtain precise data collection. This huge data collection and test records preserved at law enforcement agencies could be used to increase precision in predictive forensics modeling. Government and private employment need normal behavior candidate profiling to avoid discrepancy in organization reputation, which can be achieved by candidate test result crossmatch to this model results.

4 Conclusion

The forensic model based on statistical analysis dermatoglyphics and impulsivity can be developed which can interpret probable antisocial behavior and characteristics of potential criminal activity. Objective of digital forensic modeling is fingerprint-based predictive approach to forecast impulsive report which can catch pre-action malicious intent and historical antisocial incident cause.

Acknowledgements The statistical analysis dermatoglyphics and impulsivity project was developed and observed as a research study of Ph.D. to evaluate productive forensics outcome. Data and analysis were projected from correlation analysis of BIS-11 and IPAS tests, which is already explained in the current paper.

References

1. Barratt ES (1999) Impulsive and premeditated aggression: a factor analysis of self-reported acts. Psychiatry Res 86:163–173
2. Campbell E (2012) Fingerprints and behaviour: a textbook on fingerprints and behavioural correspondences, vol. 1. L.L.C: Amida Biometrics
3. Cattell R (1957) Handbook for the IPAT anxiety scale. Institute for Personality & Ability Testing, Champaign IL

4. Mathiasa CW, Stanford MS, Marsh DM, Frick PJ, Gerard Moeller F, Swann AC, Dougherty DM (2007) Characterizing aggressive behaviour with the impulsivity/ premeditated aggression scale among adolescent with conduct disorder. Psych Reson 151(3):231–242
5. Dougherty DM (2004) Laboratory measured behavioral impulsivity related to suicide attempt history. Suicide Life-Threat Behav 34:374–385
6. Evenden JL (1999) Varieties of impulsivity. Psychopharmacology 146(4):348–361. https://doi.org/10.1007/PL00005481
7. Hollander E, Stein DJ (1995) Impulsivity and aggression. Wiley, pp 7–24
8. Giancola PR (2006) The mediating role of executive functioning in the relation between difficult temperament and physical aggression. J Psychopathol Behav Assess 28:211–221
9. Gorynych Z (2014) What you fingerprints say about you; the link between fingerprints and personality. Europe
10. Gray JA (1981) A critique of Eyesenck's theory of personality. In: Eyesenck HJ (ed) A model for personality. Springer, Berlin, pp 246–277
11. Heilbrun AB (1976) Criminal impulsivity and violence and subsequent parole outcome. Br J Criminol 16:367–377
12. Heilbrun AB (1978) Impulsive and premeditated homicide: an analysis of subsequent parole risk of the murderer. J Crim Law Criminol 69:108–114
13. Knyazey G. G., S. H. (2006). Personality types and behavioural activation and inhibition in adolescents. Pers Individ Differ 41, 1385–1395
14. Krishnan TR (1994) Geographic and ethnic variability of finger ridge-counts: biplots of male and female Indian samples. Ann Hum Biol 21:155–169
15. Kulkarni DU, Herekar NG (2005–06) Dermatoglyphics in essential hypertension in Western Maharashtra population. J Anat Soc India 54(2)
16. Maceo AV (2011) Anatomy and physiology of adult friction ridge skin
17. Madden GJ (2010) Impulsivity: the behavioral and neurological science of discounting. Am Psychol Assoc, pp 11–37
18. Maltoni DM (2009) Handbook of fingerprint recognition. Springer Science & Business Media
19. Mandeep singh OM (2015) Dermatoglyphics: blueprints of human cognition on fingerprints. Int J Comput Sci Commun 6(2):124–146. https://doi.org/10.090592/ijcsc.2015.608
20. Moeller RG (2009) Impulse-control disorders not classified elsewhere. In: Sadock VABJ (ed) Kaplan & Sadock's comprehensive textbook of psychiatry, 9th edn. Lippincott Williams & Wilkins, Philadelphia, pp 2178–2186
21. Mustafa Mumtaz HM (2012) User psychology and behavioural effect on palmprint and fingerprint based multimodal biometric system. Am J Signal Process 2(2):1–9. https://doi.org/10.5923/j.ajsp.20120202.01
22. Patton JH, Stanford MS (1995) Factor structure of the Barratt impulsiveness scale. J Clinic Psychol 51:768–774
23. Rohit PP (2013) A novel extended biometric approach for human character recognition using fingerprints. Int J Comput Appl 77(1):0975–8887
24. Kumar S, Yadava A (2016) Exploring the relations between executive functions and personality. Int J Indian Psychol 3(2):161–171
25. Schawmann B, Alter M (1976) Dermatoglyphics in mental disorders. Springer-Verlag, pp 187–189
26. Smith PW (2006) Driving aggression in forensic and non-forensic populations: Relationships to self-reported levels of aggression, anger and impulsivity. Br J Psychol 97:387–403
27. Stanford MS, Mathias CW, Dougherty DM (2009) Fifty years of the barrat impulsiveness scale: an update and review. Personal Individ Diff 47:385–395
28. Stanford MS (1995) Irritability and impulsiveness: relationship to self-reported impulsive aggression. Pers Individ Differ 19:757–760
29. Stanford MS (1997) Neuropsychological correlates of self-reported impulsive aggression in a college sample. Pers Individ Differ 23:961–966
30. Stanford MS-P (2003) Characterizing aggressive behavior. Assessment 10(2):183–190. https://doi.org/10.1177/1073191113010002009

31. Stein DJ, Hollander E, Simeon D (1994) Impulsivity scores in patients with obsessive-compulsive disorder. J Nerv Mental Disord 182:240–241
32. Swann AC, Doughtery DM, Pazzaglia PJ (2004) Impulsivity: a link between bipolar disorder and substance abuse. Bipolar Disord 6:204–212
33. Van Del CJ, Baaré WFC (2001) Differentiating between low and high susceptibility to schizophrenia in twins: the significance of dermatoglyphic indices in relation to other determinants of brain development. Schizophr Resour 52:181–193
34. Whiteside SP (2001) The five factor model and impulsivity: using a structural model of personality to understand impulsivity. Pers Individ Differ 30:669–689
35. Shakibaei F, Asadollahi GA, Tabibi A (2011) Dermatoglyphics in patients with schizophrenia. J Res Med Sci
36. Ahmed-Popova1 FM, Mantarkov MJ, Sivkov ST, Akabaliev VH, (2014) Dermatoglyphics—a possible biomarker in the neurodevelopmental model for the origin of mental disorders. Department of Anatomy, Histology and Embryology. https://doi.org/10.2478/folmed-2014-0001
37. Hertel J, Schutz A. Emotional intelligence and mental disorder. J Clin Psychol. https://doi.org/10.1002/jclp
38. Pandey A, Vyas JM (2014) A comparative case study of fingerprint patterns in male convicts of sabarmati jail (Ahmedabad) in Gujarati population. Indian J Foren Med Toxicol, 1–5. https://doi.org/10.5958/j.0973-9130.8.1.001

A Secure Framework for File Encryption Using Base64 Encoding

Ajeet Ram Pathak, Sarita Deshpande and Mudra Panchal

Abstract File security is the practice of preventing unauthorized access, use, disclosure, disruption, modification, inspection, or destruction of files stored on a database. Nowadays, security has become a very challenging issue, as the number of attacks on cyber security has increased which has affected the confidentiality, authentication, integrity, and availability of files. Here, we are assuring confidentiality and integrity aspects. To cope up with today's security demands, we propose a secure method to provide security to all types of file formats using AES (Rijndael) and Base64 encoding, which will be later stored on MySQL database server in cloud. After thorough research, we did not find a framework which provides dynamic encryption. The model consists of three layers. First layer performs encoding and encryption. The critical aspect of using Base64 is for encoding binary data into ASCII text which can be encrypted, stored, and transmitted. Hereafter, the encoded data is encrypted by AES (Rijndael) algorithm. Second, the encrypted file is stored into MySQL Server, and finally in third step processing is done through cloud, which protects the critical information from larceny, data leakage, and deletion, thereby providing a total of three security layers.

Keywords AES (Rijndael) · Base64 · Cloud · Decryption · Encryption · Framework · File security · MySQL

A. R. Pathak (✉) · S. Deshpande · M. Panchal
Department of Information Technology, PES's Modern College of Engineering, Pune, India
e-mail: ajeet.pathak44@gmail.com

S. Deshpande
e-mail: ddsarita@rediffmail.com

M. Panchal
e-mail: mudracpanchal@gmail.com

S.-L. Peng et al. (eds.), *Computing and Network Sustainability*, Lecture Notes in Networks and Systems 75, https://doi.org/10.1007/978-981-13-7150-9_38

1 Introduction

With the rapid growth in science and technology, distributed, and parallel computing, security has become a major concern of our lives. Many secure transmission methods require a type of encryption. Various types of encryption techniques are available which differ in key size and security they provide. Along with the security, another influencing parameter of the algorithm is its energy requirement required in processing the queries. Energy consumption has a direct effect on the battery life of user's system, and consequently, on the duration and extent of mobility of user. Thus, minimal requirement of energy for a portable system is of primary importance together with security.

In our proposed framework, Base64 encoding and Rijndael encryption algorithm are used. Base64 handles the compression of varying file size so that the encryption process will be faster. This compressed file is taken as an input to the Rijndael algorithm which encrypts the file [1]. Rijndael algorithm follows block cipher strategy in iterative fashion. Following that, the encryption or decryption of a block of data is achieved by round function with specific variation. The framework is deployed on the cloud platform which provides the users to store files online so that they can access them from any location via the Internet. The execution time required to encrypt huge files can be reduced by expanding the processing power, storage, and networking. Another advantage of using cloud is you can work at scale and still stay secure as it is less vulnerable to various threats. Aspects of confidentiality and integrity for outsourced databases have been scrupulously analyzed in our previous work [2–5]. We found that providing security to the text is not the only concern, and therefore, we aim to provide security in cloud computing scenario which focuses on multiple file formats, i.e., jpg, pdf, png, etc. In this framework, we are giving focus on confidentiality and integrity aspects.

Our framework gives resistance against all known attacks to all types of file formats such as txt, doc, ppt, zip, exe, rar, png, jpg, jpeg, html, css, pdf, jar, odt, mp3, and xlsx. The proposed approach supports efficient rate of execution with compactness of code on various platforms [6].

The rest of the paper is organized as follows. Section 2 discusses about the motivation behind this framework, and Sect. 3 and 4 describe system architecture and detailed methodology, respectively. Result analysis is done in Sect. 5 with conclusion in Sect. 6.

2 Motivation

With the rapid development of digital devices, computers, and networks, digital data has become a fundamental part of our daily lives. In every case, storage and transmission of data securely are primary concern, and data have to be transmitted securely. Secure transmission is basically a method to transfer the data over a secure channel. So in order to achieve confidentiality and integrity to a higher extent, we have studied various authors' work as follows.

As per [7], author here has provided a standard encoding technique which can deal with the problems in securing the data while transmission in a network. Base64 is a block cipher algorithm that operates on a bit, but the Base64 mode is easier in its implementation than others. The Base64 encoding scheme is basically used, to encode binary data. In this method, Base64 transforms the 8-bit into 6-bit character. It is actually not used for encryption, but sometimes it seems like it is used for providing encrypted data. In order to have a perfectly well transmission, the Base64 algorithm converts the character to limited number of character. Nothing like secret code or password is used in Base64 algorithm. Still the algorithm requires improvement for better security.

A reference model of software system is proposed in [8, 9] to examine the performance of AES algorithm related to security. During this experiment, it was noted that the encrypted information is introduced with noise as well as it also shows that for the selected set of values, the AES function does not generate the correct output if the error is included. Creating and implementation of programs verified through Xilinx Project Navigator software, to verify the results for both encryption and decryption processes.

A comparative study is carried out in [10] on the most basic security mechanisms algorithms in cloud computing. Based on the results of simulation, AES performs better compared to RSA and Blowfish. AES gains more attention and is considered to be the effective available secret key cryptographic algorithm today.

Harran et al. [11] uncovered an aspect as how data about data get managed by different applications. Technique put forth in [12] allows users to store critical data on mobile cloud only after encrypting of data. The author here has proposed bio-computing solution approaches to direct security of data.

Combination of Diffie-Hellman and Blowfish methodology is proposed in [13] for images and text files encryption–decryption. An approach-based combination of cryptosystem with merging and spitting mechanisms with public key encryption techniques is put forth in [14]. It provides better results with respect to processing time and security. Public key encryption helps to provide user authorization and splitting and merging protects against various attacks. Garfinkel and McCarrin[15] presented a hash-based carving system which works efficiently for large size target files.

3 System Architecture

The architecture is shown in Fig. 1. The system is divided into three parts, namely the cloud server, database, and crypt process [16]. The cloud server provides a platform to deploy the entire system, whereby the users can access it through Internet; this serves the purpose of accomplishing one of the basic security services—availability. Another reason for deployment on cloud is that the data is centrally stored on the

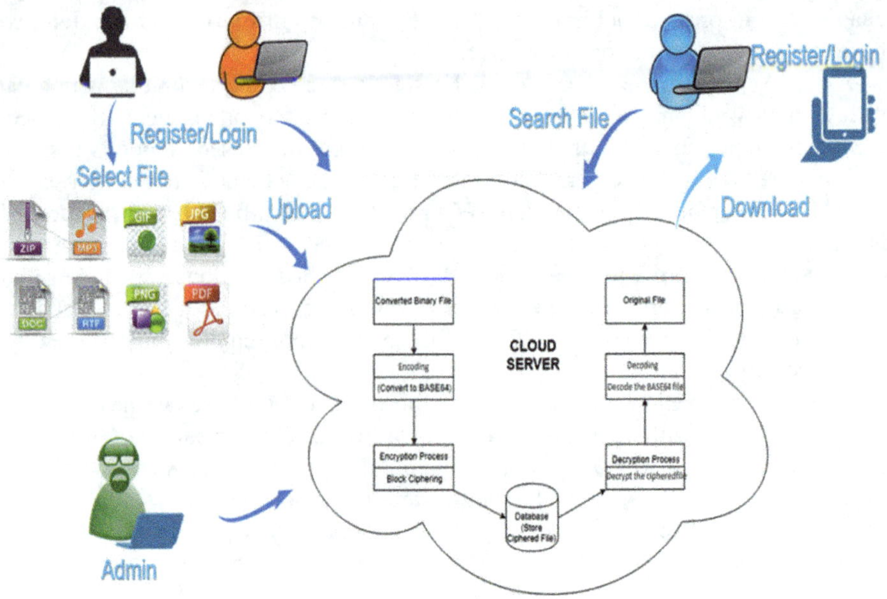

Fig. 1 System architecture

cloud server, which enables to maintain consistency furthermore helps to achieve fault tolerance [17]. The database stores the details of the file along with the actual contents of file in junk format. Junk format is obtained as a result of the entire encryption process. This is advantageous as if an intruder gets hold of the file stored in the database, he will not be able to extract the contents of the file, as it is in junk format. The crypt process consists of two phases, viz. encoding and encryption. In encoding phase, the file is encoded in Base64 format [18]. This has two benefits: (1) It reduces the size of file which results in fast execution of encryption process, and (2) query processing in database is faster as uploading to database on smaller file size is rapidly achieved. In encryption phase, Rijndael algorithm with a key size of 128 bits is used as it is AES compliant. Rijndael is an iterated block cipher, which encrypts a block of data in iterations (rounds) [19]. The number of iterations depends upon the size of the key used. In this case, as the key size is 128 bits, number of iterations will be 9. Each iteration in this algorithm has four stages: byte substitution, shift row transformation, mix column, and add round key. After nine rounds, the ciphered data will be generated and stored in the database.

The second phase of the *crypt* process is composed of decryption and decoding [20]. In the decryption process, inverse of the above four stages is performed in iterations to achieve encoded data. In the decoding process, the original lossless file is obtained.

4 Detailed Methodology of Proposed Architecture

Base64 encoding scheme takes files in binary format as an input and encodes it using the Base64 index table. Base64 encoding uses three bytes to specify four printable ASCII characters [21]. This encoding process is done in two steps as follows. The initial step changes over three bytes to four numbers of six bits. Each character in the ASCII standard comprises of seven bits. Base64 utilizes just six bits to guarantee the encoded information which is both printable and human readable formats [22]. This procedure does not utilize any of the unique characters accessible in ASCII. Base64 considers only ten digits, 26 lowercase and uppercase characters, and symbol of plus (+) and forward slash (/). In second step, these numbers are converted into ASCII characters using Base64 index table.

Advanced Encryption Standard (AES Rijndael) [19] supersedes the Data Encryption Standard (DES). It is a symmetric key-based algorithm used to encrypt and decrypt the data; hence, key having size 128, 192, or 256 bits remains unchanged during execution cycle. Here rounds, contingent upon key/block sizes. As it is a substitution linear transformation cipher, not requiring a Feistel arrangement. The process of Rijndael algorithm is made up of three primary parts: (a) cipher, (b) inverse Cipher, and (c) key expansion. (a) *Cipher* encrypts the data into cipher text, and (b) *inverse cipher* decrypts data back into its actual form. (c) *Key expansion* produces a key. Schedule used in the above two procedures and composed of particular number of rounds. Every round in the cipher and inverse cipher process has four stages: SubByte, Shift Row transformation, Mix column, and Add Round key. The methodology for designing and manufacturing phase based on aforementioned technique is mentioned below.

Methodology: Designing phase

1. User will register to the system by providing login and password.
2. User Select's any file. (mp3, zip, pdf, png, rtf, doc, etc.)
3. Upload the selected file on cloud server providing username and password to login to the system.
4. Uploaded file will be converted to binary format.
5. Binary format file is encoded using Base 64 encoding scheme.
6. Encoded file is encrypted using Rijndael symmetric key encryption technique.
7. Encrypted (ciphered) file is stored on cloud server.

Methodology: Manufacturing Phase

1. User will register to the system by providing login and password.
2. User will search for a particular file.
3. Searched file is decrypted using Rijndael Symmetric key encryption technique.
4. Decrypted file is decoded using Base 64 encoding scheme.

5. The original file is downloaded by user providing username and password to login to the system.

5 Result Analysis

Given below are the experimental results followed by the analysis of the algorithms presented here. The experiments are carried out to find the encryption time of various types of files of divergent sizes. In Table 1, the experimental results for different file size are shown. The results are based on the execution time required for processing of files after applying Base64 encoding and Rijndael encryption technique. The execution time given is in milliseconds. In the graph plotted, the file size is depicted on x-axis and encryption time in y-axis. From Figs. 2 and 3 are graph obtained after Base64 encoding and encryption process and inserting file into database. It has been observed that all the types of files of any size can be encrypted by Rijndael technique. Even the encryption time for two different file size it is sometimes observed to be same.

6 Conclusion

Rijndael algorithm has been used along with 128-bit block key to encrypt and decrypt the data, following Base64 encoding scheme for files encryption. The time required

Table 1 File sizes and time required for encryption

File size in Kb	Encryption time	File size in Kb	Encryption time
19	0.007141828	2978	0.0006761550
83	0.0102961063	2211	0.0001029968261
54	0.00939202308	57	0.00168490409
2	0.002997875	13	0.00072693824
47	0.006946086	37	0.00184106826
6360	0.0047390460	119	0.00310087203
373	0.03420400619	2616	0.00048208236
232	0.0202810764	311	0.0099430084
50	0.0055658817	100	0.0026798248
4200	0.0004808902	434	0.0387511253

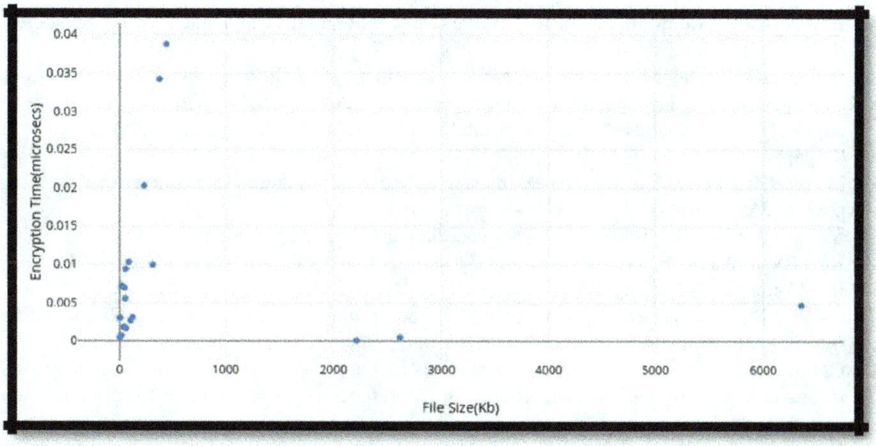

Fig. 2 Scatter plot

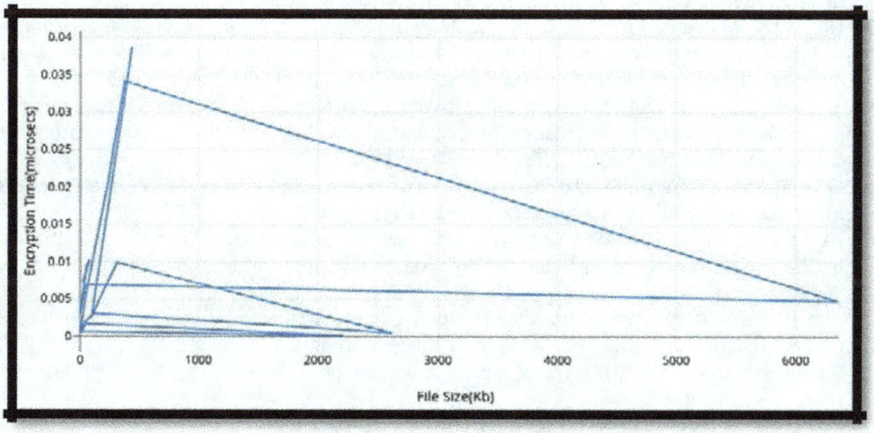

Fig. 3 Line plot

to encrypt the data after encoding in Base64 is compared with time required to complete encryption process and store data in cloud.

References

1. Yang M, Bourbakis N, Li S (2004) Data-image-video encryption. IEEE potentials 23:28–34
2. Pathak AR, Padmavathi B (2014) Analysis of security techniques applied in database outsourcing. (IJCSIT) Int J Comput Sci Inf Technol 5:665–670
3. Pathak AR, Padmavathi BA (2014) Secure threshold secret sharing framework for database outsourcing. In: IEEE International Conference ICACCCT, pp 1642–1649

4. Pathak AR, Padmavathi B. (2013) Survey of confidentiality and integrity on outsourced database. Int J Sci Eng Technol
5. Pathak AR, Pandey M, Rautaray S (2018) Assessment of integrity auditing protocols for outsourced big data. In: Bhateja V, Coello Coello C, Satapathy S, Pattnaik P (eds) Intelligent engineering informatics. Advances in intelligent systems and computing, vol 695. Springer, Singapore
6. Mao Y, Chen G (2005) Chaos-based image encryption. In: Handbook of geometric computing, pp 231–265. Springer
7. Siahaan APU (2017) Base64 character encoding and decoding modeling
8. Ravi RV, Mahalakshmi R (2014) Simulation and error analysis of Rijndael algorithm. Int J Adv Res Comp Sci Softw Eng 4(5)
9. Mankar M, Kshirsagar RV, Vyawahare MV. Encryption and decryption using Rijndael algorithm
10. Cordova RS, et al (2017) Comparative analysis on the performance of selected security algorithms in cloud computing. In International Conference on Electrical and Computing Technologies and Applications, pp 1–4
11. Harran M, Farrelly W, Curran K (2017) A method for verifying integrity & authenticating digital media. Appl Comput, Informatics
12. Sukumaran SC, Mohammed M (2018) PCR and bio-signature for data confidentiality and integrity in mobile cloud computing. J King Saud Univ Inf Sci
13. Hazra TK, Mahato A, Mandal A., Chakraborty AK (2017) A hybrid cryptosystem of image and text files using blowfish and Diffie-Hellman techniques. In: Industrial automation and electromechanical engineering conference (IEMECON), 2017 8th Annual, pp 137–141
14. Sharma M, Sharma VA (2016) Hybrid cryptosystem approach for file security by using merging mechanism. In 2016 2nd international conference on applied and theoretical computing and communication technology (iCATccT), pp 713–717
15. Garfinkel SL, McCarrin M (2015) Hash-based carving: searching media for complete files and file fragments with sector hashing and hashdb. Digit Investig 14:S95–S105
16. Kumar GM, Chandrasekaran V (2009) A novel image encryption scheme using Lorenz attractor. In: 4th IEEE Conference on Industrial Electronics and Applications, 2009. ICIEA 2009, pp 3662–3666
17. Zhu H et al (2013) A novel image encryption–compression scheme using hyper-chaos and Chinese remainder theorem. Signal Process Image Commun 28:670–680
18. Gersho A, Gray RM (2012) Vector quantization and signal compression, vol 159. Springer Science & Business Media
19. Linde Y, Buzo A, Gray R (1980) An algorithm for vector quantizer design. IEEE Trans Commun 28:84–95
20. Poonkuzhali SM, Therasa M (2015) Data hiding using visual cryptography for secure transmission. Int J Adv Res Comput Commun Eng 4
21. Siahaan APU (2016) A three-layer visual hash function using Adler-32. Int J Comput Sci Softw Eng 5:142
22. Solanki K, Vankani V, Pukle P, Iyer S. Multimedia encryption using visual cryptography

Review of Machine Learning-Based Intrusion Detection Techniques for MANETs

Fouziah Hamza and S. Maria Celestin Vigila

Abstract Mobile ad hoc network is a widely developing technology that has been used in various areas such as in health care, military, virtual classrooms and conferences. However, mobile ad hoc networks are installed in critical situations security in this network is an important issue. Many susceptible characteristics of mobile ad hoc networks make an attacker breach the system easily. So, it is important to have an intrusion detection system which can monitor mobile ad hoc networks constantly to identify any suspicious behaviour. Anomaly and misuse detection are the two widely used intrusion detection mechanisms used to analyse the attacks in mobile ad hoc networks. Anomaly intrusion detectors were proven to be more effective against unknown attacks. A number of anomaly-based intrusion detectors based on machine learning techniques were developed and tested against various attacks. In this paper, several intrusion detection techniques which used machine learning approaches for detection are reviewed and a hybrid IDS technique which combines with genetic algorithm and Bayesian game theory is proposed.

Keywords Bayesian network · Genetic algorithm · Hybrid system · Machine learning techniques · Neural network

1 Introduction

The growth of laptops and other computing devices such as pocket PCs, tablet PCs, smartphones and the advancements of wireless technology made Mobile ad hoc network (MANET) a widespread research topic since the mid-1990s. MANET is an autonomous, decentralized, self-organizing, infrastructure less, multi-hop wireless network of mobile devices. In MANETs, nodes can quickly establish a network without any pre-existing centralized server. Due to its ease of deployment, auto-

F. Hamza (✉) · S. Maria Celestin Vigila
Noorul Islam Centre for Higher Education, Kumaracoil, Kanyakumari, India
e-mail: fouzi_ajmal@yahoo.com

S. Maria Celestin Vigila
e-mail: celesleon@yahoo.com

S.-L. Peng et al. (eds.), *Computing and Network Sustainability*, Lecture Notes in Networks and Systems 75, https://doi.org/10.1007/978-981-13-7150-9_39

configuration, low cost, MANETs are increasingly being used in many applications such as battlefield communication, vehicular networks, emergency services, home networking, education, and entertainment and so on.

Security is a challenging problem for MANET because of various properties of MANETS which makes them exposed to various attacks such as black hole, flooding, passive eavesdropping, active interfering, and denial of service, spoofing, impersonate a node, modify and delete messages. Intrusion detection system (IDS) is an effective tool to classify an attack in a MANET [1]. IDS is a software or device that assists in counterattack intrusion. An IDS continuously analyses a network looking for an unusual activity. Once an IDS senses an activity which is abnormal, it in turn alerts the security administrator by generating an alarm. Additionally, IDS can also initiate a suitable response to reduce or prevent the damage caused to the system by the attack. Designing a MANET IDS scheme detects the attack with the lowest possible false positive (FP) rate while at the same time minimizing the power consumption of nodes as well as generating low traffic in the network is a dynamic research topic.

According to the detection methods, IDs can be generally divided into anomaly-, misuse- and specification-based detectors. Anomaly detection method can spot unknown attacks by comparing the network traffic with the recorded normal actions. Any action deviated from the normal pattern is labelled as intrusions. But, in misuse detection stored attack signatures or patterns are used to detect attacks. The benefit of anomaly detection is its capacity to detect the previously unknown attacks which are not documented during the training phase. So, the concept of anomaly-based detection method is best suited for the MANETs compared with misuse detection. Machine learning-based methodologies are widely applied to anomaly detectors as it requires less human intervention. Several anomaly detection techniques are newly developed, or the current existing ones are modified by researchers to improve the performance of MANET IDSs. In the proposed system, a hybrid classifier is introduced which combines genetic algorithm and Bayesian game theory to improve the performance of MANET intrusion detection system.

2 Machine Learning-Based Anomaly Detection Techniques

Machine learning algorithms adapt their behaviour according to the environment by quickly learning its surrounding properties. A major focus of machine learning research is to identify intricate patterns and make intelligent conclusions based on data. This paper provides a brief description of the classification of different machine learning-based intrusions techniques in MANETs. For example, some IDS schemes were constructed based on single learning techniques, such as genetic algorithm [2], support vector machine [3], neural network [4], fuzzy logic [5], and data mining [6] (Fig. 1).

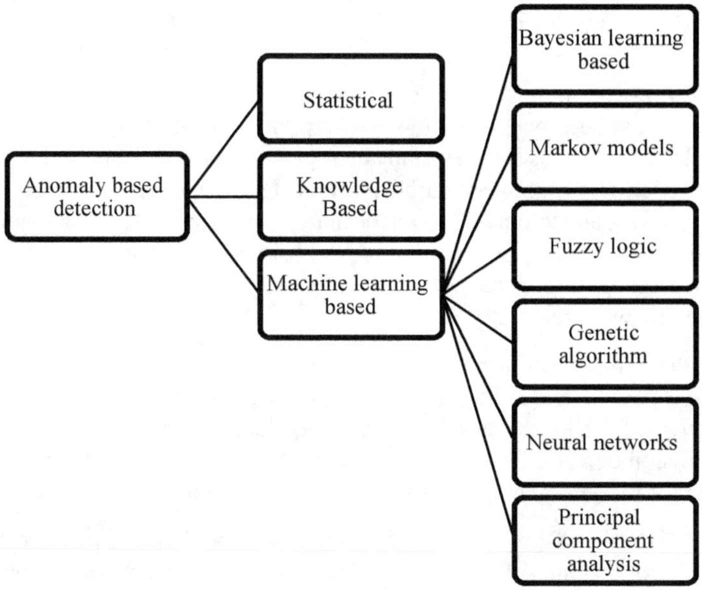

Fig. 1 Machine learning-based techniques

2.1 *Fuzzy Logic*

Fuzzy logic, also known as fuzzy set theory, is one of the prospective techniques used for reasoning. It deals with human decision making and cognitive skills. Fuzzy logic is applied in numerous fields such as engineering, business, defence, aerospace and so on. This technique is also tested against some attacks such as *probe* and *port scan* and proved to be effective. One of the main disadvantages of this logic is its high-resource consumption [7].

Yadav et al. [8] derived fuzzy-based intrusion detection systems in MANET. This scheme was not only classified the attacks correctly but also reported the extension and range of that the attack. This scheme also has an intrusion prevention system (IPS) mechanism which received the inputs from fuzzy technique and provided secure network communication.

A stationary intelligent fuzzy agent (SIFA)-based IDS for detection was proposed by Watkins [9]. This system prevented the port scanning and distributed DoS attacks in tactical MANETs. Visconti and Tahayori [10] suggested artificial immune system. This system spots and learns about the misbehaviour nodes and in turn protects the network without human interference.

2.2 Genetic Algorithm

It is an optimization strategy constructed on the natural selection and natural genetics which simulates the process of evolution. Application of genetic algorithm to the intrusion detection seems to be a promising topic. The rules are created by the genetic algorithm based on the network traffic analysis. These rules are then used to classify normal network connections from abnormal connections [11]. This algorithm can work with a huge population and produce the best possible solution.

The succeeding procedures are executed in the process of applying genetic algorithm to the network data [12].

- Firstly, the network traffic information is collected by the IDS.
- Then, the genetic algorithm develops simple rules based on the network traffic data collected and stores these rules in a knowledge base.
- The intrusion detection system then uses this collected knowledge to classify the incoming traffic as anomalous or normal based on their pattern.

Benaicha et al. [13] present an IDS combined with genetic algorithm (GA) approach to competently distinguish various types of network intrusions. With an improved population and selection operator, GA is used to enhance the search of audit files for attacks. In the testing stage, the Network Security Laboratory Knowledge Discovery and Data Mining (NSL-KDD99) benchmark dataset were employed to distinguish the misuse events occurred in the network.

2.3 Neural Network

The intrusion detector using the neural network has been identified as an appropriate technique to distinguish not only previously observed attacks but also unobserved attacks [14]. To incorporate neural network (NN) approach to intrusion detection, firstly the network is trained with data containing normal and abnormal behaviour. Then, the coefficients of the NN are adjusted according to the noted behaviour during the training. Performance tests were then conducted with real network traffic and attacks [15].

Jabbehdari et al. [16] proposed an anomaly-based intrusion detection system that used neural network to detect intrusions in mobile ad hoc networks. The simulation results of the proposed IDS result show that under different traffic and attack situations, all the DoS attacks are identified correctly. The false positive and false negative ratios were zero, thereby increasing the detection rate of IDS.

2.4 *Bayesian Network*

A Bayesian network [17] is an expert system that captures all existing knowledge to predict outcomes. The model answers the question such as "what is the probability that it is a certain type attack, constructed on the observed system variables?" This procedure is commonly used for intrusion detection in association with statistical schemes. This scheme has the capability of encoding relationships between system variables and of predicting events, as well as the ability to combine both previous knowledge and data [18].

A predictive model with naïve Bayes and decision tree classifiers was built by Amor et al. [19]. This model can segregate the attacks and normal connections of KDDCUP 99 data set. They report that the time required to train the Bayes classifiers is less than the period needed to train the decision tree classifiers, even though the decision tree classifier's performance is slightly better than the Bayes classifier.

2.5 *SVM*

Support vector machine (SVM) is one of the widely used ML algorithm techniques applied in intrusion detectors due to their good generalization nature. SVM has the ability to learn from a larger set of patterns. It also updates the training patterns during classification dynamically each and every time there is new pattern recognized [20].

3 Machine Learning Techniques—Hybrid Classifiers

Hybrid classifiers which combine one or two machine learning techniques such as SVM, GA [21], Bayesian network, hidden Markov model [22] is proposed to build a better intrusion detection system. Specifically, this approach will have two functional components. The first component produces intermediate results, and the second one generates results.

Kim and Bentley [23] proposed the artificial immune system for IDS. This approach was inspired by human immune system, and it is built on hierarchical approach. Fuzzy-based genetic algorithm was proposed by Yunwu [24]. This algorithm used fuzzy algorithm and genetic algorithm for its initial and final rules, respectively. Lee et al. [25] suggested another approach that combined genetic algorithm and selective naïve Bayesian classifiers (SNBN). To improve this method's accuracy, GA was used to extract 21 relevant features from the 41 features of KDDCUP 99 data set. The evaluation procedures are done by *selective naïve Bayesian classifiers*. The experiment results show that the performance of GA was comparatively impressive with the other IDSs which used all the features for classification, particularly for detecting unknown attacks.

Shon et al. [26] proposed machine learning framework that used genetic algorithm (GA) and SVM for feature selection and attack classification. This SVM combines the benefit of supervised and unsupervised learning. Finally, the outcomes were tested and compared using products such as Snort.

Sung et al. [27] used support vector machines (SVMs) and neural networks to categorize important features for 1998 DARPA Intrusion Detection data. These features were ranked according to their importance. It was suggested that using SVMs and neural network classifiers with only the important selected important features would produce better result than by choosing all the features. Vaishnavi Devi and Sakthi Priya [28] proposed an efficient mechanism design approach to be used in leader election-based IDS to spot the malicious actions of mobile nodes.

4 Proposed System

Most of the IDS schemes proposed for MANET has to run its own IDS continually in order to monitor the traffic in the network for any suspicious intrusion, which would be expensive not only in terms of high-resource consumption by the nodes but also affects the performance of IDS. In my proposed article, I would like to address the resource consumption issues of nodes by choosing a leader from the cluster, who will be responsible for providing intrusion services to all other nodes in that particular cluster. The hybrid IDS proposed has two components: light weight and heavy weight. Initially, only the lightweight module activated. To maximize the probability of detection, heavyweight module is activated based on the decision of cluster leader. Figure 2 shows the system architecture of the proposed system.

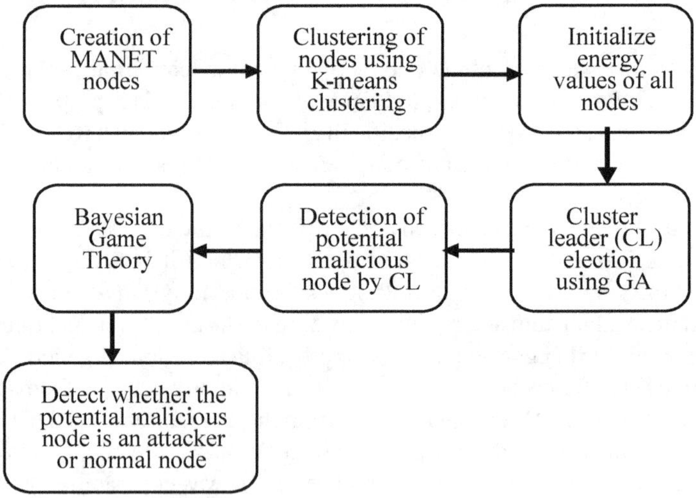

Fig. 2 Proposed IDS architecture

5 Conclusion

A great deal of work has been carried on to detect and prevent the Intrusions in MANETs. Additionally, there are many IDS systems that used machine learning techniques were also proposed. In this review paper, various types of intrusion detection have been analysed, particularly the machine learning-based IDS are discussed. The proposed hybrid system model uses genetic algorithm and Bayesian game theory concepts [29] to achieve better IDS.

References

1. Barry BI, Chan HA (2010) Intrusion detection systems. In: Handbook of information and communication security. Springer, Berlin, Heidelberg, pp 193–205
2. Sujatha KS, Dharmar V, Bhuvaneswaran RS (2012) Design of genetic algorithm based IDS for MANET. In: International conference on recent trends in information technology (ICRTIT). IEEE, pp 28–33
3. Justin V, Marathe N, Dongre N (2017) Hybrid IDS using SVM classifier for detecting DoS attack in MANET application. In: International conference on I-SMAC (IoT in social, mobile, analytics and cloud). IEEE, pp 775–778
4. James C (2010) Dynamic neural networks in the detection of distributed attacks in mobile ad-hoc networks. Int J Netw Secur Appl (IJNSA) 2:1–7
5. Balan EV, Priyan MK, Gokulnath C, Devi GU (2015) Fuzzy based intrusion detection systems in MANET. Procedia Comput Sci 50:109–114
6. Nadiammai GV, Hemalatha M (2014) Effective approach toward intrusion detection system using data mining techniques. Egypt Inf J 15:37–50
7. Kim J, Lee JH (2008) A slow port scan attack detection mechanism based on fuzzy logic and a stepwise policy. In: 4th international conference on intelligent environments (IE 08)
8. Yadav P, Gill R, Kumar N (2012) A fuzzy based approach to detect black hole attack. Int J Soft Comput Eng 2:388–391
9. Watkins D (2004) Tactical MANET attack detection based on fuzzy sets using agent communication. In: 24th army science conference
10. Visconti A, Tahayori H (2011) Artificial immune system based on interval type-2 fuzzy set paradigm. Appl Soft Comput 11:4055–4063
11. Sujatha P, Priya CS, Kannan A (2012) Network intrusion detection system using genetic network programming with support vector machine. In: Proceedings of the international conference on advances in computing, communications and informatics, Chennai, India, pp 645–649
12. Hoque M (2012) An implementation of intrusion detection system using genetic algorithm. Int J of Netw Secur Appl 4:109–120
13. Benaicha SE, Saoudi L, Guermeche SEB, Lounis O (2014) Intrusion detection system using genetic algorithm. In: Science and information conference. IEEE, pp 564–568
14. Wu HC, Huang SHS (2010) Neural networks-based detection of stepping-stone intrusion. Expert Syst Appl 37:1431–1437
15. Govindarajan M, Chandrasekaran RM (2011) Intrusion detection using neural based hybrid classification methods. Comput Netw 55:1662–1671
16. Jabbehdari S, Talari SH, Modiri N (2012) A neural network scheme for anomaly based intrusion detection system in mobile ad hoc networks. J Comput 4:61–66
17. Altwaijry H, Algarny S (2012) Bayesian based intrusion detection system. J King Saud Univ Comput Inf Sci 24:1–6

18. Scott SL (2004) A Bayesian paradigm for designing intrusion detection systems. Comput Stat Data Anal 45:69–83
19. Amor NB, Benferhat S, and Elouedi Z (2004) Naive Bayes vs decision trees in intrusion detection systems. In: Proceedings of 2004 ACM symposium on applied computing, pp 420–424
20. Deng H, Zeng QA, Agrawal DP (2003) SVM-based intrusion detection system for wireless ad hoc networks. In: 58th vehicular technology conference, vol 3. IEEE, pp 2147–2151
21. Gharaee H, Hosseinvand H (2016) A new feature selection IDS based on genetic algorithm and SVM. In: 8th international symposium on telecommunications (IST). IEEE, pp 139–144
22. Devarakonda N, Pamidi S, Kumari VV, Govardhan A (2012) Intrusion detection system using Bayesian network and hidden Markov model. Procedia Technol 4:506–514
23. Kim J, Bentley PJ (2001) The artificial immune system for network intrusion detection: an investigation of clonal selection with a negative selection operator. In: Evolutionary Computation proceedings. IEEE, pp 1244–1252
24. Yunwu W (2009) Using fuzzy expert system based on genetic algorithm for intrusion detection system. In: 2009 international forum on information technology and applications. https://doi.org/10.1109/ifita.2009.107
25. Lee CH, Chung JW, Shin SW (2006) Network intrusion detection through genetic feature selection. In: Seventh ACIS international conference on software engineering, artificial intelligence, networking, and parallel/distributed computing (SNPD'06). IEEE, pp 109–114
26. Shon T, Kim Y, Lee C, Moon J (2005) A machine learning framework for network anomaly detection using SVM and GA. In: Proceedings from the sixth annual information assurance workshop 2005, IAW'05. IEEE, pp 176–183
27. Mukkamala S, Janoski G, Sung A (2002) Intrusion detection using neural networks and support vector machines. In: Proceedings of the 2002 international joint conference on neural networks. IEEE, pp 1702–1707
28. Vaishnavi Devi B, Sakthi Priya N (2015) Intrusion detection and response systems in MANET using leader election based mechanism design approach. Int J Innov Res Comput Commun Eng 3:8670–8679
29. Kim, S (2014) Game theory applications in network design. IGI Global

Classification of Two-Class Motor Imagery EEG Signals Using Empirical Mode Decomposition and Hilbert–Huang Transformation

Ravindra Ghritlahare, Mridu Sahu and Rahul Kumar

Abstract This work proposes classification of two-class motor imagery EEG signals using instantaneous frequencies as the feature, which is extracted from intrinsic mode functions (IMFs) using Hilbert–Huang transformation (HHT). IMF is computed by using Empirical Mode Decomposition (EMD). The classification of two-class motor imagery is prerequisite for brain–computer interface (BCI). In this work, the EMD method is used for identification of the right-hand and feet imagery movements. The proposed method has been applied on the subjects 05 and 06, which are publicly available on BNCI-horizon-2020. Upon investigation of the performance of the proposed feature extraction method, the following results are obtained; we found that the BFTree turned out to be the best classifier with 59.375% accuracy with subject 05 and DecisionStump classifier with 58.75% accuracy with subject 06. From the above results, it is evident that the proposed method is reliable for classification of two-class motor imagery EEG signals.

Keywords Instantaneous frequency · Intrinsic mode function · Sifting process · Windowed sampling · Weak Learner

1 Introduction

Brain–computer interface (BCI) is a association between a brain and a EEG device that measures brain signals' activity, such as hand and feet movements [1]. The interface provides direct communication pathway from the brain to the environment surrounding a person with neuromuscular disabilities [2]. BCI involves the extraction of feature and classifying different types of motor imagery movements. Detecting motor imagery movements like right hand and feet is one of the main function of the BCI [3]. This work proposes a technique for extracting features from such type of

This work is supported by National Institute of Technology, Raipur.

R. Ghritlahare (✉) · M. Sahu · R. Kumar
National Institute of Technology, Raipur 492001, Chhattisgarh, India
e-mail: ravindraghritlahare129@gmail.com

S.-L. Peng et al. (eds.), *Computing and Network Sustainability*, Lecture Notes in Networks and Systems 75, https://doi.org/10.1007/978-981-13-7150-9_40

EEG signals using EMD. EMD is an adaptive time–space analysis method suitable for processing series that are non-stationary and nonlinear. We proposed EMD as an entirely new tool to extract features form EEG signals for application in BCI. In this work, the Hilbert–Huang transform (HHT) is used to obtain instantaneous frequency from the IMFs generated [2]. To decompose a signal into intrinsic mode functions (IMFs), the HHT uses the Empirical Mode Decomposition method and applies the Hilbert Spectral analysis method to the IMFs to acquire instantaneous frequency data. The obtained data is used as extracted data. The extracted features are then used in different classifiers to classify the imagery movements. The performance of the proposed method is tested for different classifiers and compared with other existing techniques [3].

2 EEG Database

The EEG database used in this work is acquired from the BNCI-horizon, two-class motor imagery which is provided by the Graz University of Technology [5]. The training, recording and feedback were done in the same session. Every session involves eight runs: five runs were used for training and three runs for feedback for validation. Every runs in turn had 20 trials. In every class, 50 trials were meant for training and 30 trials for validation. Participants had to perform sustained kinesthetic motor imagery for right hand and feet. Bio-signal amplifier and active Ag/AgCl electrodes were used to measure EEG at a sampling rate 512 Hz. The electrodes' placement was designed for obtaining three Laplacian derivations. The 13 participants were aged between 20 and 30 years, 8 unaffected to diseases [11].

3 Methodology

3.1 Feature Extraction

The process of feature extraction has two parts, one is EMD of EEG and the other is obtaining instantaneous frequencies from IMFs.

The Empirical Mode Decomposition EMD is a process to breakdown initial data into certain functions called as intrinsic mode functions (IMFs). An IMF satisfies the following two conditions:

1. In the entire signal, the number of extremes and the zero-crossings must be equal or differ at most by one.
2. The mean value of local maxima envelope and local minima envelope must be zero at any point of an IMF [12].

The algorithm for EMD is as following:

1. Given incoming signal $s(t)$.
2. Finding all upper extrema of $s(t)$.

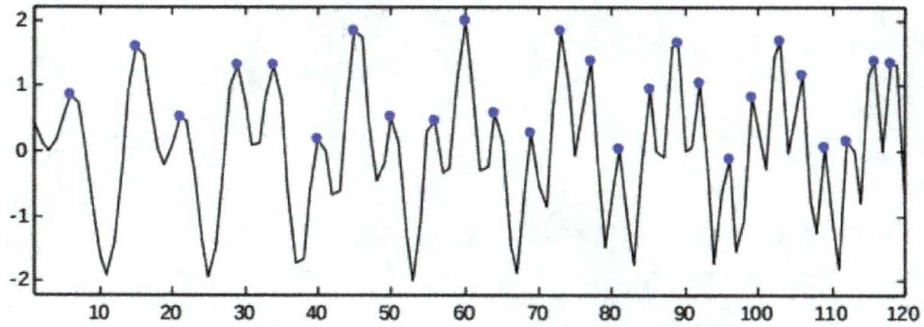

3. The local maxima is interpolated to form an upper envelope $e_1(s)$.

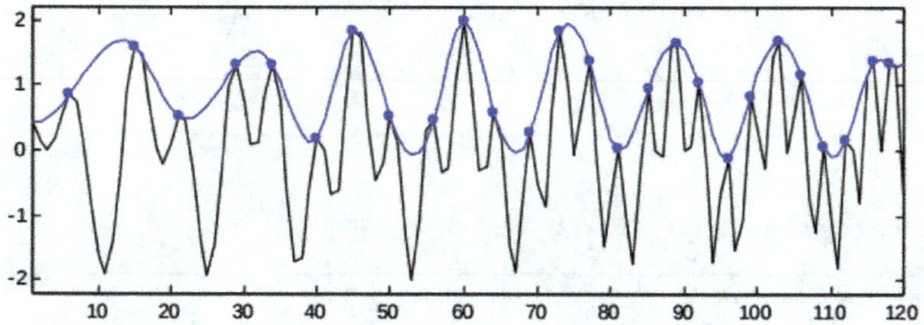

4. Find out all the lower extrema of $s(t)$.

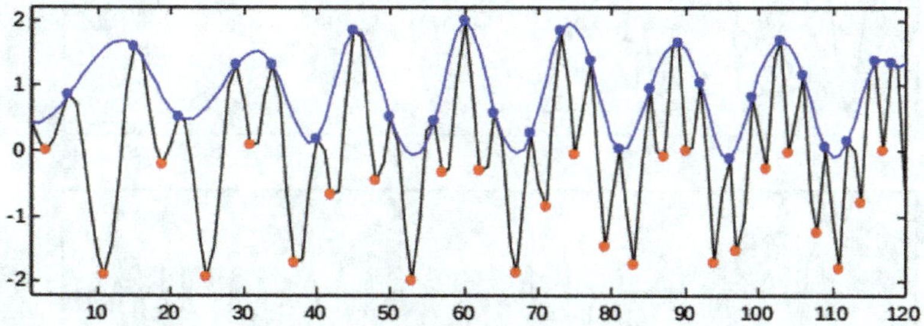

5. The local minima is interpolated to form an lower envelope $e_2(s)$.

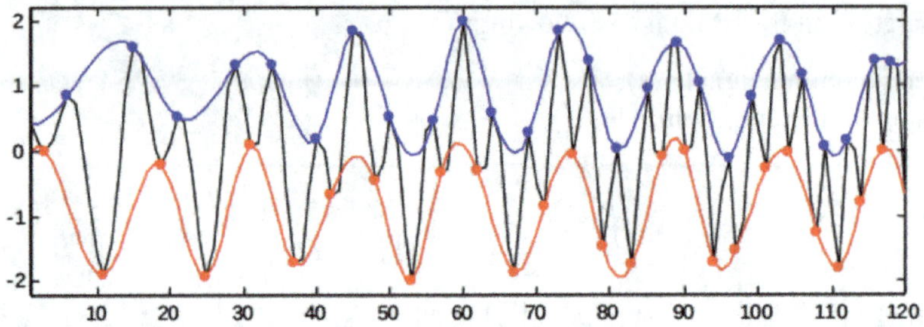

6. Find the mean envelope: $m(t) = [e_1(s) + e_2(s)]/2$.

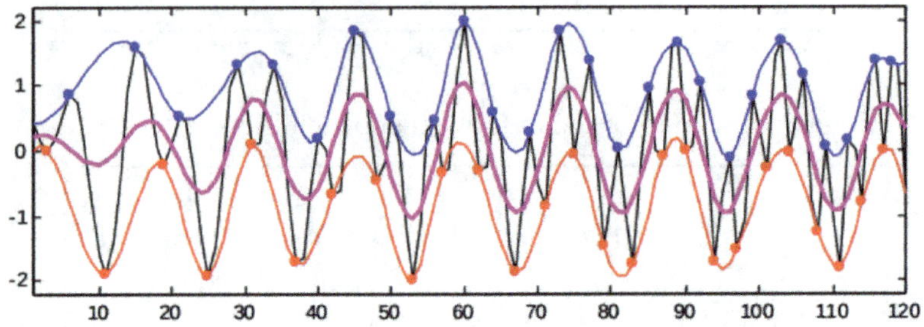

7. Find the signal mean: $g(t) = s(t) - m(t)$.

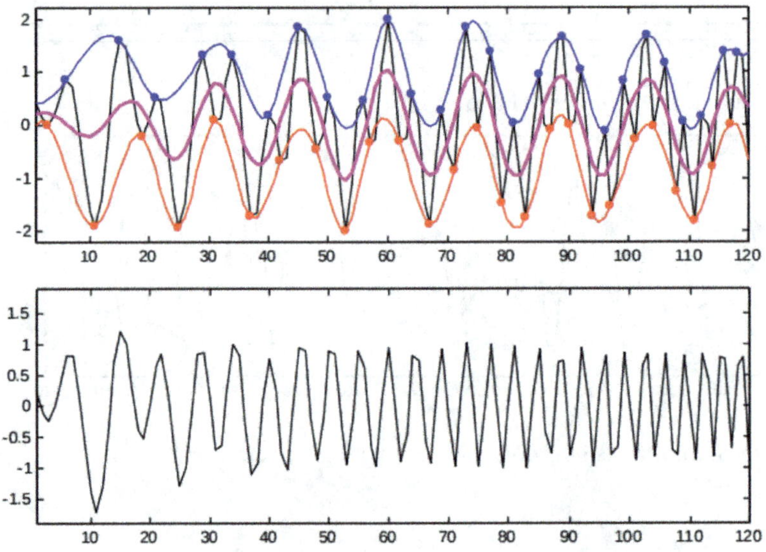

8. Check if $g(t)$ satisfies the IMF condition.
   YES: $g(t)$ is an IMF, stop sifting.
   NO: let $s(t) = g(t)$, keep sifting.

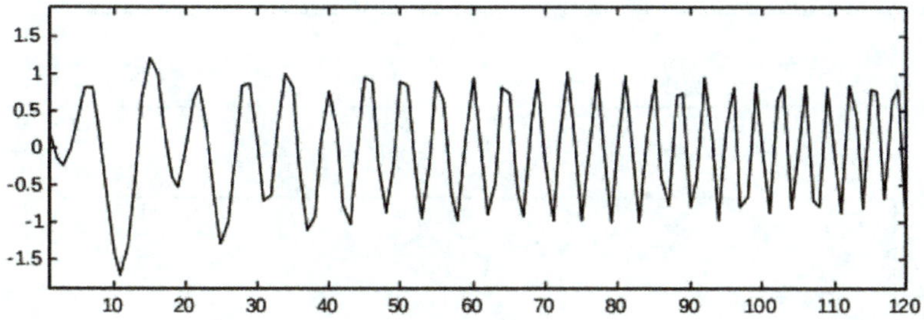

The overall sifting process will be expressed as:

$$s(t) = \sum_{i=1}^{N} g^{i}(t) + q(t)$$

The process stops when $q(t)$ is either constant or monotonic slope or function with only one extreme. The output of the EMD process is and IMF with residue signal q^{n}. Adding all the extracted IMFs and the residue will give original signal $s(t)$ (Fig. 1).

We use EMD tool box for MATLAB to find the IMFs of a given signal. The following figure shows IMFs of an EEG signal. From the figure, we can see that lower-order IMF captures the faster oscillation modes while the higher-order captures the slower oscillation modes [1]. The given figure shows that the IMF 3 and IMF 4 show frequential components that could have been removed during the filtering step, but these IMFs have amplitude 20 times lesser as compared to IMF 1 (Fig. 2).

Feature extraction from IMFs In IMFs, we have to see at instantaneous frequency components of the signals. To analyze the instantaneous frequency components, we use Hilbert–Huang transform (HHT). Instantaneous frequency is valid for non-stationary process and nonlinear signals. Instantaneous frequency allows us to avoid a local-averaging timescale altogether. We apply the HHT algorithm (HSA) technique to the IMFs to obtain instantaneous frequencies. Since the signal is decomposed in time domain and the length of the IMFs is the same as the original signal, HHT preserves the characteristics of the varying frequency. This is the main advantage of

Fig. 1 Sifting process

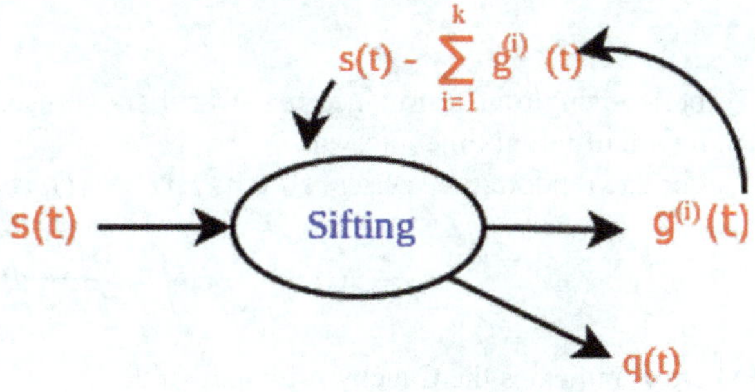

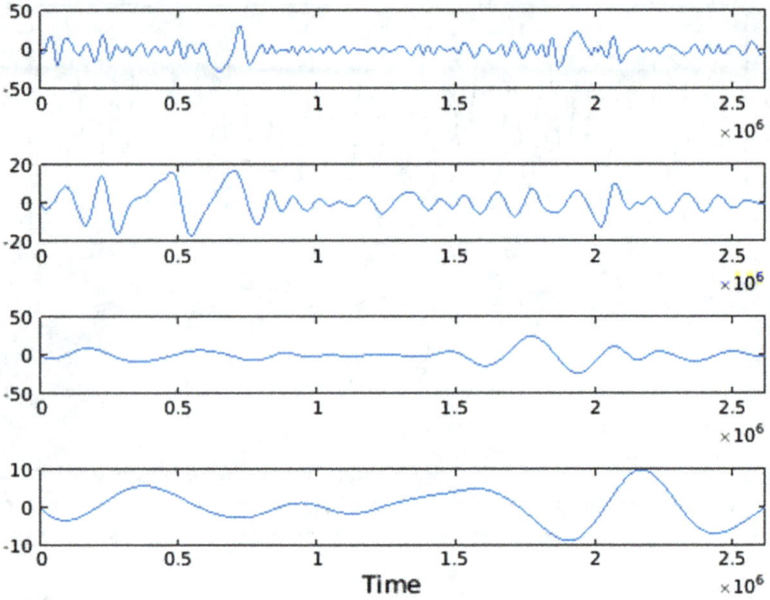

Fig. 2 IMF of Decomposed EEG segment

Algorithm 1 EMD Algorithm

$j \leftarrow 1, q_0(t) \leftarrow s(t)$ *and thresold* $= \delta$
top:
if $q_j(t) \geq 3$ **then return** false
extract local maxima and minima of $q_{j-1}(t)$
compute $e_j^1(t)$ *and* $e_j^2(t)$
compute $m_j(t) = (e_j^1(t) + e_j^2(t))/2$
$g_j(t) = q_{j-1}(t) - m_j(t)$
$q_{j-1}(t) \leftarrow g_j(t)$
    $g_{j,0}(t) = g_j(t)$ and $m_j, k(t)$; $k = 0, 1, ...$
    $g_{j,k}(t) = g_{j,k-1}(t) - m_{j,k-1}(t)$
    Until $SD(i) = \Sigma |g_{j-i-1}(t) - g_{j,i}(t)|^2 / g_{j,i-1}(t)^2$
$IMF_j(t) = g_{j,k}(t)$
$q_j(t) = q_{j-1}(t) - IMF_j(t)$
goto *top*.

Hilbert–Huang transform since real-world signal usually has multiple causes happening in different time intervals.

For any random time series, $s(t)$, its HHT, $R(t)$, is defined as

$$R(t) = \frac{1}{\pi} P \int_{-\infty}^{\infty} \frac{s(t')}{t - t'} dt'$$

where P indicates the Cauchy principal value.

Observe that the Hilbert–Huang transform, when viewed from the windowed sampling, the weight $1/(t-t)$ can be interpreted as a windowing width as the Hilbert–Huang transform is nothing but a produced function of integral of $s(t)$ with $1/t$. Thus, the Hilbert–Huang transform emphasizes the local possessions of the signal, $s(t)$.

For each of IMFs (R_j, $j = 1, \ldots., k$) and their corresponding instantaneous frequencies (ω_j, $j = 1, \ldots., k$), let their Hilbert complexification be

$$Z(t) = s(t) + iR(t) = L(t)^{i\theta(t)}$$

$$L(t) = [s^2(t) + R^2(t)]^{\frac{1}{2}}, \theta(t) = \arctan\left(\frac{R(t)}{s(t)}\right)$$

$$\omega = \frac{d\theta(t)}{dt}$$

The $Z(t)$ denotes the analytic signal, $L(t)$ and $\theta(t)$ are the instantaneous amplitude and phase functions, and ω is the instantaneous frequency which is our desired data. As a result, these data are the required feature for our work.

3.2 *Classification*

Any problem of classification consists of two main processes: feature extraction and classification. In feature extraction, we select measures which can reproduce the source signal. Even a simple classifier can give require results if feature to be extracted is proper. Among different classifiers, DecisionStump classifier with subject 06 and BFTree classifier with subject 05 perform best in our study. In this work, we will see a little description about these two classifiers below.

DecisionStump One of the Weak Learner classifier. Let us suppose that the input variable $a \in R^n$ is real valued. A DecisionStump is a function ϕ which has two parameters: a threshold v and index $k \in \{1, 2, \ldots, n\}$ and returns

$$\phi_{k,v}(a) = sign(a_k - v) = \begin{cases} 1\ if\ \ a_j \geq v \\ -1\ Otherwise \end{cases} \tag{1}$$

This is very simple classifier; we can fit it efficiently even to a weighted data set.

This classifier proceeds as follows. We start with a distribution—set of weights $w^{(1)}, \ldots, w^{(m)}$ summing to 1—on the training set, and to reduce the misconception on the training set, we wish to choose a DecisionStump of the form (1). That is, we wish to find a threshold $v \in R$ and index k such that

$$Err(\phi_{k,v}, w) = \sum_{i=1}^{m} w^{(i)} 1\{\phi_{k,v}(a^{(i)} \neq b^{(i)})\} \tag{2}$$

$$= \sum_{i=1}^{m} w^{(i)} 1\{b^{(i)}(a^{(i)} - v) \leq 0\} \tag{3}$$

is minimized. This could be a wasting of time in calculation but we can solve this problem roughly $O(nm \log m)$ time. For every feature $k = 1, 2, ..., n$, we sort the raw input features so that

$$a_k^{(i_1)} \geq a_k^{(i_2)} \geq ... \geq a_k^{(i_m)}$$

$a_k^{(i)}$ are the only values for the error of the DecisionStump can change. We compute

$$\sum_{i=1}^{m} w^{(i)} 1\{b^{(i)}(a_k^{(i)} - v) \leq 0\} = \sum_{j=1}^{m} w^{(i_j)} 1\{b^{(i_j)}(a_k^{(i_j)} - v) \leq 0\}$$

efficiently by incrementing and modifying the sum sorted order, which takes time $O(m)$ after sorted the values $a_k^{(i)}$. Thus, it takes total time $O(nm \log m)$ to perform this calculation for each of n input features. We may choose the index k and threshold v that give the best DecisionStump for error (3).

BFTree The best node is enlarged first in BFTree learner. The best node leads to maximum reduction of impurity (like Gini index or information gain) in comparison of any other nodes that can be split. The produced tree is same; only order is different in which it is built. BFTree is like binary tree, i.e. the node has exactly two child nodes. If node does not represent any homogeneous subset, then there is impurity. The impurity measures are entropy-based definition of information gain and Gini index. Entropy specifies the purity of any sample set. If there are m values of target attribute, then E entropy set corresponding to this m-wise classification is defined as

$$Entropy(E) = \sum_{i=1}^{m} -w_i \log_2 w_i \tag{4}$$

where w_i is a part of E which is a member of class i. There are some reduction in entropy caused when splitting the data set according to attribute, and this reduction is information gain. The gain is defined as

$$Gain(E, C) = Entropy(E) - \sum_{m \in Value(C)} \frac{|E_m|}{|E|} Entropy(E_m) \tag{5}$$

where $Gain(E, C)$ is information gain of attribute C, E is set of examples, $Values(C)$ is set of all possible values for C and E_m is the subset of E for which attribute C has value m. When the target feature has instances as a single value or when foremost information gain is lesser than zero, the tree ceases (Figs. 3 and 4).

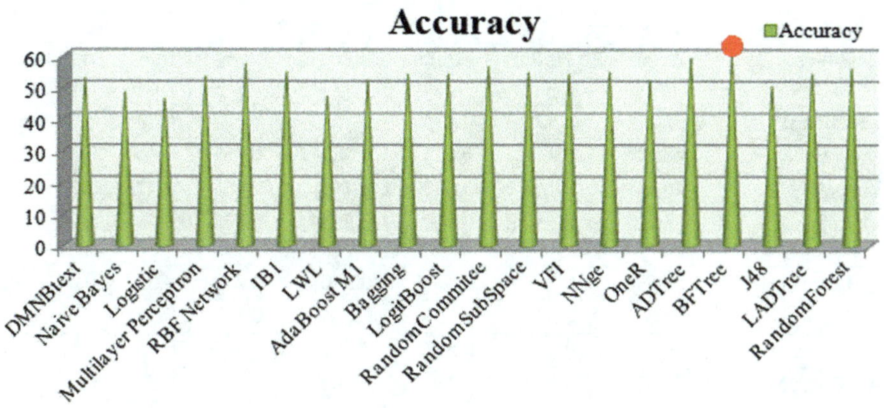

Fig. 3 Accuracy chart of subject 05

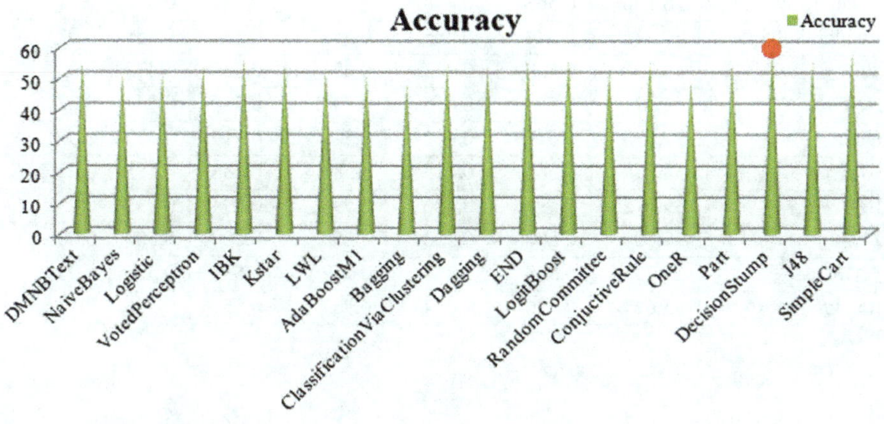

Fig. 4 Accuracy chart of subject 05

4 Results

In order to get the suitable features for our classification process, the feature extraction is an important issue. With the application of the HHT, we got the suitable and good features for our classification scheme. The result is obtained with the accuracy of 59.375% using BFTree classifier with subject 05 and with the accuracy of 58.75% using DecisionStump with subject 06. We use the tenfold cross-validation to obtain more accurate results. The results of different subjects are shown in given tables (Tables 1 and 2).

Table 1 Subject 05 results

Classifier	Accuracy (in %)	TP Rate	FP Rate	Precision	Recall	ROC Area	Build time (in sec.)
DMNBtext	52.500	0.525	0.475	0.525	0.525	0.498	0.00
NaiveBayes	48.125	0.481	0.519	0.481	0.481	0.466	0.01
Logistic	46.250	0.463	0.538	0.463	0.463	0.478	0.02
Multilayer Perceptron	53.125	0.531	0.469	0.532	0.531	0.512	0.25
RBF Network	56.875	0.569	0.431	0.570	0.569	0.537	0.04
IB1	54.375	0.544	0.456	0.544	0.544	0.544	0.00
LWL	46.875	0.469	0.531	0.461	0.469	0.462	0.00
AdaBoost M1	51.875	0.519	0.481	0.588	0.519	0.505	0.01
Bagging	53.750	0.538	0.463	0.538	0.538	0.555	0.04
LogitBoost	53.750	0.538	0.463	0.539	0.538	0.519	0.02
RandomCommittee	56.250	0.563	0.438	0.563	0.563	0.582	0.04
RandomSubSpace	54.375	0.544	0.456	0.544	0.544	0.553	0.01
VFI	53.750	0.538	0.463	0.604	0.538	0.479	0.00
NNge	54.375	0.544	0.456	0.545	0.544	0.544	0.02
OneR	51.875	0.519	0.481	0.519	0.519	0.519	0.00
ADTree	58.750	0.588	0.413	0.588	0.588	0.583	0.03
BFTree	**59.375**	0.595	0.406	0.598	0.594	0.593	0.05
J48	50.000	0.500	0.500	0.250	0.500	0.500	0.00
LADTree	53.750	0.538	0.463	0.539	0.538	0.524	0.06
RandomForest	55.625	0.556	0.444	0.557	0.556	0.584	0.11

Table 2 Subject 06 results

Classifier	Accuracy (in %)	TP Rate	FP Rate	Precision	Recall	ROC Area	Build time (in sec.)
DMNBtext	55.000	0.550	0.450	0.550	0.550	0.530	0.00
NaiveBayes	51.250	0.513	0.488	0.513	0.513	0.526	0.00
Logistic	51.250	0.513	0.488	0.513	0.513	0.519	0.02
VotedPerceptron	53.750	0.538	0.463	0.538	0.538	0.521	0.01
IBK	56.250	0.563	0.438	0.563	0.563	0.569	0.00
Kstar	54.375	0.544	0.456	0.544	0.544	0.548	0.00
LWL	52.500	0.525	0.475	0.529	0.525	0.509	0.00
AdaBoostM1	51.250	0.513	0.488	0.515	0.513	0.525	0.01
Bagging	48.125	0.481	0.519	0.481	0.481	0.491	0.04
ClassificationViaClustering	53.750	0.538	0.463	0.538	0.538	0.538	0.01

(continued)

Table 2 (continued)

Classifier	Accuracy (in %)	TP Rate	FP Rate	Precision	Recall	ROC Area	Build time (in sec.)
Dagging	53.125	0.531	0.469	0.533	0.531	0.499	0.13
END	55.000	0.550	0.450	0.639	0.550	0.507	0.01
LogitBoost	56.250	0.563	0.438	0.569	0.563	0.545	0.03
RandomCommittee	53.125	0.531	0.469	0.531	0.531	0.504	0.04
ConjuctiveRule	55.000	0.550	0.450	0.614	0.550	0.558	0.01
One	48.750	0.488	0.513	0.587	0.488	0.488	0.00
Part	54.375	0.544	0.456	0.647	0.544	0.548	0.00
DecisionStump	**58.750**	0.588	0.413	0.661	0.588	0.542	0.00
J48	55.000	0.550	0.450	0.639	0.550	0.507	0.00
SimpleCart	58.125	0.581	0.419	0.645	0.581	0.530	0.05

5 MATLAB

Also known as matrix laboratory, MATLAB is a tool which supports more than one programming paradigms and also provides technique for numerically solving mathematical problems. The MATLAB programming language was developed by MathWorks. We can perform following functions in MATLAB:

1. matrix manipulation,
2. creation of user interface,
3. function plotting,
4. data plotting,
5. implementation of algorithms
6. and also we can interface with another programming languages, like Java, C, C++, Python and Fortran.

In MATLAB, there is a package, named as Simulink. This package provides the graphical analysis in MATLAB.

MATLAB tool is mostly used in the field of engineering, economics and science.

The MATLAB system consists of five main parts:

1. the MATLAB language,
2. the MATLAB working environment,
3. handle graphics,
4. the MATLAB mathematical function library
5. and the MATLAB Application Program Interface (API).

In this work, we have used MATLAB for implementing our methods.

6 Conclusions

This work proposes a method for extracting features by analysis of EEG signals and classifying mental task and estimation of the instantaneous frequencies from each of the IMFs. Different classifiers are used, but BFTree gives the best result with accuracy 59.375% among all of them for subject 05 and for subject 06, classifier DecisionStump gives the best result with 58.75% among all of them. We compared all the different classifiers.

References

1. Diez PF, Mut V, Laciar E, Torres A, Avila E (2009) Application of the empirical mode decomposition to the extraction of features from EEG signals for mental task classification. In: 2009 annual international conference of the IEEE engineering in medicine and biology society, 2009
2. Bashar SK, Bhuiyan MIH (2016) Classification of motor imagery movements using multivariate empirical mode decomposition and short time Fourier transform based hybrid method. Eng Sci Technol Int J
3. McFarland DJ, Anderson CW, Mller K-R, Schlgl A, Krusienski DJ (2006) BCI meeting 2005 workshop on BCI signal processing: feature extraction and translation. IEEE Trans Neural Syst Rehabil Eng 14(2):135–138
4. Huang NE, Shen Z, Long SR, Wu ML, Shih HH, Zheng Q, Yen NC, Tung CC, Liu HH (1998) The empirical mode decomposition and Hilbert spectrum for nonlinear and nonstationary time series analysis. Proc R Soc Lond A 454:903–995
5. Matlab 2015. http://www.mathworks.com/products/matlab/
6. Lotte F, Congedo M, Lcuyer A, Lamarche F (2007) A review of classification algorithms for eeg-based braincomputer interfaces. J Neural Eng 4
7. </http://bnci-horizon-2020.eu/database/data-sets/>
8. </https://en.wikipedia.org/wiki/MATLAB>
9. Khan YU, Sepulveda F (2010) Braincomputer interface for single-trial eeg classification for wrist movement imagery using spatial filtering in the gamma band. IET Signal Process 4(5):510–517
10. Yi W, Qiu S, Qi H, Zhang L, Wan B, Ming D (2013) EEG feature comparison and classification of simple and compound limb motor imagery. J Neuroeng Rehabil 10
11. Abibullaev, B (2017) Learning suite of kernel feature spaces enhances SMR-based EEG-BCI classification. In: 2017 5th international winter conference on brain-computer interface (BCI), 2017
12. Orosco L, Laciar E, Garces Correa A, Torres A, Graffigna JP, An epileptic seizures

Cloud-Based Secured QR Code for Self-service Access Control System at Resort and Hotels

M. Anitha, Anurag Babel, Anushil Kumar, Aman Rauniyar and Kashif Zahid

Abstract Self-service technologies (SSTs) are increasingly becoming popular with the advancement of mobile technologies, cheaper Internet prices and the growing popularity of mobile devices in daily life. Popular hotels/resorts tend to adopt SSTs to enhance their service quality standards, consumer experience, loyalty and most importantly overall customer satisfaction. The long queue in hotels and resorts has been the major problem for causing dissatisfaction for customers, especially for minors and senior citizen. Customer satisfaction has been the main criteria for the business to grow. With this goal, we propose a cloud-based secure access control for hotel and resort rooms using the QR code technique. To gain access to the hotel room, the customer needs just to scan the QR code sent from the secure cloud. The user details are encrypted using MD5 algorithm and stored as checksum in database to avoid various attacks such as phishing, malware, code injections and others possible in Web services. When the user scans the QR and decodes, it compares the checksum stored in database. In case of any discrepancies, further process is stopped and hence makes our approach more secure from other techniques. Also, cloud provides additional security for the Web services.

Keywords Self-service technologies · QR code · Access control

1 Introduction

Business and market strategy change as the technology changes to provide better quality to the customers. Self-service technology has become a popular marketing strategy, as the service is faster with the latest technologies. Many SSTs are popular such as automated teller machines (ATMs), mobile banking, online booking, self-check-in machines at airports and online shopping. Self-service technologies also serve as an alternative for human resource and have added advantages of potential

M. Anitha (✉) · A. Babel · A. Kumar · A. Rauniyar · K. Zahid
Dayananda Sagar College of Engineering, Kumaraswamy Layout, Bangalore, India
e-mail: anitha-cs@dayanandasagar.edu

S.-L. Peng et al. (eds.), *Computing and Network Sustainability*, Lecture Notes in Networks and Systems 75, https://doi.org/10.1007/978-981-13-7150-9_41

increase in revenue, efficiency in process, flexible control for customers and less investment in labor resources [1, 2].

Moreover, the customer satisfaction is important in their growth of business. The smarter technologies are constantly added to increase their customers which in turn increase their profit margin. With the reduced prices in smarter devices such as mobile phones and iPads, Internet helps in providing services to customers more efficiently. Besides these added benefits, SSTs also help them to sustain their top place in the service industry. Hotels and resorts being part of the service industry, they constantly invest in SSTs for improving their service quality and reducing overall cost [3].

Varieties of secure access control systems are available such as password protected ones, RFID card technologies, fingerprint, biometric protected systems, OTP-based, cryptography-based and many more. Each system is applicable to different application zones depending upon their technical usage [4–7]. We propose to use secure cloud-based QR access control system, which is more efficient and convenient than other approaches.

Quick Response (QR) code is a two-dimensional barcode image which can store more information than barcodes and can be easily encrypted making it convenient to be used in various applications. Usually, it is used to store data such as personal information, secret key providing privacy to users. It can be easily decoded by just scanning the image and relives the user from remembering the secret key or the password. The popularity of this technique relies on the advantage that it can be scanned regardless of its position and angle of orientation of the camera [8].

Due to the versatile nature of QR, it used many applications such as online banking, attendance management, health care, urban parking and tourism and also helps minors, senior citizens, visually impaired people. Hence, we use QR code technique to provide an access control to hotel rooms.

The ease with which one can create and distribute QR codes has not only attracted businesses, but also scammers seeking to direct people to phishing Web sites [9]. Attackers misuse QR codes to encode malicious links that lead, e.g., to phishing sites or to the execution of malicious code. These malicious QR codes can be printed on small stickers and pasted over preexisting QR codes on billboard advertisements [10]. To avoid malicious QR attacks, we propose to use cloud to provide more secure access control for the customers.

Since the service, storage, related sources and communication are on the cloud, our approach has significant advantages for the reason that the cloud platform is technically stable, taking many security elements under their own control. Our solution is built on such a trustable platform and makes best use of it to enhance the security of QR code.

2 Literature Survey

This section presents a survey of various access control systems used in different applications and compares their techniques with others. With the trend of digital technology usage becoming popular in all places [11], QR code technique of an access control is used in many applications because of its low complexity and with its added advantage of providing privacy since it is used in the encoded form.

Alshattnawi et al. [4] propose a QR code usage for Muslim pilgrimage activity in Mecca, Saudi Arabia, where QR codes are used as a label or tag is placed at important places where pilgrims can refer to it to get directions to follow. Bechini et al. [5] suggest an efficient technique of using QR codes in urban parking. Charoensiriwath et al. [6] propose another useful usage where overcrowding and bottleneck problems, due to less resources in hospitals, are solved using QR code. Ishak et al. [7] suggest an application where QR code is used as a tag on plants to give information to the tourists. Because of very convenient way of usage to customers and being popular in all places, we propose to use QR code technique as an access control mechanism to hotel rooms.

Chow et al. [12] propose a unique challenge response visual one-time password (OTP) authentication scheme which obtains a session key from public or the user end rather than generated by system and is sent to the user. This scheme shows that visual-based encryption techniques are better than OTP and short message services (SMS) which suffer from problems such as time synchronization, hash chains complexity making it complex and security attacks such as key loggers and phishing. OTP can be used only for one session. Multiple OTPs create annoyance among customers leading to less business.

Khandal et al. present an innovative registration mechanism using the QR code to fill up the registration form which contains the basic information of the user [13]. Another advantage of QR code helps in automatic filling of forms, thereby helping the minors and the senior citizens from waiting in long queue [14]. Haron et al. suggest an automated form-filling mechanism using smart card readers in government offices making it a time-efficient method [15].

Shafin et al. [16] proposed an RFID-based authentication system where only authorized person was allowed access to the protected area. Other approaches [17–19] also make use of RFID along with different encryption techniques to provide secure access control. Even though it uses good encryption techniques, RFID and smart cards are not cost effective and are likely prone to be misused, to be tampered or to electronic data theft.

Fu et al. [14] present an access-based control system using fingerprint with an additional authentication process using OTP and QR code. It also provides extra authorization scheme to assist a remote manager to grant temporary access to an unauthorized person based on certain emergency situations. But this paper uses three different types of security techniques which lead to complexity in managing the access control in the operation and also make the process slow which can be the major drawback for business industry.

But for more secure approach, we choose a cloud-based QR access control using smart locks [11] to hotel rooms to provide efficient, cost-effective, secure and faster approach compared to other techniques.

3 System Design

3.1 Web Service Module

Customer who needs to book a hotel room needs to visit the hotel management Web site and registers their details such as username, password, phone number and other personal details for proper user authentication as shown in Fig. 1. After validating the data entered (correct phone number, password complex enough, etc.), a unique identification number (UID) is generated for each user, and a checksum (CK) value is created using MD5 encryption algorithm [20] to protect against various malware, phishing attacks and well-performed code injection filters and is stored in the database.

Once the validation is over, a message is sent to the customer on confirmation for further process. Then, the customer can visit the Web site for booking the room based on the availability. Once the room is booked, QR code is generated in the cloud and sent to the customer's registered phone number. After this, the customer can proceed to download QR code, scan the code and access the door to unlock using Wi-fi.

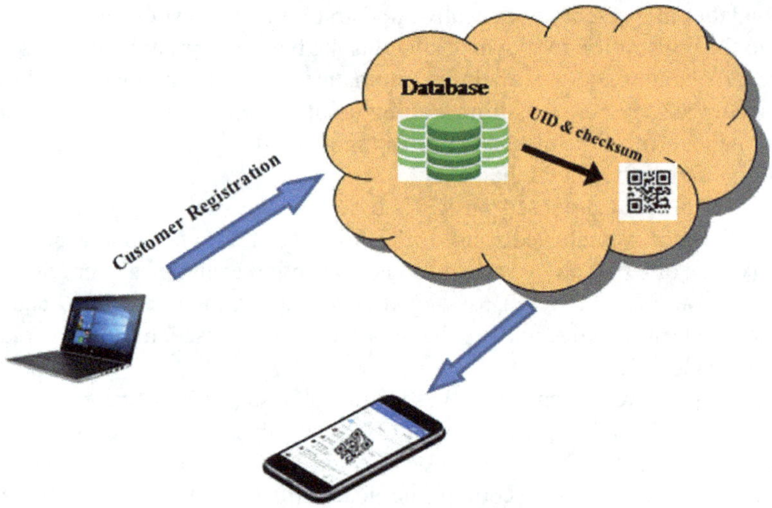

Fig. 1 Schematic diagram of QR generation

3.2 Cloud Module

Cloud module contains the database and the QR generator services. When a customer tries to book a room, a request is sent to the cloud. The URL that is generated is encrypted by MD5 encryption algorithm [20], and a checksum (CK) value is generated and stored in database. Next, the UID along with CK value, instead of the real data pertaining to the user, is encoded in the generation of QR code and sent to the customer. Every time the customer makes an access to the room, time entry and exit timing detail entry is made to the database for future use of security and other statistics measures.

There is also a provision for an admin having the direct access to cloud, who can monitor QR generation in cloud, Web services and the database. The admin has the full privileges to view the content of database and make modifications. In case of emergencies, the admin will have the right to reject a customer or protect from fraud customers by not releasing the QR code.

3.3 Android App Module

An Android app is used for the communication between the customer and the Web service connected to cloud. First, the customer has to log in the app before for accessing the QR code. First, the user is authenticated based on the username and password given during the registration process. Later, it uses the QR scanning and decoding process. The QR code received is first scanned and then decoded.

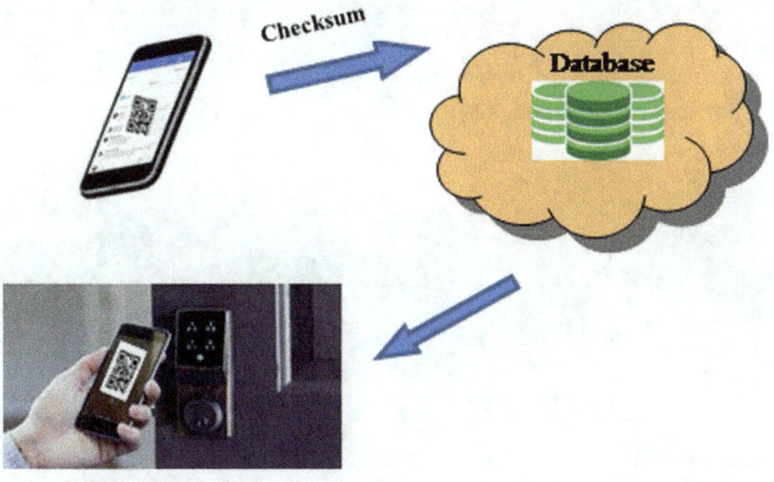

Fig. 2 Schematic diagram of decoding QR code

As shown in Fig. 2, the decoding part contains mainly three tasks: Decodes the QR code, verifies the CK value for secure authentication, and unlocks the room lock. The decoder decodes the QR code to get UID, CK value. This CK value is compared with the CK stored in the database to avoid attacks such as code injections and others. The decoded information contains the smart lock access key which is sent over the Wi-fi to unlock the room.

4 Implementation

The system implemented is comprised of three parts: cloud-based database, Web service and Android application. Android application is developed in Eclipse, while the Web service is programmed in .NET using C# language, with the support of Windows Azure Cloud, and the database is created in Microsoft SQL Server which is stored on cloud. C# has the advantages of having its own countermeasures to mitigate certain kind of code injection such as parameterized statement, escaping, pattern checking and so on [23]. The snapshot results of the modules created are as shown in Figs. 3, 4 and 5.

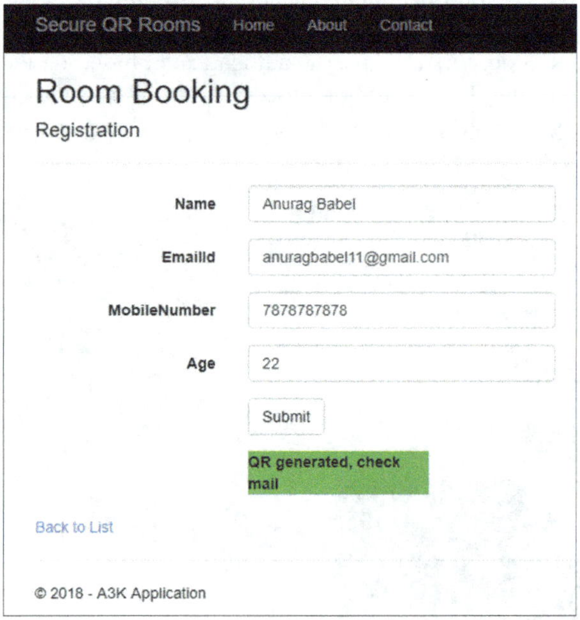

Fig. 3 Registration form

Fig. 4 Decoding of QR image

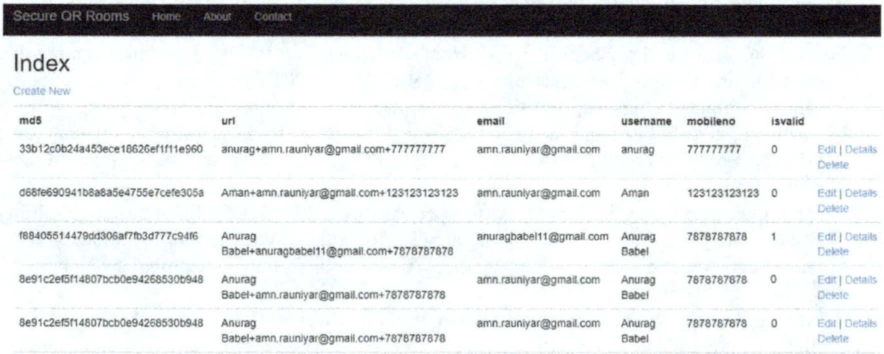

Secure QR Rooms Home About Contact

Index

Create New

md5	url	email	username	mobileno	isvalid	
33b12c0b24a453ece18626ef1f11e960	anurag+amn.rauniyar@gmail.com+777777777	amn.rauniyar@gmail.com	anurag	777777777	0	Edit | Details Delete
d68fe690941b8a8a5e4755e7cefe305a	Aman+amn.rauniyar@gmail.com+123123123123	amn.rauniyar@gmail.com	Aman	123123123123	0	Edit | Details Delete
f88405514479dd306af7fb3d777c94f6	Anurag Babel+anuragbabel11@gmail.com+7878787878	anuragbabel11@gmail.com	Anurag Babel	7878787878	1	Edit | Details Delete
8e91c2ef5f14807bcb0e94268530b948	Anurag Babel+amn.rauniyar@gmail.com+7878787878	amn.rauniyar@gmail.com	Anurag Babel	7878787878	0	Edit | Details Delete
8e91c2ef5f14807bcb0e94268530b948	Anurag Babel+amn.rauniyar@gmail.com+7878787878	amn.rauniyar@gmail.com	Anurag Babel	7878787878	0	Edit | Details Delete

Fig. 5 Snapshot of database details

5 Conclusion

With the technology invading into the private life of humans, there is a need to reduce the complexity of usage of technology to make life hassle-free for a comfortable living. We propose an idea which makes a traveler have an easy accommodation without any delay and have a wonderful experience without any chaos. The idea presented in this paper reduces the number of keys required to manage with the increasing number of gadgets in the current days. The cloud-based QR approach has less complexity than other approaches making it user-friendly for the customers. This idea also helps the hotel industry to increase the customer by providing a luxury based on hassle-free environment to have a great relaxation during their vacation and making their travel experience a pleasant one.

References

1. Kaushik AK, Agrawal AK, Rahman Z (2015) Tourist behaviour towards self-service hotel technology adoption: trust and subjective norm as key antecedents. Tour Manag Perspect 16:278–289. ISSN 2211-9736
2. Oh H, Jeong M, Baloglu S (2013) Tourists adoption of self-service technologies at resort hotels. J Bus Res 66(6):692–699
3. Iqbal MS, Ul Hassan M, Habibah U (2018) Impact of self-service technology (SST) service quality on customer loyalty and behavioral intention: the mediating role of customer satisfaction. Cogent Bus Manage 5
4. Alshattnawi S (2012) Effective use of QR codes in religious tourism. Int Conf Adv Comput Sci Appl Technol (ACSAT), pp 497–451
5. Bechini A, Marcelloni F, Segatori (2013) A mobile application leveraging QR-codes to support efficient urban parking. In: 2013 sustainable internet and ICT for sustainability (SustainIT), Palermo, pp 1–3
6. Charoensiriwath C, Surasvadi N, Pongnumkul S, Pholprasit T (2015) Applying QR code and mobile application to improve service process in Thai hospital. In: 2015 12th international joint conference on computer science and software engineering (JCSSE), Songkhla, pp 114–119
7. Ishak I, Sidi F, Affendey LS, Sani NFM, Hamzah AS, Bawon P (2013) Mobile plant tagging system for urban forest eco-tourism using QR code. In: 2013 international conference on advanced computer science applications and technologies, Kuching, pp 37–41
8. Mohamed K, Sidi F, Jabar MA, Ishak I (2013) A novel watermarking technique in data transmission between QR codes and database. In: 2013 IEEE conference on open systems (ICOS), Kuching, pp 95–99
9. Vidas T, Owusu E, Wang S, Zeng C, Cranor LF, Christin N (2013) QRishing: the susceptibility of smartphone users to QR code phishing attacks. In: Adams AA, Brenner M, Smith M (eds) Financial cryptography and data security. FC 2013. Lecture Notes in Computer Science, vol 7862. Springer, Berlin, Heidelberg
10. Kieseberg P Schrittwieser S, Leithner M, Mulazzani M, Weippl E, Munroe L, Sinha M (2012) Malicious pixels using QR codes as attack vector. In: Trustworthy ubiquitous computing. Atlantis Press, pp 21–38
11. https://www.owasp.org/index.php/Top_10–2017_A1-Injection
12. Chow Y, Susilo W, Yang G, Phillips JG, Pranata I, Barmawi AM (2016) Exploiting the error correction mechanism in QR codes for secret sharing. In: Liu JK, Steinfeld R (eds) ACISP 2016. Lecture Notes in Computer Science, vol 9722. Springer, pp 1–17
13. Khandal D, Somwanshi D (2015) A novel cost effective access control and auto filling form system using QR code. In: 2015 international conference on advances in computing, communications and informatics (ICACCI), Kochi, pp 1–5
14. Fu X, Feng K, Wang C. Zhang J (2015) Improving fingerprint based access control system using quick response code. In: 2015 IEEE international conference on computational intelligence and virtual environments for measurement systems and applications (CIVEMSA), Shenzhen, pp 1–5
15. Haron NZ, Mohter MB, Yunus SAMJ, Isa MSM, Rahim YA (2013) SAFe: a secure and fast auto filling form system. In: IEEE symposium on industrial electronics and applications (ISiEA), pp 206–209, 22–25 Sept 2013
16. Shafin MK et al (2015) Development of an RFID based access control system in the context of Bangladesh. In: 2015 international conference on innovations in information, embedded and communication systems (ICIIECS), Coimbatore, pp 1–5
17. Mathew M, Divya RS (2017) Super secure door lock system for critical zones. In: 2017 international conference on networks & advances in computational technologies (NetACT), Thiruvanthapuram, pp 242–245
18. Cho JS, Yeo S-S, Kim SK (2011) Securing against brute-force attack: a hash-based RFID mutual authentication protocol using a secret value. Comput Commun 34(3):391–397

19. Song B, Mitchell CJ (2008) RFID authentication protocol for low-cost tags. In: Proceedings of the first ACM conference on Wireless network security. ACM, pp 140–147
20. Yong-Xia Z, Ge Z (2010) MD5 research. In: 2010 second international conference on multimedia and information technology, Kaifeng, pp 271–273. https://doi.org/10.1109/mmit.2010.186
21. Wang Z, Cao L (2013) Implementation and comparison of two hash algorithms. In: 2013 international conference on computational and information sciences, Shiyang, pp 721–725
22. Ning LW, Siah YK, Khalid M, Yusof R (2000) Design of an automated data entry system for hand-filled fonns. In: Proceedings of TENCON 2000, vol l, pp 162–166
23. Tsudik G (2006) YA-TRAP: yet another trivial RFID authentication protocol. In: Fourth annual IEEE international conference on pervasive computing and communications workshops (PERCOMW'06), Pisa, pp 4, 643
24. Kassem A, Murr SE, Jamous G, Saad E, Geagea M (2016) A smart lock system using Wi-Fi security. In: 2016 3rd international conference on advances in computational tools for engineering applications (ACTEA), Beirut, pp 222–225

Palm-Leaf Manuscript Character Recognition and Classification Using Convolutional Neural Networks

R. S. Sabeenian, M. E. Paramasivam, R. Anand and P. M. Dinesh

Abstract In this paper, a character recognition approach using convolutional neural networks with specific focus on Tamil palm-leaf characters has been presented. The convolutional neural network (CNN) used in this paper has utilized around five layers viz., convolution layer, pooling layer, activation layer, fully connected layer, and softmax classifier. The database of character set has been created using scanned images of palm-leaf manuscripts. The database comprises of 15 variety of classes and each class contains around 1000 different samples. The recognition of CNN Classifier if found to be around 96.1% to 100%. The prediction rate is found to be higher due to the large quantum of features extracted for each of the CNN layers. A comparison of the proposed method with other machine learning algorithms has also been presented in the paper.

Keywords Convolutional neural network · Convolution layer · Pooling layer · Activation layer · Tamil palm-leaf manuscript · Handwritten character recognition · Prediction rate

R. S. Sabeenian · M. E. Paramasivam · R. Anand (✉) · P. M. Dinesh
Department of Electronics and Communication Engineering, Sona SIPRO, Salem, India
e-mail: anand.r@sonatech.ac.in

R. S. Sabeenian
e-mail: sabeenian@sonatech.ac.in

M. E. Paramasivam
e-mail: sivam@sonatech.ac.in

P. M. Dinesh
e-mail: dinesh@sonatech.ac.in

R. Anand · P. M. Dinesh
Sona Signal and Image PROcessing Research Center, Advanced Research Centre Block, Sona College of Technology, Salem 636 005, Tamil Nadu, India

S.-L. Peng et al. (eds.), *Computing and Network Sustainability*, Lecture Notes in Networks and Systems 75, https://doi.org/10.1007/978-981-13-7150-9_42

1 Introduction

Centuries before Egyptians invented '*Papyrus*', knowledge was shared by inscribing on [1] tree barks, skin hides, rocks, and leaves. Under the leaf category, palm leaves were widely used due to its enormous availability and capability to withstand rigorous conditioning [2]. Organizations [3–9] have taken efforts on war-footing basis to preserve these manuscripts as most of them contain information varying from traditional medicines, land documents, astrology, astronomy, and many more. Despite such conditionings, the organic nature of palm leaves restricted its life to an average of 300–400 years.

Convolutional neural network (CNN) has been a strong machine learning approach in the area of deep learning. CNN learns feature representation from a wide range of images. This paper has utilized CNN for classifying and recognizing Tamil palm-leaf manuscript characters from isolated character images. The paper has also involved studying the end-to-end mapping for each character utilized for recognition. The proposed approach shows that the computational complexity and speed of classification is comparatively low when compared to existing approaches. The results obtained for the proposed CNN architecture have been compared with support-vector machine (SVM), *k*-NN classifier [10], feed-forward neural network classifier, and convolution neural network (LeNET- Architecture) [11].

2 Proposed Dataset

Bharathidasan University, Digital Repository [3] has made available the digital versions of ancient manuscripts on the Internet. '*Agathiyar Vaithiyam*' a treatise on traditional medicine (*Siddha*) written by a saint *Agasthiyar*, centuries ago is one among the digitalized manuscripts. Manuscript characters recognition require efficient recognition algorithms to address the constraints viz., low contrast between foreground and background, uneven background color, touching/overlapping in closed loops of character structures and many more thereby creating confusion among character symbols.

The dataset used in this paper has been created using 50 manuscript images. The images were randomly chosen such that the characters were better visible despite the degradation of the manuscript. Figure 1 shows a sample palm-leaf manuscript embedded with artifacts. Expert advice was also sought to ensure that the manuscript images were of the same scriber. Isolated character symbols (20,000 images) were segmented and in turn classified into 60 different classes.

An analysis of various binarization algorithms on palm-leaf manuscripts was carried out by [12–14]. We have utilized the widely exploited global thresholding method proposed by Otsu [15]. Geometric features [16] were applied on a $M = 9\,(3 \times 3)$ uniform grid for the created database. A group of 20 symbols were chosen viz., ள, ன, ◌ை, ◌ெ, ◌ே, க, ஈ, ச, ண, ட, ய, ப, ம, ர, ◌ா, ல, வ, த, ந and உ.

Fig. 1 Sample palm-leaf manuscript with noise artifacts

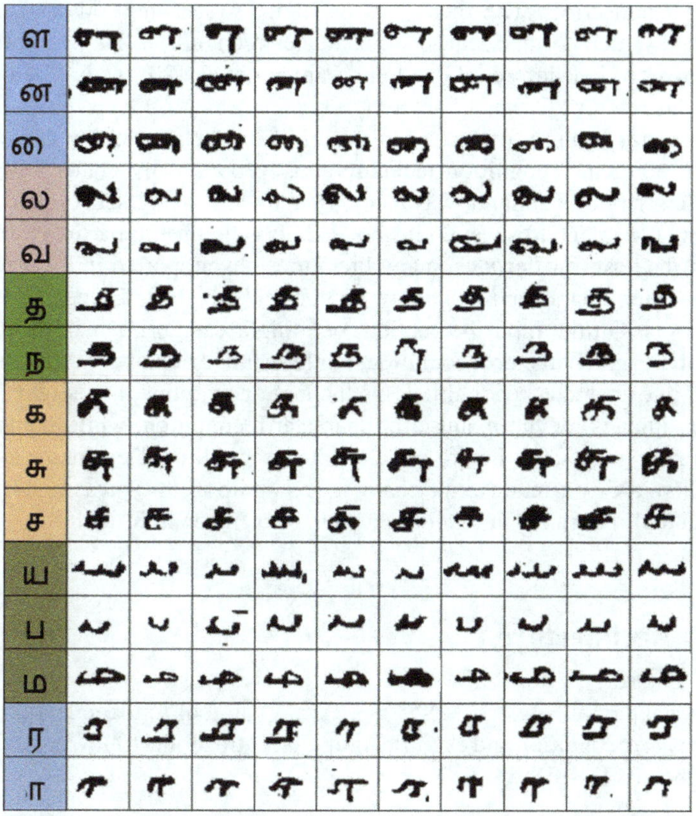

Fig. 2 Samples of binarized character samples utilized for experimentation

The choice for these selected character symbols was due to its predominant usage in palm-leaf manuscripts (nominal structural complexity), thereby fairly increasing the investigation and training samples. Figure 2 shows a few samples of segmented and binarized manuscript characters utilized in this experimentation.

3 Related Works

For more than a decade, optical character recognition had been a wide area of research for scientist around the globe. Gatos et al. [17] have made an OCR system to recognize printed Latin text from ancient manuscripts. Thammano and Pravesjit [18] have invented an apparatus to recognize text in images of an ancient printed books and manuscripts. The focus was on Latin, Arabic, and other characters in printed/handwritten form. Manuscript character recognition has been a emerging area of research. Lakshmi et al. [19] and Panyam et al. [20] have developed a 3D modeling system for '*Telugu*' characters.

The character ligatures of ancient manuscript Greek characters were investigated and a fast, efficient recognition technique was proposed by Gatos Lanna script, an obsolete script of Thailand, was recognized by [18] using self-organizing maps for dividing the image to several clusters and classified using particle swarm algorithm. For the past few decades, many literatures have reported in the areas of off-line handwritten character recognition of numeric, English, Chinese, Indian, and Arabic scripts. On the input perception, one might categorize manuscript images for the manuscript character recognition, while scanned document character images for the off-line character recognition. On the input perception, one might categorize manuscript images for the manuscript character recognition, while scanned document character images for the off-line character recognition. Despite the availability of such literatures on manuscript character recognition, the presence of complex curves and multiple joints in '*Tamil*' characters has made it difficult to recognize.

4 CNN Architecture

Convolutional neural network (CNN) is an important tool for deep learning, widely used for image recognition and classification. CNN utilizes the following steps toward recognition and classification:

1. Collecting the data;
2. Design the CNN architecture;
3. Construct the CNN network;
4. Compute the weights and biases;
5. Train the designed network;
6. Evaluate the designed network for identifying different classes.

One of the unique features of CNN is the presence of '*Hidden Layers*'. Each of these hidden layers plays a vital role in extracting additional characteristic features, thereby enhancing the performance of classification. A variety of CNN architectures has been reported in the literature viz., LeNET [21], Alexnet [22], GoogleNet [23], etc. It has also been reported in these literatures that LeNET (Fig. 3), which composes of minimal layers is the most sought architecture for character recognition and

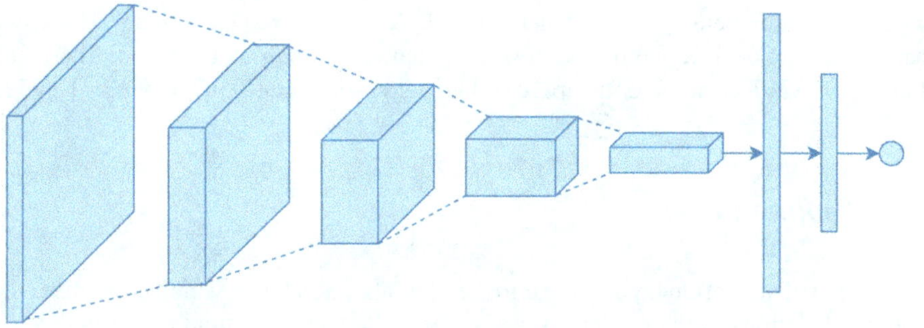

Fig. 3 LeNET CNN architecture

classification. This paper has focused on recognition of palm-leaf manuscript characters and hence the modified LeNET architecture was utilized for the same.

4.1 Convolution Layer

The segmented manuscript character images were ensured that each is of size [48 × 48 × 3]. Each of the images is convolved with 32 different filters [3 × 3 × 32], as a result of which we obtain the output of the first layer [48 × 48 × 32] of CNN architecture. This paper has also utilized different types of convolution layers viz., normal convolution layer, separable convolution layer, and convolution transpose layer. If X_1 is the size of the input image and K_1 is the size of the input filter, then the size of the convolved image can be defined as:

$$\text{Number of Rows / Columns in the convolved image} = \frac{(X_1 - K_1 + 2P)}{S} + 1 \quad (1)$$

where 'P' refers to the Zero Padding and 'S' is the Striding fashion to be followed.

4.2 Activation Layer

The next layer in CNN is the activation layer. In this paper, rectified linear unit (ReLU) [24] has been utilized, which performs the thresholding operation for each element in the convolution layer. The threshold value has been fixed as

$$f(x) = \begin{cases} x(\lambda + 1) & , x \geq 0 \\ \alpha e^x - \alpha(\lambda + 1) & , x < 0 \end{cases} \quad (2)$$

where ‘x’ is the intensity value of an image, ‘λ’ & ‘α’ are fixed parameters. These two parameters are derived from input. For a standard scaled input, the mean is zero and standard deviation is 1 and the approximate value for $\lambda = 1.0507$ and $\alpha = 1.6732$.

4.3 Pooling Layer

The output of activation layer is fed into the normalization layer, which normalizes the output activations of the earlier layer after every batch. The output of normalization layer is in turn fed into pooling layer. This layer plays a significant role in reducing the unwanted pixels in the image. In the created CNN architecture for the palm-leaf manuscript character images, 2,238,441 parameters were learned and from this totally 2,235,945 trainable parameter and 2,496 non-trainable parameters were identified. These parameters in turn are fed to the classification layer.

5 Results and Discussion

The performance of the proposed CNN architecture for palm-leaf manuscript character recognition has been compared with other machine learning approaches viz., support-vector machine (SVM), *k*-NN, and fast artificial neural network (FANN). Table 1 shows the prediction rate and mis-classification rate along with the prediction time for classifying the character present in the image database (Sect. 2) utilizing machine learning approaches. For the developed CNN -rchitecture, results show that the prediction rate of 96.21%, mis-prediction rate of 3.79% and prediction time for a single character is 0.64 s. The obtained results are comparatively better to other existing machine learning approaches.

Table 1 Classification rate for various machine learning approaches and its corresponding prediction time

Classification method	Prediction rate (%)	Mis prediction rate (%)	Prediction time (s)
SVM	85.46	14.55	1.06
k-NN	77.21	22.79	1.04
FFNN	89.21	10.79	0.95
CNN	96.21	3.79	0.64

6 Conclusion and Future Work

A palm-leaf manuscript character image database has been created by in this work. This paper has also developed a CNN architecture and evaluated the performance of the same on the developed database. Other machine learning approaches have also been subjected with the same developed database for comparing the prediction rate. The authors wish to extend the database with additional character images and evaluate the performance of the developed architecture. Characters from different scribes are to be included in the character database, thereby increasing the size of the database. The presence of other scribe character images in the database would be challenge to the machine learning approaches for reduced mis-classification.

References

1. Diskalkar DB (1979) Materials used for Indian epigraphical records. Bhandarkar Oriental Research Institute, Poona
2. Krishnamurthi S, Hikosaka S, Samuel JG (1994) A history of Telugu literature, vol 1. Institute of Asian studies
3. மின்னணுச் சுவடிக் காட்சியகக் குழு Digital Manuscript Gallery, Bharathidasan University, Tiruchirappalli, India. http://www.bdu.ac.in/suvadi/1.1/. Accessed 12 March 2019
4. Institute of Asian Studies, Chennai, Tamil Nadu, India. http://www.instituteofasianstudies.com. Accessed 12 March 2019
5. Memory of the World Programme—UNESCO. http://www.unesco.org/new/en/communication-and-information/flagship-project-activities/memory-of-the-world/homepage/. Accessed 12 March 2019
6. National Mission for Manuscript—The Republic of India. http://www.namami.org/. Accessed 12 March 2019
7. Project Madurai. http://www.projectmadurai.org/pmworks.html. Accessed 12 March 2019
8. Tamil Virtual Academy (Erstwhile Tamil Virtual University). http://www.tamilvu.org/library/suvadi/html/index.htm. Accessed 12 March 2019
9. Tara Prakashana. http://www.taraprakashana.org/. Accessed 12 March 2019
10. Anand R, Veni S, Aravinth J (2016) An application of image processing techniques for detection of diseases on brinjal leaves using k-means clustering method. In: 2016 international conference on recent trends in information technology (ICRTIT). pp 1–6
11. Yann L (2015) LeNet-5, convolutional neural networks. 20. http://yann.lecun.com/exdb/lenet
12. Fung CC, Chamchong R (2010) A review of evaluation of optimal binarization technique for character segmentation in historical manuscripts. In: Third international conference on knowledge discovery and data mining WKDD '10. pp 236–240
13. Chamchong R, Fung CC, Wong KW (2010) Comparing binarisation techniques for the processing of ancient manuscripts. Springer, Berlin, Heidelberg, pp 55–64
14. Sabeenian RS, Paramasivam ME, Dinesh PM (2016) Appraisal of localized binarization methods on Tamil palm-leaf manuscripts. In: IEEE international conference on wireless communications, signal processing and networking (WiSPNET)
15. Otsu N (1979) A threshold selection method from gray-level histograms. IEEE Trans Syst Man Cybern 9(1):62–66
16. M Blumenstein, B Verma, H Basli (2003) A novel feature extraction technique for recognition of segmented handwritten characters. In: Proceedings of the 7th international conference on document analysis and recognition, ICDAR '03

17. Gatos B et al (2006) An efficient segmentation-free approach to assist old Greek handwritten manuscript OCR. Pattern Anal Appl 8(4):305–320
18. Thammano A, Pravesjit S (2015) Recognition of archaic Lanna handwritten manuscripts using a hybrid bio-inspired algorithm. Memet Comput 7(1):3–17
19. Lakshmi TVR, Sastry PN, Rajinikanth TV (2016) A novel 3D approach to recognize Telugu palm leaf text. Eng Sci Technol Int J, In Press
20. Panyam NS, Lakshmi TRV, Krishnan RR, Koteswara Rao NV (2016) Modeling of palm leaf character recognition system using transform based techniques. Pattern Recognit Lett 84: 29–34
21. Arandjelovic R et al (2016) NetVLAD: CNN architecture for weakly supervised place recognition. In: Proceedings of the IEEE conference on computer vision and pattern recognition
22. Han X et al (2017) Pre-trained AlexNet architecture with pyramid pooling and supervision for high spatial resolution remote sensing image scene classification. Remote Sens 9(8):848
23. Peng C et al (2017) Large kernel matters improve semantic segmentation by global convolutional network. In: 2017 IEEE conference on computer vision and pattern recognition (CVPR)
24. Hecht-Nielsen R (1992) Theory of the back-propagation neural network. In: Neural networks for perception. pp 65–93
25. Howse J (2013) OpenCV computer vision with python. Packt Publishing Ltd
26. Sastry PN et al (2015) Analysis of Telugu palm leaf characters using multi-level recognition approach. J Appl Eng Sci 10(20):9258–9264
27. Yahya SR et al (2009) Review on image enhancement methods of old manuscript with the damaged background. In: International Conference on Electrical Engineering and Informatics, ICEEI'09, vol 1. IEEE
28. Chamchong R, Fung CC (2011) Character segmentation from ancient palm leaf manuscripts in Thailand. In: Proceedings of the 2011 workshop on historical document imaging and processing. ACM
29. Sahiner B et al (1996) Classification of mass and normal breast tissue: a convolution neural network classifier with spatial domain and texture images. IEEE Trans Med Imaging 15(5):598–610
30. Shin HC et al (2016) Deep convolutional neural networks for computer-aided detection: CNN architectures, dataset characteristics and transfer learning. IEEE Trans Med Imaging 35(5):1285
31. Lin M, Chen Q, Yan S (2013) Network in network. arXiv preprint arXiv:1312.4400
32. Girshick R (2015) Fast R-CNN. In: Proceedings of the IEEE international conference on computer vision
33. Chinmaya International Foundation. http://www.chinfo.org. Accessed 12 March 2019

Forecasting of Indian Stock Market Using Time-Series Models

Sourabh Yadav and Nonita Sharma

Abstract In the present era, stock market has become the storyteller of all the financial activities of any country. Therefore, stock market has become the place of high risks, but even then it is attracting the mass because of its high return value. Stock market tells about the economy of any country and has become one of the biggest investment places for the general public. In this manuscript, we present the various forecasting approaches and linear regression algorithm to successfully predict the Bombay Stock Exchange (BSE) SENSEX value with high accuracy. Depending upon the analysis performed, it can be said successfully that linear regression in combination with different mathematical functions produces the best results. This model gives the best output with BSE SENSEX values and gross domestic product (GDP) values as it shows the least p-value as 5.382e−10 when compared with other model's p-values.

Keywords Stock market · Forecasting · Time series · Univariate analysis · Multivariate analysis · Regression · Linear regression

1 Introduction

Stock market prediction has been an active area of research as it has huge applications in financial domain. Stock market trend analysis is one of the difficult tasks because of the daily ups and downs in prices of the stock. Hence, it is important to build an accurate and precise prediction model for predicting the stock prices. Further, there are various approaches to analyze the stock prices, but the statistical approach for analyzing the prices is one of the most widely used approaches [1]. Furthermore, if time-series approach is used, it will provide the better accuracy and precise prediction

S. Yadav (✉)
Gautam Buddha University, Greater Noida, Uttar Pradesh, India
e-mail: sy9643391664@gmail.com

N. Sharma
Dr. B. R. Ambedkar National Institute of Technology, Jalandhar, Punjab, India
e-mail: nonita@nitj.ac.in

S.-L. Peng et al. (eds.), *Computing and Network Sustainability*, Lecture Notes in Networks and Systems 75, https://doi.org/10.1007/978-981-13-7150-9_43

model [2]. There are various parameters of stock market, and BSE SENSEX is one of them. Moreover, there are many other additional factors which affect the BSE SENSEX, like gross domestic product (GDP), inflation, exchange rates like the value of US dollar in Indian rupee, and many other [3].

This manuscript specifically targets for predicting the BSE SENSEX depending upon the historical values [4] and factors affecting the BSE SENSEX. Initially, univariate analysis or understanding the historical trends in the dataset is performed to provide a model for predicting the stock prices depending upon past values [5]. To increase the accuracy of the results found in univariate analysis, the next step performed is of multivariate analysis. Multivariate analysis involves determining the correlation values among the BSE SENSEX vector and all the factors affecting BSE SENSEX. Depending upon the correlation values, the correlation matrix is prepared to judge highly affecting factor. Moreover, multivariate analysis of the dataset provides a mathematical relation between highly affecting factors and prices of stock. Hence, the next target was to create a mathematical relation between BSE SENSEX values and additional factors affecting the BSE SENSEX. Further, ensemble is created to improve the accuracy and precision of the model.

2 Proposed Method

Proposed method involves creating the ensemble of the various regression techniques applied in the dataset. Ensemble, also known as data combiner, is a data mining approach that converges the strength of multiple models to achieve better accuracy in prediction.

2.1 Univariate Analysis

Step 1 for preparing the prediction model is to create a forecasting model depending upon the previous trends in the dataset or univariate analysis. For performing analysis on any dataset, univariate analysis acts as a basic step. 'Uni' means one, 'variate' here means variable, so one variable analysis is known as univariate analysis. Under this step, the first step is data collection. Data collection is the process of collection of data from all the relevant sources in a systematic fashion that enables one to answer the relevant questions and evaluate outcomes [6]. After collecting the data, data cleaning is the next step. Data cleaning refers to the process of removing invalid data points from the dataset [7]. After data cleaning, the next step is an exploratory analysis of the dataset. For exploratory analysis, data is loaded in the statistical environment for performing the different statistical functions on the dataset. Further, the dataset is converted into time series. This means that data exists over a continuous time interval with equal spacing between every two consecutive measurements. Converting the dataset into time series always proves to be an effective method for the analysis of any dataset, especially in the stock analysis [8]. The next step involves checking the time series for stationarity which can be done by performing Ljung–Box test and augmented Dickey–Fuller test.

Next step involves testing for stationarity under which ADF test is performed on the dataset. Moreover, the null hypothesis states that large *p*-value indicates non-stationarity and smaller *p*-values indicate stationarity [9]. The next step is the decomposition of the dataset which involves breaking down the dataset into parameters that are observed, trend, seasonal, random [10]. The next step is performing the model estimation. Firstly, auto-correlation function (ACF) plot and partial auto-correlation function (PACF) plot are prepared. These ACF and PACF plots tell about the correlation factor of statistical analysis and in turn help to judge covariance of the dataset. Next step is to build the model, which means deducing that which particular model applies best on our dataset depending upon our statistical results. Different models are:

Autoregressive Integrated Moving Average (ARIMA) Model: ARIMA is a forecasting technique that projects the future values of a series entirely based on its own inertia. Its main application is in the area of short-term forecasting requiring at least 40 historical data points. It offers great flexibility to work upon univariate time series [11].
BoxCox Transformation: BoxCox transformations are generally used to transform non-normally distributed data to become approximately normal.
Exponential Smoothing Forecast: This forecasting method relies on weighted averages of past observations where the most recent observations hold higher weight. This method is suitable for forecasting data with no trend or seasonal pattern.
Mean Forecast: This forecasting method relies on the mean of the historical data.
Naive Forecast: The naive forecasting method which gives an output as ARIMA (0, 1, 0) with a random walk model that is applied to time-series object.
Seasonal Naive Forecast: This forecasting method works almost on the same principles as the naive method, but works better when the data is seasonal.
Neural Network: Neural networks are forecasting methods that are based on simple mathematical models of the brain. They allow complex nonlinear relationships between the response variable and its predictors. This model is very helpful when combined with the statistical computational approach for forecasting of stock market [12].

The model which has the least error or has the higher accuracy will be the best fit model for the dataset. Moreover, error analysis suggests the improvements that can be made in the results in the future [13].

2.2 Multivariate Analysis

Step 2 involves multivariate analysis for improving the results of Step 1. Multivariate analysis is a statistical approach in which dataset is analyzed on the basis of different factors and the main objective is to prepare a combined model for better performance, analysis, and accuracy. Results from Step 1 can be improved if a relationship analysis is carried out among the dataset vector and factors affecting the dataset. For relationship analysis, the statistical approach used is regression. Regression is the statistical approach which is used to build a model in terms of mathematical equations for determining the relationship among the different factors with the main variable. In regression, one of the variables is known as a predictor variable whose value is carried out by performing different experiments, and another variable is response variable.

The linear regression approach is preferred over other regression approaches as all other regression approaches are built by understanding the working of linear regression [14]. A key requirement for linear regression is linearity among the variables. Moreover, correlation values also help to judge the dependability of any response variable upon the predictor variable. The correlation values have the range of -1 to 1. So, larger the absolute value of the correlation coefficient, more the dependability of variables upon each other and more is the linearity among them. After determining the correlation value, the most influencing factor will be extracted. Furthermore, model fitting is done by applying different mathematical functions like logarithmic function or exponential function, on both response variable and predictor variable, for making model estimation simple and easier. Moreover, instead of passing a single factor, i.e., most influencing factor, one can pass all the factors at the same time as an argument to the regression algorithm. Then, whichever model performs better will be the best fit model. For determining the accuracy, steps will be the same, i.e., summarization of regression model.

2.3 *Ensemble Technique*

Step 3 involves the building of ensemble for the dataset which involves combining the result from Step 2, i.e., after multivariate analysis, and hence creating an equation which provides the best accuracy. The Ensemble can be built by taking the numerical total of all the factors affecting BSE SENSEX, i.e., GDP, inflation, USD value. In this firstly, the absolute value of the correlation coefficients needs to be more than 0.5 for effective linearity. The correlation coefficient suggests that a linear regression model can be constructed as an experiment to improve the accuracy. For applying regression, open vector is used as the response variable and the total vector as a predictor variable. As a result, there will be a linear equation between the open vector and the total vector which is represented by Eq. 1.

$$\text{Open} = 714.2 * \text{Total} - 27557.8 \quad (1)$$

Table 1 Model estimation for open vector

	ME	RMSE	MAE	MASE	ACF1
ARIMA	−2.808	59.618	48.379	0.7	0.05
BoxCox	0.834	59.938	48.878	0.707	0.06
ETS	0.068	59.931	48.867	0.707	0.06
Meanf	0	59.928	48.865	0.707	0.06
Naïve	−0.18	82.312	64.539	0.934	−0.457
Snaive	0.468	86.49	69.132	1	0.13
Neural network	**−0.022**	16.769	12.939	0.187	−0.034

Table 2 Correlation coefficients

Vector	Correlation coefficients
Inflation	0.4155946
GDP	0.9675431
Exchange rates	0.6650287

3 Results and Discussion

The tool which is used for forecasting is *R*. Various packages related to the various functionalities described in Sect. 3 are included as: forecast package and tseries package. Datasets used in our analysis are BSE SENSEX collected from official Web site of BSE India [15], GDP of India collected from official Web site of World Bank [16], USD prices in rupee collected from official Web site of Reserve Bank of India [17], and inflation collected from official Web site of European Union [18]. BSE SENSEX dataset contains variable that is open. The open variable represents the opening price of the stock market. Results after applying the procedure mentioned are detailed below.

In Step 1, open vector is analyzed by applying different model functions depending upon which error matrix is prepared.

As seen from Table 1, all the models have performed differently. Considering every model, best results are given by neural network model as its mean error value is comparatively low.

Further in Step 2, multivariate analysis is performed to improve the accuracy. For performing multivariate analysis, linearity among the BSE SENSEX value and GDP values, inflation, and exchange rates is determined, respectively. So for that purpose, correlation coefficients are determined among the different vectors with the open vector of the dataset. Table 2 depicts the different correlation values of different vectors with the open vector.

After interpreting Table 2, it is clearly observable that GDP vector is highly correlated with BSE SENSEX open feature.

Table 3 Summary of linear regression model object with log (GDP) vector

Parameters	Values
p-value for intercept	<2e−16
p-value for logarithmic GDP coefficient	5.38e−10
Net p-value	5.382e−10
Multiple R-squared values	0.9527
Adjusted R-squared value	0.6650287

Table 4 Summary of a combined linear regression model with improvements

Parameters	Values
p-value for intercept	2.61e−08
p-value for exp (exp(inflation)) coefficient	0.399
p-value for log (GDP) coefficient	4.11e−06
p-value for USD value coefficient	0.472
Net p-value	5.213e−08
Multiple R-squared values	0.9606
Adjusted R-squared value	0.9499

So, next step is to build a linear regression model between open vector and GDP vector. The linear equation obtained by decreasing the residuals value is given in Eq. 2.

$$\log(\text{Open}) = 1.35712\log(\text{GDP}) + 9.15080 \tag{2}$$

To check the accuracy of the equation, the regression object can be summarized as given in Table 3.

Then for better comparison among the models, next step is to build a combined model, i.e., using all the factors that affect the BSE SENSEX, i.e., inflation, exchange rates, GDP value, as a predictor variable and response variable will remain the same, i.e., open vector. Further to increase the accuracy, logarithm of open and GDP, exponential of reciprocal of inflation can be used for determining the equation. As a result, the linear equation between open as a response variable and GDP, inflation, USD value as a predictor variable is constructed, which is quoted in Eq. 3.

$$\log(\text{Open}) = 1.318037\log(\text{GDP}) - 0.208758\mathrm{e}^{\mathrm{e}^{\frac{1}{(\text{Inflation})}}} - 0.006184\ \text{USD Value} \tag{3}$$

For checking the accuracy, the regression model object is summarized, and Table 4 depicts the different p-values and R-squared values.

Next step is to build an ensemble. The ensemble can be built by taking the numerical total of all the factors affecting BSE SENSEX, i.e., GDP, inflation, USD value. In this firstly, the absolute value of the correlation coefficients needs to be more than 0.5

Table 5 Summary of ensemble

Parameters	Values
p-value for intercept	0.011661
p-value for total coefficient	0.000688
Net p-value	0.0006876
Multiple R-squared values	0.6008
Adjusted R-squared value	0.5701

Table 6 Net p-values for all the above regression models

Parameters	Values
Net p-value for open – GDP model	3.828e−09
Net p-value for log (open) – log (GDP) model	5.382e−10
Net p-value for open – GDP + inflation + USD value model	5.966e−07
Net p-value for log (open) – log (GDP) + exp (exp(1/inflation)) + USD value model	5.213e−08
Net p-value for ensemble	0.0006876

for effective linearity. Correlation coefficients when calculated between open vector and total vector, it comes out to be 0.7751132. The correlation coefficient suggests that a linear regression model can be constructed as an experiment to improve the accuracy. For applying regression, open vector is used as the response variable and the total vector as a predictor variable. As a result, there will be a linear equation between the open vector and the total vector which is represented by Eq. 4.

$$\text{Open} = 714.2 * \text{Total} - 27557.8 \tag{4}$$

For checking the accuracy of the result, summarization of regression object is required and Table 5 shows the summary of the model.

The next and final step is to analyze the results of all the above linear regression models and find the best and suitable model for forecasting which can closely predict the value of BSE SENSEX. Table 6 shows all the net p-values of all the above regression models, through which we can easily compare the results.

It is clearly observable that open-GDP model with a mathematical function gives the least p-value when compared with all other models.

4 Conclusion

In this manuscript, there is a research performed on the dataset of BSE SENSEX from January 1997 to January 2016, and the dataset of GDP of India (in trillions), inflation (in %), USD values (in rupees), which is interpreted annually from the year 2001 to

2015. On applying different forecasting models in the beginning, then applying linear regression techniques, it has been found that each and every model results differently and can be analyzed on the basis of mean error and the net p-values of the models. After analyzing the different mean errors of forecasting model in the beginning, it has been concluded that exponential smoothing and neural network give the consistently less mean error, with the small exception in the low vector where the mean error of neural network is comparatively high. Moreover, when linear regression algorithm is applied in the datasets for the improvements, it has been concluded that linear regression model of logarithmic open values of BSE SENSEX and logarithmic GDP values of India gives the best accuracy and precision, from all other quoted models.

References

1. Armstrong JS (2001) Combining forecasts. In: Principles of forecasting. Springer, Boston, MA, pp 417–439
2. Frick RW (1996) The appropriate use of null hypothesis testing. Psychol Method 1(4):379
3. Cleveland WP, Tiao GC (1976) Decomposition of seasonal time series: a model for the census X-11 program. J Am Stat Assoc 71(355):581–587
4. Sharma N, Juneja A (2017) Combining of random forest estimates using LSboost for stock market index prediction. In: 2nd international conference for convergence in technology (I2CT). IEEE
5. Mondal P, Shit L, Goswami S (2014) Study of effectiveness of time series modeling (ARIMA) in forecasting stock prices. Int J Comput Sci Eng Appl 4(2):13
6. Rao A, et al (2015) Survey: stock market prediction using statistical computational methodologies and artificial neural networks
7. Cole R (1969) Data errors and forecasting accuracy. In: Economic forecasts and expectations: analysis of forecasting behavior and performance. NBER, pp 47–82
8. Litterman RB (1986) A statistical approach to economic forecasting. J Bus Econ Stat 4(1):1–4
9. Alam P (2016) Factors affecting stock market in India. Spl Int J Prof 3(9):7
10. Montgomery DC, Peck EA, Vining GG (2012) Introduction to linear regression analysis vol. 821. Wiley
11. Sapankevych NI, Sankar R (2009) Time series prediction using support vector machines: a survey. IEEE Comput Intell Mag 4(2)
12. Devers KJ, Frankel RM (2000) Study design in qualitative research–2: sampling and data collection strategies. Educ Health 13(2):263
13. Rahm E, Do HH (2000) Data cleaning: problems and current approaches. IEEE Data Eng Bull 23(4):3–13
14. Angadi MC, Kulkarni AP (2015) Time series data analysis for stock market prediction using data mining techniques with R. Int J Adv Res Comput Sci 6(6)
15. BSE Homepage, http://www.bseindia.com. Last accessed 2018/07/05
16. The Worldwide Inflation Data Homepage, http://www.inflation.eu. Last accessed 2018/07/10
17. The World Bank Homepage, http://www.worldbank.org. Last accessed 2018/07/10
18. The Reserve Bank of India Homepage, https://www.rbi.org.in. Last accessed 2018/07/10

Automatic Generation of Description for Images Using Recurrent Neural Network

G. S. Veena, Savitri Patil and T. N. R. Kumar

Abstract In this paper, a generative long short-term memory (LSTM) model for generating description of the image is implemented. Automatic generation of description that describes the content of a given image is a fundamental problem in artificial intelligence. This kind of work is achieved by connecting two different domains like computer vision and natural language processing. The solution proposed here makes use of deep learning. A deep learning framework known as Keras is used which uses TensorFlow for the backend process. TensorFlow is a framework used to do a series of operations in a chain. The general technique is to feed the features of an image to the model, which is capable of generating text of length less than or equal to a predefined caption length. The dataset Flickr30 K is used to train the model. The InceptionV3 is used to extract features of the images. BLEU metric is used to measure the accuracy of the description that is generated for that image using LSTM model.

Keywords Artificial intelligence · Consciousness · Deep learning · Intelligent agent · Long short-term memory · InceptionV3 · Flickr30k · Bilingual evaluation understudy

1 Introduction

Cognition is defined as the mental action or process of acquiring knowledge and understanding through thought, experience, and the senses. It encompasses tasks such as perception, problem solving, decision making, reasoning, computation, attention, memory, and production of language [1]. Cognitive processes involve the use of

G. S. Veena · S. Patil (✉) · T. N. R. Kumar
Ramaiah Institute of Technology, Bengaluru, India
e-mail: savitribp41@gmail.com

G. S. Veena
e-mail: veenags@msrit.edu

T. N. R. Kumar
e-mail: tnrkumar@msrit.edu

S.-L. Peng et al. (eds.), *Computing and Network Sustainability*, Lecture Notes in Networks and Systems 75, https://doi.org/10.1007/978-981-13-7150-9_44

existing knowledge to generate new knowledge. Cognitive science is the interdisciplinary study of the mind and its processes [2]. Artificial intelligence (AI) involves the study of cognitive phenomenon in machines. The primary goal of AI is to implement the processes of human thinking into machines. Describing and understanding the contents of images are a simple and easiest task for humans, and it can be accomplished with maximum accuracy. Humans can perceive, recognize, depict, and distinguish many objects present in the images. The task of automatic generation of description can be applied to robotic vision research [1]. Computer vision, machine translation, and object detection are all areas of machine learning that have been growing rapidly in the past decade [1]. As such, there now exist several different frameworks that provide for caption generation implementations. In this project, we provide an image captioning implementation using one such framework—Keras [3]. To develop the image captioning model, there were three parts to be considered: extracting image features for use in the model, training the model on those features, and finally using the trained model to generate caption text when given an input image's features. How these three parts were eventually accomplished was through the use of two different techniques: a InceptionV3 for the feature extraction and a recurrent neural network (RNN) [4] to train and generate caption text. After this was accomplished, we then evaluated the model using the same scoring standard—a bilingual evaluation understudy (BLEU) score—as our baseline for comparison [5]. In the end, there was a significant increase, over the baseline, in BLEU score accuracy, with the added advantage that this model can also truly generate image captions.

2 Related Work

Karpathy [2] done a project titled 'Deep Visual-Semantic Alignments for Generating Image Descriptions.' This project is involving the model which depends on a combination of different convolutional neural networks for the identification of areas in the image and bidirectional recurrent neural networks for generation of sentences. They at that point depicted a multimodal recurrent neural network model that makes use of inferred alignments to figure out how to produce descriptions for images by considering all objects and relationship between them. They experimented this model with datasets like Flickr8K, Flickr30K, and MSCOCO.

Yuille et al. [6] had proposed an end-to-end system for the purpose of image captioning. The system consists of three parts, namely a language model part, image part, and the multimodal part. The language model learns at the time of training. The training is done by providing the dense feature embedding for each and every word in the dictionary as input to recurrent neural network (RNN) units. The image part is used to obtain the low-level features of the image given as input. The multimodal part connects the image part and the language part. The output of the RNN is combined with the image features obtained from the image part and is fed to the multimodal part. The output of the multimodal part is the next word in the caption. Hence, the previous word and the image feature are used to generate the next word in the caption.

3 Tools

To build the model, we leveraged and learned several different modern machine learning tools/frameworks such as Keras and Google's TensorFlow

3.1 Keras

Keras Python library that provides an appropriate way to develop range of deep learning models using Theano or TensorFlow as backend. It is one of the mostly used deep learning framework. Keras allows for simple and fast prototyping. Keras is having so many useful characteristics such as extensibility, modularity, and friendliness. Keras is supporting both convolutional neural network (CNN) and recurrent neural network (RNN); it also supports combination of both CNN and RNN. It runs effortlessly CPU and GPU. Keras is utilized in python. It can utilize Theano or tensor as low backend. As defined by developer of Keras, Keras is a high-level neural network API implemented using Python.

3.2 Tensor Flow

TensorFlow is easily available open-source library. It is specially applied for numerical calculation using data flowgraphs. Generally, mathematical operations are represented by nodes in the graph. The communication between the multidimensional arrays called tensors is represented by edges. We can install computation to one more CPUs or GPUs in server, mobile devices, and desktop with single API. TensorFlow is developed by researchers working on the Google. Actually, TensorFlow is developed with the intention of carrying the machine learning and deep learning research. It is sufficient to be applicable in different domains as well. Actually, it has been designed with deep learning in mind, but it can be applied to varieties of problems.

4 Methodology

In our implementation, we use the Flickr30K image dataset to train our model. The model we trained to do our predictions is called a long short-term memory (LSTM) neural network [5, 6]. We use a combination of the features extracted from the image along with the sentences that are captions for our data—all provided in the Flickr30k dataset. In prediction of an image, the most important parts are then InceptionV3 feature extraction and caption prediction using the trained LSTM.

4.1 Dataset: Flickr30k

Flickr30k dataset comprises of 31,930 images concentrating mostly on people and animals and 158,919 English captions. There are five captions for each image. We used Flickr30k dataset to train our model to generate automatic description for images. This dataset is the standard dataset for text generation for images. The dataset Flickr30k is extended version of Flickr 8k dataset new images and captions emphasis on people involved in day-to-day events.

4.2 Feature Extraction Using InceptionV3

The first step is extracting the features from images. We did so using a pre-trained InceptionV3. Extraction of features is done by using a pre-trained InceptionV3 neural network model. The pre-trained InceptionV3 neural network model takes image as input; once the image has been read, image is resized to 224 * 224 and then passed into InceptionV3 neural network. After that, features are extracted and stored as numpy array. We choose InceptionV3 model for feature extraction based on accuracy of classification of object. After carried out the practical work on pre-trained model for image feature extraction, we got results as shown in given Table 2. Based on results of our work only, we choose InceptionV3 to extract features of image. To run it through Inception-v3, when giving image as input to the pre-trained model Inception-v3, we did pre-processing of image. The pre-processing is, For CNN, we have subtracted the mean estimation of each image from each pixel of that image. At that point, we have subtracted the mean estimation of all images from each pixel of each image and divided the value with its variance. The comparison of top three feature extraction model has been done. Among InceptionV3, VGG16, and ResNet50, the accuracy of InceptionV3 is having good accuracy in classification of object. So InceptionV3 has been used for feature extraction (Table 1).

The Inception v3 model is one of the pre-trained models for ImageNet Large Visual Recognition Challenge. Inception v3 can able differentiate around 1000 classes such as cat, dog, and plane. The inception v3 is giving high performance better than VGGNet. This model is having the less computational cost. Inception v3 is having the layers for each feature extraction of image. For examples, network has been

Table 1 Comparison of pre-trained feature extraction models

Pre-trained feature extraction model	Number of parameters	Accuracy of object classification (soccer ball)
IneptionV3	23,587,712	99.97
VGG16	134,260,544	99.10
ResNet50	21,802,784	91.98

trained to identify faces, the first layer detects edges, the second layer detects whole design, the third layer detects eyes, and the fourth layer detects mouth, and so on. The main purpose of the inception model behaves as 'multi-level feature extractor.' This is can be developed by computing 1 × 1, 3 × 3, and 5 × 5 convolutions in the same model of the network. The output all these three filters then loaded along with the channel dimension and beforehand being served into the next layer in the network. The inception V3 is having less number of weights compared to weights of VGGNet and ResNet. Inception V3's weights are available in 96 MB.

4.3 Building LSTM Model

Long short-term memory (LSTM) networks are a kind of recurrent neural network suitable for learning order dependency in sequence prediction problems. This is a performance required in complex problem fields like machine translation, speech recognition, and so on. Long short-term memory (LSTM) systems are an addition for recurrent neural networks systems, which fundamentally increase their memory. In this way, it is appropriate to learn from experiences that have long time lags in between. The units of a LSTM are utilized as building units for the layers of a RNN, which is then regularly called a LSTM network. LSTMs empowers RNNs to recall their inputs over a long period of time. This is on the grounds that LSTMs contain their data in a memory that is much similar to the memory of a computer in light of the fact that the LSTM can read, compose, and erase data from its memory.

The Structure of LSTM cell.

The structure of long short-term memory cell is shown in Fig. 1.

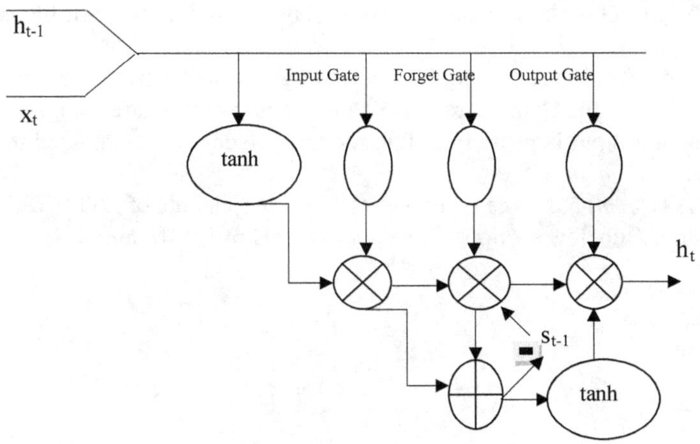

Fig. 1 LSTM cell structure

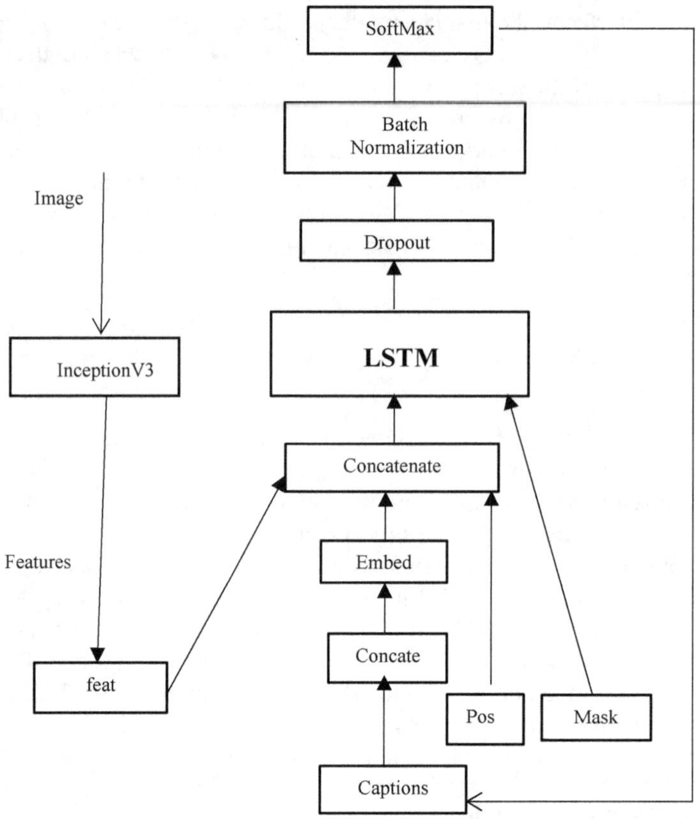

Fig. 2 CNN-LSTM model

The LSTM network is recurrent neural network having the cell blocks instead of neural network layers. Figure 2 shows the structure of LSTM cell. As shown in Fig. 2, the information is flowing from left to right, where x_t is the current input and h_{t-1} is previous output. The current input and previous output are combined together. This combined input is passed through the tanh layer and then passed through the input gate.

Each LSTM cell has three components. The components of LSTM cell are listed and explained in below section. Three components of LSTM are:

i. Input gate
ii. Forget gate
iii. Output gate.

i. **Input gate**

First, the input is squashed between -1 and 1 using a tan$h$ activation function. This can be represented using equation:

$$g = \tan h\left(b_g + x_t U_g + h_t - 1V_g\right). \tag{1}$$

where

U_g and V_g are the weights for the input, and b_g is the input
V_g is the weights previous cell output
b_g is the input bias.

Each element of output of the input gate is multiplied with each element of squashed input. The input gate is nothing but one of the hidden layer of sigmoid activation node. These sigmoid activation nodes have weights x_t and h_{t-1} values. T he output of sigmoid activation node a value in the range of 0–1. When these output values are multiplied element-wise by the input which takes the decision what all inputs have to switched on and off. It's nothing but s type of input filter. The representation of input gate in terms of expression is

$$i = \sigma\left(b_f + x_t U_i h_t - 1V_i\right) \tag{2}$$

The expression for the output of the input gate of long short-term memory is given below:

$$g \circ i \tag{3}$$

where
$\circ$ represents the operator used for element-wise multiplication, '*i*' is the weights, and '*g*' is the squashed input.

ii. **Forget gate**

This phase in the LSTM is the place the vast majority of the enchantment happens. As can be seen, there is another variable s_t which is the internal condition of the LSTM cell. This state is postponed by one-time step and is ultimately added to the g *o* i input to give an inside recurrence loop to take in the relationship between inputs isolated by time. Two things to see—to start with, there is an forget gate here; this forget gate is again a sigmoid activated set of nodes which is element-wise multiplied by $s_t - 1$ to figure out which past states should be remembered that is output of the forget gate close to 1 and which should be forgotten that is output of forget gate close to 0. This enables the LSTM cell to learn appropriate context. Consider the sentence 'Sinam offers the full-time job to Vitul, he was very gratefull'—for the LSTM cell to realize who 'he' refers to, it needs to forget the subject 'Sinam' and supplant it with the subject 'Vitul.' The forget gate can facilitate such operations and is expressed as:

$$f = \sigma(b_f + x_t U_f + h_t - 1V_f) \tag{4}$$

The output of the element-wise product of the previous state and the forget gate is expressed as

$$s_{t-1} \circ f \tag{5}$$

where
s_{t-1} is the previous state, f is the output of the forget gate

Once more, the forget gate goes about as weights for the internal state. The second is to see about this stage is that the forget gate 'filtered' state is just added to the information, as rather than multiplied by it or mixed with it by means of weights and a sigmoid activation function as happens in a recurrent neural network. This is critical to lessen the issue of vanishing gradients. The yield from this stage, s_t, is expressed by

$$s_t = s_{t-1} \circ g \circ i \tag{6}$$

The last phase of the LSTM cell is the output gate. The output gate has two segments—another tanh squashing function and an output sigmoid gating function. The output sigmoid gating function, similar to the next gating functions in the cell, is multiplied by the squashed state s_t to figure out which estimations of the state are output from the cell. As should be obvious, the LSTM cell is exceptionally adaptable. LSTM cell containing three gating functions to control the input, internal state, and output. These gating function have control over what should be input, what should be remembered in internal state, and what should be output od LSTM cell. The output gate is expressed as:

$$O = \sigma(b_o + x_t U_o + h_t - 1 V_o) \tag{7}$$

So the final output produced by LSTM cell can be expressed as:

$$\text{ht} = \tan h(s_t) \circ o \tag{8}$$

4.4 Training the Predictive Model

The first step was to preprocess our input captions and build a dictionary. Then, we present all the words in our caption as index list-mapping words into an index. For captioning, using Keras, we create a single LSTM cell with 256 neurons. For this cell, we have four inputs: image features, captions, a mask, and a current position. For first, the caption input and position input are merged (using concatenate), and then, it goes through a word embedding layer. Then, the image features and embedded words are also merged (using concatenate) with the mask input. Together, they all go through the LSTM cell. The output of LSTM cell then goes through dropout and batch normalization layer to avoid overfitting of the model. Finally, the activation (softmax) layer is applied, and we get the result. The result itself is a vector with each entry representing the possibility of every word in the dictionary. The word with the largest probability would be our current 'best word.' Along with pre-built dictionary, this vector is used to 'interpret' the generated next word—which can be considered a type of ground truth for training in the true caption.

The mask plays an important role in all of this, 'recording' the previous words used in captions, so that the model knows the words before the current word and inputs the model with current position of the sentence so that it will not fall into a loop. To be more specific, loop refers to caption like 'man in white shirt in a garden' may look like 'man in white shirt in white shirt in white shirt in…'. Because, after the model generates the second 'in', the word input, feature input, and the mask input for the model would all be exactly the same. Then, the model would generate 'in white shirt' again which will create a loop and generate a caption that does not fully represent the image. Therefore, it is necessary for us to input the current position of the sentence.

4.5 Generation of Description for Image

Like training, we also need to get the features for each image to be predicted. So, the images go through the VGG16 network first, to generate the features. For captioning, we used the same LSTM model. The first word input for this model is the '#start#' tag, and the following input is the prediction result from the previous iteration. We also need to set the mask and input current position of the sentence as we did in training process. When we reach a prediction which produces '.', we have our final prediction! Figure 3 gives an example of generated caption—'a dog is running in the yard', for image. So, the first word is '#start#', the position is 0, and mask input is empty. The model would generate 'a' which is used as input for the next iteration. Then, the word 'a' is marked by the mask input, and the position input moves forward. By doing this until our model predict '.', we stop the model and get the final caption (Fig. 4).

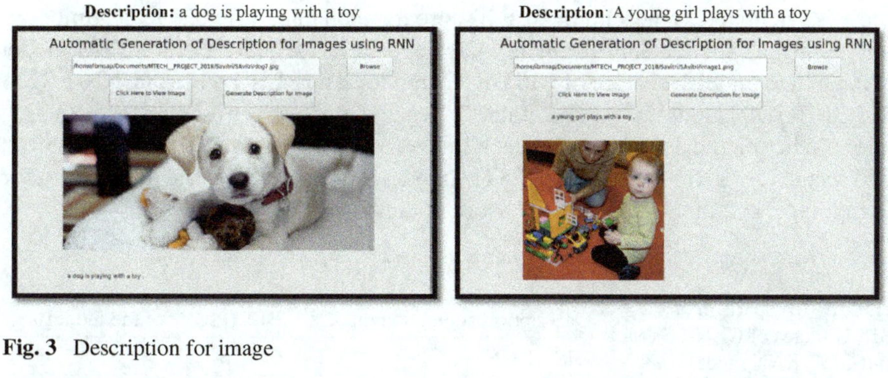

Fig. 3 Description for image

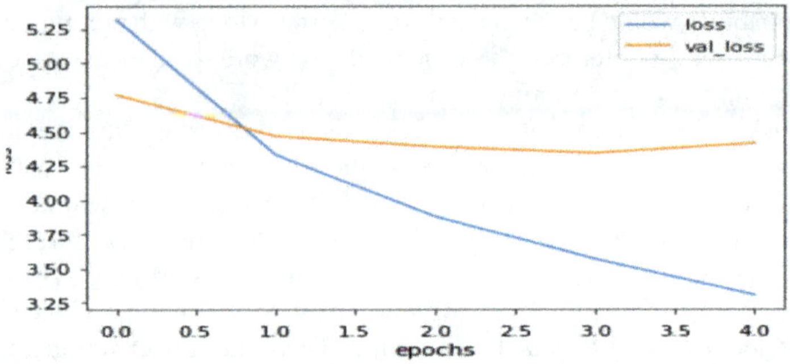

Fig. 4 Graph representing the number of epochs versus loss

5 Evaluation

Bilingual evaluation understudy (BLEU) [7] is an algorithm that was originally designed to evaluate the quality of machine-translated natural language text, doing so using n-grams to compare and assign sentences an appropriate score. The bilingual evaluation understudy score, or BLEU for short, is a metric for measuring a generated sentence to a reference sentence. Correct coordinate outcomes in a score of 1.0, though a is there is mismatch results in a score of 0.0. The score was produced for assessing the predictions made through programmed machine translations frameworks.

The approach works by including coordinating n-grams the candidate translation to n-grams in the reference content, where 1-gram or unigram would be every token and a bigram examination would be each word pair. The comparison is made with respect to word order. The checking of matching n-grams is altered to guarantee that it considers the occurrence of the words in the reference content, not rewarding a candidate translation that produces a many numbers of sensible words [8]. BLEU can be utilized as a part of, generation of description for image, speech recognition. The CNN-LSTM model achieved average BLEU score of 74%. The average BLEU score is calculated as the sum of BLEU score for each image divided by a number of images given. There is improvement in average BLEU score of CNN-LSTM model compared to the BLEU score of GIST-KNN model. The BLEU score of CNN-LSTM and GIST-KNN model is shown in Table 2. Ten images are given as input to the model, after predicting the description for each image, and the average BLEU score is 50%; 250 images are given to the CNN-LSTM model, after predicting the description for each image, and the average BLEU score obtained is 74%.

Table 2 Comparison of BLEU scores of CNN-LSTM and GIST-KNN models

Description generation models	BLEU score of the model
GIST-KNN	0.5076
CNN-LSTM	0.7445

6 Conclusion and Future Scope

The project briefs on the development of a model to generate natural language description for images automatically. A deep learning neural network model is preferred to generate description of the images with accurate result. An extensive hyper-parameter search over CNN-LSTM model architecture is conducted. The CNN, i.e., VGG16, is used to extract features of images. LSTM is used to generate descriptions for images. This model is trained especially on a dataset Flick30 k. The BLEU score is calculated using Flickr8k dataset as reference. Average BLEU score of 74% is achieved. GIST-KNN model achieved average BLEU score of 51% [8], but using CNN-LSTM model, there is improvement in the average BLEU score, i.e., 74%.

To achieve better accuracy, one could develop attention models to generate description for image. LSTM can replace by gate recurrent unit (GRU). To make our prediction process better, beam search would be a promising choice. Beam search would consider the set of n best sentences and keep track of resulting n of the candidates, which would help generate overall most probable caption.

References

1. Kinghorn P, Zhang L, Shao L (2017) Deep learning based image description generation. IEEE
2. Karpathy A, Fei-Fei L (2017) Deep visual-semantic alignments for generating image descriptions. Department of Computer Science, Stanford University
3. Liang M, Hu X (2015) Recurrent CNNs for object recognition. In: CVPR, 2015; R. Girshick, Fast R-CNN. In IEEE International Conference on Computer Vision (ICCV)
4. Vinyals O et al (2015) Show and tell: a neural image caption generator. In: Proceedings of the IEEE conference on computer vision and pattern recognition
5. Donahue J, Hendricks LA, Guadarrama S, Rohrbach M, Venugopalan S, Saenko K, Darrell T (2015). Long-term recurrent convolutional networks for visual recognition and description. In: CVPR
6. Mao J, Xu W, Yang Y, Wang J, Yuille A (2015) Explain images with multimodal recurrent neural networks (m-RNN). In: ICLR
7. Papineni K et al (2002) BLEU: a method for automatic evaluation of machine translation. In: Proceedings of the 40th annual meeting on association for computational linguistics. Association for Computational Linguistics
8. Sutskever I, Vinyals O, Le QV (2015) Sequence to sequence learning with neural networks. In: Proceedings advances in neural information processing systems, vol 27. pp 3104–3112
9. Simonyan K, Zisserman A (2015) Very deep convolutional networks for largescale image recognition. In: ICLR
10. Arjovsky M, Bottou L (2017) Towards principled methods for training generative adversarial networks. In: Proceedings of the international conference on learning representations, ICLR
11. Farhadi A, Hejrati M, Sadeghi MA, Young P, Rashtchian C, Hockenmaier J, Forsyth D (2010) Every picture tells a story: generating sentences from images. In: ECCV
12. Ren S, He K, Girshick R, Sun J (2015) Faster R-CNN: towards real-time object detection with region proposal networks. In: NIPS

Intrusion Detection in Wireless Network Using Fuzzy Logic Implemented with Genetic Algorithm

S. Sai Satyanarayana Reddy, Priyadarshini Chatterjee and Ch Mamatha

Abstract This paper suggests how fuzzy logic can be applied to route traffic between two nodes in a wireless network. The main focus of this paper is also to use fuzzy logic, implemented with genetic algorithm, for intrusion detection in the network. As wireless network has unparalleled mobility, it is important for the nodes to re-arrange themselves frequently. Not only this, there are multiple nodes between source and destination that helps to route packets to the destination. So every node acts as a router. There may be delay and packet loss during the transit. Fuzzy logic helps to reduce these drawbacks by clearly configuring the route. Apart from this, fuzzy logic can be implemented using GA for intrusion detection in the network. Intrusion detection broadly deals with how to be aware of network-based and Internet attacks. Firewalls could not properly handle nefarious attacks. This has proved to be very robust and effective with 80% success rate. This paper also suggests how fuzzy logic can be used to address some of the cybersecurity challenges.

Keywords Wireless network · Fuzzy logic · Genetic algorithm · Intrusion detection

1 Introduction

A wireless network comprises of independent devices that can communicate with each other. As these devices have unlimited mobility, there is always a necessity that they should keep on reconfiguring frequently. This is to be done in order to

S. Sai Satyanarayana Reddy (✉) · P. Chatterjee
Department of Computer Science and Engineering, Vardhaman College of Engineering, Hyderabad, India
e-mail: saisn90@gmail.com

P. Chatterjee
e-mail: jinipriya@gmail.com

C. Mamatha
Department of Information Technology, Vardhaman College of Engineering, Hyderabad, India
e-mail: mittumamatha03@gmail.com

S.-L. Peng et al. (eds.), *Computing and Network Sustainability*, Lecture Notes in Networks and Systems 75, https://doi.org/10.1007/978-981-13-7150-9_45

maintain the route between nodes. There exists an intermediate node that helps in retransmitting the packet from the source to the destination which also calls for reworking on the existing route [1]. Therefore, we can say that every intermediate node in the wireless network will provide an immense functionality of a router. These independent routers can move the packet to any direction with any speed based on the requirement. This nature of the router demands for a continuous change in the links [2].

Apart from civil functions like defense communication, the wireless communication can be extended to entertainment also [3]. These civil functionalities demand the wireless network which can be used for external use, especially with respect to Internet. The connection between the two networks, i.e., the wireless intranet and the Internet, requires a gateway [4]. Our paper in the initial part describes how to identify the gateway. The gateways are identified based on their nature of emission of messages called connection request messages. The proposed algorithm also suggests how to control the emission rate of the CR messages by the gateway. Our paper is about implementing a fuzzy system that is rule-based. As the gateways connect to the internet, they are prone to cyberthreats. The second part of our proposed system helps to identify intrusion detection with the help of a fuzzy system.

2 Related Work

There are other hybrid networks that consist of two parameters, i.e., the circumference of the area in which modified router adjustment messages are emitted and the time gap between two emissions [5]. There are schemes where the gateway sends messages with the time-to-live parameter set to minimum hops. These messages reach all sources from the gateway that are required to communicate with the Internet. It requires low overhead and is scalable. It is called as low overhead and scalable proxied algorithm [6].

Our paper is implemented by using the concepts of fuzzy logic to deal with uncertainties. In fact, fuzzy system is the best way to cope up with the uncertainties. This is the reason fuzzy logic is widely used in telecommunication and networking applications. Fuzzy logic can be also applied to select more than one path between the source and the destination [7]. Fuzzy logic chooses the paths based on the quality of service measure. Fuzzy logic can be also applied to systems that have minimum sensor devices. In this paper, fuzzy logic is used to optimize the emission of CR messages. Fuzzy logic also controls the time of emission of two CR messages [8].

Another broad and one of the most important areas of application of fuzzy logic is in intrusion detection [9]. Intrusion detection is a major area in networking and communication. Intrusion detection can be broadly classified as network-based intrusion detection and host-based network detection [10]. Host-based intrusion detection is used to analyze information on a single host, more than one host and also including the standalone operating system [11]. On the other hand, network-based intrusion detection is used to comprehend network attacks by analyzing the packets that flow across the network [12].

Speaking about the functional components of an intrusion detection system, there are broadly components [13]

- Source (generator)
- Evaluation engine
- Communication manager.

The source or the generator is the event generator. The second component of the intrusion detection system takes the information from the source and analyzes that information. Based on the systems, it can say whether it is an attack, and if it is an attack, then what kind of attack it is [14]. The communication manager only acts if there is any anomaly. It communicates with the part of the system that can handle the anomaly [15].

The proposed system that this paper deals with is based on fuzzy system implemented using GA. Genetic algorithm technique uses biological evaluation as the main concept [16]. The proposed GA-based fuzzy evaluation system works on two parts. Each part works on two different stages. As part of the training, few classification rules are established from the network accounting data [17]. These rules are applied in the intrusion detection stage in order to classify the network traffic in a real-time frame. Based on these rules, the intrusion detection system becomes efficient and experienced [18].

Genetic algorithm uses some functions such as [19]

- Mutation
- Crossover
- Selection.

Each of these functions works on a randomly generated population of genes and after repeating use of the above function reaches a candidate solution [20].

3 Proposed System

We are proposing a wireless network that requires a special node to connect between the network and the Internet.

This node has to execute the following activities:

1. Gateways are to be identified. Gateways are used to send specific messages. Other than from sending particular messages, gateways also route the packets to the destination (wireless network node) in hops.
2. Gateways also provide a viable IP address.
3. They also detect duplicate IP addresses.

This paper proposes a system for the identification of a gateway. There are two messages we need to consider, i.e., changed router and changed router request. The

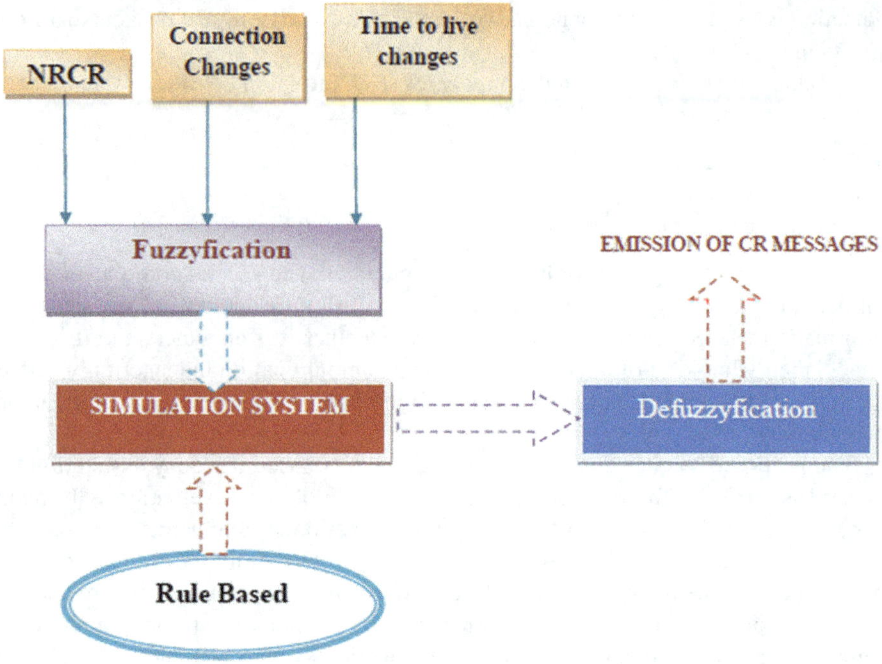

Fig. 1 Flow diagram of the proposed system (first part)

fuzzy system that we will be using informs about the wireless network condition. The fuzzy system is based on the following parameters:

- The number of received changed router is defined as: number of CR messages/number of message sources.

  This is the ratio between a number of live sources and the number of CR messages.

- The number of connection changes (how active is the Internet gateway) = number of connection changes/number of message sources.
- Time-to-live changes (represents frequency of changes from the source to the gateways) = number of time-to-live changes/number of message sources.

In the fuzzy system, these three parameters are updated and provided as input. Apart from this, these three parameters are fuzzified. The three input variables that are used are low, high, and moderate. We are using a triangle-based function to increase the computational efficiency. The output of the fuzzy system is used as an input efficiency to calculate the number of broadcasted CR messages. Once the output is computed, the gateway compares the calculated output with a benchmark. If the calculated output increases the benchmark, then the CR message is considered essential and a new one is allowed to be emitted. The proposed fuzzy system is used to control the broadcast of CR messages (Fig. 1).

In the simulation phase of the algorithm, 25 groups of chromosomes are made from the data set [21]. The number of chromosomes varies in each group depending on the relationship of data in the group. In the detection phase, an initial population is computed for each data set. This population is then compared with the population of the training data set. Some of the part of the test data population that closely represents training data population is removed. Mutation and crossover are implemented on the leftover population. This is used as the newly generated population. This process is repeated till generation size becomes 1.

Second part of the paper deals with making use of the fuzzy system to deal with cybersecurity in the wireless network. The proposed intrusion detection system contains two parts that work in different stages. The first stage, i.e., the training stage, consists of some classification rules that are generated from network. In the second stage, i.e., the intrusion detection stage, few rules are used to classify the incoming traffic of the network connection during real-time execution environment. The chromosome is randomly generated. This population of chromosomes represents all possible solutions of problems. Among the chromosomes, various positions are encoded as characters, numbers, or bits. In a broad sense, these encodings can also be also referred to as genes. We are using a fitness function that calculates the working of each chromosome. Crossover mechanism is used to calculate original reproduction, and "mutation" mechanism is used for the mutation of the species (Fig. 2).

We are using a confusion matrix to simulate the fitness of the chromosomes. The fuzzy negative operator and the fuzzy truth logic are represented by the chromosome. The fitness of a chromosome is calculated using few equations. The equation gives the value of the following parameters: true negative, false positive, true positive, and false negative.

4 Result Analysis

For simulation of the initial part of the proposed algorithm (selecting the gateway), we have used ns-2.31 simulator. This simulator tool is widely used as a network simulator. We have also compared the proposed system with LOSP and FGD. These algorithms are described in the related work. The proposed system is much more efficient in controlling the emission of CR messages. It is efficient in regulating the interval between the emissions of two CR messages. The proposed approach is initially evaluated on the following two parameters, i.e., normalized routing overhead and packet loss rate. For the second part of the algorithm, we are using KDD 99 intrusion detection data set [22]. Each data set is set to have eight network features. These eight network features are used in GA. There is one record type. The record type identifies the connection to be an intrusion or a normal connection. The system is trained with the training data set. The fitness values that were used are $w1 = 0.3$, $w2 = 0.6$, $w3 = 0.4$. Twelve numbers of chromosomes, 1995 generations, 260 initial rules, and 045 crossover, and 0.01 mutation rate are used. With the completion of

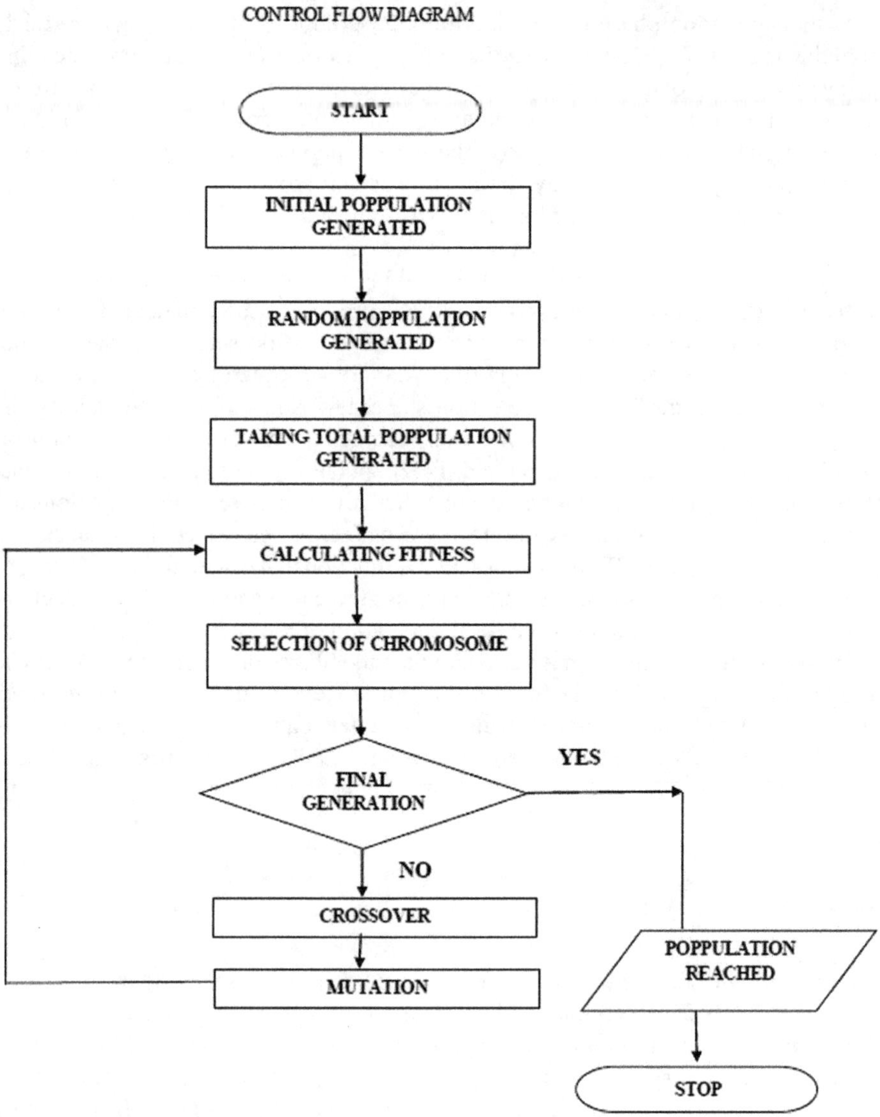

Fig. 2 Control flow diagram

the training process, top 20 rules are used for the final classification. These rules are used to classify the testing and the training data eventually.

5 Conclusion

Intrusion detection is the process of monitoring any anomalous behavior in the computer network. It also deals with detection of threats and how to overcome those threats. Intrusion detection occurs if there is a attack on integrity and confidentiality. Our paper deals with intrusion detection with fuzzy logic implemented with genetic algorithm. The data set that was used is KDD CUP 99 benchmark. This data set is very efficient having a high detection rate. In order to simulate the fitness of a chromosome, the fuzzy confusion matrix is used. The proposed system can update the fuzzy rules and can use the same rules, if there is new intrusion detection in the network. This was related to the second topic of the paper.

As the first part of the proposed system is concerned, it deals with the recognition of the Internet gateway and the emission of CR messages. The proposed system using fuzzy logic controls the emission of CR messages.

References

1. Gomez J, Dasgupta D (2005) Evolving fuzzy classifiers for intrusion detection. In: Proceedings of the IEEE
2. Shabadkar R, Reddy SSS, Vijay Bhaskar Reddy V (2017) Sensor based vehicle motion recognition in portable computing machines for wireless networks. Int J Mech Eng Technol (IJMET) 8(12):465–468
3. Xia T, Qu G, Hariri S, Yousif M (2005) An efficient network intrusion detection method based on information theory and genetic algorithm. In: Proceedings of the 24th IEEE international performance computing and communications conference (IPCCC'05), Phoenix, AZ, USA
4. Reddy SSS, Mamatha C, Chatterjee P, Nagarjuana Reddy S (2017) Contemporary supply chain and inventory data management using data analytics. Int J Mech Eng Technol (IJMET) 8(12): 290–294
5. Yao, JT, Zhao SL, Saxton LV (2005) A study on fuzzy intrusion detection. In: Proceedings of the data mining, intrusion detection, information assurance, and data networks security, SPIE, vol. 5812. Orlando, Florida, USA, pp. 23–30
6. Reddy SS, Shabadkar R, Vijaya Bhaskar Reddy Y, Ranjith Kumar A (2017) Sensor based spontaneous water level recognition in smart cities environment. Int J Civ Eng Technol (IJCIET) 8(8)
7. Chatterjee P, Reddy SSS, Prasada Reddy GS, Hemanth PV (2012) Train time scheduling using genetic algorithm. Int J Civ Eng Technol (IJCIET) 8(12): 410–413
8. Reddy SSS, Nagarjuna Reddy S, Subba Reddy N, Srinivasa Rao M (2017) Framework for efficient authentication mechanism for genuine user's payload transmission in cybernetic locations. Int J Civ Eng Technol (IJCIET) 8(8): 898–904
9. Neog TJ, Sut DK (2011) Complement of an extended fuzzy set. Int J Comput Appl 29(3): 39–45

10. Shahabadkar R, Reddy SSS, Manjunath C, Channabasava U, Shahabadkar KR (2017) Secure framework of authentication mechanism over cloud environment, Springer International Publishing AG 2017. In: Silhavy R et al. (ed), Software engineering trends and techniques in intelligent systems, advances in intelligent systems and computing, vol 575. pp 285–291
11. Reddy SSS, Prasada Reddy GS, Balakrishna B, Sirisha N (2017) Drone control vehicle for prospective protected shipping. Int J Civ Eng Technol (IJCIET) 8(8): 905–909
12. Lu W, Traore I (2004) Detecting new forms of network intrusion using genetic programming. Comput Intel 20(3), 475–494 (BlackIll Publishing, Malden)
13. Reddy SSS, Jabbar MA, Shahabadkar R (2017) Framework for safe direction-finding for protected multimedia contents in cloud computing environment. International Conference on Cognitive Informatics & Soft Computing (CISC-2017)
14. Pushpalatha N, Reddy SSS (2017) Towards an extensible web usage mining framework for actionable knowledge. In: Proceedings in international conference on inventive communication and computational technologies (ICICCT), pp 39–40
15. Sung A, Mukkamala S (2003) Identifying important features for intrusion detection using support vector machines and neural networks. In: Symposium on applications and the internet, pp 209–216
16. Reddy SSS, Shabadkar R, Mamatha C, Chatterjee P (2017) Protection Policy in networked locations using machine learning and data science approach, ICATCCT-2017, pp 237–240
17. Reddy SSS, Reddy LSS, Khanna V, Lavanya A (2008) Advanced techniques for scientific datawarehouses. In: Proceedings of IEEE international conference on advanced computer control, pp 576–580
18. Kumar S, Spafford E (1995) A software architecture to support misuse intrusion detection. In: The 18th national information security conference, pp 194–204
19. Ilgun K, Kemmerer R, Porras PA (1995) State transition analysis: a rule-based intrusion detection approach. IEEE Trans Softw Eng 181–199
20. Bobor V (2006) Efficient intrusion detection system architecture based on neural networks and genetic algorithms. Department of Computer and Systems Sciences, Stockholm University/Royal Institute of Technology, KTH/DSV
21. KDD-CUP (1999) Task description http://kdd.ics.uci.edu/databases/kddcup99/task.html
22. KDD Cup (1999) Tasks http://www.kdd.org/kddcup/index.php?section=1999&method=task

Securing Network Using Software-Defined Networking in Control and Data Planes

Nishant S. Pawar, A. Arunvel, Gardas Naresh Kumar and Aditya Kumar Sinha

Abstract Software-defined networking (SDN) security has look into subject in network security area because of its hugeness in correspondence architecture. SDN architecture mainly focuses on control plane and data plane to enhance security for whole network. SDN approaches depend on the broadly utilized OpenFlow protocol to provide communication among controllers and network element. In this research paper, we use controller Rest API which controls and manages network device and host machine in data plane and also create VLAN topology for security to create small network in data plane. Load balancing is used in data plane for feasible communication. Also firewall application is used for flow control between nodes and fetching network information like port number, IP and MAC address, port mapping, etc.

Keywords Software-defined networking · Traditional networking · OVS · Network attack · OpenFlow · Controller · Data plane · Control plane · Access control list

1 Introduction

Nowadays, software-defined networking becomes advanced network technology. In traditional networking, it consists of management plane, control plane, and data plane. All planes are referred as static configuration of network. They have dis-

N. S. Pawar (✉)
Gujarat Technological University, Ahmadabad, India
e-mail: nish.pawar.25@gmail.com

A. Arunvel · G. N. Kumar
CDAC, Pune, India
e-mail: arunvela@cdac.in

G. N. Kumar
e-mail: guru.jasper@gmail.com

A. K. Sinha
DAC, Pune, India
e-mail: saditya@cdac.in

S.-L. Peng et al. (eds.), *Computing and Network Sustainability*, Lecture Notes in Networks and Systems 75, https://doi.org/10.1007/978-981-13-7150-9_46

tributed control plane, so it is difficult to manage network. Software-defined networking (SDN) is a network structure approach that enables the network to be centrally and intelligently controlled or programmed using software applications. Network administrator manages the entire network reliably and comprehensively, regardless of the underlying network technology. Software-defined networking (SDN) is an example of advanced network infrastructure which overcomes the imperfection of the traditional networking infrastructure. SDN has benefit to decoupling of control plane and data plane; network infrastructure is abstracted from application, so that whole network is managed in centrally as well as logically because of centralized control plane. Traditional network have network device which is use forwarding table and control traffic using forwarding table which is manage traffic by routing table. Network functionality is mainly implemented on network devices as software or hardware in traditional network. Software-defined networking manages and configures all network elements using centralized configuration control.

1.1 SDN Architecture

SDN creates dynamic and flexible network architecture that can be improved as business requirement changes. The purpose of SDN is to reduce cost and enhance user experiment by automating whole network services, from end users to network elements and decoupling of control from network traffic processing and forwarding, centralized control and mobility of customer and application to interact with network control. Software-defined networking architecture is described as three planes as shown in Fig. 1 [1].

Application Plane: This layer contains application like business application, controlling application, networking management, analytics, datacenter management, etc. In application plane, application collects information and data from control layer. This application builds abstracted view of network for decision-making purposes. Interface between control plane and application plane is referred as northbound interfaces.

Control Plane: This layer is useful for programming and managing data plane. It gives information from data plane and also gives command or instruction to data plane. One or more software controller communicates with data plane, and it uses standardized interface known as southbound interface. In controller plane, network controllers configure the network element with forwarding rules based on request from the application and security policy. Control plane contains the forwarding logic and additional routing logic.

Data Plane: This plane is responsible for forwarding data, observing local data or information, and gathering statistics. Data layer is built up by using network element and providing connectivity. The data plane provided the resources that deal directly with external traffic, along with the necessary supporting resources to ensure proper virtualization, connectivity, security, availability, and quality.

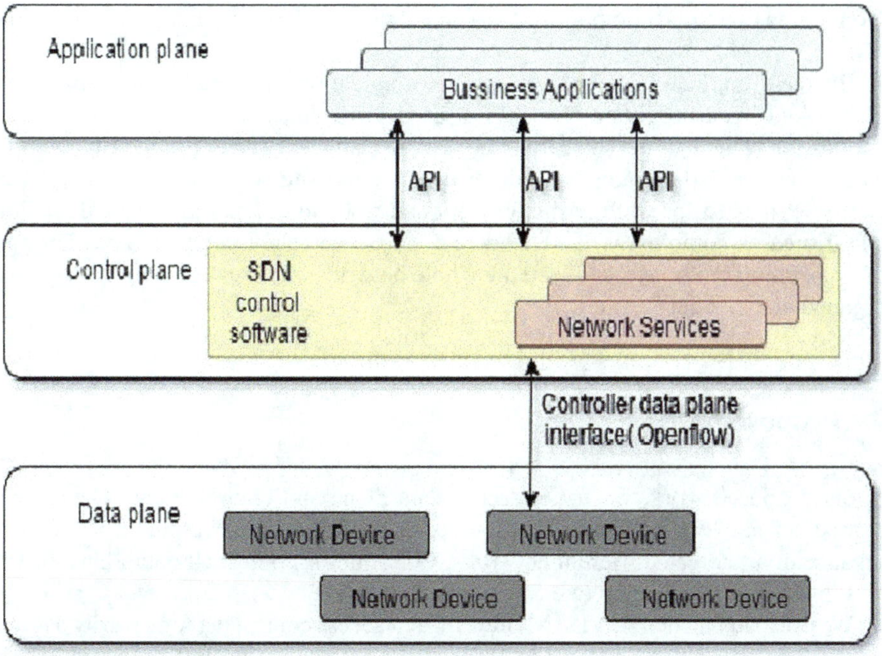

Fig. 1 SDN architecture

Nowadays, network development is mainly focused on security to secure network element, configuration of device, etc. Under the condition that some of the switches in a network are under attack, attacker can effectively catch all data in the packet sent by the switches. Here, we proposed to enhance the security in SDN for data plane and control plane.

2 Problem Identification and Research Scenario

SDN is network architecture that provides unprecedented programmability, automation, and network control. SDN controller provides centralized control and monitoring in network and also protects against malicious traffics. The basic properties of a secure communication network for information are confidentiality, integrity, and availability [2, 3].

These are supported by techniques of authorization, authentication, and encryption. There are some issues in software-defined network such as unauthorized access, data leakage, data modification, malicious/compromised applications, denial of service, configuration issues, and system-level SDN security. Sometimes, attacker finds or exploits the system vulnerability. So this causes security issue. SDN is used in different areas like cloud computing, data center, wireless network, and cellular network [4].

Business application and management application are running in application plane. And open programmable API is used for business application and service application. This may be security issue, because attacker easily found vulnerability and use as injection malicious code or use as cross site scripting (XSS) [5]. SDN controller is used for configuring network and fast and efficient routing calculation and decreases complexity of network. So it may be easily target by Attacker and take unauthorized access and lead the whole network which cause a big damage for organization and data center [6].

3 Proposed Work

In our proposed work, we try to secure data plane and control Plane. Data plane consists of network switch, host machine, and SDN switch. We create VLAN for organization to create different networks. Controller is control all data Plane means all Flows, network element and SDN Switch(OVS) (Fig. 2).

We use load balancing, VLAN, firewall, and access control list for security as well as feasible performance. We use ACL for secure authentication and authorization for controlling whole network. Here, access control list allows only authorized machine and network element. We create virtualization network using Mininet as well as physical network. We can use virtualization network in data center using software-defined networking.

Using security application, we try to fetch network information and also number of nodes connected in our network. We use Postman API to fetch virtualized network information and also create Python script to fetch network information like MAC address and IP address and NODE information like port number and port mapping.

We use firewall concept in VLAN network so that in VLAN network host communicates with authorized host. For example, Webserver communicates with only authorized database server. We try to provide secure OpenFlow between controller and data plane using OVS in L2switch and L3switch layer.

We use ODL controller for software-defined networking. ODL controller supports wireless network connection, so we try to use in virtualization network. We can create network with a number of machines.

As shown in Fig. 3, we create VLAN in Mininet for small network separation. In VLAN, host machine communicates with only same not for other networks. Here, we create VLAN networks such as VLAN1 and VLAN2. VLAN1 cannot communicate with VLAN2. So here, security is implemented using VLAN. Controller allows packet flow from authorized host and network element.

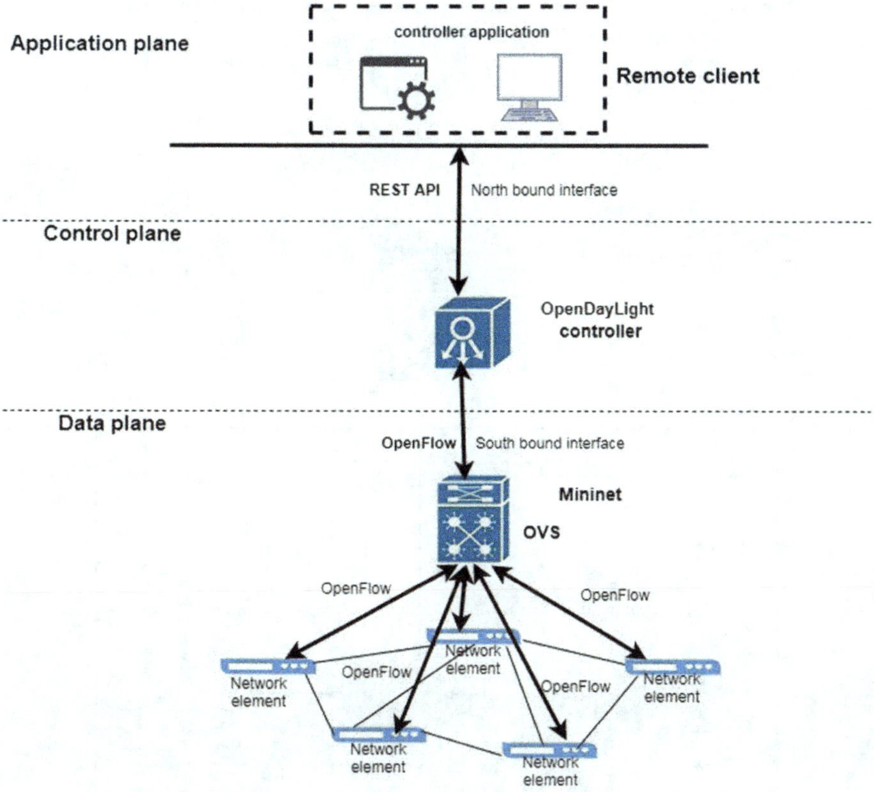

Fig. 2 SDN architecture using security API

4 Related Work

In our proposal, network admin can access and control the data plane using security application. We try to provide automation application to secure the data plane. But our focus is on controller and data plane for security. We also try to provide security using L2switch and L3switch layer. Also we try to provide secure OpenFlow between controller and data plane using OVS. Our goal is to provide security in SDN as controller and data plane. Mainly attacker attacks networking device, so that we try to develop security application for network admin who accesses data plane and gathers information from data plane like MAC address and IP address of networking devices. So we can overcome some security attacks like DDOS attack and MITM attack.

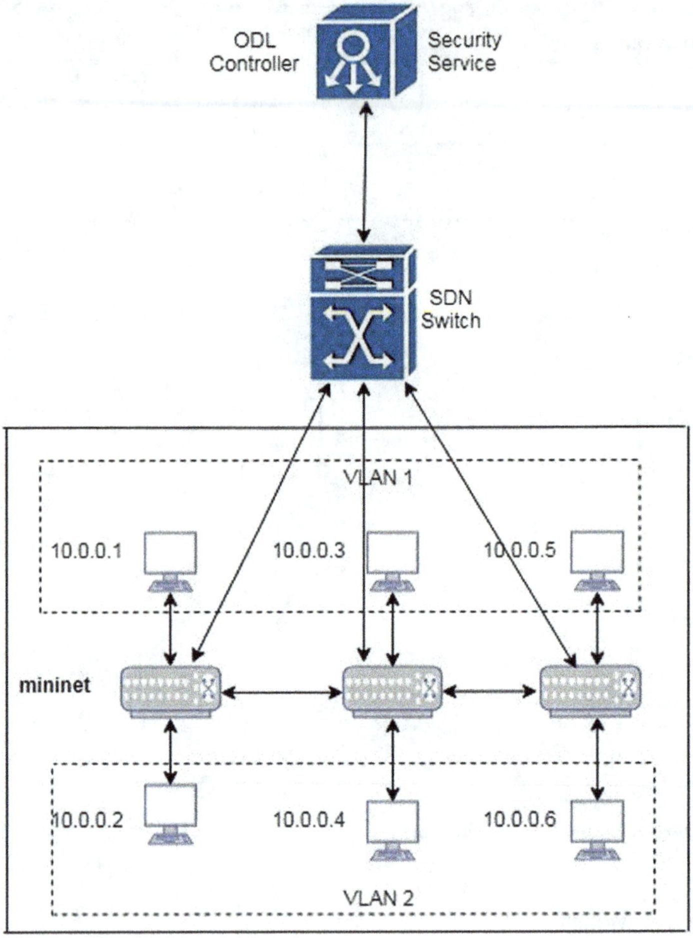

Fig. 3 SDN architecture with VLAN network

5 Implementation Work

In our implementation work, we use OpenDaylight controller with the use of ODL feature like AAA, MD-SAL, BGP, topology, etc. We use OVS as SDN switch in data plane. We use data plane as Mininet network (Fig. 4).

Show in ODL GUI, create Vlan Topology with MAC and IP address and also use OVS in Mininet (Fig. 5).

Using Python script, we fetch IP and MAC addresses from network (Fig. 6).

We also use Python script for load balancing for node and find shortest path from one node to another node (Fig. 7).

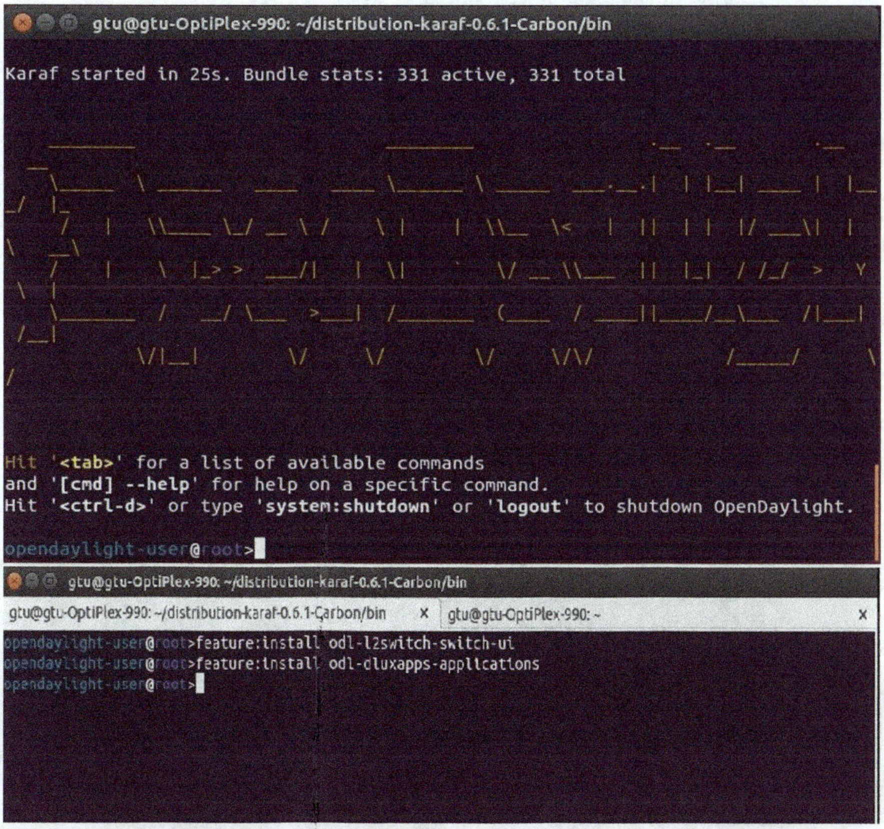

Fig. 4 ODL controller terminal

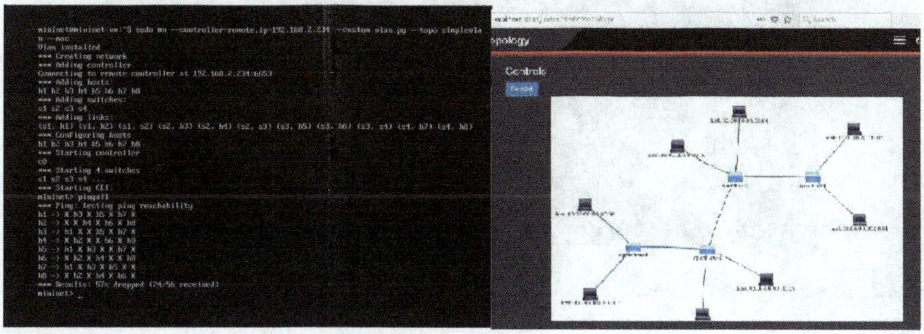

Fig. 5 Mininet using VLAN

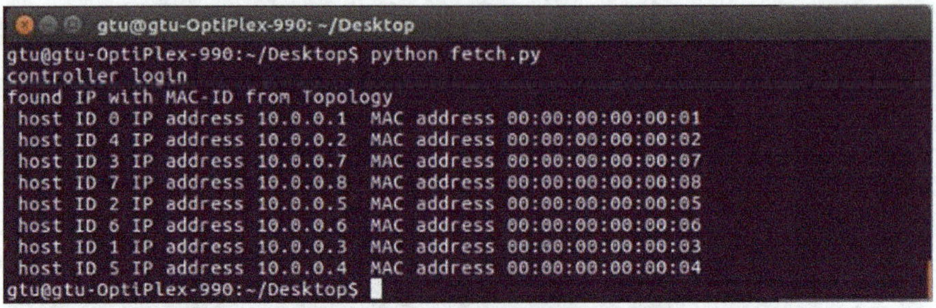

Fig. 6 Fetched MAC and IP address

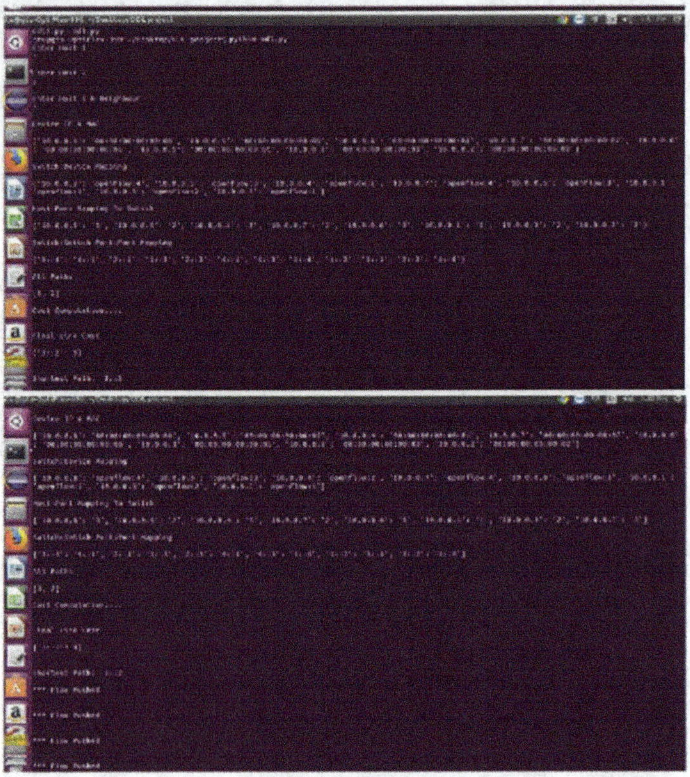

Fig. 7 Managing load balancing in network

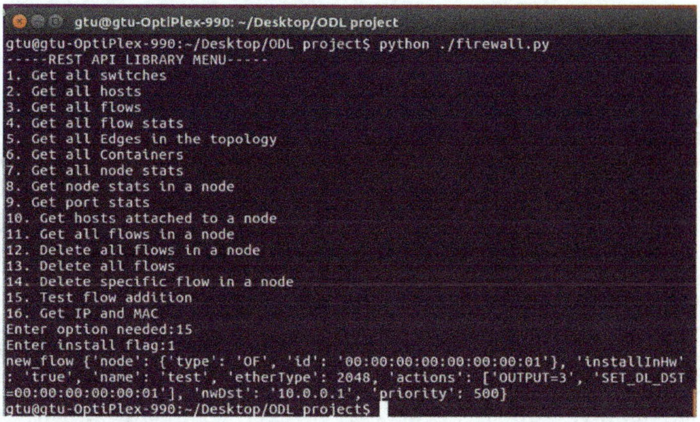

Fig. 8 Rest API for controller

Another script is use as Rest API which is use for node detail, network information, port Mapping, finding shortest Path and also use allow or drop flow packet in network (Fig. 8).

6 Result

Virtualized network: We create virtualized network using SDN which is controlled by controller.

Network information: We create the network using Mininet and also implement it in physical network.

OVS Switch: We create OVS switch in Mininet and also implement in Ubuntu machine. This switch controls virtualized network using controller, and we use Open-Daylight controller.

Fetch Network Information: We use Postman API to fetch virtualized network information and also create Python script to fetch network information like MAC address and IP address and node information like port number and port mapping.

Load Balancing: In virtualized network, we create Python script to balance load of data forwarding between one host to other node and also find shortest path in virtualized network. And we drop all longest paths in network.

VLAN: We create VLAN in virtualized network. This is used as small different networks and all these networks are separated by VLAN network. This is use as Security Proposed.

Firewall: We use firewall concept in VLAN network so that in VLAN network, host communicates with authorized host. For example, Webserver communicates with only authorized database server.

7 Conclusion and Future Work

Our main objective is improving security in SDN architecture. We can provide the security layer in data plane and control Plane. VLAN is used as security proposed to provide authentication in small network and also use firewall to control the flow in virtualized network. We can also use AAA concept in software-defined network. Using security Python API, we fetch network information like MAC address and IP address so that we identify valid and authorized host device. Firewall is also used to block unauthorized device in virtualized network and also used to allow device to communicate with each other.

We use as virtualized network so that we could implement in virtualized data center. NSX allows network virtualization in data center to host users to create and manage multiple virtual networks without physical hardware. We could also implement security technique as micro-segmentation which allows security policies to be assigned in data center application and also decreases workload using load balancing. Security policies are synchronized with all virtual network devices in virtualization data center. We could use data fetching using deep learning algorithm, and using this algorithm, we could automatically fetch all data plane information and the controller controls the packet forwarding in virtualized network.

References

1. Pawar N, Sinha AAA (2018) A survey paper on SDN security in controller and data plane. [online] Doi.one. Available at: http://doi.one/10.1729/IJCRT.17444 Accessed 28 May 2018
2. El Moussaid N, Toumanari A, El Azhari M (2017) Security analysis as software-defined security for SDN environment. In: 2017 4th international conference on software defined systems (SDS)
3. Saxena M, Kumar R A (2016) Recent trends in software defined networking (SDN) security. IEEE
4. Anon (2018) [online] Available at: https://en.wikipedia.org/wiki/Software-defined-networking
5. Xue N, Huang X, Zhang J (2016) S2Net: A security framework for software defined intelligent building networks. In: 2016 IEEE Trustcom/BigDataSE/ISPA
6. Ammar M, Rizk M, Abdel-Hamid A, Aboul-Seoud A (2016) A framework for security enhancement in SDN-based datacenters. 2016 8th IFIP international conference on new technologies, Mobility and security (NTMS)
7. Karmakar K, Varadharajan V, Tupakula U (2017) Mitigating attacks in software defined network (SDN). In: 2017 4th international conference on software defined systems (SDS)
8. Song C, Park Y, Golani K, Kim Y, Bhatt K, Goswami K (2017) Machine-learning based threat-aware system in software defined networks. In: 2017 26th international conference on computer communication and networks (ICCCN)
9. Dacier M, Konig H, Cwalinski R, Kargl F, Dietrich S (2017) Security challenges and opportunities of software-defined networking. IEEE Secur Priv 15(2):96–100
10. Shin S, Xu L, Hong S, Gu G (2016) Enhancing network security through software defined networking (SDN). In: 2016 25th international conference on computer communication and networks (ICCCN)
11. Satasiya D, Raviya R, Kumar H (2016) Enhanced SDN security using firewall in a distributed scenario. In: 2016 international conference on advanced communication control and computing technologies (ICACCCT)

12. Hogg S (2018) SDN security attack vectors and SDN hardening. [online] network world. Available at: https://www.networkworld.com/article/2840273/sdn/sdn-security-attack-vectors-andsdn-hardening.html
13. Team M (2018) Mininet overview—mininet. [online] Mininet.org. Available at: http://mininet.org/overview/
14. Openvswitch.org (2018) Open vSwitch. [online] Available at: http://openvswitch.org/
15. Docs.us-phoenix-1.oraclecloud.com (2018) Scenario A. Public Subnets. [online] Available at: https://docs.us-phoenix-1.oraclecloud.com/Content/Network/Tasks/scenarioa.htm
16. Blackhat.com (2018) [online] Available at: https://www.blackhat.com/docs/us-16/materials/us-16-Yoon-Attacking-SDN-Infrastructure-Are-We-Ready-For-The-Next-Gen-Networking.pdf
17. OpenDaylight (2018) Home—OpenDaylight. [online] Available at: https://www.opendaylight.org/

LAN Security Management System

V. Rama Krishna and R. Subhasini

Abstract In today's competitive era, the success of an organization solely depends on its protection of resources involved in computing. Intrusions are silent enemies that can strike any organization without warning. While these threats can go unnoticed until they cause operational disasters, there is no single "silver bullet" cure for organizational security threats. There are numerous intrusion detection systems (IDS) that work at network level which could not trace all the intrusions. So, it is required to develop integrated IDS which works at both host level and network level and reacts proactively to the threats. To develop an effective network security system for LAN includes the following components.

- To develop a network management system (NMS) which analyzes the traffic and throughput, predicts the link failure and monitors the availability of the device.
- To develop a hybrid intrusion detection system (IDS) which identifies the known and unknown malicious activities (Signature and Anomaly) and alarms the administrator.
- To perform port scanning which scans the activities of network devices (routes, switches, and host computer services) and report abnormal services.

Keywords IDS · NMS · Fault management

1 Methodology:

The objective of LAN security management system (LSMS) is to propose a distributed/host IDS which serves to find new anomalies that intrudes the network. The proposed work is categorized into three phases namely distributed/host intrusion detection system (IDS), network management system, (NMS) and analyzer which

V. R. Krishna (✉) · R. Subhasini
Sathyabama Institute of Science and Technology, Chennai, India
e-mail: rvankama09@gmail.com

R. Subhasini
e-mail: subhaagopi@gmail.com

S.-L. Peng et al. (eds.), *Computing and Network Sustainability*, Lecture Notes in Networks and Systems 75, https://doi.org/10.1007/978-981-13-7150-9_47

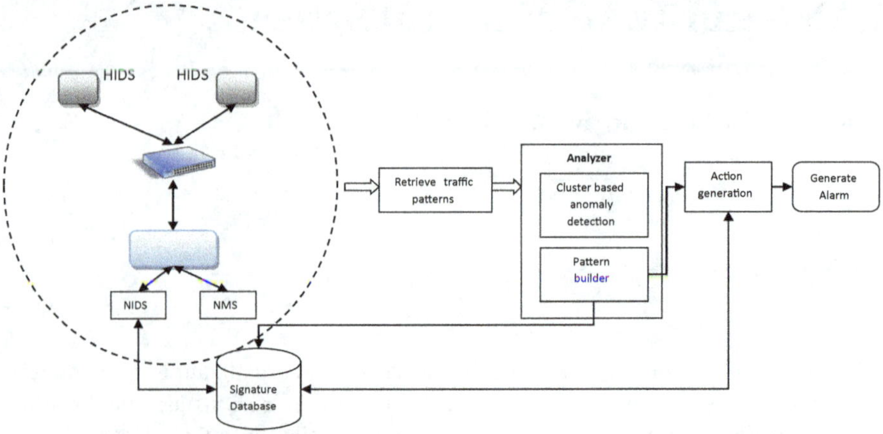

Fig. 1 Architecture of LSMS

runs the cluster-based anomaly detection algorithm to categorize the new anomalies and initiates the pattern builder to develop a new pattern on occurrence of an anomaly. This pattern is updated in the signature database. Figure 1 below shows the overall architecture of the proposed distributed IDS.

The LSMS is designed with the following components namely distributed/host IDS, NMS, and analyzer. The IDS monitors the basic intrusions, NMS manages the network, and analyzer runs a cluster-based anomaly detection algorithm to categorize the unusual anomalies that occurs in LAN.

2 Distributed/Host Intrusion Detection System (IDS)

The IDS is the first component of LSMS. Security in private networks is an important concern as internal hackers and intruders try to capture the confidential information in the network. The intrusion detection system (IDS) helps to identify the suspicious activity within a network. IDS can be deployed in both host level as well as network level and is capable of handling signature-based and anomaly-based detection. HIDS evaluate information found on a single or multiple host systems, including contents of operating systems, system, and application files. NIDSs capture the information traveling in the network, and NIDS analyzes packets and streams of flow to identify the threats. This component hosts the following levels of detection.

- Signature-based detection
  This is used to identify the known patterns of attacks. Well-known pattern of vulnerabilities signature was identified. NIDS mainly uses these signatures to identify the known attacks. The main draw back of this method is we can identify know attacks only using this. New attacks cannot be detected by using this.

- Anomaly detection
  Anomaly detection will work on phenomena that something deviation form the usual traffic pattern is an attack. This mainly depends on statistics of the network traffic, events, and pattern of activities. Things of deviations like huge volume of traffic, suspicious event, and pattern of activities are considered as attack pattern. The main disadvantage is the analysis takes lot of time and cycles, and it was not straight forward way to detect attack. Need to identify when attack is taking place it was expensive.

In network, SNORT NIDS is deployed in every network segment to monitor the known signatures and in host OSSEC is deployed for log analysis, integrity checking, rootkit detection; time-based alerting, and active response.

SNORT NIDS also performs port scanning in order to make the system more effective. The description about port scanning is given below:

2.1 *Port Scanning:*

Systematically scans the computer ports and identifies open ports. It is one of the most popular techniques used by hackers to exploit a system. Port scanning used to find what services specifically system was using, what user running those services. By using this can identify any common logins are permitted, authentications are required (or) not. We cannot completely eliminate this feature as this can identify the security vulnerabilities in the system effectively.

2.2 *Steps in Port Scanning:*

(1) Port scanner sends request to host or range of hosts that you want to scan.
(2) It waits for reply from available hosts.
(3) It probes available hosts for all the possible ports depending upon requirement.

2.3 *Information Provided by Port Scans:*

(1) Hosts that are active and reachable.
(2) Network addresses of hosts found.
(3) Services/Applications that hosts may be running.

3 Network Management System (NMS)

NMS is the second component in LSMS. NMS is widely used scalable and flexible tool that can be adopted for wired and wireless networks equally. It can span from hundreds to thousands of nodes and can be administrated with less cost that can control, monitor, debug, and report of problem on wired and wireless networks from one single location. NMS periodically polls and retrieve information about the network elements. NMS can configure parameters. NMS polls and collects the data periodically and analyzes the collected information. If any thresholds reaching (or) service downs were there that was intimated to the administrator. Usage graphs it can provide When there was threshold limit (or) critical event identified this can be intimated to the administrator immediately by SMS (or) mail. So that administrator can take action immediately.

Proper management improves efficiency of the network and provides cost-effective network utilization. The functions of NMS taken into consideration for this work include fault management and accounting management as shown in Fig. 2.

3.1 Fault Management

The efficiency of fault management lies in the manner in which the fault is located, diagnosed, and rectified. It deals with device monitoring and prediction of link failure. Initially, the devices in local area network are probed at regular intervals say 50 ms. The information collected during probing is stored in f-basket. The f-basket keeps the record of the devices that are available, unavailable, and the duration of downtime of

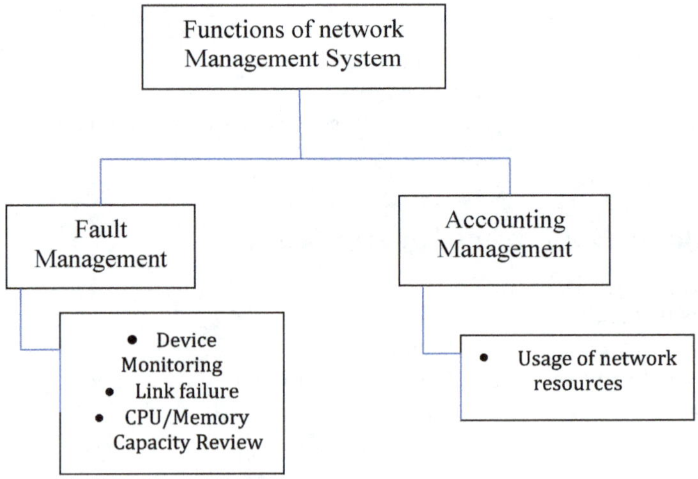

Fig. 2 Block diagram for functions of NMS

Table 1 Linguistic variables for link failure

Input parameters		Output parameters
Link capacity	Synchronization loss	Probability of link failure
Small	Small	Small
Medium	Medium	Medium
High	High	High

the unavailable devices. If the device is unavailable for more than 2 s, then an alarm is triggered to the administrator and also a popup message indicating the unavailability and duration of downtime of system is shown in nearby located available devices. Thus by the use of popup menu, even the users working nearby the devices can also alert the administrator.

Link failures are quite common in large networks and the software error, hardware problem, or link disconnection may lead to link failures. Proactive approach reserves a backup path before a link failure occurs. But in the reactive method, the backup paths are identified only when the failure happens. Link failure has got a great impact on link capacity and synchronization loss. Link capacity is the amount of traffic that the link can carry. If the capacity of the link is small then there is a chance for link failure and also if the synchronization loss is high then there is high risk of link failure. Table 1 shows the input, output parameters, and their respective linguistic variables.

3.2 Accounting Management

With changes in network traffic, connection log varies from time to time. Using the connection log, number of systems in connection and the number of transactions, and amount of packets transmitted are determined. CPU overloads and the consumption of memory greatly depends on the amount of packets transmitted between the systems. A threshold is set for the packets and if the threshold is exceeded, then an alarm signal is sent to the administrator stating the current load of the CPU and the space in memory. In addition, SMS and e-mail alerts regarding the details will also be sent.

4 Analyzer

This is the third component of LSMS. For analysis, the input to this component includes the normal traffic after the basic analysis of signature and anomaly detection is made and the alerts from NMS. The analyzer runs a cluster-based anomaly detection algorithm namely the k-medoids algorithm to categorize the anomalies.

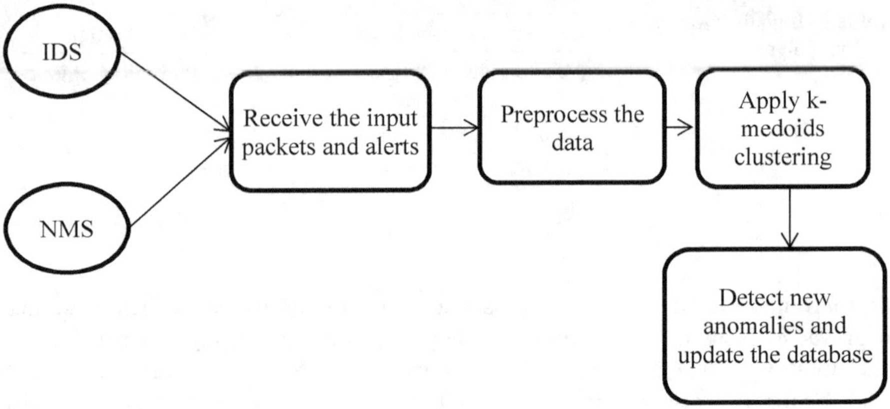

Fig. 3 Block diagram of analyzer

Based on the new anomalies detected, the pattern builder module creates a new signature pattern and updates in the database for further signature detection. Figure 3 shows the block diagram of analyzer. The analyzer performs the following functions namely the cluster-based anomaly detection and pattern building. It receives the input from both the previous components IDS and NMS. The data received is preprocessed and the k-medoids algorithm is applied for categorization. This algorithm provides the lowest false alarm rate while clustering data.

4.1 Analyzer with NMS Data

Analyzer will correlate both data and traffic patterns to see any malicious activity playing role. Some of the use cases as below

High CPU in network elements will see the traffic flooding from specific host (if so block hosts)
Link going unreachable because of flooding from group of hosts (if activity repeating block the specific services)
Hosts crossing the thresholds limit more frequently (check validity by port scan block malicious hosts/services will be blocked)
More control traffic is flowing form specific hosts (send challenges to validate)
Same traffic pattern from multiple hosts (validate the service are bot/malicious)
Frequent configuration changes taking place (check user validity by challenging)
Non-working hours' hosts generating more traffic (validate background process, close them on need basis)
Connection staying idle for long (Will send RST to close them)

4.2 Analyzer with Port Scan data

Port scan data will be given to analyzer. Non-standard ports used across many systems/used for long time/doing more activities(can get this from NMS) are suspicious.

Analyzer initiates three-way handshake to identify malicious services, complete hand shake may take more bandwidth

TCP flows

Analyzer sends SYN packet.
Listening target respond with a SYN+ACK.
Non-listening target respond with a RST (can close if it responded in multiple iterations).

Close the ports who is not responding

UDP flows

In order to find UDP suspicious ports, the analyzer sends empty UDP packets. If
The port is listening, the service will send back an error message
The port is closed, "ICMP Port Unreachable" message.
Neither UDP packets nor the ICMP errors are generated they are suspicious
Through NMS audit records need to ensure packets have not reached end host if needed need to retransmit
With multiple iterations UDP response (or) ICMP error not coming consider them as malicious service.

4.3 Analyzer with IDS data:

HIDS sends large volumes of transfer (to/from), traffic coming in fixed time interval of time, specific protocol/user traffic based on usage, and control connection and their protocols. Similarly, NIDS sends traffic to analyzer large volumes with host/destination/protocol pair, protocol traffic coming more with control flows, idle connections stayed for long duration, and same volume of traffic with same time intervals. By using K-mean cluster algorithm, anomaly patterns will be identified.

4.4 k-medoids algorithm

The k-medoids algorithm is relevant to k-means algorithm and medoid shift algorithm. This is a partitioning algorithm which attempts to minimize the distance between points labeled in a cluster.

Steps:

1. Select in random k of the n data points as medoids.
2. Associate each data point to the closest medoid.
3. Do the following steps as the cost decreases.
   a. For each medoid m and non-medoid data point o
      i. Swap m and o, recompute the cost.
      ii. If the total cost increases, undo the swap.

5 Conclusion

Currently, we are working on the analyzer system which will get the inputs from both IDS and NMS and format them in meaningful way for the clustering algorithm. Currently, we are starting with K-means, and in future, we are planning to enhance this to X-means clustering algorithms. Currently working Botnet detection by analyzer some of the works were published and given in reference [18, 19] We are planning to make LSMS as open source management system can very will provide all NMS activates and additionally can provide best security protection.

References

1. Jian Y (2009) An improved intrusion detection algorithm based on DBSCAN. Micro Comput Inf 25:1008–0570(2009)01-3-0058-03, 58-60
2. Yu S, Guo S, Stojmenovic I (2015) Fool me if you can: mimicking attacks and anti-attacks in cyberspace. IEEE Tran Comput 64(1): 139–151
3. Shue CA, Kalafut AJ, Gupta M (2012) Abnormally malicious autonomous systems and their internet connectivity. IEEE/ACM Trans. Netw 20(1): 220–230
4. Zeng Y, Hu X, Shin KG (2010) Detection of Botnets using combined host- and network-level information. In: IEEE international symposium on network computing and applications. NCA, 2010, pp. 148–152
5. Li L, Yang DZ, Shen FC (2010) A novel rule-based intrusion detection system using data mining. In ICCSIT, IEEE international conference
6. L. Dongxia, Z. Yongbo (2012) An intrusion detection system based on honeypot technology. In: 2012 international conference on IEEE computer science and electronics engineering (ICCSEE), vol 1
7. Chandrasekhar AM, Raghuveer K (2013) Intrusion detection technique by using K-means, fuzzy neural network and SVM classifiers. In: ICCCI, 2013
8. Aydin et al (2009) A hybrid intrusion detection system design for computer network security. Comput Electr Eng 35:517–526
9. Huang W et al (2010) A multi-agent-based distributed intrusion detection System. In: 2010 3rd international conference on advanced computer theory and engineering (ICACTE)
10. Xue Q et al (2003) The design of a distributed network intrusion detection system IA-NIDS. In: Proceedings of the second international conference on machine learning and cybernetics, 2–5 Nov 2003

11. Chen YM, Yang Y (2004) Policy management for network-based intrusion detection and prevention. In: IEEE NOMS 2004, Seoul
12. Khan NY et al (2010) Comparative study of intrusion detection system and its recovery mechanism. IEEE, 2010
13. PAN Z et al (2005) An integrated model of intrusion detection based on neural network and expert system. In: Proceedings of the 17th IEEE international conference on tools with artificial intelligence (ICTAI'05)
14. Kim G et al (2014) A novel hybrid intrusion detection method integrating anomaly detection with misuse detection. Expert Syst Appl 41:1690–1700
15. SanieeAbadeh M et al (2011) Design and analysis of genetic fuzzy systems for intrusion detection in computer networks. Expert Syst Appl 38:7067–7075
16. Hubballi N et al (2011) LAN attack detection using discrete event systems. ISA Trans 50:119–130
17. Elhag S et al (2015) On the combination of genetic fuzzy systems and pairwise learning for improving detection rates on intrusion detection systems. Expert Syst Appl 42:193–202
18. Rama Krishna V, Subhashini R (2018) BOTNET algorithm adoptabulity for mimicking attack and inducing mimicking attack. J Adv Res Dyn Control Syst 06:977–991
19. Rama Krishna V, Subhashini R Detecting HTTP based mimicking attacks at HTTP server. Int J Eng Technol 0975–4024

Evaluating Three-Dimensional Directions of Focuses on a Target and Utilizing Estimations in Two Photographic Frames

Manish Sharma and Varun Gupta

Abstract 'Z' data in 3D sensing plays a vital role in 3D image acquisition. Third dimension can significantly improve the reliability of scene analysis. Therefore, using a 3D imaging device, an obstacle can be reliably detected and interpreted for various applications. In this paper, different techniques are compared and reviewed for 3D data estimation from two photographic frames. High-precision 3D data can be obtained and analyzed by digital signal processors or multi-core scalar processors. Stereoscopic vision is cost-effective which can fit in a small system making them a good choice for 3D data-based devices. The use of the present application outcome can be adapted for further advancement in 3D imaging methodologies and applications. It would be helpful in the development of cost-effective laser-based 3D imaging systems for the generation of image models of various strategic targets, which will be very useful in depicting hot spot identifications, geometrical profiling, etc. The objective of this literature review is to outline the present techniques of the 3D imaging and to assess the applications.

Keywords Image processing · Geometry · 3D image · Laser · Camera

1 Introduction

The determination of three-dimensional coordinates of a point on an object employing measurement made in two or more photographic images taken from different positions is commonly known as stereo photogrammetry. The past work done related to the determination of three-dimensional coordinates has high percentage of error due to various factors in which camera parameters play a major role. The camera parameters include extrinsic and intrinsic parameters. The camera calibration is also done with the help of following parameters: focal length, image sensor, size of the pixel, position of principle point. Target detection is one of the most important assignments in the area of defense; hence, computer vision is used for determining

M. Sharma (✉) · V. Gupta
Department of Electronics and Communication, Amity University, Uttar Pradesh, India
e-mail: manish.nsit07@gmail.com

S.-L. Peng et al. (eds.), *Computing and Network Sustainability*, Lecture Notes in Networks and Systems 75, https://doi.org/10.1007/978-981-13-7150-9_48

the various data of an image and utilizing data of an image captured from camera. The main property of camera model is that the relationship between world and pixel coordinate is linear.

Three-dimensional coordinates are retrieved in a given two-dimensional frame which can be further utilized for object tracking and also the reconstruction of 3D image which can be implemented by various algorithms. In this paper, we have reviewed the techniques for finding the three-dimensional data from the two-dimensional frame by the two cameras. However, the mathematical geometry for all the techniques has much in common, but due to different target structures, the implementation and error percentage vary.

Geometrical calculations need to be focused on the image projection from two different positions of camera. 3D location in an image taken from two or more cameras is depending on the Doppler shift, the distance of the points; i.e., 'z' data of the image may be retrieved. The implementation of algorithm can be done in the MATLAB, and it gives the results based on the geometry.

Complexity for finding the three-dimensional coordinates of target is constantly growing. The current algorithm and tools have reached limits of adaptability. For detection and data analysis, we need new algorithm and approaches; hence, comparison of different algorithms developed with data analysis is studied for less error system. This review paper focuses on the tracking and detection of the target, based on their characteristics to distinguish a target from other areas on the captured frames. A few most generally utilized active and dynamic optical range imaging systems are surveyed with their qualities and shortcomings. The objective of this literature review is to outline the present condition of the 3D imaging methods and to assess the applications in defense.

2 Related Work

Stereoscopic measurement—Stereoscopy is a technique used for recording and representing stereoscopic (3D) image. Stereoscopic picture created with the two cameras just like our own eye. The important limitations to take a pair of stereoscopic pictures are: Camera should be horizontally categorized, and second thing is at same instant picture should be taken. With the help of stereoscopic pictures, we can calculate the distance between the camera and the chosen object in the picture.

Object recognition—Location of both the picture is selected automatically. The author describes the location of an object and clicking of image.

This describes the method of noninvasive distance of an object and pictures of an object taken from two horizontally displaced cameras. An algorithm is created for two cameras which are placed horizontally.

Stancik et al. [2] describes the design of special software for 3D image reconstruction. 3D point representation and its transformation into 2D image space by mathematical formula. Three-dimensional data is important for the fields of computer vision and photogrammetry. With the help of stereo vision, image data will be

achieved. Stereo vision is used for achieving three-dimensional data from image data. The main characteristics for 3D reconstruction are the camera calibration and stereo correspondence problem. The main motive of camera calibration is to establish the projection from 3D world coordinates to the 2D image coordinates. Camera calibration needs intrinsic and extrinsic parameters. So the determination of these parameters takes problem. This method is used for designing special software for these problems. This software design establishes the projection from three-dimensional coordinates to the two-dimensional image coordinates.
Geometric camera model—In this section, full geometrical calculation is described. The relation between three dimension and two dimension is described by pinhole camera. When the picture is captured, the origin of the camera lies on the upper left corner of the image. So with the help of matrix calculation, transformation of the coordinates will be done.

$$X = [X,\ Y,\ Z]$$

Outline is given by an unbending body change as,

$$X = RX' + T$$

where
R = Orthogonal 3 * 3 rotation matrix
The directions of a similar guide relative toward camera
T = translation vector
Focal length is applied with point X into image plane. The homogeneous equation will be formed with rotational and translation matrices. The author also described the epipolar geometry.
Reconstruction from two calibrated views—The epipolar constraint is well known in stereo vision. In this, the author tells about photogrammetry and 3D scene reconstruction. For 3D reconstruction, stereo matching is also important task. This task is difficult to apply, but when it is applied, the depth could be recovered. The matching algorithm can be divided into two main groups: one is feature-based stereo, and second is area-based stereo. The software RECON 3D for three-dimensional measurement is designed. The software is worked with digital camera. The software communicates with the camera and saves the scene into a file in computer memory, and after that, automatic detection of object in image occurs. For above process, it is important to get the intrinsic and extrinsic parameters of the camera for the proper three-dimensional construction. If the program directly communicates with camera, it recovers the rotation matrix and translation vector. If the camera calibration is done and images get matched, then the three-dimensional coordinates of point are reconstructed.

Lee et al. [3] describes that with the help of image processing technique, the coordinates in three dimension can be done by using laser range finder. In this, the algorithm can be done for 2D-to-3D transformation. Three point coordinates are measured in advance. The algorithm was verified through the experiment. Sensors are used to perform local mapping. The monocular camera algorithm method gives 3D measurement data via a much simpler computational scheme. Here, the monocular camera method, however, relies on the structured light such as CMM or SL; both have different techniques. These technologies have quite expensive devices and complicated processes. In this study, 3D coordinates will be determined by 2D image. The camera position is assumed as origin, and with the help of three known points, the relationship between 2D image and 3D points will be obtained.

The author describes that 3D coordinate measurement can be done by using laser range finders or stereo vision. The algorithm can be used in modern industry measurement systems and the development of weapon system.

The applications are machine perception, pattern recognition, computer vision, robotic vision, etc. Here, the assumption is made that the coordinates of origin and three points are known. With the help of vector phenomenon, the whole mathematical calculation is made. The four points are selected for the experiment, and these four points were on the same plane. The camera position was measured from a plane $z = 0$. Pixel coordinates were measured from camera image. The pixel values and coordinates of all points were the input to the equations, and the 3D coordinates were obtained.

Experimental results showed that the measurement has about 1.7% error.

Saulo et al. [4] describe the algorithm for finding distance using image processing technique and laser triangulation distance measurement technique. With the help of embedded system, the processing of device is good as compared to other devices because of less processing power.

A CMOS camera, laser, and vibration motor are used in this method for an electrical virtual white cane. In this, the algorithm is used for laser detection on camera image. This procedure applies to a minimal effort white cane design for helping outwardly weakened individuals. The loss of vision is one of the most difficult disabilities. So this device is developed for visually impaired people for getting any hindrance around them. MATLAB software is used to apply the algorithm and get good result. With the help of algorithm, a microcontroller communicates with camera and laser which are working together. ADC is used for checking the battery level, and buzzer gets the information from it and is also used for indicating the problem comes in detecting the laser point. The micromotor transmits mechanical vibration which shifts its power as per the separation figured by image processing. For developing the model, these steps are followed—distance detection using a laser, hardware assembly, image processing algorithm validation, and system validation tests.

The laser detection is done by looking for pixels with the laser fringe shading. The laser relocation model is based on snag distance. For finding the distance between laser point and obstacle from the camera, mathematical model has been introduced. Minimum distance of an obstacle from the camera (dA), maximum distance of an obstacle from the camera (dB), $dA < d < dB$, auxiliary variable (p), real size of the image in a distance d (l) are used for finding distance. The limitation of the system is surfaces as glass and mirrors.

3 Acknowledgment

The review is preliminary work done in 3D data calculations to find the best geometry for developing the algorithm which can be implemented on DSP. The work is supported by ER & IPR Department of DRDO, India.

4 Conclusion

In this paper, we compare image processing algorithms for the detection of the target point, i.e., depth of the point from camera. The area of 3D imaging and handling is wide and complex where certain issues need to be resolved. The computing technologies are focused on soft behavior of the systems. Based on various algorithms, it is concluded that:

- The depth of the image depends upon the baseline of the cameras.
- The disparity calculated is dependent on camera external and internal parameters.
- The value of the x, y coordinates depends upon the threshold value of the color and the pixels.
- Procedure can be followed for targets with different geometries and wavelengths.

References

1. Mrovlje J, Vrancic D (2008) Distance measuring based on stereoscopic pictures. In: 9th international PhD workshop on systems and control: young generation viewpoint, 6
2. Stancik P, Ricny V (2009) Software for camera calibration and 3D points reconstruction in stereophotogrammetry. In: EUROCON 2009, EUROCON'09, IEEE
3. Lee HM, and Choi WC (2015) Algorithm of 3D spatial coordinates measurement using a camera image. J Image Graphics 3(1)
4. Barreto SVF, Sant'Anna RE, Feitosa MAF (2013) A method for image processing and distance measuring based on laser distance triangulation. In: 2013 IEEE 20th international conference on electronics, circuits, and systems (ICECS), IEEE
5. Ge W, Collins RT, Barry RR (2012) Vision-based analysis of small groups in pedestrian crowds. IEEE Trans Pattern Anal Mach Intell 34(5): 1003–1016
6. Boleček L, Říčný V, Slanina M (2019) Fast method for reconstruction of 3D coordinates. In: 2012 35th international conference on telecommunications and signal processing (TSP), IEEE
7. Matsubara H et al (2003) A three-dimensional position measurement method using two pan-tilt cameras. R&D Rev Toyota CRDL 38(2): 43–49
8. Saxena A, Chung SH, Ng AY (2008) 3-d depth reconstruction from a single still image. Int J Comput Vision 76(1):53–69
9. Francis SLX, Anavatti SG, Garratt M (2011) Reconstructing the geometry of an object using 3D TOF Camera. In: 2011 IEEE workshop on merging fields of computational intelligence and sensor technology (CompSens), IEEE

ICT Enable Artificial Intelligence in Healthcare Management in India

Rajeev Pandya, R. S. Shaktawat and Namita Pandya

Abstract In the present scenario, healthcare management plays an important role to develop strong bonding between peoples and hospital management. Rapid enhancement in medical and healthcare supporting equipment and technology is big motivation factor to develop the area. Nowadays we are observing that the development in the healthcare sector is so fast due to use of latest technology and technology-oriented services. The credit for motivation behind better healthcare services goes to several sectors; the information communication technology is one of them. The ICT supported artificial intelligence services is one of the crucial motivational factors to shape the current healthcare services. We can say that the ICT enable artificial intelligence is motivational pinnacle in healthcare sector at much instance driven by the information technology. Automation in equipment and atomized services are now backbone of health care. The present study tries to provide an insight of information communication technology enabled artificial intelligence in the favor of motivational factors for the development of healthcare industries in the Indian scenario.

Keywords Motivation · Health care · Services · Automation · ICT

1 Introduction

Nowadays worldwide information technology plays a critical part in functioning of all industries and has made our lives totally different than before. Innovations and advancements are taking place all across the world in medical science behind which information technology is playing a major role. The healthcare professionals of today are wiser then the healthcare professionals of twentieth century. Technological advancements have reduced the treatment and recovery time of patients by using

R. Pandya (✉) · R. S. Shaktawat
B N University, Udaipur, India
e-mail: pandya.rajeev@gmail.com

N. Pandya
Department of Human Resource, Geetanjali University, Udaipur, India
e-mail: pandya.namita@gmail.com

S.-L. Peng et al. (eds.), *Computing and Network Sustainability*, Lecture Notes in Networks and Systems 75, https://doi.org/10.1007/978-981-13-7150-9_49

latest system and communication advancements. It has motivated maximum number of people to make their career in healthcare sector which is booming like anything in India and is becoming hub for medical inventions and innovations. Whatever developments have happened in past decade has made the life of people more comfortable and it has helped in increasing average life of human beings in India. Hence, contribution of information technology in motivating healthcare professionals cannot be ignored. In the same series, the ICT-enabled artificial intelligence services are performing commendable in the development of equipment and services. This section also helps the sector by the cognitive, predictive, and descriptive kind of prospects.

2 Use of Artificial Intelligence in Health care

The ICT-enabled artificial intelligence is one of the major sub-sectors in the healthcare services. In the way of understanding all those things, there is need to understand artificial intelligence, computerized and automated section of health care. The utilization of artificial intelligence for healthcare sector is assorted in sub-sectors. To know the sort of ICT-enabled 'artificial intelligence' in health care, we have to understand the different category of the activity which is descriptive, predictive and perspective18:

- **Descriptive**

The descriptive artificial intelligence is very widely employed in healthcare equipment and machineries nowadays and clutches the most promising parts in terms of short-term latent. It measures actions that have previously take place and utilize this information and data to expand further insights. It works as detecting the patterns of the trends and trying to find out the miner change which is cause of the problems. This is generally escapes by the professionals. So we can say that this technology is atomized the pattern matching and recording the miss matches in the pattern and concern things.

- **Predictive**

The role of predictive artificial intelligence is to ensure suitable utilization of data received by the descriptive intelligence. So predictive artificial intelligence uses ICT technology to make prediction which gives imminent and suggest actions in a predictive manner. 21 The ICT-enabled artificial intelligence can play an important part in predictive-analytical healthcare tools and in the management of hospital (Fig. 1).

- **Prescriptive**

The ICT-enabled perspective artificial intelligence furthers the purpose of predictive artificial intelligence, and it is not simply detected pattern which is not be predicted, but also recommend promising dealing. It is basically one of the processes of decision making. This type of ability makes prescriptive artificial intelligence motivating and

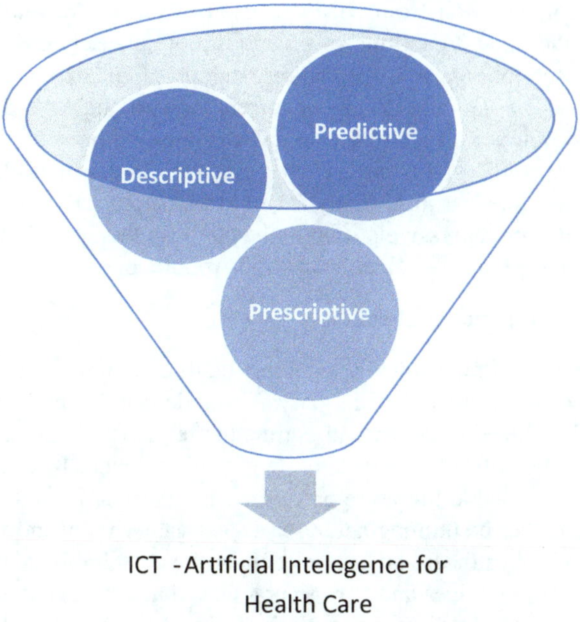

Fig. 1 ICT-enabled AI services

complex problem-solving tools. The prescriptive artificial intelligence may potentially be mobilized as cognitive mediators which mimic the mind to diminish cognitive freight on patients. The ICT-enabled artificial intelligence powered little agents to search and apply clinical information and knowledge with staff and researchers to improve the efficiency, capacity, and quality of working.

3 ICT-Enabled Artificial Intelligence and Healthcare Segments in India

The role of ICT in context of atomization is very positive for the healthcare industries in India. Nowadays the healthcare industry plays a big role in developing variety of automated and digitized equipment and services. As per assessment of Industries developing ICT enable solutions for health care and opinion of healthcare practitioners along with researchers, it was found that atomization and digitization are employed in a variety of ways in the industries. Some of the important segments are as given below:

1. **Hospitals**

Nowadays all types of hospitals are working with the support of ICT and concern equipment and support services. Hospitals in India are employing descriptive and predictive atomization. To illustrate, we can take Manipal Group of Hospitals has collaboration from IBM's Watson in support of oncology to assist doctors in the finding and handling of seven kind of cancer [1]. Here, automation intelligence is used for data analysis and research proof and enhance the quality of the report.

The "Google Braine" is currently served the Arvind Eye-Care, earlier Google also developed a system for retinal transmission bat y the help of images and associated "parsing" Algorithm. In that series, various products like MS Azure for cloud service, machine learning, Analytics, online Customer Relationship Management and Office 365 are regressively used in the big hospitals like LVPEI "L V Prasad Eye Institute" Narayana, Max Healthcare, and Appolo, etc. [2].

2. **Pharmaceuticals**

The pharmaceutical are comprised of creating, manufacturing, sanitizing the chemicals and medicines in general. The descriptive and predictive artificial intelligence is utilized to create and outline the tests. The popular work of automated testing is utilized for drug revelation and testing, where ICT-AI is prepared to scan through all available literature on a particular particle for a drug which would some way or another be unimaginable for even a gathering of individuals to manually carry out. Also healthcare businesses deliver novel technological innovations like apps for the heart and liver and vertigo practices that take a shot at the idea of the virtual world utilize amplified and implicit reality) [3]. Pharmarack50 is a software-as-a service (SaaS) app deals with the location that uses AI to computerize the pharmaceutical supply chain.

3. **Diagnostics**

It involves organizations and laboratories which proffer methodical or investigative services. For example, Google and IBM, India is congregation to startup industries that specialize in connecting artificial intelligence to investigate syndrome. Since a survey of arrangements take on it emerge that investigation are utilizing descriptive and predictive artificial intelligence. In India, AI is being utilized through chat bots, for example, Wysa that gives mental health bolster.

4. **Medical Equipment and Supplies**

It incorporates establishments with built-up medical things and supply, e.g., surgical, dental, orthopedic like laboratory gadget, and so on. As of an audit of arrangements take on it appears that industries producing medical apparatus and distributing in nation are utilizing ICT enable artificial intelligence (Fig. 2).

5. **Medical Insurance**

This incorporates health cover and medical repayment amenities, hospitalization costs brought about because of disorder. Since a survey of arrangements take on it come into view that industries offer medical cover utilizing various form of ICT enable artificial intelligence. Machine learning can mechanize claims administration by analyzing huge sum of data in a smaller amount instance, which diminishes preparing moment and managing costs and enhances client encounter. Recognizing

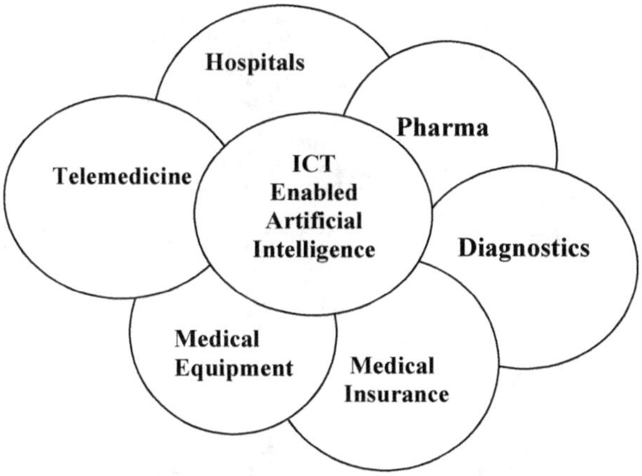

Fig. 2 Major ICT-enabled artificial intelligence and healthcare segments in India

doubtful trends in data know how to distinguish fake claims, which could accelerate agreement of honest to goodness claims. By consolidating huge information with artificial intelligence, back up plans can distinguish the way of life behavior of clients to give them redid contributions. Enormous information may be exploited to make possible safety net providers to recognize pre-time sicknesses and diminish the dangers concern to treatment [4]. Some of the companies working in the area are "Bajaj Allianz" with general Insurance uses Boing, a chatbot that deals with client queries inquiries on engine. The "ICICI Lombard" uses its chatbot stand MyRA to offer insurance policies. The "HDFC Life's" system claims to be pioneer in nation which is automatically deal with client emails [5] (Fig. 3).

6. **Telemedicine**

In the telemedicine system, the client served by automatic machine by means of e-communications. It provides clinical services to the patient with the help of virtual world. In the normal sense, we can say that it reduces the visiting time, appointment problems and other transportation problems. Sometimes parallel processing is not possible by the practitioners; also the human efforts are get failed here. But due to utilization of artificial intelligence, all kinds of scheduling and continuity of prescription till date are automatically managed by the telemedicine system.

These activities consist of audio and video conferencing about medical treatments and follow-up. Again the records of the patient are managed for lifetime [6].

This approach bypasses the complex process traditional consultancy-type process and provides well-defined knowledge-based treatment in a continuity without any delay. It is low cost and repetitive approach [7].

In the various surveys, it is found that telemedicine approach is better healthcare solution for backward and isolated area; it also minimizes the efforts of training and action management [8]. But the telemedicine approach facing problem regarding infrastructure, ICT-enabled artificial intelligence services are able to minimize the

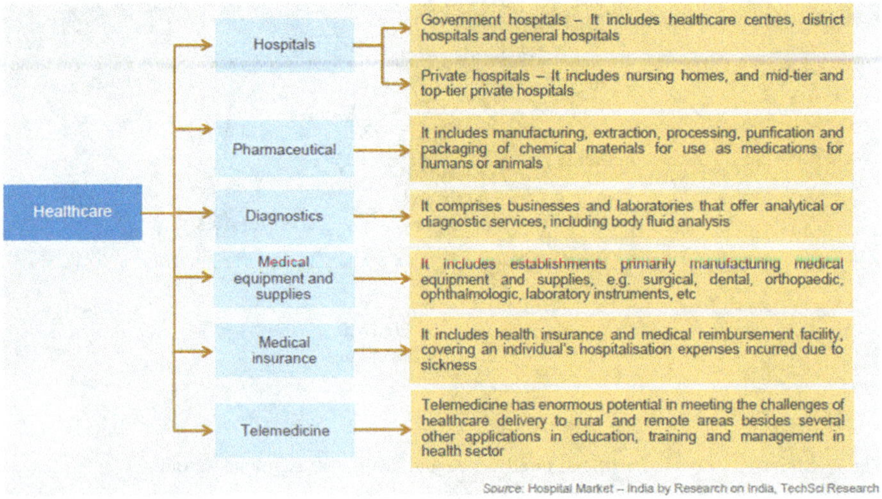

Fig. 3 Major ICT-enabled healthcare segments in India

resources efforts and needs. The SigTuple may analyze "blood slides" and produce a pathology description with no support of any qualified person. Such kind of approach is used in isolated region with very low cost, usually as no cost [9].

The Telangana government ties up with Microsoft cloud analytics services for "Rastriya Bal Swasthya Kalyan" program by using MINE "Microsoft Intelligent Network for Eyecare", this is one of the successful ICT-enabled artificial intelligence service [10].

4 Growth Scenario in Healthcare

Health care has turned out to be one of India's biggest segments mutually in the form of income and business. The business is developing at a huge speed inferable from its reinforcing scope, administrations what's more, expanding use by open also private players.

During the year 2008–2020, bazaar is relied upon to evidence a CAGR of 16.5%. The aggregate business measure is required to contact USD160 billion by 2017 and USD280 billion through the year 2020. According to the Health ministry, advancement of 50 innovations has been focused in the financial year 2016, for curing of syndrome as TB and Cancer.

Administration is underlining the electronic health activities, for example, "Mother and Child Tracking System" (MCTS) and "Facilitation Focus" (MCTFC) (Fig. 4).

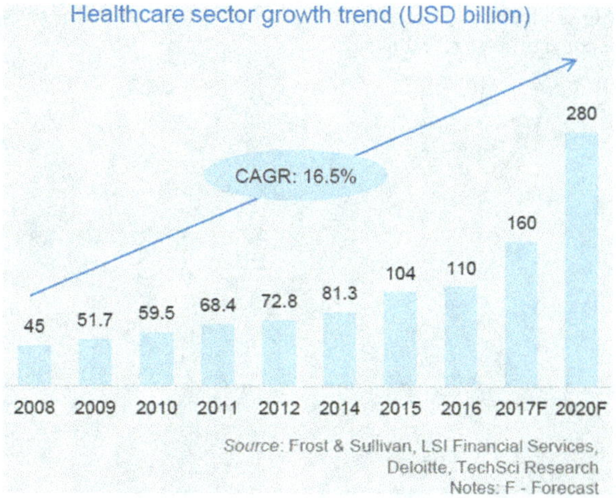

Fig. 4 Healthcare sector growth current and expected in India

Indian organizations are going for unification and attainments with residential and remote organizations to oblige development and pick up novel bazaar. By the date 12/16/2016, “Sun Pharm” a finished the procurement of US supported organization, Ocular Technologies Sarl.

As per NASSCOM, the Indian healthcare showcase is esteemed at USD1 billion in 2016 and is required to increment by 1.5 times by 2020.

- **Rising in the sector**

Rising income means a consistent development in the capacity to get to health care and related administrations.

During the year 2015–2019, per capita income growth is accepted to boost at a CAGR of 8.09%.

Per capita consumption on health care in India is USD68.6 billion.

In addition, changing socioeconomics will likewise contribute to more noteworthy healthcare expenditure. It is probably going to proceed with the span of the mature populace set to ascend from the present 98.9 million to around 168 million by the year 2026 (Fig. 5).

- **Milestone Achieved**

In financial year 2016, net direct premium pay from health cover remained at 25.4% of entire gross direct premium income for non-extra security section

- Health cover is picking up energy innovation; seeing development at a CAGR of 15.36%, during the financial year 2008–16, net healthcare cover expense in 2016 remained at USD2.8 billion
- The pattern is probably going to keep, profiting the nation’s healthcare sector versatile-based health conveyance.

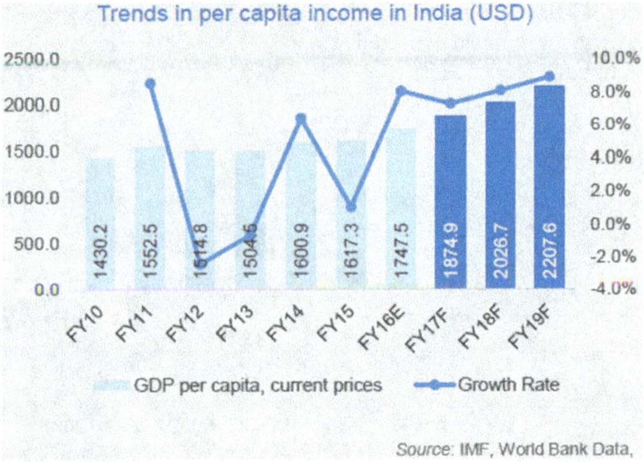

Fig. 5 Trend in per capita income in India (Health care)

- Strong mobile technology infrastructure and dispatch of 4G is relied upon to drive mobile health activities in the nation
- Cycle tel Humsafar is a SMS-based portable administration intended for ladies, it empowers ladies to design their family betterly.
- Currently, there are more than 20 mobile health benefit activities in the nation for spreading mindfulness about family arranging and different infirmities.
- Mobile health industry in India is achieve USD0.6 billion by 2017
- To institutionalize the nature of administration conveyance, control cost, and upgrades understanding commitment, healthcare suppliers are concentrating on the mechanical part of healthcare conveyance.
- Digital Health Knowledge Resources, Electronic Medical Record, Mobile Healthcare, Electronic Health Record, Hospital Information System, PRACTO, Technology-empowered care, telemedicine, and Hospital Management Information Systems are a portion of the innovations increasing wide acknowledgment in the area.
- All India Institute of Medical Sciences (AIIMS) to change over all its installment exchange cashless, for which it has related with portable wallet organization, MobiKwik, in January 2017.

5 Outcomes

- Earlier approach to rural areas for clinicians was not their. Now ICT has made it easy to provide healthcare advice in through information technology.
- Healthcare training has become so easy by expert clinicians at rural areas.
- It has become so easy to collect data and records through the easy approach of ICT in rural areas which helps in making quick decisions.

- ICT provides quick advice on positive and negative impacts of drugs and medicines with content.
- Radiodiagnosis images can be transmitted anywhere and anytime which has reduced treat time TAT.
- Quality of patient care has improved.
- It has been easy in tracking and monitor status of various diseases in community.
- Surveillance, tracking, and monitoring of ANM, Asha and medical officers have become so easy.
- Early diagnosis reduced morbidity and mortality.
- Early detection of epidemic situation.
- Fast reporting of patient's disease and timely treatment in real-time situation.

6 Conclusion

Usage of information and communication technology in healthcare sector has immense potential, and of course, it has motivated healthcare professionals to provide the better results comparative to past decade. It has helped people in easy connectivity, patient's mobilization, and developing fast response system and in forecasting of diseases in advance. It is a need of an hour to continue further inventions and innovations in this segment to provide better healthcare system to nation. ICT is going to create more number of jobs in the country as it is churning demand from the government. No doubt, we need to change the infrastructure with changing need of the market. The study proved that ICT-enabled services like artificial intelligence with information technology and computerized equipment and services are the face of the future India. The healthcare services like e-health record, telemedicine atomized and informatics are now backbone of Indian health care.

References

1. Simonite T (2017, June 7) Google's AI eye doctor gets ready to go to work in India. Retrieved 5 Jan 2018, from https://www.wired.com/2017/06/googles-ai-eye-doctor-gets-ready-go-work-india/
2. Jaiswal S (2016, September 23) Healthcare & BFSI companies lead Microsoft local cloud adoption. Retrieved 5 Jan 2018, from https://news.microsoft.com/en-in/healthcare-bfsi-companies-leadmicrosoft-local-cloud-adoption/
3. Raghavan P (2016, October 31) Indian pharma sector going digital at a fast pace. Retrieved 5 Jan 2018, from https://economictimes.indiatimes.com/industry/healthcare/biotech/pharmaceuticals/indian-pharma-going-digital-at-a-fast-pace/articleshow/55146271.cms
4. Mantri S (2017, September 25) Artificial intelligence and healthcare. Retrieved 5 Jan 2018, from https://www.icicilombard.com/experts-blogs/story/artificial-intelligence-and-healthcare
5. Saleem S (2017, June 20) Bots are welcome in the insurance sector. Retrieved 5 Jan 2018, from http://www.livemint.com/Money/BuEJg3dxFvAR66lgd2rfTO/Bots-are-welcome-in-the-insurancesector.Html

6. Chiron Health, What is telemedicine? Retrieved 5 Jan 2018, from https://chironhealth.com/telemedicine/what-is-telemedicine/
7. Dalmia V (2013, February 12) India: Telemedicine in India—legal analysis. Retrieved 5 Jan 2018, from http://www.mondaq.com/india/x/221258/food+drugs+law/Telemedicine+In+IndiaLegal+Analysis
8. Vision 2K+Inc (2015) What is telehealth? Retrieved 5 Jan 2018, from http://www.v2k.in/Telehealth.html
9. Chakraberty S (2017, February 13) An artificially intelligent pathologist bags India's biggest funding in healthcare AI. Retrieved 5 Jan 2018, from https://www.techinasia.com/artificially-intelligentpathologist-bags-indias-biggest-funding-healthcare-ai
10. Nagpal J (2017, August 3) Government of Telangana adopts Microsoft Cloud and becomes the firststate to use artificial intelligence for eye care screening for children. Retrieved 5 Jan 2018, from https://news.microsoft.com/en-in/government-telangana-adopts-microsoft-cloud-becomes-firststate-use-artificial-intelligence-eye-care-screening-children/
11. Paul Y, Hickok E, Sinha A, Tiwari U (2018) Artificial intelligence in the healthcare industry in India. The Centre for Internet and Society, India
12. www.ima-india.org
13. India Brand Equity Foundation www.ibef.org

A Study on the Impact of Cognitive Radio Network in Cloud Computing

Keren Lois Daniel and Ankur Meena

Abstract The functions of cloud computing are basically to use remote assistance in providing the best service to the users through networks. In such cases, the concept of mesh network has to be taken into consideration. Growing trends of cloud computing in today's scenarios has brought the use of CRNs. The impact of cognitive radios is on an increasing scale due to the role of radio nodes which are formed with mesh topology. The mesh network improves the network performance in case of balancing throughput and the high traffic as it creates its own network with little hops for transmitting data. The main goal is to efficiently use the spectrum without any interference. The influence of CRNs in the cloud computing can bring a better output because cloud has the functionality of its super computational resources and the extent of its highest usage. The use of mesh network can reduce the cost-effectiveness with high bandwidth in network, and the focus of this paper is to find out how effective is CRNs in cloud computing.

Keywords CRNs · SU · PU · WMN · Cloud computing

1 Introduction

The growth of wireless network has led us to the use of spectrum efficiently. A "cognitive radio" is a radio that can change its transmitter parameters based on interaction with the environment in which it operates [1]. The main aim of the cognitive radio is to reconfigure the wireless transceiver automatically and adapt to the parameters which are required by the users. It is an intelligent program that intellectually and dynamically configures. The Federal Communications Commission (FCC) has distributed spectrums among the users [2]. The users are classified as primary users

K. L. Daniel (✉)
St. Xavier's College, Jaipur, India
e-mail: kerendaniel@stxaviersjaipur.org

A. Meena
B N University, Udaipur, India
e-mail: ankurmeena8777@gmail.com

S.-L. Peng et al. (eds.), *Computing and Network Sustainability*, Lecture Notes in Networks and Systems 75, https://doi.org/10.1007/978-981-13-7150-9_50

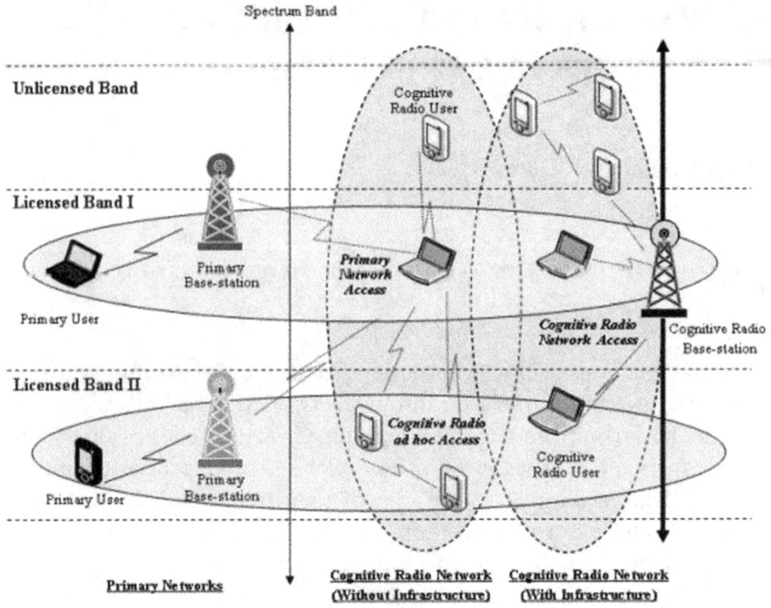

Fig. 1 Cognitive radio network architecture

and secondary users who are the un-licensed spectrum holders. The large amount of spectrum is assigned to primary users which are mostly unused which paved the way for many more dynamic accesses. The cognitive radio tries to minimize the interference of the external users by giving the radio frequency (RF) spectrum.

These radios help in detecting the available channels and then change accordingly the transmissions or receiving of packets from source to destination. The function of CR is to determine the geographical location to identify and authorize the users [3]. This is a dynamic form of spectrum management. But on the other hand, to find the unutilized spectrum band is quite difficult. So dynamically using the spectrum is the best way. Thus, a cognitive radio (CR) allocates spectrum in an opportunistic way for maximum use of spectrum (Fig. 1).

2 Characteristics of Cognitive Radio

- *Cognitive Capability*

It is the ability of the radio technology to capture or sense the information from its radio environment. It cannot be just realized by just monitoring the power of the frequency bands that it needs more advanced techniques through which we can capture the spatial variations in the environment and avoid interfering the other users that are present. With this capability, it is easy to identify the unused spectrum at a

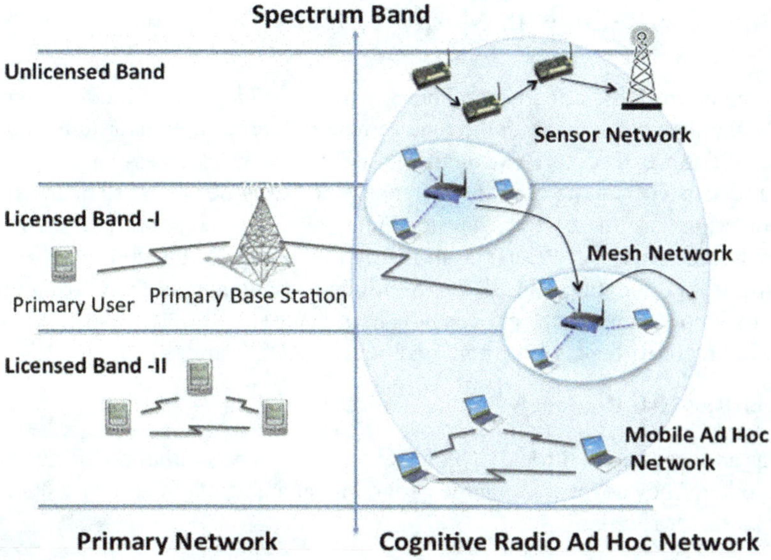

Fig. 2 Use of spectrum by CRAHN

particular location. So correctly, the unbeaten spectrum and sufficiently employed variables can be chosen.

- *Reconfigurability*

The reconfigurability helps us to dynamically utilize the radio environment. It can transmit and receive on various frequencies to access the technology present in the system. The term cognitive radio can be considered as very fragmented share of materialistic world for its utilization and contribution by collecting information from the surrounding. The fundamental agenda of cognitive radio will be acquiring the optimal obtain ability spectrum by cognitive capability and reconfigurability way as defined above. As the way spectrum is previously allocated, the main thing will be how they will portion the licensed spectrum cracking disturbing the scattering of primary users as shown in figure.

The cognitive radio allows the consumption in secular way to unallocated spectrum that would be either termed as spectrum hole/white space [4]. As the band seems acquired by licensed user, the cognitive radio switches other spectrum hole either remains in the unchanged spectrum band, amending the way of transmission power level or modulation type so that interference can be evaded (Fig. 2).

3 Cognitive Radio with Mesh Networks

It is an emerging cost-effective technology that provides high broadband connection. As the network increases, the requirement for throughput and higher capacity functionalities are needed. This cognitive radio allows us to access larger amount of spectrum. This CR can be used in mesh network that is dense in urban areas where the contentions are more. To meet the challenges and new opportunities, wireless mesh network moves very faster from its traditional way of allocation system to an open spectrum allocation and access. Significant challenges in the CRMN are that in real-time communication services how to give QoS.

The main constituents of the cognitive mesh network are as follows:

- **Cognitive Mesh Router**:

It acts as an access point through which many users in a particular region can connect to each other. They cover a very large geographical area and dynamically select end-to-end paths for routing.

- **Cognitive Mesh Client**:

The mesh clients are free to access the mesh routers any time that they require or have their own ad hoc network. In ad hoc network, they are connected to peer that is very close to them.

- **Gateway**:

They help in connecting to any wired or wireless network.

The area where the cognitive radio network functions can have an increased output when they mesh with the help of the cognitive access point and fixed cognitive relay nodes. The cognitive radio network has an ability to add fixed spectrum or temporary spectrum to the infrastructure when there is heavy traffic load in the links.

4 Challenges in Cognitive Mesh Network

The use of energy to transmit resources efficiently, with the available spectrum without the interference in the network, is a challenge. The dynamic allocation of channels with different characteristics securely is the major aspect to be considered while creating a routing algorithm. It can be classified

- Security
- Portability
- Reliability
- Bandwidth cost.

5 Significance of Cloud Computing

Cloud computing is on a rising edge to a destination where there is no return. It gives a challenging opportunity. It represents the future world of Internet. It is a combination of internet, resources, information, software, hardware that are shared without any hiding [5]. The computational part of the data is done in the cloud where optimal utilization of resources is done. Cloud is further classified into four models—Private, Public and Hybrid. A public cloud is like an open source that is available to all through Internet but is less secure. But a private cloud gives service through a single owner architecture designed for their desired purpose, whereas the hybrid is a mixture of both which give the best result when data is shared. The role of networks is the heart of intelligence that plays a vital role in controlling the security of the data and with minimal cost, effectively delivering end-to-end dealings of cloud with a balance between the users. When

6 Role of Cognitive Networks in Cloud

This helps the users to take the best action for analyzing and recognizing the related data by adapting to the alternatives provided with optimal decision. To respond to the ever-changing demand in businesses network can be best suited with cloud for reliability, security and complying with the needs of cloud or else it can be adverse. Since the data needs to be critically analyzed by both the external and internal users, the efficient use of network keeps these together intact (Fig. 3).

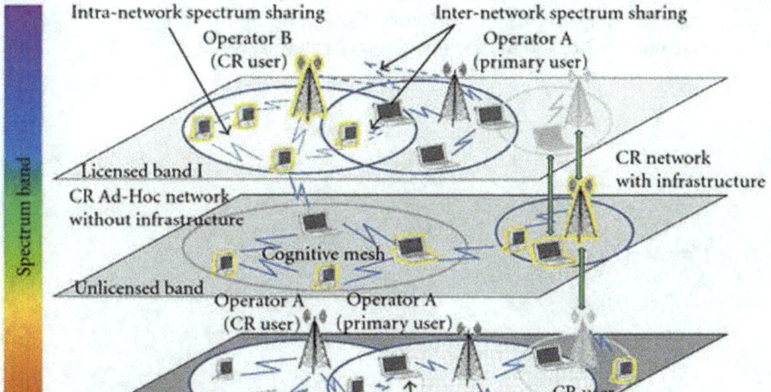

Fig. 3 Cloud with CRMN [6]

7 Conclusion

To gain the transformation in the competitive world of cloud computing, we need to have cognitive network in place to control large amount of data that are being transmitted to many users. With cognitive radio and cloud computing working together would meet the demand for services that are virtual storage and productivity of time management efficiently with the available energy resources. When the data is extracted from various networks, the energy utilization is more and the optimal use of the spectrum without interference makes a perfect combination for a cloud management. It is considered to be the most intelligent and effective communication where the opportunistic use of bandwidth in the spectrum that is unused. Therefore, a cognitive radio mesh network can be considered as bridge to cloud computing services in network.

References

1. Chowdhury KR, Akyildiz IF (2011) CRP—A routing protocol for cognitive radio ad hoc network, member, IEEE, Fellow, IEEE. IEEE J Sel Areas Commun 29(4)
2. Fcc E (2003) Docket no 03222. Notice of proposed rulemaking and order
3. Satyakonda PNV, Pavankumar T, Suresh Babu E (2015) Routing on cognitive radio wireless mesh networks survey. Int J Adv Res Comput Sci Softw Eng 5(2)
4. Mitola III J (1999) Cognitive radio for flexible mobile multimedia communication. In Proceedings of IEEE international workshop on mobile multimedia communications (MoMuC)
5. Namboodir V (2010) Towards sustainability in portable computing through cloud computing and cognitive radios. In: 39th international conference on parallel processing workshops, pp 468–475
6. Irianto KD, Kouvatsos DD (2014) An investigation of performance versus security in CRNs with supporting cloud platforms. Int J Electron Commun Eng
7. Furht B, Escalante A (2010) Handbook of cloud computing. Springer

Influence of Information and Communication Technology on Telecommunication Industries

Chetan Mali and Pooja Shrimali

Abstract Nowadays, the Indian telecom sector is one of the huge and driving businesses on the planet interfacing distinctive parts of the nation through different modes of communication. Now the connecting services like phone, satellite, and Web are in lead role due to development of Internet. The Telecom Regulatory Authority of India administers this sector by giving an administrative structure and ideal condition for its productive task. The Indian telecom sector remains as second biggest on the planet because of its fast progression. Services provided by this industry are effortlessly open at moderate costs to the clients of urban and rural zones. The present position of telecommunication industries in India is fully supported by the information communication technology (ICT), and the live evidence is the availability of internet facilities. The development of use of smartphone and web-based application is due to Internet. The current study has subsequently been attempted the development of Indian telecom industry in the special context of information communication technology (ICT) with the telecom policies

Keywords Telecommunication sector · Government · Policies · Growth · ICT · Internet

1 Introduction

It is quite obvious that India is holding second position in the worldwide telecommunication market. According to the data over 1.418 billion subscribers are part of this sector which is noticed in April 2018. For the duration of financial year 2007–2018, the wireless user contribution gives evidence of a CAGR of 19.62% to achieve 1,183.41 Mn.

C. Mali (✉)
Department of CS & IT, JRN Rajasthan Vidyapeeth, University, Udaipur, Rajasthan, India
e-mail: mali.chetan05@gmail.com

P. Shrimali
Informatics & Computational Sciences, Mohanlal Sukhadia University, Udaipur, Rajasthan, India
e-mail: pooja_shrimali@rediffmail.com

S.-L. Peng et al. (eds.), *Computing and Network Sustainability*, Lecture Notes in Networks and Systems 75, https://doi.org/10.1007/978-981-13-7150-9_51

One of another fact which is very supportive to modern telecommunication system is that India is also holding the tag of second-leading nation of Internet users by means of 493.96 million Internet users, on March 2018. So it is very clear that the information communication technology (ICT) boosts the telecommunication sector. Another factor which is associated with both of the above-mentioned factors is that in the present scenario, India is the worldwide second-leading smartphone user marketplace. It is also predicted that in India, there would be approximately 01 billion exclusive cell phone uses by the year 2020.

If we talk in terms of revenue generation by the telecom accessories and equipment sector, it is expected to cultivate to US$ 26.38 billion by 2020. Due to this success ration, the NTP 2018 has imagined to catch the attention of investments of US$ 100 billion in the sector till the year 2022. Nowadays, our telecom world is probable for high growth, empowered by development of data income and privileged in rural market. It is noticeable that at present status of rural area goes to 56.68% in April 2018. These all are due to the use of Internet by the middle class of the Indian population. So Internet use by the common person is the chief reason for the development of telecommunication industries. Powerful policy for ICT support from the government has been vital to the sector's growth and expansions. FDI ceiling in the telecommunication area gets to rise to 100% from 74%. Furthermore, now the Indian government is resultant with a novel "National Telecom Policy 2018" in fast technical progression in the sector over the last times.

2 Objectives

Major objectives for the study:

- To identify trends in Indian telecom sector.
- To know the progress and development of telecom industries in the context of ICT.
- To know about growth and prospect in the telecom sector.
- To understand the role of ICT in telecom industries.

3 Methodology

The current research is empowered by the secondary data taken from the "Telecom Regulatory Authority of India," Dept. of Telecommunication (DoT) and the reports from Govt. of India and similar sources. Also research paper and different telecom magazines were consulted for congregation of information. Data concern to development of Internet is also considered to make conclusion of the study. The primary data was also collected through survey and interviews with intellectuals which are employed in the sector. In the way of achieving the objectives of the research, the development of telecom industries was considered with the development of ICT.

4 Discussion

This is the study to identify the role of information communication technology (ICT) in the development of telecommunication industries in Indian scenario. The study not much focused on the historical background of the telecommunications. It tries to depict the actual scenario in the present world and actual performance of telecommunication industries with support of Internet. The wireless communication system with the web-based application gives new heights to the Indian telecom industries. An assortment of landmark of the Indian telecom industry like public and private sector telecom operators, the progress of telephones (landline and cell phone), tele-density in rural and urban areas, the wireless and wire-line communication along with their market share has also been considered and also discussed the relation between ICT and telecom industries developments.

- IT-enabled Services

At present, the Indian mobile financial system is rising speedily and will put in substantially to gross domestic product (GDP), as per the report of "GSM Association" (GSMA) in association with the "Boston Consulting Group" (BCG). Downloading of app in the nation raises just about 215% during the year 2015–2017.

Moderate and reformist policies have been involved with well-built user demand in the fast expansion. The concept of easy market access with business supportive policies is accepted. Also reasonable and down to business control structure has made sure the development of telecom services to user at reasonable cost. For all those "foreign direct investment" (FDI) norms have given opportunity to remain stay in top five employment generator sector for nation.

Also it is noticeable that directly or indirectly approximately four million jobs will be generated by the sector in the upcoming five years. The job scenarios for the prospect are probable from the efforts and liberty given by government to develop telecom sector in the rural areas. The demand for smartphone and rising Internet usage is also considerable for the same.

- Market size and ICT support

In the context of market size, cell phone company is projected to produce an overall monetary value of Rs 14 trillion (US$ 217.37 billion) at the end of year 2020. There would produce about three million straight job prospect and two million circumlocutory jobs.

The growth of smartphone in Indian market is approximately 14% per year. In the year 2017, the growth reached to total of 124 million. Growth in cell phone reaches to everyone, and downfall the price of data will introduce 500 million fresh Internet users for nation in the upcoming five years. This would be big occasion for up-to-the-minute trade. Also it would be considerable that the data usage may be increased up to 3.9 GB in 2017 to 18 GB by 2023 monthly.

5 Present Scenario of Indian Telecom Industry and ICT

According to the data of month July 2018, by the telecom regularity authority of India, the development of telecom industries is very much supported by the Internet and Internet-enabled facilities. Some of the noticeable points are:

- The number of telephone users in India crossed the figure of 1,175.01 million by January 2018.
- Almost 98.04% users are associated with wireless services in India by January 2018.
- Number of wireless users in the urban areas are now crossed the figure of 56.76% in the nation (Fig. 1).

As per Fig. 2, it is clear that we are holding third leading position as Internet users in the world at present. We posses number two positions worldwide as Internet user during the year 2017. The details of data depict that 7 out of 8 users accessing Internet from their mobile phones. Since 2012, the share of time spent on watching videos on mobile device has grown 200 tears a year. These pictures clarify the role of ICT "internet" to develop the telecom industries in India.

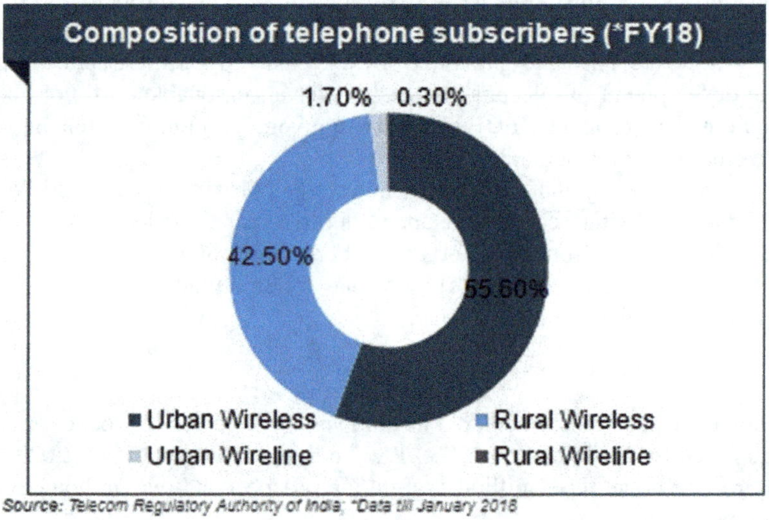

Fig. 1 Share of wireless and wire-line in 2018

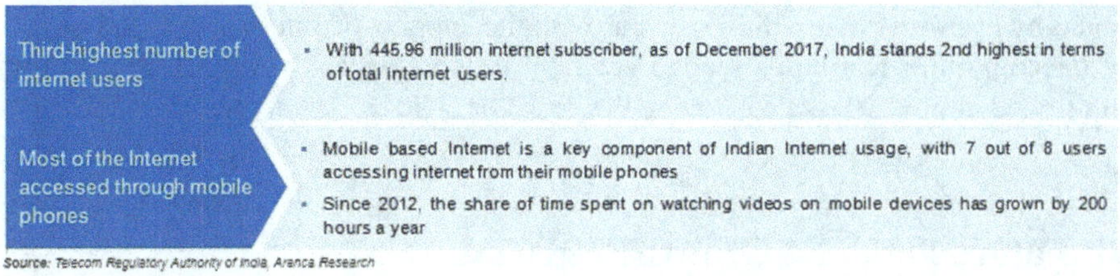

Fig. 2 Internet and mobile phone users

At present, the subscribers' base of the Indian telecom industries is of 1,206.22 million. This figure is very challenging for any other country. With this huge user base, another reality is:

- World's fastest growing app market
- World's second largest telecom network
- World's third highest number of Internet user (earlier in 2017 we possess second positions).

6 Sector Trends and ICT Role

According to sector trends, it is found that the wireless subscription captures remarkable growth, in the year 2007 the wireless subscription was 165 (Mn) which is increased drastically, and till now it is 1183.41 (Mn). Such speedy development takes place with the help of Internet support. The figure shows the use of Internet from the year 2006–2018 (Fig. 3).

The year 2012 is the year from there the Internet comes in the use with the launching of the smartphones for common public. The fast development may be easily

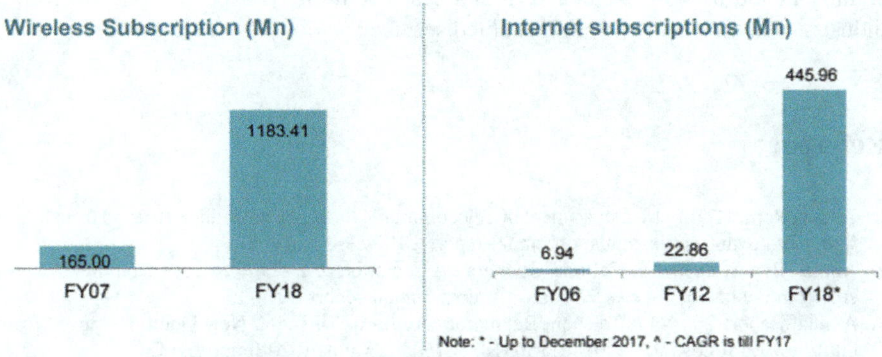

Fig. 3 Comparative and growth scenario of Internet (2006–2018)

noted by everyone that within six year with the support of Internet/ICT, the growth of telecom industries holds second position on the planet. The use of Internet also multiplied almost 20 times between the year 2012–2018.

7 Future Path and Growth Scenario

In the present and future prospective, India will come into view as a top performer for virtual world with 700 million Internet consumers of the 4.7 billion worldwide users by 2025, according to a Microsoft report. The scenario if ICT shows that Internet economy may touch Rs 10 trillion (US$ 155 billion) by the end of 2018, contributing about 5% to the country's GDP. By means of favorable regulation for telecom sector and 4G services striking the bazaar, the nation sector is projected to eyewitness rapid expansion in the coming times. The Indian govt. tries to launch 5G spectrums in bands 3,300 MHz and 3,400 MHz to endorse new technology and its associated applications like Internet of Things (IoT), high-definition (HD) video transfer, and the smart cities initiative. The production of supporting accessories like mobile phone battery chargers will result in setting up of 365 factories, thereby generating 800,000 jobs by 2025.

8 Conclusion

From the evidence received by the authentic and authorized Web sites of Indian government, in general it may be concluded that telecom sector gives significant site in the development of social and economic area of the nation. The role of ICT with special reference of Internet service is too positive in the present time. It is necessary to support for technological development of the country. Now providers of telecom services place forward voice and data services to the user in the whole country including both urban and rural areas. The future prospective of the telecom industry in association with ICT-enabled services is too bright in the nation.

References

1. Annual Report 2017–18, Department of Telecommunications, New Delhi. Retrieved from http://www.dot.gov.in/reports-statistics/annual-report-2017-18-0
2. Annual Report 2014–15, Cellular Operators Association of India. Retrieved from http://www.coai.com/media-room/news…./coai-in…/coai-annual-report-2014-15
3. Annual Report 2011–12, Telecom Regulatory Authority of India, New Delhi. Retrieved from http://www.trai.gov.in/…./20130115031838678 0062Annual%20Report%20En
4. Annual Report 2012–13, Telecom Regulatory Authority of India, New Delhi. Retrieved from http://www.trai.gov.in/…/AnnualReports/TRAI-English-Annual-Report-10032

5. Annual Report 2013–14, Telecom Regulatory Authority of India, New Delhi. Retrieved from http://www.trai.gov.in/…/AnnualReports/TRAI-Annual-Report-(English)=010420
6. Gopika GG (2014) Growth and development of telecom sector in India—an overview. IOSR J Bus Manag (IOSR-JBM) 16(9): 25–36. e-ISSN: 2278-487X, p-ISSN: 2319-7668 (Ver. I)
7. op.cit. Anjana, Prachi et al (2009) Telecom industry-business environment domain study. http://www.scribd.com/, 5.08 pm, 6 Jan 2012
8. Pritish and Saxena T (2015) An analysis of the Indian Telecom Industry. IOSR J Bus Manag (IOSR-JBM) 17(10):35–42 (Ver. II)
9. Media Reports and Press Releases, Cellular Operators Authority of India (COAI), Telecom Regulatory Authority of India (TRAI), Department of Telecommunication (DoT), Department of Industrial Policy and Promotion (DIPP), India Services Sector Report by Deloitte

A Perception of ICT for Social Media Marketing in India

Sanjay Gaur, Lokesh Sharma and Darshanaben D. Pandya

Abstract In the present scenario, social media marketing plays an important role to develop business activity. Enhancements in social media marketing are supported by information communication technology which gives a pace to develop a new era of marketing and business. The combination of ICT and SMM provides a comfort to show positive involvement in the marketing and acts according to the scenario. The development of Internet and 4G technologies boosts social media in the direction of price control in respect of specified product and services. These circumstances are a firm base for small-scale industries and entrepreneurship development. The present study tries to provide an insight of information communication technology in the favor of development of social media marketing in India.

Keywords SMM · ICT · Digital marketing · Social media · Internet

1 Introduction

The growth of information communication technology boosts up various sectors of daily life. Similarly, online social networking gets empowered through the development of ICT. Nowadays, social media marketing is an autonomous and self-sufficient platform, which is developed by the support of ICT. In general, the term social media marketing (SMM) refers to a method that targets social networks and its applications to extend brand consciousness or endorse particular products and services. The social media marketing works in the form of campaigns and online advertisements. The

S. Gaur (✉)
Jaipur Engineering College & Research Center, Jaipur, India
e-mail: sanjay.since@gmail.com

L. Sharma
JRN Rajasthan Vidyapeeth University, Udaipur, India
e-mail: lokeshsharma.sharma190@gmail.com

D. D. Pandya
Madhav University Abu Road, Sirohi, India
e-mail: pandya_darshana@rediffmail.com

S.-L. Peng et al. (eds.), *Computing and Network Sustainability*, Lecture Notes in Networks and Systems 75, https://doi.org/10.1007/978-981-13-7150-9_52

word campaign in the context of social media marketing usually focuses on the following points:

- Set up a social media existence on various platforms;
- Producing shareable information in the form of advertisements;
- Cultivating feedback of campaign in the course of surveys.

As per the above-mentioned points, we can say that social media marketing is a regressively targeted type of promotion and advertisements. Therefore, it is more effective marketing to create brand awareness than traditional marketing. The network of online media exposed the product and services very frequent as one cannot amaze. The prosperous data is accessible via social media, and also, media focuses specific audiences. In the future, the specific audience would be the potential customer with better marketing outcomes.

2 ICT-Empowered Social Media Marketing

It is obvious that the backbone of social media marketing is Internet. Strong foundation of information communication technology definitely empowers dependent services. Therefore, one can easily say that social media marketing is empowered by ICT. Social media destinations are valuable for building social and business systems along with trading belief. It is part of a pattern, which is known as Web 2.0 in the technical form. It is a collective utilization of the Webs that improves inventiveness and learning trade. It is a perceptive and client-driven approach to enable clients for the utilization of applications and administrations. It is basically a wide area of exposure and linked together to share information and further activity. Various advancements that are open and reasonable like Ajax (a technique for building intelligent applications for the Web that procedures client asks for immediately) and Really Simple Syndication (RSS). Social media marketing utilizes Webcasts, wikis, blogs, folksonomies, and online recordings, photograph sharing, news, message, and posts to achieve focused people. The subsequent diagram highlights different types of social media and corresponding (Fig. 1).

3 Motives to Use Social Media

The ICT-enabled social media marketing is now backbone for boosting medium- and small-scale industries. Such marketing takes less financial assistance and produces more as desired. The main reason behind the adoption of social media marketing is:

- One can achieve an extensive number of traffics without paying huge advertising expenses.

Fig. 1 ICT-enabled social media

- The utilization of blogs and business organizing destinations can expand traffic to your Web site from other social media. This may expand your page rank, bringing about expanded traffic from driving search engines.
- Social media supplements and other marketing procedures, for example, a paid advertising effort.
- One can earn goodwill by valid contribution in applicable zone.
- Social media destinations have data, for example, client profile information, which can be utilized to focus on a particular arrangement of clients for advertising.

Above given advantages creates an impression that there are no drawbacks in social media marketing. It gives a targeted audience marketing which requires very low investment.

4 ICT-Enabled Marketing Strategies

The ICT-enabled social media marketing takes place by the help of various tools and application. These applications and tools are the way by which strategies of the organization may be applied on the social media. Operations which are committed SEO and SEM is the central points of digital marketing are boosted by ICT. Search engine provides tools and application to promote the digital marketing and concern activities:

- **Search Engine Optimization**

Search engine optimization (SEO) includes the approaches with Web environment that improve the traffic to your Web site during the search. The searching names as SEO include various procedures to expand Web site traffic. These procedures are available in the various forms which are managed by the various internal algorithms. The organic form of such types of searching is known as "search engine optimization" (SEO), whereas the paid form is known as the "search engine marketing" (SEM). The basic purpose of both the searching is traffic diversion.

- Affiliate Marketing

A business contracts with different people or organizations ("affiliates") that showcase the organization's items for a commission (on deals or leads) are accountable under the affiliated marketing. Affiliate marketing covers with other Internet marketing techniques since affiliates can utilize an assortment of marketing systems. Those strategies incorporate SEO, SEM, SMM, e-mail marketing and show advertising. A typical sort of affiliate marketing happens where a Web publicist or trader initiates Web site administrations to put the vendor's flag advertisements or catches individually. Web site administrations will get a referral expense or commission from deals when the client has tapped the affiliate interface.

- **Online Display Advertisements**

Online display advertisements are concern from the promotion of the products with the help of images and videos on the Internet through various Web sites. The Web site pages are one of the resultant pages of search engine keywords. These promotional pages are almost paid by the business units and corporate house. The online display advertisements are also promoted by the Google AdWords and other campaign units (Fig. 2).

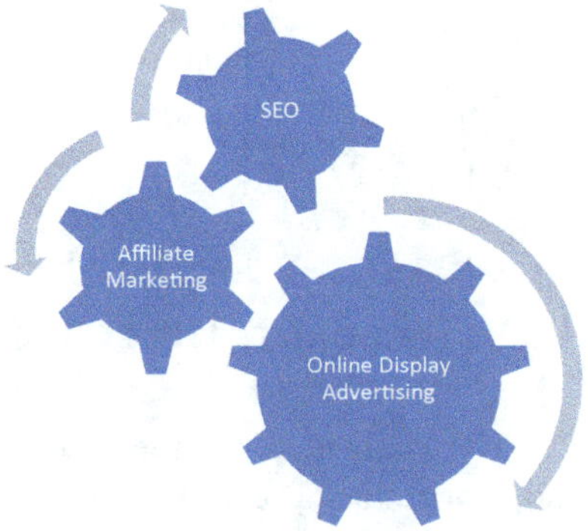

Fig. 2 ICT-enabled base marketing

5 Scope and Objectives of Social Media

The main objective of the social media strategy is as follows:

1. To know about the brand awareness and its development on social network
2. To form social communities and find out the targeted audience
3. To enhance customer loyalty and belief
4. To examine and monitor the customer feedback
5. To manage the social followers for intellectual world and business activity.

Scope of Social Media Marketing in India

Social media is one of the fastest catching up activities among all marketing activities available now. But, this trend is not apparent visible in India. The scope of social media marketing in India is very wide, but the understanding of media in Indian people is very low. They are mainly looking for entertainment on the social media. There is need to understand the commercial role of bout "digital marketing", that it **need of time and one have to understand the significance of social media** over traditional media. To understand the scope of social media marketing, one has to understand the meaning of return of investments.

6 Social Media Marketing Network Users in India

As per the "Statista", the position of social media marketing is going very strong day by day in the whole world as well as in India. As per one of the assumption, it is predicted that by 2019, there will be around 258.27 million social network users in India. It is noticeable that the figure for the same in 2016 was approximately 168 million. There is drastic growth recorded in last three years. The prediction does not stops here; it is assumed that by 2022, it will be around 370 million (Fig. 3; Table 1).

7 Global Social Media Research Summary 2018

The worldwide scenario of social media users and ICT-enabled technology users depicts that in the year 2018, the Internet user will be 4.021 billion with the 7% growth from the year 2017. In the same way, mobile phone users are 5.135, and it records approximately 5% growth from the year 2017. Similarly, the figure for the social media users is 3.196 billion and the growth recorded here is approximately 13% on the basis of the year 2017. So, we can say that ICT-enabled technologies like Internet and mobile phone are having continues growth, but the growth of social media usage is just double than other similar technologies. So, it is very clear that the development of ICT-enabled technologies supports the development of social media marketing worldwide (Fig. 4; Table 2).

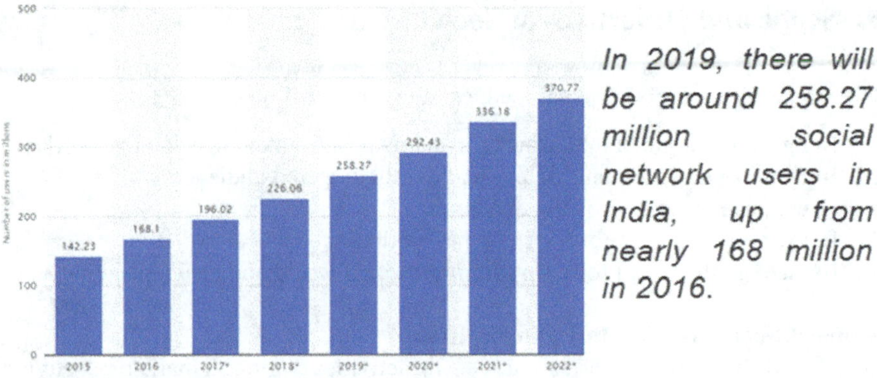

Fig. 3 Data about social network users in India (actual and prediction). *Source* Statista

Table 1 Data about social network users in India (today, yesterday, and tomorrow)

Social media network users in India								
Year	2015	2016	2017	2018	2019*	2020*	2021*	2022*
Social network user (million)	142.25	168.1	196.02	226.06	258.27	292.43	336.18	370.77

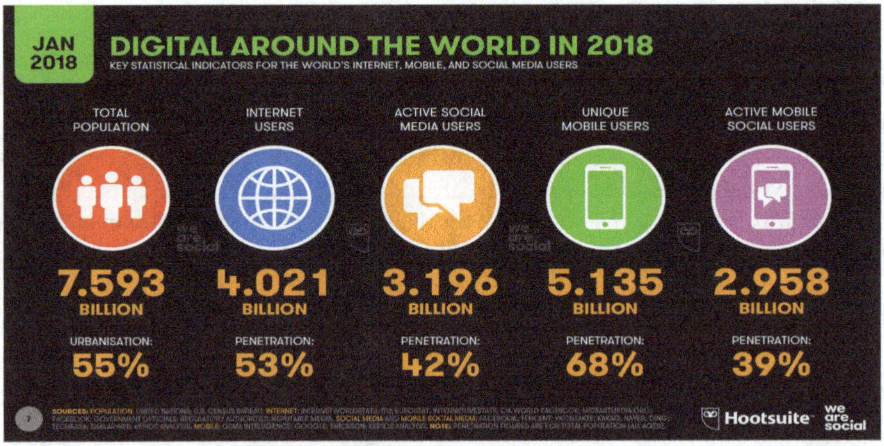

Fig. 4 Digital word summary 2018. Reference: https://digitalreport.wearesocial.com/

Table 2 Worldwide ICT and social media users (2018)

Worldwide social media research summary 2018		
User type	Number (billion)	Growth per year (%)
Internet users	4.021	7
Social media users	3.196	13
Mobile phone user	5.135	4–5

It is observed that in the commencement of the year 2018, according to US census bureau, the worldwide urban population was approximately 7.593 billion, in which 4.021 billion are Internet users. Whereas 3.135 billion are mobile phone users and 2.958 billion are the users who are using social media by the help of mobile phone where as total 3.196 people are social media users. So, we can say that the support of ICT plays an important role to develop the social media uses.

Yearly expansion is continued, particularly by the users using mobile phone to access social media is 39% penetration up 5% from 2017. Contribute to of Web traffic by gadget extremely favors mobile at 52% (+4% year-on-year change), at the same time as standby PC remains in 2nd position by means of only 43% of gadget contribute to entire Web pages, downward by 3% year-on-year.

8 Internet Penetration Worldwide

If we are looking toward Internet penetration region-wise, it is found that the Northern, Western Europe and Northern America encompass the biggest Internet saturation. At the second stage, Eastern, South Europe, Central America, Oceania, Western Asia, and South America have the middle-level Internet saturation with 77–61% Internet client measure up to entire populace (Fig. 5).

The regions of Caribbean, Northern Africa, Central Asia, Southern Africa, Eastern, and South East Asia have average penetration with around 50% Internet users as

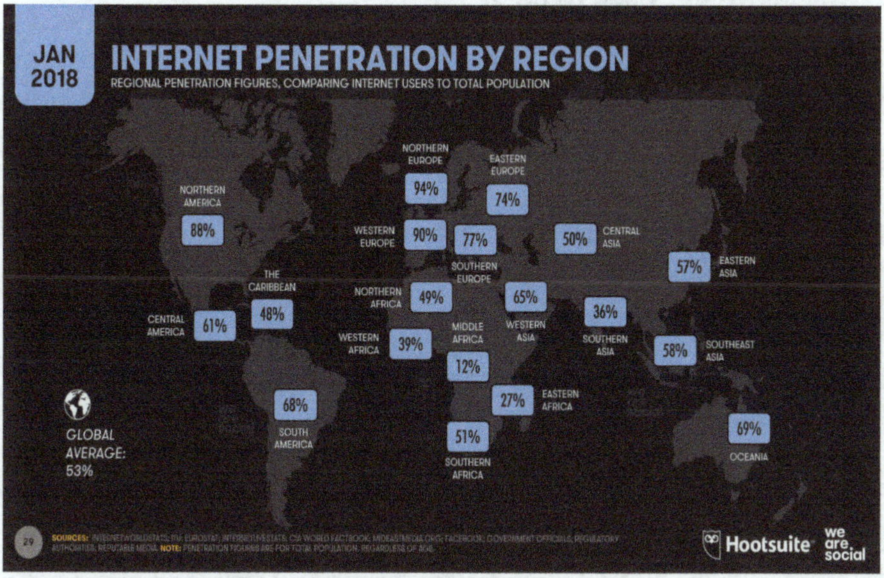

Fig. 5 Worldwide Internet penetration 2018. Reference: https://digitalreport.wearesocial.com/

Table 3 Worldwide Internet penetration 2018

Worldwide Internet penetration 2018					
WSN	Region	Penetration (%)	SN	Region	Penetration (%)
1	Northern America	88	11	Western Asia	65
2	Central America	61	12	Southern Asia	36
3	The Caribbean	48	13	South East Asia	58
4	South America	68	14	Northern Africa	49
5	Northern Europe	94	15	Western Africa	39
6	Eastern Europe	74	16	Middle Africa	12
7	Western Europe	90	17	Eastern Africa	27
8	Southern Europe	77	18	Southern Africa	51
9	Central Asia	50	19	Oceania	69
10	Eastern Asia	57	20	Global average	53

Reference: https://digitalreport.wearesocial.com/

compared to total population. Rest of regions is having very less penetration as compare to others. The global penetration average is 53% of the worldwide population (Table 3).

9 Social Media User Growth

The worldwide annual growth of social media users for the year 2018 is shown. Growth is shown on the basis of the data of 2017. Here, it is found that the worldwide growth is 13% as compared to the last year. The growth percentage for the India is 31% annually (Fig. 6).

The highest growth percentage for the year 2018 is 32% recorded by the Saudi Arabia. So, it is observed that as compared to other countries, the growth rate of social media users in India is much rapid than the average growth of the world.

10 Time Spent Per Day on Internet

As per the survey report, the average numbers of hours spent on the Internet per day are depicted in the graph. From the figured graph, it is noticed that one common Indian user spend at least 07 h and 25 min per day on the Internet (Fig. 7).

The minimum amount of time spending on Internet is 02 h and 35 min, whereas the maximum amount of time spending on Internet is 09 h and 38 min in a day. The utilization of the Internet per day by the users shows the association of the

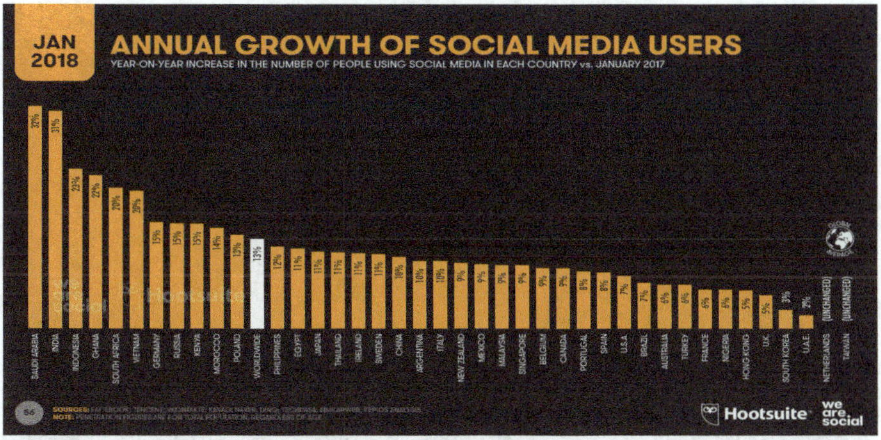

Fig. 6 Annual growth of social media users 2018. Reference: https://digitalreport.wearesocial.com/

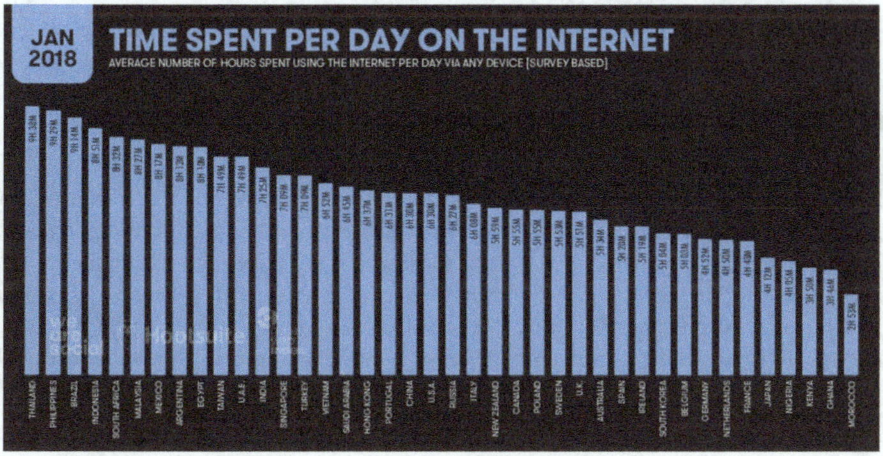

Fig. 7 Time spent by Internet users in a day, 2018. Reference: https://digitalreport.wearesocial.com/

technology. The association of Internet tool gives us occasions to interact with social media and other networking sites. So, it may conclude that the use of Internet by a user is directly concern to the use of the social media.

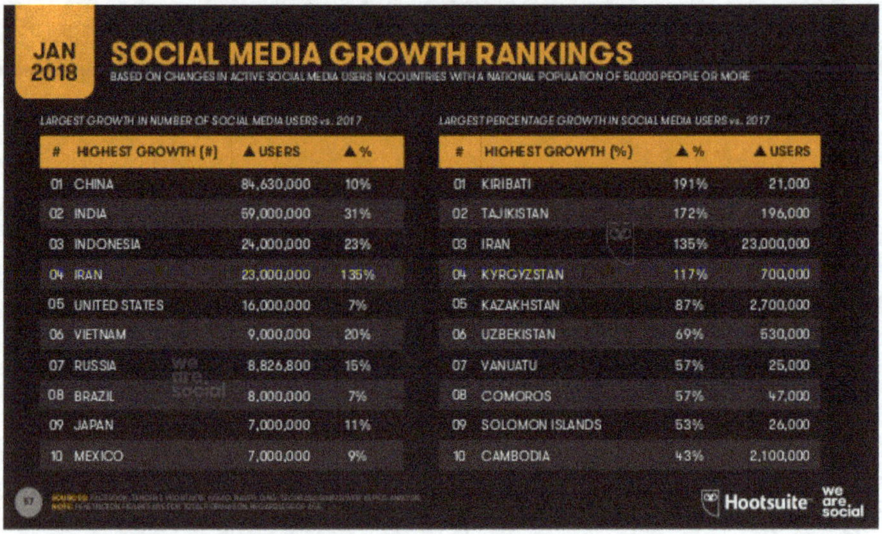

Fig. 8 Social media growth ranking, 2018. Reference: https://digitalreport.wearesocial.com/

11 Social Media Growth Ranking

The social media growth is one of the prime outcomes in the context of the use of Internet per day. It gives a relation between the ICT-enabled services and social media marketing and its expansion. According to Fig. 8, it is noticed that China holds the leading position in the social media growth and India leads with the second position for the same. At present, the figure of the social media user in India is 59,000,000, and it is 31% growth in comparison of the year 2017. So, one can be easily observed that the users of the social media in the India are possessing good growth as compared to the year 2017.

12 Conclusion

From the facts produce in the study, it is found that the use of social media marketing in India is rapidly growing day by day. The analysis of data and scenario of the year 2018 show that there is justified growth recorded in the favor of development of social media awareness. Similarly, the use of ICT tools and supporting technologies was depicted in the productive manner. By the help of data and evidences, it is found that there is a big role of ICT in the development of social media marketing in the India. So, we can say that the growth of ICT tools and services plays a favorable role in the development of social media marketing and it is proved in the Indian context by the present study.

References

1. Namrata A, Vandana K (2014) Growth of apparel industry in India: present and future aspects. Scholars World-Int Refereed Multidisciplinary J Contemp Res 2(1):64–70
2. Vanmathi A, Thangavel N (2016) An investigation of media marketing on Online consumer behaviour. I J A BE R 14(4):2465–2475
3. Richa Dahiya (2012) Impact of demographic factors of consumers on online shopping behaviour: a study of consumers in India. Int J Eng Manage Sci 3(1):43–52
4. Ioanas E, Stoica I (2014) Social media and its impact on consumers behavior. Int J Econ Pract Theor 4(2):295–303
5. Ranawat HS, Panwar A, Gour S (2018) Influence of social media on consumer purchase behaviour in Udaipur City. Int J Res Culture Soc 2(4):484–487
6. Mangold G, Faulds D (2009) Social media: the new hybrid element of the promotion mix. Bus Horizons 52:357–365
7. Davis MR, Malthouse E, Calder B (2010) Engagement with media. J Media Bus Stud 7(2):39–56
8. Anukrati S (2013) Study on e-commerce and online shopping: issues and influences. Int J Comput Eng Technol 4(1):364–376
9. https://digitalreport.wearesocial.com/
10. https://www.statista.com/

Particle Swarm Optimization for Flexible Job Scheduling Problem with Mutation Strategy

Kirti Choudhary, Geetika Gautam, Neha Bharti and Vijay Singh Rathore

Abstract FJSP provides the better solution for job scheduling at the given set of machines, but as the rapidly increasing environment of technology, it becomes complex to maintain and schedule the machines according to the optimal solution in a particular amount of time. In this paper, FJSP (flexible job shops scheduling problem) for the given set of the machine with the PSO (Particle Swarm Optimization) for optimization of it with mutation operator; the target is to reduce the makespan. Particle swarm optimization rule of the flexible job shop scheduling problem with mutation operator is introducing dissimilarity within the searching and scheduling procedure. Once the modification of the total records tends to decrease, the mutation method can begin. The proposed algorithm is implemented on DP data which shows a better result.

Keywords Flexible job shop scheduling problem · Job shop scheduling problem · Mutation operator · Particle swarm optimization

1 Introduction

In comparison with job shop scheduling problem (JSP), flexible job shop scheduling problem (FJSP) provides a much better result. On the available set of machines, FJSP operation is processed. In [1], all the machines are combined parallelly by having "WC" possible work centers. An operation or a task can be performed by any machine from any work center while consisting of a set of "m" machines. FJSP is considerably complex as compared to JSP, as it [2] has "n" number of "m" machines which work on flexible mode. This scheduling is the [3] solution to bottleneck problem. Figure 1 shows the layout of working of the FJSP where each work center (WC) has the corresponding machines and each machine is located in equidistant fashion. And the machines are labeled as $M_{x,\, y1-y2}$, where "x" denotes the job number, "$y1$" represents the "WC" number, and "$y2$" indicates the number of machines in each "WC".

K. Choudhary · G. Gautam (✉) · N. Bharti · V. S. Rathore
Department of CSE, Jaipur Engineering College and Research Center, Jaipur, India
e-mail: geetikagautam.cs@jecrc.ac.in

S.-L. Peng et al. (eds.), *Computing and Network Sustainability*, Lecture Notes in Networks and Systems 75, https://doi.org/10.1007/978-981-13-7150-9_53

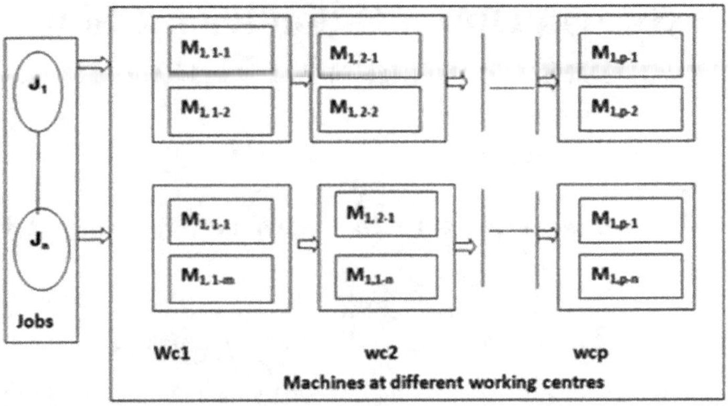

Fig. 1 Generalized flexible job shop problem

2 Literature Review

In the 1950s, the researchers were to introduce a several number of methods to solve the scheduling problems; heuristic procedure is one of the best among them. This literature analysis arises after 1990 because it came in focus for researchers.

Table 1 describes the review of flexible job shop scheduling (FJSP). Firstly, this algorithm was proposed by Brucker and Schile in the year 1990. The table below shows the ample techniques which are used to solve the FJSP problem by using heuristic techniques.

3 Proposed Methodology

Particle swarm optimization starts the search for particles at a limited diversity so it trapped after a fixed amount of time, so diversity has lessened within the population [4]. To overcome the loss of diversity, mutation is introduced in the proposed FJSP-PSO algorithm. Once the modification is decreased in the given set of archive, then the mutation process [5] begins. In order to overcome the scheduling problems of the traditional FJSP, the below flowchart shows the process of proposed FJSP-PSO algorithm with mutation operation.

3.1 Flowchart of Proposed FJSP-PSO Algorithm

See Fig. 2.

Table 1 Literature review on FJSP

Sl. no.	Year of publication	Author	Findings
1	1990	Brucker and Schlie	Introduces a technique for FJSP where machines processed those jobs which have the same processing time. If numbers of the jobs are more than two, then two approaches came into development: integrated approaches and hierarchical approaches
2	1993	Brandimarte	Tabu search has been used to decompose the problem of FJSP using the past literature dispatching rules and then concentrated on the resulting job shop subproblems
3	2002	Kacem, Hammadi, and Borne.	Proposed GA using two kinds of flexible approach and described the performance of FJSP. Total flexibility used an ideal assignment for solving the resource allocation, and partial flexibility used an assignment model
4	2006	L. Gao, C. Y. Peng, C. Zhou, P. G Li.	The author used a bottleneck shifting method along with the genetic algorithm (GA) for solving the FJSP
5	2007	M. Saidimehrabad and P. Fattahi	Used tabu search method in flexible job shop problem where two important factors come into consideration, i.e., sequence-dependent setups and alternative operation. Author comprises the proposed algorithm in two parts; firstly, it searches the finest sequence of jobs, and the researcher found the method to choose the best machine
6	2009	G. H. Zhang, P. G. Li, X. Y, Shao, and L. Gao	A hybrid algorithm combining PSO with tabu search (TS) algorithm. The researcher used PSO to assimilate local search and global search, and TS has been used to find a near-optimal solution for FJSP
7	2012	Rakesh Kumar, Ajai Jain	Introduce a simulation-based GA, where four different manufacturing circumstances have been considered to enhance the working performance of a job shop. And the main objective is to minimize the makespan, tardiness, and flow time
8	2014	Ye Xun, Ling Wang, Sheng-Yao Wang, Min Liu	Proposed to solve the flexible job shop problem with fuzzy processing time (FJSPF); this is used to represent an encoding scheme to signify the solution, and the decoding procedure has been used to get the feasible schedule in a fuzzy sense. Researcher incorporated teaching–learning mechanism and special local search operators to make a balance between exploration and exploitation

Fig. 2 Flowchart of proposed FJSP-PSO algorithm

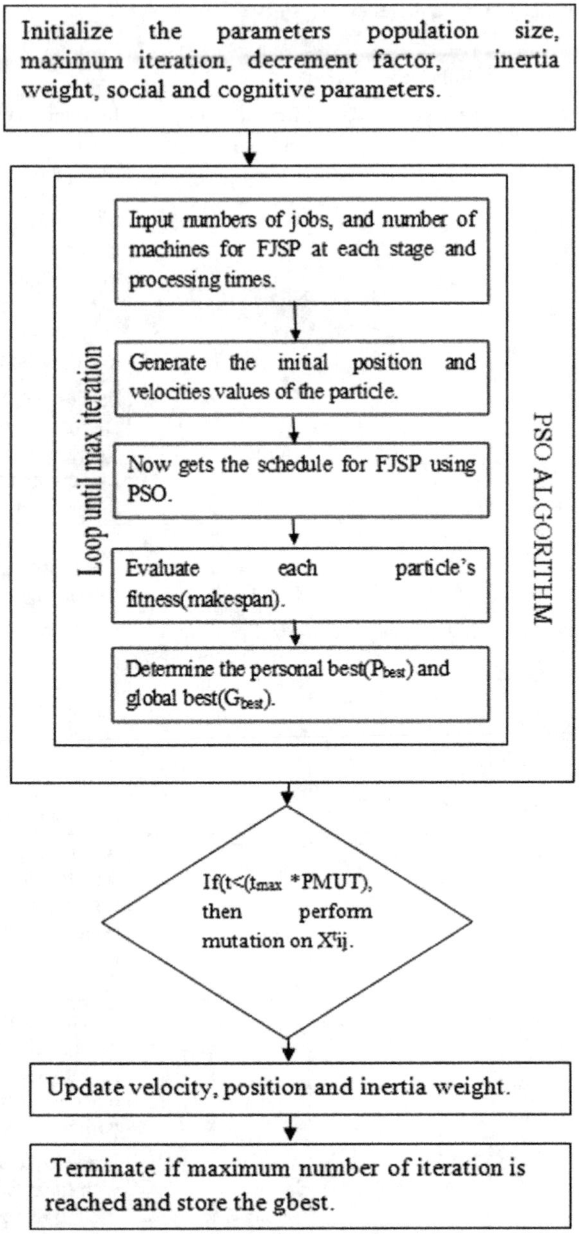

4 Result and Discussion

This study shows the analytical performance of FJSP with the core concept of the PSO algorithm to minimize the makespan for the given set of machines. The proposed algorithm is implemented in MATLAB 13 on the Windows 7 operating system, and it is tested on Dauzere-Peres [6] (DP dataset of 18 problems), which is universally approved benchmark instances in the FJSP. The range of operation varies from 15 to 25 for each individual job by using a number of machine ranges from 5 to 10 and the number of job ranges from 10 to 20. To compare the result of the proposed PSO algorithm with the other literature algorithms, relative deviation criterion is used. The relative deviation is obtained as follows:

$$\mathrm{Dev}(\%) = \frac{(C_{\max})_i - C_{\max}}{(C_{\max})_i} \times 100 \tag{1}$$

where makespan of the past literature algorithm is $(C_{\max})_i$ and the makespan of the proposed PSO algorithm is $C_{\max}$.

Table 2 illustrates the comparisons of the computational results of the DP dataset (Dauzere-Peres dataset) in reference to resultant makespan of the proposed FJSP-PSO algorithm. The proposed algorithm was also compared with best resultant makespan of the hybrid genetic algorithm (hGA) [13], effective genetic algorithm (eGA) [14], and TS [12] (Mastrolilli and Gambardella 2000), hybrid genetic algorithm (hGA) [13] and effective genetic algorithm (eGA) [14] (Zhang et al. 2011). The proposed FJSP-PSO algorithm produces a better result than hGA but less than eGA. The research experiment is done on six test cases by using a number of problems, name, and size of the machine and the value of lower bound (LB) and upper bound (UB).

5 Conclusion and Future Scope

From the result, the paper shows that the proposed algorithm of FJSP and PSO provides the better and nearest optimal scheduling solution for job scheduling for the given set of machines. To decrease the computational burden and impulsive convergence and to improve the solution density, the mutation operation is used with one of the genetic algorithms, i.e., particle swarm optimization (PSO). In this paper, the proposed algorithm was investigated on the given set of machine as well as with the given set of problems, which were generated from Kacem et al. [3] and Dauzere-Peres [6] (DP data). The results indicate that PSO produces a much better result than the well-known solution algorithms, which are described in the literature.

A more optimum schedule can be produced by considering other performance measures like machine loading capacity and workload on a critical machine. The setup time cannot be bounded or integrated with the processing time, but if the setup

Table 2 Results of the Dauzere-Peres instance

Problem	$n \times m$	LB, UB	Proposed PSO	eGA		hGA		TS	
				C_{max}	Dev (%)	C_{max}	Dev (%)	C_{max}	Dev (%)
1a	10 × 5	25,052,530	2505	2516	0.44	2518	0.52	2518	0.52
2a	10 × 5	22,282,244	2230	2231	0.04	2231	0.04	2231	0.04
3a	10 × 5	22,282,235	2229	2232	0.13	2229	0	2229	0
4a	10 × 5	25,032,565	2503	2515	0.68	2515	0.68	2503	0.2
5a	10 × 5	21,892,229	2207	2208	0.05	2217	0.45	2216	0.41
6a	10 × 5	21,622,216	2170	2174	0.18	2196	1.18	2203	1.5
7a	15 × 8	21,872,408	2264	2217	−2.12	2307	1.86	2283	0.83
8a	15 × 8	20,612,093	2073	2073	0	2073	0	2069	−0.19
9a	15 × 8	20,612,074	2066	2066	0	2066	0	2066	0
10a	15 × 8	21,782,362	2205	2189	−0.73	2315	4.75	2291	3.75
11a	15 × 8	20,172,078	2050	2063	0.63	2071	1.01	2063	0.63
12a	15 × 8	19,692,047	2019	2019	0	2030	0.54	2034	0.74
13a	20 × 10	21,612,302	2253	2194	−2.69	2257	0.18	2260	0.31
14a	20 × 10	21,612,183	2167	2167	0	2167	0	2167	0
15a	20 × 10	21,612,171	2165	2165	0	2165	0	2167	0.09
16a	20 × 10	21,482,301	2252	2211	−1.85	2256	0.18	2255	0.13
17a	20 × 10	20,882,168	2134	2109	−1.19	2140	0.28	2141	0.33
18a	20 × 10	20,572,139	2123	2089	−1.63	2127	0.19	2137	0.66

time depends on the execution time of current, previous and the next operations which are going to be performed on the machine, then the proposed PSO algorithm will be beneficial. Hence, setup time should be integrated into the scheduling problem as a critical parameter.

References

1. Muth JF, Thompson GL (1963) Industrial scheduling. New Jersey, Englewood Cliffs
2. Dave M, Choudhary K (2016) Job shop scheduling algorithms—a shift from traditional techniques to non-traditional techniques. In: Proceedings of the INDIA Com-2016, IEEE conference ID: 37465, ISSN 0973-7529, 3 rd international conference on "computing for sustainable global development, Mar 2016
3. Kacem I, Hammadi S, Borne P (2002) Approach by localization and multi-objective evolutionary optimization for flexible job-shop scheduling problems. IEEE Trans Syst, Man, Cybern 32:1–13
4. Xing LN, Chen YW, Wang P, Zhao QS, Xiong J (2014) A knowledge-based ant colony optimization for flexible job shop scheduling problems. Appl Soft Comput:888–96
5. Dave M, Choudhary K (2009) Particle swarm optimization with the mutation for flexible job shop scheduling problem. IJEE 9:16–21. ISSN-09737383
6. Dauzere P, Pauli E (1997) An integrated approach for modelling and solving the general multiprocessor job shop scheduling problem using tabu search. Ann Oper Res:281–306
7. Brandimarte P (1993) Routing and scheduling in a flexible job shop by tabu search. Ann Oper Res:157–183
8. Ruiz R, Maroto C (2006) A genetic algorithm for hybrid flow-shops with sequence dependent setup times and machine eligibility. Eur J Oper Res:781–800
9. Saidi M, Fattahi P (2007) Flexible job shop scheduling with tabu search algorithms. Int J Adv Manuf Technol:563–570
10. Zhang G, Shao X, Li P, Gao L (2009) An effective hybrid particle swarm optimization algorithm for multi-objective flexible job-shop scheduling problem. Comput Ind Eng:1309–1318
11. Garey MR, Johnson DS, Sethi R (1976) Complexity of fellowship and job shop scheduling. Math Oper Res:117–129
12. Mastrolilli M, Gambardella M (2000) Effective neighbourhood functions for the flexible job shop problem. J Scheduling:3–20
13. Gao J, Sun L, Gen M (2008) A hybrid genetic and variable neighbourhood descent algorithm for flexible job shop scheduling problems. Comput Oper Res: 2892–2907
14. Zhang GH, Gao L, Shi Y (2001) An effective genetic algorithm for the flexible job-shop scheduling problem. Expert Syst Appl:3563–3573
15. Kennedy J, Eberhart R (1995) Particle swarm optimization. In: Proceedings of IEEE international conference on neural network, pp 1942–1948. Washington, USA
16. Phanden RK, Jain A, Verma R (2012) A genetic algorithm-based approach for flexible job shop scheduling. Appl Mech Mater:3930–3937

An Empirical Study on Implementing Efficient Cashless System in Southern Rajasthan

Chandresh Kumar Chhatlaniand and Hemant Sahu

Abstract This paper reveals insights of rural and urban people of Southern Rajasthan on cashless system. Not only urban but also rural people put considerable impact on development of the country economically and socially. With the up-and-coming developments of the technology rural people can also go upwards. The study was conducted in Southern Rajasthan to understand whether the cashless system can be adopted by people of Southern Rajasthan or not. All the data have been collected with the help of a structured questionnaire and analyzed. Respondents' answers reveal that cashless economy can become a powerful tool in reducing black money and forged money and stealing of money. Many of them were disagreed that cashless economy system can fight with terrorism. However, it is general truth that, after adopting any software-based system not only economy but efficiency of getting various financial reports instantly increases. Many rural people rely on cash-based system rather cashless due to unawareness of computer system, bad hands in Internet, afraid from cyber fraud. Few challenges are very slow Internet connections, less or almost nil hardware support agencies, electricity shortage, etc. The result of the hypothesis testing using Pearson chi-square was showed that there is a significant relationship in the predicted attributes on the users' intention to use cashless system.

Keywords Cashless economy · Online transaction · Internet-based economy · Southern Rajasthan · Rural area

1 Introduction

Cashless economy has all its transactions through electronic mediums viz. Debit cards, Credit cards, Net Banking, National Electronic Funds Transfer (NEFT), Unstructured Supplementary Structured Data (USSD), Aadhar Enabled Pay-

C. K. Chhatlaniand (✉)
JRN Rajasthan Vidyapeeth, University, Udaipur, Rajasthan, India
e-mail: chandresh.chhatlani@gmail.com

H. Sahu
Geetanjali Institute of Technical Studies, Udaipur, Rajasthan, India
e-mail: hemantsahu@gmail.com

S.-L. Peng et al. (eds.), *Computing and Network Sustainability*, Lecture Notes in Networks and Systems 75, https://doi.org/10.1007/978-981-13-7150-9_54

ment System (AEPS), Unified Payment Interface (UPI), Immediate Payment Service (IMPS), Mobile Wallets, Point of Sale, Mobile Banking, Real Time Gross Settlement (RTGS) and Micro ATMs. The Government of India is also promoting Cashless Economy with the aim of minimizing cash robbery cases, reducing expenditure of various money-related processes, reduced black money and fake money, to fight with terrorism, among others. As per government "Faceless, Paperless, Cashless" is one of professed role of Digital India [1]. A survey on money transaction through Internet, banks and bank-related cards expresses that government of any nation, which wants cashless economy, have to done homework "not less than excellent" before launching cashless system, otherwise most of the people of that nation will suffer [2]. As the payment structure has a critical and important position in every financial system whether it is of a nation or a small business. Efficient payment structure is one the foundations for development of a nation. Policies related to payment structure also aims to improve the usefulness of money, economical augments and stability in various other systems. With digital payment systems, government of a nation has to think about stable infrastructure in terms of Internet, software and computer hardware, proper education of digital payments, security enabled payments, capacity building, awareness, and encouragement of digital payments among its people and among government systems. Government of India has also tried attracting people to use cashless system by introducing Lucky Grahak Yojna and Digi Dhan Vyapaar Yojna [7]. Few more schemes have come into force like zero Excise Duty on POS Devices and all goods for manufacture of POS devices, lesser charges for IMPS, NEFT, USSD & UPI, toll-free helpline for digital payments and BHIM Mobile App [1].

Cashless payment system is the momentous provider to the efficiency and strength of an economy [4]. Innovations in technology and financial models are constantly required for the efficiency of a secured cashless system [5]. As developed countries like USA having only 7 percent cash transactions of their total transactions and in India at first sight, the cashless system might seem to be an incredible initiative, the rural community of India is not accepting transactions without using cash-based money, using of credit and debit cards and/or net banking due to traditional culture of physical saving, illiteracy, unawareness, less support in terms of Internet, software and hardware, less confidence and not relying in adoption of new technologies [3].

2 Literature Review

Alvares, Cliford (2009) in their reports—there is a huge quantity of fake currency available in India. The battle against such currency is easier. In this report, it is also conveyed that however new technology's printing facilities can found fake currency but there should be more steps to be taken.

As per a **report by Price Water House Coopers in 2015**—people in India's who are not using the facilities of banking is about 233 million. Even the people, who are using banking facility, are not able to use bank cards efficiently. Efficient users of bank cards are only about 1.46 million.

In their study, **Maitanmi Olusola et al. (2013)** have concluded—that a number of tasks have already been initiated to implement cashless economy and numerous people agreed with the Indian government's vision in terms of loss of cash, reducing theft and armed robbery.

Vidhi P. Kakkad (2017) conveyed in their study—there is lack of awareness in many people, technology uneducated people are facing difficulties on using cashless system in India. Therefore, it is impossible to develop 100% cashless societies; therefore, the idea of less cash system seems better than cashless system. In any case, Indian people strongly believe that cashless systems are good sings of development of India.

Menariya P.M. (2017) investigated—the real-life users of bank cards and Internet banking in Rajasthan. He has used Statistical Package for Social Sciences (SPSS) to get the conclusions. Their result shows that acceptance of bank cards and Internet banking is significantly influenced by region, urban or rural, age, education, income, awareness, ease of use and political background.

Jeet Amrutiya and Aakash Antala (2016) expressed that—the first and the foremost pre-requisite for building an economy having no cash is to have every single entity, whether an individual or a small-scale or a large-scale firm, to be registered under unique IDs. This can be achieved biometrically, as has already been done in India with the advent of the Government's UID scheme named "Aadhar." And already, nearly 42 million bank accounts in India have been linked with Aadhar. Such feasible and low-cost biometric systems could easily support electronic payment systems which could replace the current hand-to-hand currency system. According to a RTI reply by RBI before demonetization, India's 86% cash was in the form of 1000 and 500 notes. It cost the central bank Rs. 3,917 crore to print Rs. 500 notes in circulation, and Rs. 2,000 crore to print the Rs. 1,000 notes in circulation currently. Denomination Cost of printing is (Rs.) 2000 4.72 1000(Old) 4.06 500(Old) 3.58 50 1.80 20 1.5 5 .50.

From the literature review, it is revealed that there are problems of discontinued electricity and slow Internet connection exists in rural areas; therefore, it is very hard to implement digital payments in rural and tribal areas. Even schools and colleges in such areas are fighting with these crucial challenges. There is no reliability and instability of point of sale networks. This is biggest challenge which may take years to be resolved. However, Government of India has started regular technical audit and taking measures for risk mitigations, hackers are also clever and becoming successful in breaking the security. The charges of a cashless system should be determined on the basis of location, age and economic conditions. Otherwise old, rural and tribal people may not able to make payment of such charges. Stability of software applications is also a big fear among rural people. If the system crashes, it would really hard to implement that system again, even if it is having best technology and other measures. Rural people would immediately deny that. Not all targeted people are literate and aware. Some of them even do not know how to use of net banking and debit cards. Educating people and making them aware is one of the biggest challenges in implementing cashless system.

3 Objectives of the Study

The major objectives of this study are:

1. To know whether the Cashless Transaction can be adopted among people of Southern Rajasthan or not.
2. To know if the people of Southern Rajasthan understand impact, issues and importance of Cashless Transaction System.
3. To assess the preparation of Indian and State Government to implement Cashless System in Southern Rajasthan.
4. To calculate the methodology and character of transactions made by the rural and urban users for the different uses of Cashless System.
5. To observe the risk associated with Indian Cashless System.
6. To identify the attributes having significant impact on Cashless System.

4 Research Methodology

This study is performed to acquire information about Cashless economy. The study is conducted in Southern Rajasthan. Other than few urban people, including the rural and tribal people, a sample size of 100 was selected using the convenience sampling procedure out of which 94 have been responded. This means response rate is 94%. The responses have been recorded through the reliability of survey. Survey method is used to collect primary data. To collect the data a survey instrument (questionnaire) was designed, which consists of 27 questions like The cashless system will minimize the fake currency, the country could avoid money laundering schemes, it can stop corruption in banking system, it can stop corruption in government system, this system can rise economic development, cyber crime will increase by using of Cashless System, literacy is very required to implement this system, transparency would be increased, efficiency would be increased, Internet fraud will be increased, etc. Respondents were asked to put their opinion on a Likert scale, which is from 1 to 5, or say from "strongly disagree" to "strongly agree."

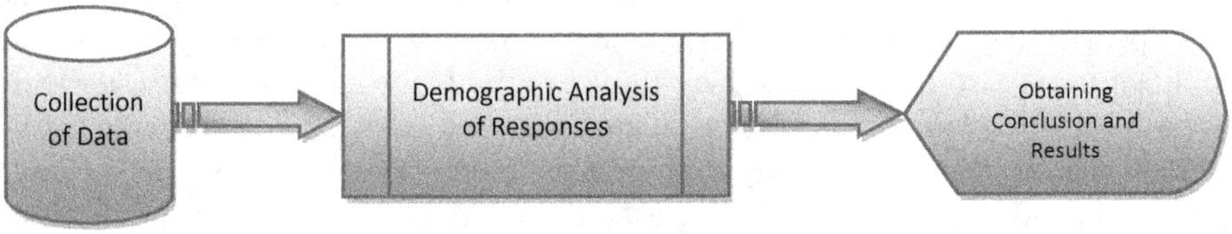

Table 1 Demographic profile of samples

Area			Age		
Group	Frequency	Percentage	Group	Frequency	Percentage
Urban	42	44.68	<25	16	17.02
Rural	52	55.32	25–35	46	48.94
			35–45	22	23.40
			>45	10	10.64
Total	94	100		94	100
Education			Cashless system's user		
Group	Frequency	Percentage	Group	Frequency	Percentage
Less than UG	26	27.66	Bank Cards	49	52.13
UG	42	44.68	Internet Banking	33	35.10
PG or above	26	27.66	Mobile Banking	12	12.77
Total	94	100		94	100

The major attributes which have been used to prepare the questionnaire are convenience, subjective norms, quality of technology, quality of basic resources, service quality and system accessibility according to area (urban or rural), literacy and age. Under these attributes, minor attributes have been mapped like Ease of Use, Usefulness, Creditability, and Trust. The research believes that above major and minor attributes can develop an mind-set then an intention to accept digital payments. The analysis of the responses is done by the Pearson chi-square method.

Following are the Hypotheses:

Null Hypothesis H0: $\mu d = 0$
H1: Intention has a positive significant effect on actual behavior.
H2: Mind-set has a positive significant effect on intention.
H3: Quality has a positive significant effect on Trust.
H4: Usefulness has a positive significant effect on intention.
H5: Creditability has a positive significant effect on Trust.
H6: Ease of use has a positive significant effect on mind-set.
H6: Age has a positive significant effect on mind-set
H7: Location has a positive significant effect on mind-set and Intention
H8: Literacy has a positive significant effect on mind-set and intention (Table 1).

5 Results and Discussions

As per the Table 2, it was analyzed that all means having values greater than 3.0, which clearly shows an overall positive response to the attributes that are calculated in this research. Standard deviations for all constructs was found less than one, therefore, it can be said that the item scores were around the mean scores.

Table 2 Descriptive statistics of the attributes

Attribute	Mean	Standard deviation
Intention	4.44	0.65
Mind-set	4.13	0.59
Trust	3.96	0.61
Actual behavior	3.79	0.57

Hypotheses were also evaluated by verifying the statistics and data. The null hypothesis is such that H0: $\mu d = 0$, which means usefulness, mind-set, ease of use, credibility, accessibility, location and literacy are not significantly associated with intention and age, location and literacy are not significantly associated with mind-set. On the other hand, H1 to H8: $\mu d \neq 0$ means that usefulness, mind-set, ease of use, credibility, accessibility, location and literacy are significantly associated with intention and age, location and literacy are significantly associated with the mind-set.

The result of the hypothesis testing by Pearson chi-square was found very similar to previous research that there is a significant relationship of the predicted attributes on the users' intention to use cashless system.

6 Conclusion

To achieve the fully implemented Cashless Transaction System in Southern Rajasthan, government has taken few steps, but these are not sufficient. It is found that age, location and literacy are significantly associated with use of Cashless System. People of old age, from rural and less literate are not having intention to use the Cashless System in Southern Rajasthan. The central and state governments need to make this system easier and safer. This system should be updated regularly, as the hackers are updating themselves continuously. There are a number of challenges in implementing Cashless System of Payment in Southern Rajasthan, which includes frequently loss of Internet connectivity, discontinued electricity, inadequate infrastructure, Internet frauds and other security issues, charges, instability of software and mobile applications, illiteracy and unawareness, etc. From the study, it is also found that if people understands the usefulness of Cashless Transaction, if it easy to use, if it is to be implemented by organizations having credibility, if it is easily accessible in every location, the mind-set of people can be diverted toward this system and ultimately they have plan to use this.

Acknowledgements The research was supported by the Janardan Rai Nagar Rajasthan Vidyapeeth Deemed to be University and Geetanjali Institute of Technical Studies. The authors acknowledge to Dr. Manish Shrimali, Director, Dept. of Computer Science and I.T., JRNRV Deemed to be University, Prof. Dr. Manish Saraswat, Professor, MCA, GITS and Prof.(Dr.) Amit Sinhal, Professor, Dept. of Computer Sc., GITS who have provided academic support and ethical approval to collect and analyze data. The full project design, methods and findings are reported elsewhere. Qualitative data from the focus groups and other meetings were transcribed and thematically analyzed and informed consent was secured from the parents/LAR of participants and participants themselves and continually re-negotiated throughout the project.

References

1. http://cashlessindia.gov.in/. Accessed on 07 Jan 2017
2. David S (2016) An empirical investigation of the level of user's acceptance of e-banking on technology acceptance model. Proceedings of international conference on e services and cashless economy, International Journal of Business and Communication Technology, volume II, issue II, pp 37–46
3. Srinivas N (2016) An analysis of the defaults in bank card payments. Econ Today 19–21
4. Lee J, Abdul-Rahman F, Kim H (2007) Debit card usage: an examination of its impact on household debt. Financ Serv Rev 16(1):73
5. Shenbagavalli R, Shanmugapriya AR, Lokeshwara Chowdary Y (2012) Risk analysis of credit card holders. Int J Trade, Econ Financ 3(3)
6. Odior ES, Banuso FB (2012) Cashless banking in Nigeria: challenges, benefits & policy implications. Eur Sci J 8:12–16
7. Ravi CS (2017) Digital payments system and rural India: a review of transaction to cashless economy. Int J Commer Manage Res 3(5):169–173. ISSN: 2455-1627
8. Boianovsky M (2017) Users observations on security pointers in internet banking systems. J Internet Bank Commer 24(1):15–19
9. Patel B, Amin U (2012) Plastic money: roadmay towards cash less society. Indian J Res 1(11). ISSN: 2250-1991
10. Kumar P (2015) An analysis of growth pattern of cashless transaction system. Int J Res Bus Manage 3(9):37–44. ISSN(E)2321-886X; ISSN(P): 2347-4572
11. Das A, Agarwal R (2010) Cashless payment system in India—a roadmap technical report 2010

An ICT Insight of Digitization of Banking in India

Sanjay Gaur, Lavika Jain, Garima Ojha and Nilam Choudhary

Abstract Digitization is basically the procedure in which all the manual or machine-driven exercises are brought out by the electronic gadgets in association with web-enabled services. In these days, the financial and banking area has a major influence on the economy of the country. Now, banks are taking control of supply of cash through the cashless, online transactions and payment gateway. These exercises are the central inspiration for digitize banking evolution of the country. It is a fact that well-constructed and developed banking area is a foundation for the development of a country. The role of ICT is very important for monetary movement. Indeed, even the digitization in banking is financial movement and is the part of information communication technology. So, ICT in banking is essential for the advancement of computerized banking. Presently, point of convergence is changing from traditional banking to smart banking. Electronic banking, mobile banking, electronic fund transfer, and other comparative items are now easily acceptable. This paper endeavors to give a genuine image of the digitization in economical segment with fundamental information study and most recent advancement.

Keywords Digitization · ICT · Banking · Electronic · Payment · Mobile

S. Gaur (✉) · G. Ojha · N. Choudhary
Jaipur Engineering College and Research Centre, Jaipur, India
e-mail: sanjay.since@gmail.com

G. Ojha
e-mail: garimaojha.cse@jecrc.ac.in

N. Choudhary
e-mail: nilamchoudhary.cse@jecrc.ac.in

L. Jain
Aishwarya Institute of Management and Information Technology, Udaipur, India
e-mail: lavijain_fms@yahoo.com

S.-L. Peng et al. (eds.), *Computing and Network Sustainability*, Lecture Notes in Networks and Systems 75, https://doi.org/10.1007/978-981-13-7150-9_55

1 Introduction

"Digital Banking" is the new trendy expression in the banking sector, with banks all around the world moving toward digitalization. All types of banks and associated areas are now influenced by digital activities. This is a target to make convinced to client for transaction and continue the present banking services. Furthermore, digitalization prompts strong banking and trust, which encourages banks to draw nearer to clients and stand in the competition. In general terms, digitization in banking industry basically implies making banking smooth and consistent for the customers.

The greater part of the private banks and open-sector banks is centered on offering new technology-based services to its customers like cloud banking, mobile banking, and e-wallets.

The greatest favorable position of digital banking is its capacity to give new facilities and customer with particular plans of action by investigating customer value. Therefore, banking makes the customer valuable with latest technology and less physical or manual transaction.

Now, information and communication technology (ICT) has turned into an essential instrument for an effective banking framework, and Indian banks have set up a genuinely solid infrastructure to use its advantages. Digitization is not only a possibility for banking industry, rather it is a truth in light of the fact that each industry is being digitized and banking sector is no special case for that. With the support of ICT, Indians are prepared to accept the cashless economy. In the present time, it is an easy decision that banks are the foundation of the economy of any country. The "Digital India" campaign can possibly change the Indian banking industry. While featuring the advancement of "Digital India," there are more than 12,000 rural branches have been connected from this movement.

The Digital India vision aims to change our nation into a digital economy. Over 190 million accounts have been opened under the consideration, in which around 38% of these being zero-balance accounts. It aims to accomplish transparent transaction all over the nation. Now, Indians are moving toward cashless economy. This is an economical consideration strategy. "Digital India" campaign is the policy of the government to control the stream of black money.

2 Customer Dimension for Product and Services

Nowadays, people in the whole world are living with digital facilities at work, at home, and anywhere. The separation between the computerized and physical cosmos is looking clearly and up progressively. Innovation energizes this duality by astutely utilizing an ever-increasing number of signs as shoppers fill their lives with new devices and services.

The computerized behavior is requiring mapping as per the need of the customer and their satisfaction. Banks, furthermore, need to look past mobility and payment

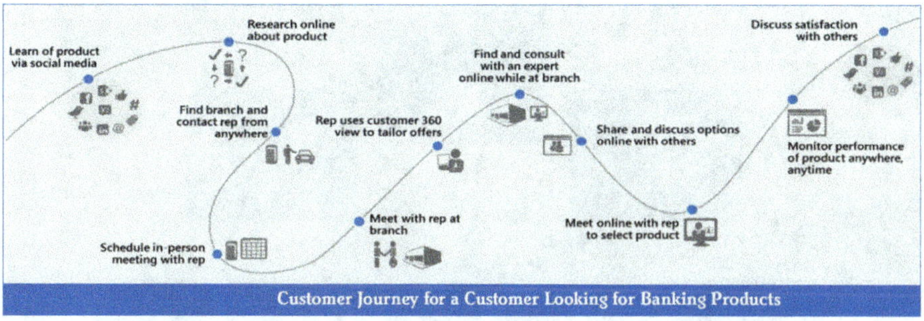

Fig. 1 Journey for looking banking products

and figure out the client activity and guide their advanced strategy. The customer dimension for products and services regarding banking services has multiphase transaction, as depicted in the figure.

Figure 1 portrays an example of client traveling for a searching of banking products and services. The major steps for the sake of this are as follows:

- Learn of product via social/digital media
- Research online about product
- Find branch and contact rep from anywhere
- Schedule in person meeting with rep
- Rep uses customer 360-degree view to tailor offers
- Meet with rep at branch
- Find and consult with an expert online while at branch
- Share and discuss options online with others
- Meet online with rep to select product
- Monitor performance of product anywhere, anytime
- Discussion satisfaction with others.

3 Novel Though and Approach with Digital Banking

Nowadays, bank is one of the important units for the people. It is not only a place where people can deposit their money to keep safe from any miss-happenings. People are now looking toward banks as an advisor for certain financial issues. So, people have to change his/her thought about digital banking, also banks have to accept the thought of people in the age of ICT. No doubt, the entry of private banks makes things easy for digitization in financial sector (Fig. 2).

Similarly, the best method to figure out and bring the connection from conventional banking to digital banking is Omni channel approach. This approach is a multichannel way to deal with customer benefits, where everyone of the channels is firmly incorporated, keeping customer in the focal point of the coordination. As

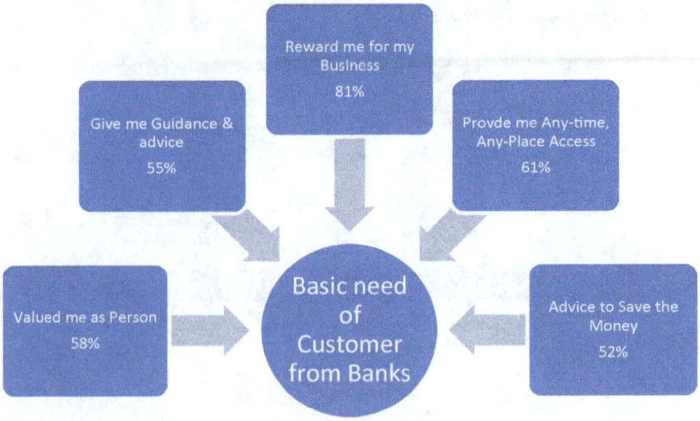

Fig. 2 Need of potential customer from banking system

customers keep on changing their channel, banks and credit firms need to focus on consistent activity and behavior of the customer. It gives a transparent feedback about the banking in association with customer services. Something other than a saying, Omni-channel banking is a prospect to take primary concern on higher note by picking up experiences from customers' channels, conduct, and inclinations. In these days, customers are more complex and technically knowledgeable, and to oblige their particular needs, every customer needs a kind affair from banking. They need the organizations to comprehend their implicit needs and additionally their preferences. It is also applicable for modern banking system also. In this way, it should not take as that these customers are expecting comparative sort of reaction and services from banking establishments as well.

From the investigation of new services, opening an account, checking balance, directing exchanges, advances, credits, customer bolster, and conveying an Omni-channel encounter has turned into a key to achievement in this focused commercial center.

Omni-channel facility is crucial to guarantee that customer encounter is brought together, fused and bolsters customer at the ideal time, in the opportune place as the correct way.

It must be according to their portable and digital way of life (Fig. 3).

In the current banking transformation involves profound of customers' needs, and requests. It, likewise, requires the exact blend of ICT infrastructure and creative new advances to ensure that one stays ahead in current market to drive top and in addition main concerns upward.

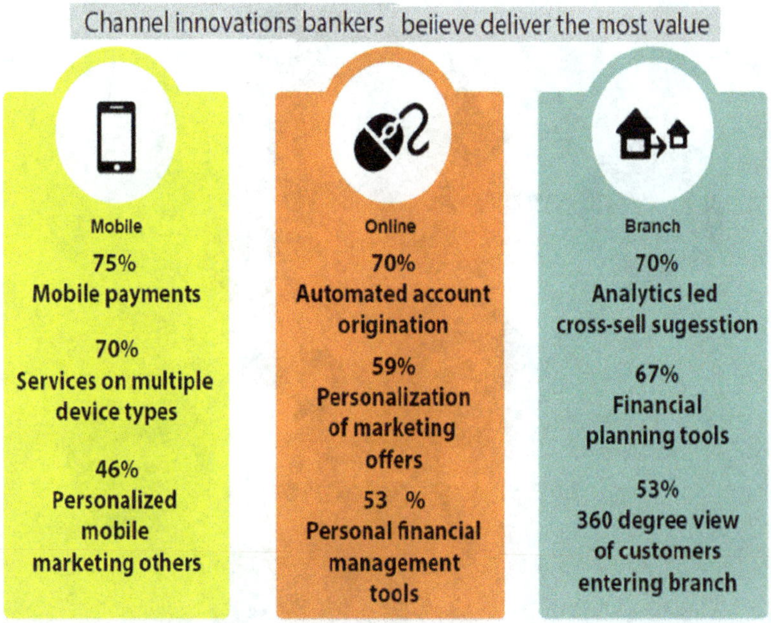

Fig. 3 Behavior of value-added potential customer

4 Major Challenges in Digitization in India

The major challenges in the digital banking in India are associated with the environment and adoption of technology by the people and bankers. Sometimes, infrastructure and technology implementation are also a matters (Fig. 4).

Legacy of existing model is a common challenge; this is type of shaping work time to time so inheritance is slowing adoptable. The cost of ICT-enabled system is too high and lot of installment requires with. With the implementation of hardware and technical instruments, there is a need for skilled men power to operate the system. To produce suitable men power, there is a need for expert, and experts are rare and expensive. After this, the transformation of data according to new technology is a tedious task. Dealing with pervasive data is again an assignment. In India, we know that policy is formed by ministers and they are lacking in technical knowledge. So, incomplete strategy is one of the major factors in real-time implementation; due to this factor, a lot of projects are getting failed. Despite all those, as per the opinion of the responders for study, we find following hurdles.

- Legacy of existing model and practices
- High cost of implementation
- Lack of technology expertise and skilled workforce
- Dealing with humongous pervasive data
- Complying with legislation

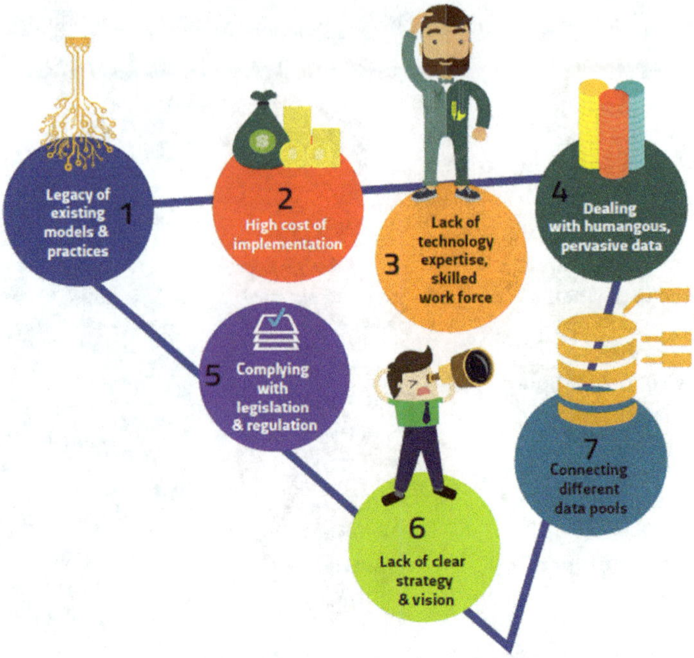

Fig. 4 Major challenges to implement digital banking (www.happiestminds.com)

- Lack of clear strategy and vision
- Connecting different data pools.

5 Bridging the Digital Delivery Gap

To bridging the gap between normal banking delivery and digital delivery, there is a need to understand about target to achieve or need to achieve and the actual achieved position till now.

We are looking for the matter connected from the main challenge. There is a need to understand the target. Also, there is a need to look toward what we achieved till now. After that, we can calculate the required efforts for bridging the gap (Fig. 5).

The digital delivery gap is concerned from the year 2017 around; although the digitization expanding in good pace, the full-flashed achievements are very far as per the statistical data (Fig. 6).

According to graph, it is depicted that the target is not achieved up to requirements. Even each and every field is not at the required level. The gap between market lag and current capabilities is too high. The negative value of the market leg shows that the target is not achieved. Despite that, we are able to stand in the digital market.

Digital Delivery Gap				
NO.	Type of Activity	Important to Achieve (%)	Current Capability (%)	Market Lag (%)
P1	Real Time Synchronization	94	24	-70
P2	Location Driven Services	82	19	-63
P3	Customer Analytics	92	30	-62
P4	Social Media Account	78	34	-44
P5	Multichannel Integration	92	36	-56
P6	Mobile Channel	96	44	-52
P7	Personnel recommendation assistance	88	28	-55
P8	Cognitive Intelligence	76	14	-62
P9	Digital Advisory Services	83	28	-55
P10	Comparison Services based on financial Profile	76	26	-50

Fig. 5 Digital delivery gap scenario in India (www.happiestminds.com)

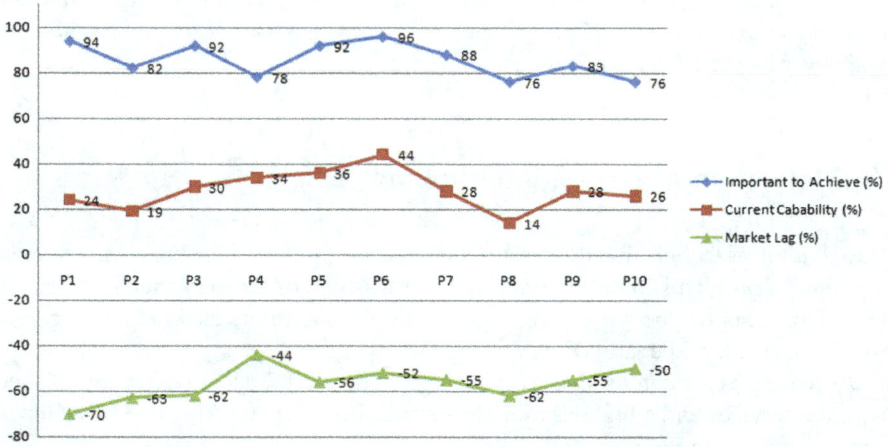

Fig. 6 Graphical representation "Digital Delivery Gap"

The red line shows our capabilities. It is good that with the occurrence of a lot of challenges, we achieved enough to run the digital banking in the agriculture-based nation.

6 Reducing the Gap and Alternate

As mentioned in the above part that a digital delivery gap exists in the implementation of digital banking. It is due to the lack of knowledge and proper interface availability. From a common overview, it is found that 41% of users are satisfied with the digital services and they are availing services, 17% are not aware of services, 07% are

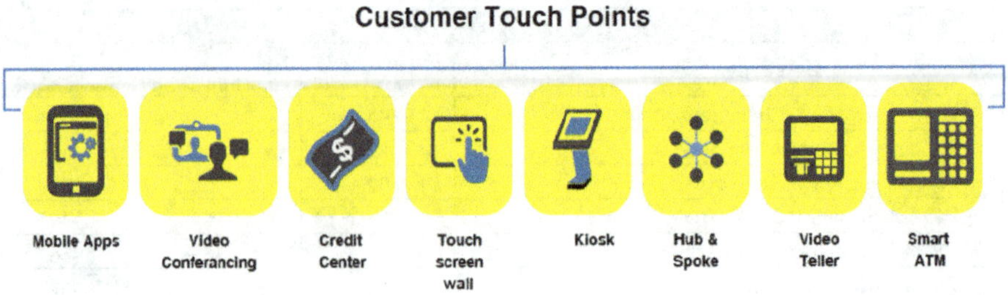

Fig. 7 Customer touching point

dissatisfied from services, and 35% people are aware but not interested to use digital services.

To overcome from such type of situation, there is a need to increase customer touch points. Some of the customer touching points are performing very well in the present context are depicted in Fig. 7. So, we can say that later or sooner the situation will cover by the advanced facilities.

7 Motivation About Digital Banking

Due to a lot of technical and infrastructure-related problems, including the operating kind of problems in digital banking, still it is one of the very much successful initiatives. This will be work very smoothly in the near future also functioning very well in the present scenario (Fig. 8).

As per Fig. 9, we can see that during the year 2012–2017 a lot of users are satisfied with the services and a big mass shifted toward the digital banking and behavior of customers are same.

Also as per the global electronic payment scenario and ration, both are increasing rapidly and it would be definitely reflected on the Indian digital banking market.

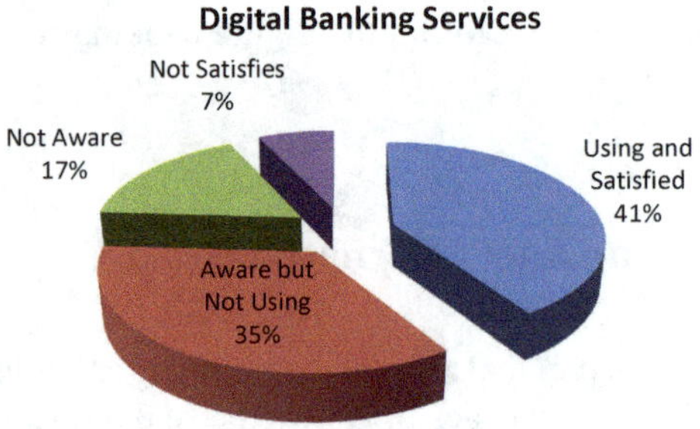

Fig. 8 Scenario of using digital banking services in India

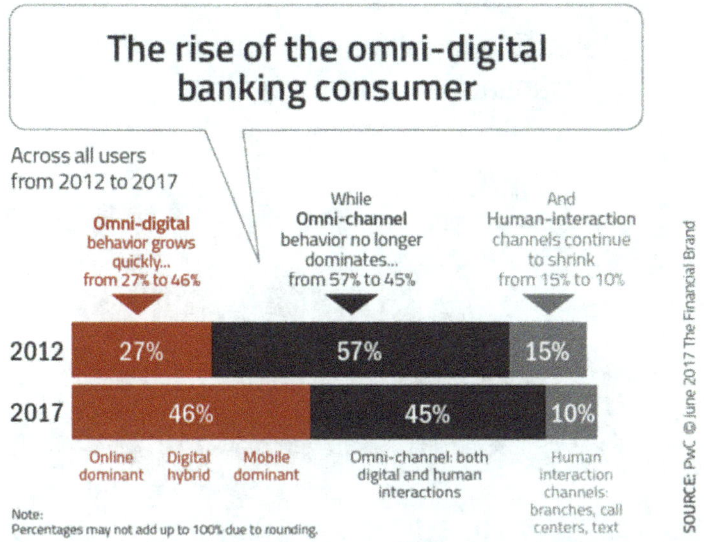

Fig. 9 Behavior of customer toward digital banking

Growth ration of e-payments' and m-payments' transaction is very positive, and it is a big motivation for the digital banking in the current scenario. Figure 10 shows a clear picture of the upcoming and running time about the use of electronic and mobile media for the electronic payments system.

8 Conclusion

The present study is based on the user views and secondary data available on the authentic web sites. It is observed that in the present scenario digital banking is the backbone of the banking system. Available facts and figures are also supporting digitization of banking in the nation. Only the lacking part is human behavior about adopting new technologies, sometimes the infrastructure and skilled men power are also a challenging situation. There is a need to bridge gap between technology implementation strategy and human acceptable situation. So, if we are working from both the sides simultaneously, then we will achieve complete digitize banking system in India very soon.

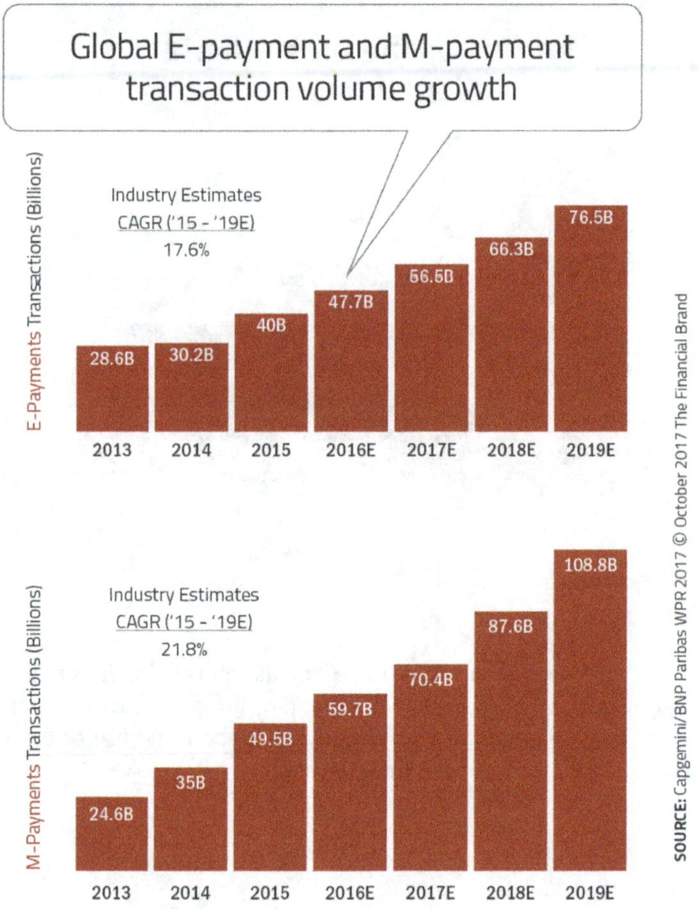

Fig. 10 E-payment and m-payment

References

1. Chellaswamy P (2010) Modern banking management? Himalaya Publishing House, India
2. Gupta K, Arya R, Goel V (2017) Emerging trends in banking sector in India: with special reference to digitization. Int J Sci Technol Manage 6(3):260–268
3. Shukla R (2014) Banking digital revolution—a way forward. Happiest Minds Technology Pvt. Ltd, India
4. The Indian banking sector: recent developments, growth and prospects (2013)
5. RBI discussion paper on banking structure in India (2013)
6. http://www.digitalindia.gov.in/content/ekranti-electronic-delivery-services
7. https://www.intelligenthq.com/finance/banking-technology-trends-and-vision-for-2016-in-india/
8. www.happiestminds.com

Author Index

S.-L. Peng et al. (eds.), *Computing and Network Sustainability*, Lecture Notes in Networks and Systems 75,
https://doi.org/10.1007/978-981-13-7150-9

CPSIA information can be obtained
at www.ICGtesting.com
Printed in the USA
LVHW080121070519
616813LV00002B/54/P

9 789811 371493